Advances in Intelligent Systems and Computing

Volume 254

Series editor

Janusz Kacprzyk, Polish Academy of Sciences, Warsaw, Poland
e-mail: kacprzyk@ibspan.waw.pl

For further volumes:
http://www.springer.com/series/11156

About this Series

The series "Advances in Intelligent Systems and Computing" contains publications on theory, applications, and design methods of Intelligent Systems and Intelligent Computing. Virtually all disciplines such as engineering, natural sciences, computer and information science, ICT, economics, business, e-commerce, environment, healthcare, life science are covered. The list of topics spans all the areas of modern intelligent systems and computing.

The publications within "Advances in Intelligent Systems and Computing" are primarily textbooks and proceedings of important conferences, symposia and congresses. They cover significant recent developments in the field, both of a foundational and applicable character. An important characteristic feature of the series is the short publication time and world-wide distribution. This permits a rapid and broad dissemination of research results.

Bing-Yuan Cao · Sheng-Quan Ma
Hu-hua Cao
Editors

Ecosystem Assessment and Fuzzy Systems Management

Editors
Bing-Yuan Cao
School of Mathematics and Information Science, Key Laboratory of Mathematics and Interdisciplinary Sciences of Guangdong Higher Education Institutes
Guangzhou University
Guangzhou
People's Republic of China

Hu-hua Cao
Department of Geography
University of Ottawa
Ottawa
Canada

Sheng-Quan Ma
School of Information Science and Technology
Hainan Normal University
Haikou
People's Republic of China

ISSN 2194-5357 ISSN 2194-5365 (electronic)
ISBN 978-3-319-03448-5 ISBN 978-3-319-03449-2 (eBook)
DOI 10.1007/978-3-319-03449-2
Springer Cham Heidelberg New York Dordrecht London

Library of Congress Control Number: 2013958224

Printed on acid-free paper

Springer is part of Springer Science+Business Media (www.springer.com)

Preface

This book is a monograph with submissions from the 3rd International Conference on Ecosystem Assessment Management and the Workshop on the Construction of an Early Warning Platform for Eco-Tourism in Hainan on May 5–12, 2013 in Haikou, China. The monograph is published by *Advances in Intelligent Systems and Computing* (AISC), Springer, ISSN: 2194-5357.

The conference received more than 120 submissions. Each chapter of it has undergone a rigorous review process. Only high-quality papers are included in it.

The book, containing 47 papers, is divided into five parts:

I. Thesis on "Ecosystem Assessment, Management and Information."
II. Subjects on "Intelligent Algorithm, Fuzzy Optimization and Engineering Application."
III. Topics are discussed on "Spatial Data Analysis and Intelligent Information Processing" appearance.
IV. Ideas circus around "Tourism Culture, Development and Planning."
V. Focus on "Application of Operations Research and Fuzzy Systems."

We appreciate organizations sponsored by Hainan Normal University, China; University of Ottawa, Canada; Department of Land Environment and Resources of Hainan Province, China.

Gratitude is shown to Fuzzy Information and Engineering Branch in ORSC and Operations Research Society of Guangdong Province for Co-sponsorships.

We wish to express our heartfelt appreciation to the Editorial Committee, reviewers, and our students. We appreciate all the authors and participants for their great contributions that made these conferences possible and all the hard work worthwhile.

Finally, thanks to the publisher, Springer, for publishing the AISC (Notes: Our series of conference proceedings by Springer, like *Advances in Soft Computing* (ASC), AISC, (ASC 40, ASC 54, AISC 62, AISC 78, AISC 82, and AISC 147, have been included into EI and all are indexed by Thomson Reuters Conference Proceedings Citation Index (ISTP)), and also thanks to the supports coming from international magazine *Fuzzy Information and Engineering* by Springer and supports by International Science and Technology Cooperation Program of China (2012DFA11270), and Hainan International Cooperation Key Project (GJXM201105).

May 2013

Bing-Yuan Cao
Sheng-Quan Ma
Hu-hua Cao

Editorial Committee

Organizing Committee of ICEAM and WCEWPET

Conference Chairs: Sheng-Quan Ma (China)
Hu-hua Cao (Canada)

Honorary Chair: Feng-min Li (China)

Local Arrangements Chair: Cheng-yi Zhang

Co-Chair: Li-hua Wu

Secretary: Jun-kuo Cao

Members:
Jun-kuo Cao (China)
Mou-song Fu (China)
Rui-bo Han (Canada)
Li-jun Bai (China)
Wen-juan Jiang (China)
Song Lin (China)
Hong-yan Lin (China)
Xiao-wen Liu (China)
Zi-wei Liu (Canada)
Jing-le Liu (Canada)
Zhuang-jian Mo (China)
Li-hua Wu (China)
Hong-li Wu (China)
Jian-xin Wang (China)
Cheng-yi Zhang (China)
Xue-ping Zhang (China)
Xian-feng Zhang (China)
Li-qin Zhang (Canana)
Zhi-qing Zhao (China)
Yu-ping Zhou (China)

Publicity Chair: Sheng-Quan Ma

Co-Chair: Hu-hua Cao

Member: Xiao-yuan Geng

Publication Chair: Bing-Yuan Cao (China)

Co-Chairs: Sheng-Quan Ma (China)
Hu-hua Cao (Canada)

Members: Jun-kuo Cao
Jing Ma
De-jun Peng

Contents

Introduction: Ecosystem Assessment and Management

Huhua Cao, Sheng-Quan Ma and Fengmin Li

Abstract This article provides an overview of the origin and evolution of the concept of Ecosystem Assessment Management (EAM), including increased use of the framework among academics. EAM is a tool that uses information technology to assess the relationship between human and natural factors of a given eco-system and provide strategies for improved management in the future.

Keywords Ecosystem Assessment Management (EAM) · Environmental sustainability · Ecosystem · Information technology · Geospatial technology

1 Origin of EAM

The concept of Ecosystem Assessment Management (EAM) was first proposed and created in 2003 when a joint Canadian–Chinese team was building a collaborative project with the Canadian International Development Agency (CIDA). This project focused on rural and agricultural development in minority regions of Gansu, China. This CIDA project included partners from the University of Ottawa (Huhua Cao), Lanzhou University (Professor Fengmin Li), and Northwest Minorities University (Professor Shengqun Ma).

H. Cao (✉)
Department of Geography, University of Ottawa, Ottawa, Canada
e-mail: caohuhua@uottawa.ca

H. Cao · S.-Q. Ma
School of Information and Technology, Hainan Normal University, Haikou 571158, China

F. Li
MOE Key Laboratory of Arid and Grassland Ecology, School of Life Science, Lanzhou University, Lanzhou 730000, China

B.-Y. Cao et al. (eds.), *Ecosystem Assessment and Fuzzy Systems Management*, Advances in Intelligent Systems and Computing 254, DOI: 10.1007/978-3-319-03449-2_1,

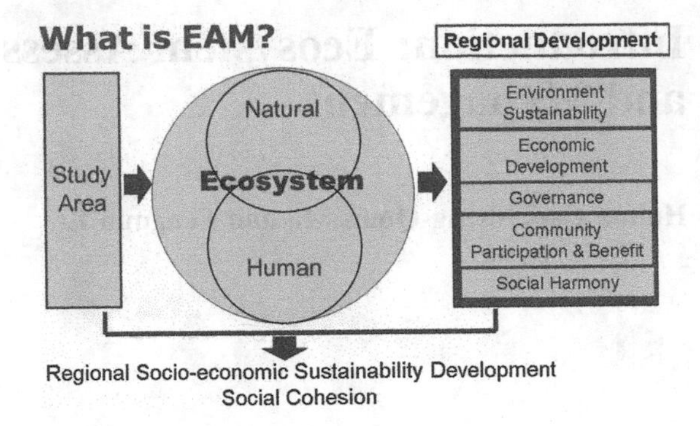

Fig. 1 Diagrammatic representation of the EAM concept

The project was created in response to challenges that China was facing in its development. After joining the WTO in 2001, it became increasingly evident that China was experiencing enormous environmental problems, but at the time most Chinese academics focused their efforts solely on the natural causes of environmental issues. EAM aims to bring new perspectives to Chinese stakeholders and emphasize the study of the region as a complete ecosystem. EAM examines all subcomponents in that system, considering two main aspects, human and nature, using geospatial technology.

Figure 1 describes the concept on which this system is based. Any given study area can be chosen, and the EAM system approaches the study area by conceptualizing ecosystems as both human and natural. With this combined social–ecological viewpoint, regional development can be broken down into multiple human and natural factors.

2 What is EAM?

EAM is a relatively new concept which uses information technology (IT) tools such as geospatial technology [geographic information systems (GIS), remote sensing (RS), and global positioning systems (GPS)] and 'what-if' scenarios to reveal the relationships and interactions between specific ecosystem components. These components include physical factors such as contaminant concentrations, species occurrence and diversity, habitat characteristics, and land use patterns. In addition, they include the social, economic, and cultural conditions of the locality in question, with emphasis on the local community's knowledge, skills, and values, its expectations of gender roles, and its ethnic composition (Fig. 2).

EAM is an ideal tool for exploring the significance of differences and discrimination on the basis of gender and minority identities and how these factors affect the prospects for various environment-related development issues. By integrating these human factors into the analysis, EAM produces more accurate and reliable assessments of what measures are likely to achieve environmental, economic, and social goals. Another feature of EAM is its emphasis on long-term

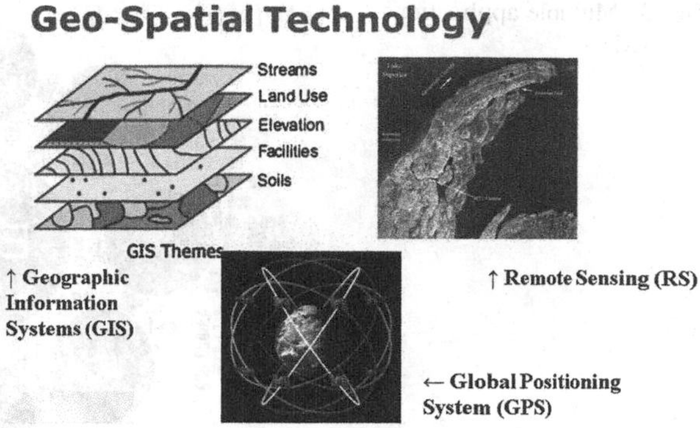

Fig. 2 Compositions of geospatial technology

Geo-Spatial Technology

Streams
Land Use
Elevation
Facilities
Soils
GIS Themes
↑ Geographic Information Systems (GIS)
↑ Remote Sensing (RS)
← Global Positioning System (GPS)

management, which entails ongoing assessment and adjustment to the evolving physical and human factors. Finally, EAM takes a bottom–up approach to assessment and management. It begins with the physical and human conditions of the specific locality and develops solutions for those specific conditions. At every stage, the participation and cooperation of the local community is a priority.

3 Multiple Applications of EAM

In recent years, the EAM concept has become more popular among Chinese academics. Through Canadian–Chinese collaboration, our partners at Lanzhou University first used the EAM framework in the agricultural field and received two significant grants of more than 2 million Yuan RMB from the Ministry of Science and Technology of the Government of China for the project entitled 'Construction of an Information Platform/module in ECOAGRICULTURAL Assessment and Management (EAM)' for 2006–2009 and 2010–2013. After transferring to Hainan Normal University, Professor Shengquan Ma and his team collaborated with Canadian partners to successfully use EAM in tourism development and received another important funding of 2.25 million Yuan RMB from the Ministry of Science and Technology of China for the project entitled 'Construction of an Early Warning Information Platform/Module based on Ecotourism Assessment and Management (EAM)—Its Application in the Building of Hainan International Tourism Island of China' for 2012–2015. Professor Yanfan Li from Nanjing University has also applied the EAM framework for his work on coastal areas (Eco-coastal Assessment and Management). In collaboration with Anwar from Xinjiang Normal University, the University of Ottawa team has obtained funding from the Canadian International Development Research Centre (IDRC) to apply EAM to urban development (Eco-city). Professor Cao and Zhang from Hainan Normal University are building another important project using EAM in Nanhai for the project entitled 'Construction of China NanHai Eco-political Assessment and Management (EAM) Platform for Policy-Making' (Fig. 3).

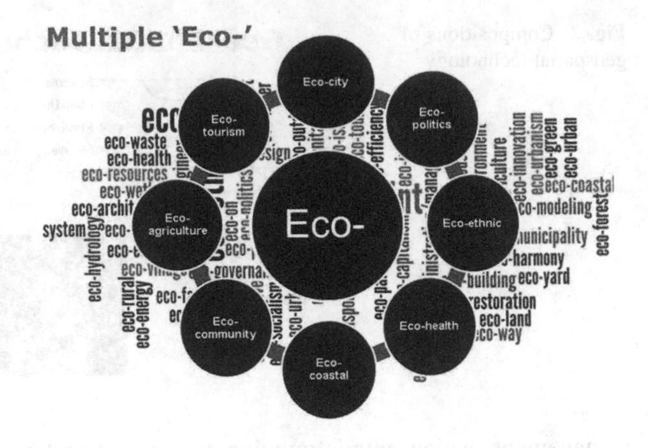

Fig. 3 Multiple applications of EAM

In fact, EAM has since been applied in multiple contexts in China and internationally. In this volume, many examples of its application can be seen and will be discussed in more detail throughout this volume. Over the years, the concept of EAM has become much more detailed since it has been applied in practice. For example, the eco-agricultural application of EAM describes landscapes that support both agricultural production and biodiversity conservation, working in harmony together to improve the livelihoods of rural communities. And an eco-political framework integrates the role of political institutions and local community participation in regional sustainable development and concentrates on the interactions between different management levels of institutions, between NGOs, local communities, and governments.

EAM systems are now not only a clear framework, but also have a clear methodology through the use of intelligent assessment methods such as fuzzy mathematics, neural network model, and others, which have been introduced into the data process. As such, EAM systems are now much more technical. It has been integrated with advanced technologies such as RS, GIS, and GPS. These technologies not only create more accurate information for EAM research, but also make EAM a practical platform for assessment and early warning.

4 Volume Contents

This volume includes five EAM-related parts: The first part is for 'Ecosystem Assessment, Management and Information,' focusing on multiple applications of EAM on ecotourism early warning, eco-safety early warning, economic structure analysis, energy footprint analysis, water resource ecological carrying capacity assessment, and so on. The case areas are ranged from the southern province of China Hainan to the northwest province Gansu. The second part is for 'Intelligent Algorithm, Fuzzy Optimization and Engineering Application,' focusing on mathematical methodology and intelligent algorithm for EAM. The third part is for 'Spatial Data Analysis and Intelligent Information Process,' focusing on virtual tourism, ecotourism resources planning, and eco-urban transition analysis. The fourth part is for 'Tourism Culture, Development and Planning,' focusing on sustainable tourism planning from local perspectives and international perspectives, especially for island ecotourism development. The fifth part is 'Application of Operations Research and Fuzzy Systems,' focusing on application of extensional operational and fuzzy systems.

Acknowledgments In addition, the authors would like to thank the many collaborators who have been involved in developing and applying EAM research and practice over the last 15 years.

Part I
Ecosystem Assessment, Management and Information

Geomatics in Ecosystem Assessment and Management

Rui-bo Han, Jun-kuo Cao and Sheng-Quan Ma

Abstract The development of Geomatics in recent years has been revolutionized by rapid advances in computer hardware and software. Geomatics has been successfully applied in biodiversity assessment, wetland mapping, drylands degradation assessment, measurements of land surface and marine attribute, and many other applications. Geomatics technologies benefit the ecosystem assessment and management process at all stages, including data collection, data management, data analysis and modeling, and presentation of results. The chapter conducts a review of the Geomatics technologies and their application in a practical framework for ecosystem assessment and management.

Keywords Geomatics · Ecosystem assessment and management · GIS · Multi-criteria evaluation

1 Introduction

With the transition of the ecosystem studies from ecosystem structure, condition, and function assessment to ecological procedure assessment, the facing problems are becoming more complicated and synthesized: The research scale is long termed and globalized; the research method is quantified; the research objective is management oriented. Obviously, the traditional statistical method are not capable to finish the job any more, a new technology—Geomatics technology [including

R. Han
Department of Geography, University of Ottawa, Ottawa, ON K1N 6N5, Canada
e-mail: han@live.ca

J. Cao (✉) · S.-Q. Ma
Department of Computer Science and Technology, Hainan Normal University, Haikou, China
e-mail: junkuocao@gmail.com

B.-Y. Cao et al. (eds.), *Ecosystem Assessment and Fuzzy Systems Management*,
Advances in Intelligent Systems and Computing 254, DOI: 10.1007/978-3-319-03449-2_2,

tools and techniques used in Remote Sensing, Geographical Information System (GIS), and Global Positioning System]—has become the most efficient and convenient method to do research. Geomatics is "a powerful set of tools for collecting, storing, retrieving at will, transforming, and displaying spatial data from the real world for a particular set of purposes [1]." It has been widely used to support the process of environmental assessment and management.

The term "remote sensing" is broadly defined as the technique(s) for collecting data about an object from a distance from the object and the recorded data are usually saved as images. Based on the information being collected and the technology used to collect data, remote sensing products can refer to satellite imagery, aerial photographs, or radar data from active or passive microwaves [2]. Remote sensing technology has been increasingly used in the past several decades to conduct research from local to global scales. The evolvement of technology has enabled the data collection from pure visual imagery (e.g., aerial photographs) to multi-spectral imagery (e.g., Landsat products). The spatial and temporal resolution has improved over time and reached a level at which high-quality spaceborne imagery of any location on earth can be acquired in a timely manner.

Geographical information systems are widely used as tools to collect, store, manage, analyze, and display geographical data. GIS can be used to build the inventory of any type of data with spatial attributes, which have seen an expanding usage in both human topics (e.g., demographic databases) and natural studies (e.g., distribution of environmental elements and factors). GIS also provides a growing and large number of tools that can be used to analyze and assess the characteristics of data over data space, temporal space, and spatial space. More importantly, GIS offers a practical environment to manage multiple types of database, which ensures a platform for a sustainable management system. GIS can also provide inputs to both static and dynamic ecosystem models [3]. For example, a static model may be used to estimate soil erosion based on soil type, meteorological data, and terrain characteristics, whereas a dynamic GIS model could be used to represent a spatial landscape transition pattern at different time periods. Another impressive characteristic of GIS is that it can visualize and present of simulation results in maps, even in three-dimensional view.

The role of Geomatics in ecosystem studies is expanding as researchers exploit the increasingly sophisticated capabilities of technologies in GIS and remote sensing. Recent advances include the ability to store and manage Big Data (large datasets) and to perform more spatial and statistical analyses. In particular, GIS has relevance to the system modeling process that is such an essential component of the usual assessment and management procedure. Therefore, Geomatics will continue to be applied as an essential toolset for data acquisition, analysis, and management in the fields of environmental assessment, planning, and management.

2 Application of Geomatics in Ecosystem Assessment and Management

Geomatics technologies benefit the ecosystem assessment and management process at all stages, including data collection, data management, data analysis and modeling, and presentation of results.

2.1 Data Collecting, Preprocessing, Storage, and Management

Large amounts and diverse sorts of data may be required for environmental modeling. Due to the nature of the environmental management problems, much of these data have spatial characteristics. For example, land use and land cover data, digital elevation models, and remote sensing imagery provide useful information to those attempting to model environmental and ecological processes. Geomatics provides a convenient means of collecting, storing, and managing such data. Steyaert and Goodchild [3] noted that Geomatics has automatic "housekeeping" functions, such as builds the inventory of data layers, and provides consistent access to diverse data that have been integrated into the system.

GIS is able to build the inventory of data for modeling the ecosystem, whether the modeling procedure is in a GIS or an external environment. GIS has the function of integrating historic, socioeconomic, and environmental data, which allows users to identify properties to conduct ecosystem assessments and rank properties and determine priorities. A well-defined geodatabase can record multiple types of information about an ecosystem, such as geographical location, area, zoning, functioning factors, development history, and infrastructure. GIS also provides compatible data services for external packages (e.g., simulation modeling packages) by reformatting and exporting data into desired format. GPS provides biologists and managers with the ability to collect accurate locational information in the field, which can be related and integrated with other spatial data using GIS.

Remote sensing has been used as the primary data source for detecting land cover and land use condition, mapping the extent and providing ancillary data for ecosystems. Moreover, remote sensing provides consistent measurements over the entire area that is not subject to varying data collection methods in different locations. The quality of remote sensing data depends on its spatial, spectral, radiometric, and temporal resolutions. Millennium Ecosystem Assessment [4] lists most remote sensing data that are useful to assess ecosystem conditions from various sensors on satellites (see Table 1).

Table 1 Remote sensing platforms for monitoring land [4]

Platform	Sensor	Spatial resolution at Nadir	Date of observations
Coarse resolution satellite sensors (>1 km)			
NOAA-TIROS (National Oceanic and Atmospheric Administration-Television and Infrared Observation Satellite)	AVHRR (Advanced Very High Resolution Radiometer)	1.1 km (local area coverage); 8 km (global area coverage)	1978–present
SPOT (Systeme Probatoire pour la Observation de la Terre)	VEGETATION	1.15 km	1998–present
ADEOS-II (Advanced Earth Observing Satellite)	POLDER (Polarization and Directionality of the Earth's Reflectances)	7 × 6 km	2002–present
SeaStar	SeaWIFS (Sea viewing Wide Field of View)	1 km (local coverage); 4 km (global coverage)	1997–present
Moderate resolution satellite sensors (250 m–1 km)			
ADEOS-II (Advanced Earth Observing Satellite)	GLI (Global Imager)	250 m–1 km	2002–present
EOS AM and PM (Earth Observing System)	MODIS (Moderate Resolution Spectra radio meter)	250–1,000 m	1999–present
EOS AH and PM (Earth Observing System)	MISR (Multi-angle Imaging Spectra Radiometer)	275 m	1999–present
Envisat	MERIS (Medium Resolution Imaging Spectra radio meter)	350–1,200 m	2002–present
Envisat	ASAR (Advanced Synthetic Aperture Radar)	150–1,000 m	2002–present
High resolution satellite sensors (20–250 m)[a]			
SPOT (Systeme Probatoire pour la Observation de la Terre)	HRV (High Resolution Visible Imaging System)	20 m; 10 m (panchromatic)	1986–present
ERS (European Remote Sensing Satellite)	SAR (Synthetic Aperture Radar)	30 m	1995–present
Radarsat		10–100 m	1995–present
Landsat (Land Satellite)	MSS (Multi-spectral Scanner)	83 m	1972–1997
Landsat (Land Satellite)	TM (Thematic Mapper)	30 m (120 m thermal–infrared band)	1984–present
Landsat (Land Satellite)	ETM+ (Enhanced Thematic Mapper)	30 m	1999–present

(continued)

Table 1 (continued)

Platform	Sensor	Spatial resolution at Nadir	Date of observations
EOS AM and PM (Earth Observing System)	ASTER (Advanced Spaceborne Thermal Emission and Reflection Radiometer)	15–90 m	1999–present
RS (Indian Remote Sensing)	LISS 3 (Linear Imaging Self-scanner)	23 m; 5.8 m (panchromatic)	1995–present
Very high resolution satellite sensors (<20 m)[a]			
JERS (Japanese Earth Resources Satellite)	SAR (Synthetic Aperture Radar)	18 m	1992–1998
JERS (Japanese Earth Resources Satellite)	OPS	18 × 24 m	1992–1998
IKONOS		1 m panchromatic; 4 m multi-spectral	1999–present
QuickBird		0.61 m panchromatic; 2.44 m multi-spectral	2001–present
SPOT-5	HRG-HRS	10 m; 2.5 m (panchromatic)	2002–present

2.2 Spatial Analysis

What make geospatial data more important than non-spatial data is the spatial attribute, based on which the spatial modeling of environment assessment and management can be established. Based on this unique property, GIS can take into consideration of the spatial scale and can integrate the scale theory into ecosystem studies. In addition, the spatial dimension of GIS data can be used to simulate and test spatial hypotheses, which are difficulties most ecologists face while address ecosystem assessment issues.

The spatial analytical functions in GIS consist of a combination of spatial analysis tools, which can work with data in both vector and raster data. For example, data layers in GIS for the same area can be imposed on top of each other to be analyzed for change detection over time. In raster-based GIS, remote sensing products can also be integrated into the analysis procedure. For example, Dahdouh-Guebas [2] analyzed how remote sensing technology and other scientific tools can be integrated in long-term studies, both retrospective and predictive, in order to anticipate degradation and to take mitigating measures at an early stage of a sustainable management of tropical coastal ecosystems. Additionally, GIS can be used to analyze the results of a modeling process. For example, the use of Boolean logical operators and weighted reclassification functions, which are generic to GIS, would facilitate the identification of suitable areas in a multi-criteria analysis.

However, GIS as a stand-alone system is sometimes not sufficient for environmental assessment or management modeling. GIS is commonly combined with other environmental software, but the passing of data from GIS to other software often results in bulky data conversion procedures [3]. On the other hand, more and more GIS packages are equipped with application programming interfaces (APIs). These APIs make it possible for users with programming capabilities to customize or design generic models.

2.3 Visualization and Presentation

A final category of the potential contributions of Geomatics to the ecosystem assessment and management process is the visualization and presentation of results. A chart is worth a 1,000 words, but in geographical studies a map is worth a 1,000 charts. As an important component of Geomatics, cartographic mapping is capable of adding more explanatory and intuitive power to the traditional tabular and graphical reporting formats. The importance of this capability in ecosystem assessment and management is reflected by the persistent reference in the adaptive management literature to the significant communicative role of clear visual presentation of results [5–7].

With the prevalence of Internet and mobile devices, the visualization and presentation of geographical data are not limited to paper-based maps any more.

GIS Web services are the software components that host spatial data and GIS functionalities that can be accessed and integrated into customized GIS applications through the Internet. Developers utilize GIS Web services for custom applications that process geographical information without having to maintain a full GIS system or the associated spatial data [8]. Users can tap into Web-based GIS distribution systems through their Web browsers without having any specialized GIS software on the desktop system. Web-based GIS technologies have also enabled the possibility of using Internet to publish the data from the inventory and GIS database. The focus in this mode of GIS use is not necessary on its geodetic or analytic capabilities (although they do play a major role), but rather on the visual and contextual exploration of the problem situation and issues connected to it.

Web-based GIS also facilitates the procedure of collecting inputs from public in the process of ecosystem assessment and management. The integration of user requirement and user feedbacks is now indispensable in general information systems design [9] and in GIS design [10]. Public Participation GIS (PPGIS) has emerged as a test bed for techniques, methodologies, ideas, and discussion about the social implication of GIS technology. PPGIS enables users to benefit from GIS' ability to bring together many different data sources into comprehensive and manageable format making it an excellent tool for data management. For instance, community groups and citizens can contribute information such as historic land uses, old photographs, or other data that completes the inventory of an ecosystem.

Two key benefits of Web-based GIS distribution systems are the increased interaction with users and connections to a wider audience [11] and its advanced data integration capabilities [12]. Thus, there is potential for more people over a broader area to be reached through the Internet than other forum options and certainly at a lower cost compared to traditional methods—i.e., printing or public forums. In addition, any updates to data can be made on the Web server and are immediately available to users with little or no printing costs.

The second key benefit discussed in the literature is the capability of Web-based GIS distribution systems to relate a wide range of spatial and non-spatial data sets. The systems discussed are used as public forums and as decision support tools for projects from environmental assessments to transportation infrastructure and mass transit routing [8]. These systems can integrate spatially referenced shape files with tabular attribute data, satellite imagery, and aerial photographs. In addition, other photographs, images, and documents along with links to additional Web resources can be incorporated. Common analysis tools also allow users to extract, overlay, and join spatially related data, create buffers and service areas around features, and perform advanced spatial and network analyses.

This Web-based GIS thus allows users to search the repository through selection criteria with a series of menus and query functions to retrieve data results for ecosystem assessment or management. Output is displayed through a combination of text, maps, and digital orthophotographs. This system is thus a comprehensive data delivery tool that can assist policymakers' and stakeholders' with development decisions to encourage sustainable ecosystem management.

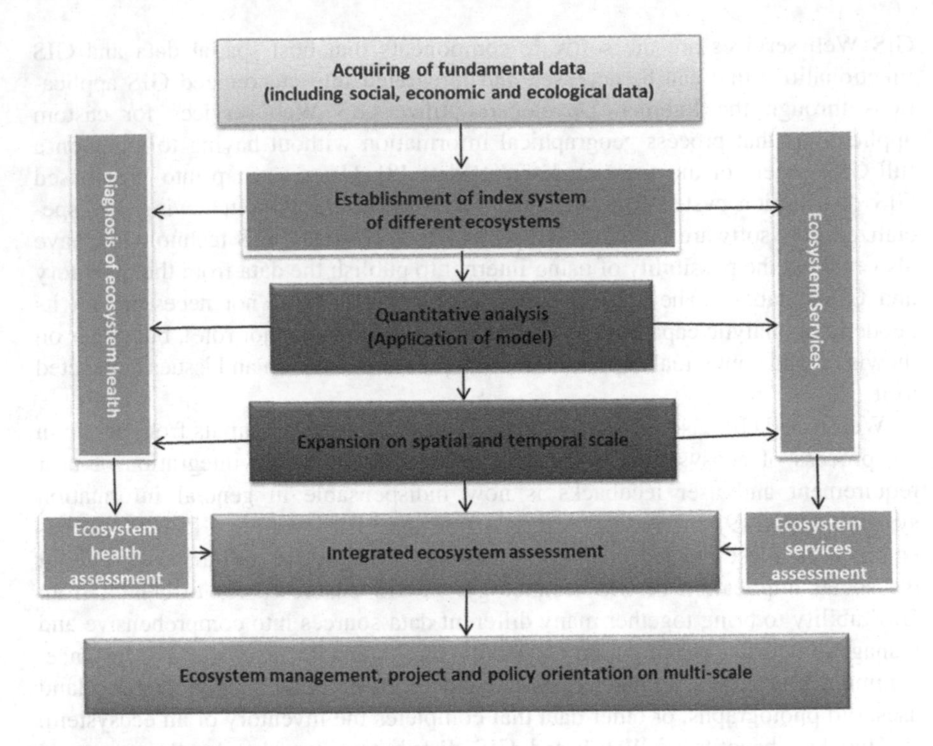

Fig. 1 Framework for ecosystem assessment

3 Practical Framework for Ecosystem Assessment and Management

Ecosystem assessment is a complicated, comprehensive, multi-scale, and dynamic process, which covers knowledge of ecology, geography, economics, and many other related disciplines. In order to find a way to integrate the entire ecosystem attributes from natural, social, and cultural scopes, an integrated ecosystem assessment model is recommended to synthesize the effects of multiple drivers on all ecosystems. Figure 1 servers as a standard procedure of an ecosystem assessment and management project.

Since ecosystem health assessment somewhat includes risk assessment, ecosystem health assessment and service assessment are taken as the subsets of integrated ecosystem assessment. Evaluating different indicators of ecosystem sustainability from two separated perspective, the result of these two methods is required to be integrated to demonstrate a synthesized condition of the ecosystem.

3.1 Required Geomatics Techniques of Practicing the Framework

3.1.1 Graph-Theoretic Analysis Methodology

Since drivers or indicators may interact with each other and have a combined effect on ecosystems, it is important to understand not only the impacts of drivers upon the ecosystem itself, but also the interactions among drivers. Therefore, it is essential to figure out a way to identify and weigh the relationships among drivers.

A graph-theoretic analysis methodology is developed by Wenger et al. [13] to identify the relationships among ecosystem change drivers. It is required at first to identify a list of drivers affecting the ecosystem under study. A matrix is then constructed in which the relationship between each pair of drivers in this list is identified. Specifically, a determination must be made to measure whether the first stressor in a given pair has an augmenting, a diminishing, or no impact on the second stressor in the pair. Both the construction of the list of drivers and the determination of the relationships between each pair of drivers must be reinforced by the best information contained in scientific journals and reports. An effective way to interpret the available scientific information required to make these decisions is to employ, for example in a workshop setting, the scientific expertise of persons acquainted with the ecosystem under study.

3.1.2 Multi-Criteria Evaluation

In an integration process, different indicators are evaluated according to weighted criteria, resulting in a ranking on a suitability scale. Usually, several criteria will be required and be evaluated all together in order to meet a specific objective. Such a procedure is called Index Overlying or multi-criteria evaluation (MCE). MCE is in fact a weighted linear combination of all indicators or factors by applying a weight to each followed by a summation of the results to yield a suitability or risk assessment (see Fig. 2).

The most commonly used method is multi-linear weighted model. The basic model is

$$I = \sum_{i=1}^{m} W_i \left(\sum_{j=1}^{m} W_{ij} P_{ij} \right),$$

where W_i is the weight of the ith factor, W_{ij} is the weight of jth indicator in the ith factor, P_{ij} is the standardization value of the weight of jth indicator in the ith factor, and I is the synthesis index of ecosystem condition.

The basic procedure is as follows: firstly, to (1) establish the indicators system according to the object of assessment, then (2) determine the weight of each indicator and standardize these indicators, and finally (3) assess the ecosystem

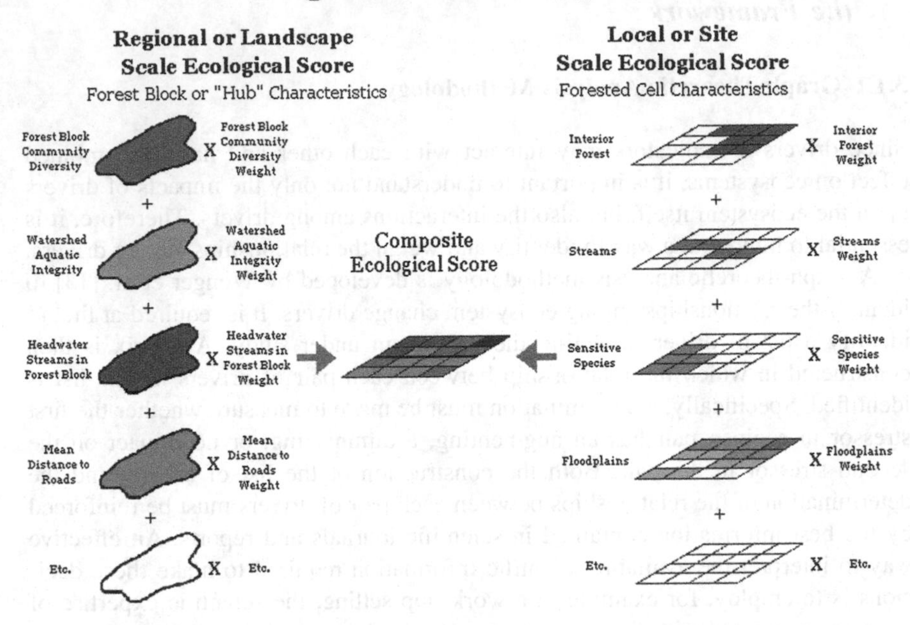

Fig. 2 Example of the process of index overlay. *Source* http://www.dnr.state.md.us/forests/planning/sfla/images/ecol_apph.gif

synthetically with respect to the assessment model. Based on fundamental model, several ameliorated models have been generated with the introduction of fuzzy sets and multi-criteria theories [14–16].

3.1.3 Fuzzy Logic Approach

Conventional approaches for assessing ecosystems are based on crisp sets for indicators of ecosystem health or services, such as population, regional GDP, biodiversity, and environmental quality. Defining a crisp sets for attributes of an ecosystem is based on the assumption that it is possible to make a sharp, unambiguous distinction between an ecosystem that is healthy and one that is comparably not so healthy. However, due to the uncertainties inherent in ecosystem assessments, defining thresholds for those attributes is arbitrary and could give rise to faulty or misleading conclusions. An alternative approach for evaluating strong sustainability is proposed by Prato [16] that uses fuzzy sets to develop fuzzy propositions about ecosystem attributes and strong sustainability and applies fuzzy logic to evaluate those propositions. Due to fuzzy logic's ability to resolve ambiguity, it is able to process an input space to an output space through a mechanism of if–then inference rules. Fuzzy logic techniques in the form of approximate reasoning offer powerful reasoning capabilities for decision support and expert systems [17].

According to Prato [16], evaluation of the an ecosystem in terms of sustainability requires the manager to (1) specify the prior probability the ecosystem is strongly sustainable; (2) define fuzzy sets for combinations of attribute values and probability qualifiers; (3) estimate the joint frequency distribution for ecosystem attributes; and (4) develop a rule for inferring strong sustainability from fuzzy propositions.

4 Conclusion

In conclusion, studies of both ecosystem assessment and ecosystem management can benefit from the very rapid advances in geospatial technology in the past decades. Geomatics has been successfully applied in biodiversity assessment, wetland mapping, drylands degradation assessment, measurements of land surface and marine attribute, and many other applications, as inputs to ecosystem assessment and management models. Wang et al. [18] and Yang [19] applied remote sensing and GIS in the spatiotemporal dynamic analysis of land use/cover change by using post-classification comparison and GIS overlay techniques. Bydekerke et al. [20], Ceballos-Silva and Lopez-Blanco [21], and Kalogirou [22] all have used GIS techniques to identify suitable areas for crops, MCE approach and fuzzy membership function were also used to generate standardized factor maps. RS and GIS environment was also applied successfully to anticipate degradation and to take mitigating measures at an early stage [2]. The quality of life (QOL) of could also be assessed by integrating environmental variables extracted from Landsat thematic mapper data with socioeconomic variables obtained from the Census data, in which principal components analysis (PCA) method was used [23].

The development of Geomatics in recent years has been revolutionized by rapid advances in computer hardware and software. Despite the technical advances in Geomatics and the expanding applications in research, there are still a number of fundamental issues that remain unaddressed when using Geomatics techniques. For example, there is little theoretic consideration of finding the most appropriate spatial resolution (scale) for remote data while considering the domain of the study area. In addition, GIS still lacks many statistical functions. Users often have to use external statistical software in order to run some statistical analysis.

Acknowledgments The authors would like to thank International Science and Technology Cooperation Program of China (2012DFA11270) and Hainan International Cooperation Key Project (GJXM201105) for the support given to the study.

References

1. Burrough, P.A.: Principles of Geographical Information Systems for Land Resources Assessment. Clarendon Press, Oxford (1986)
2. Dahdouh-Guebas, F.: The use of remote sensing and GIS in the sustainable management of tropical coastal ecosystems. Environ. Dev. Sustain. **4**, 93–112 (2002)
3. Steyaert, L.T., Goodchild, M.F.: Integrating geographic information systems and environmental simulation models: a status review. In: Michener, W.K., Brunt, J.W., Stafford, S.G. (eds.) Environmental Information Management: Ecosystem to Global Scales. Taylor and Francis Ltd., London (1994)
4. Millennium Ecosystem Assessment: Ecosystems and Human Well-Being. Island Press, Washington D. C. (2005)
5. Holling, C.S.: Adaptive Environmental Assessment and Management. Wiley, London (1978)
6. Environmental and Social Systems Analysts Ltd. Review and Evaluation of Adaptive Environmental Assessment and Management. Environment Canada, Vancouver (1982)
7. Walters, C.: Adaptive Management of Renewable Resources. Macmillan Publishing Company, New York (1986)
8. Li, S., Dragicevic, S., Veenendaal, B.: Advances in Web-based GIS, Mapping Services and Applications. CRC Press Inc., Boca Raton (2011)
9. Onwuegbuzie, A.J., Leech, N.L., Collins, K.M.T.: Innovative data collection strategies in qualitative research. The qualitative report, 15, pp. 696–726 (2010)
10. Brown, G.: An empirical evaluation of the spatial accuracy of public participation GIS (PPGIS) data. Appl. Geogr. **34**, 289–294 (2012)
11. Kyem, P.A.K., Saku, J.C.: Web-based GIS and the future of participatory GIS applications within local and indigenous communities. Electron. J. Inf. Syst. Dev. Countries **38**, 1–16 (2009)
12. Kulawiak, M., Prospathopoulos, A., Perivoliotis, L., łuba, M., Kioroglou, S., Stepnowski, A.: Interactive visualization of marine pollution monitoring and forecasting data via a web-based GIS. Comput. Geosci. **36**, 1069–1080 (2010)
13. Wenger, R., Harris, H., Sivanpillai, R., DeVault, D.: A graph-theoretic analysis of relationships among ecosystem stressors. J. Environ. Manage. **57**, 109–122 (1999)
14. Cornelissen, A., Van den Berg, J., Koops, W., Grossman, M., Udo, H.: Assessment of the contribution of sustainability indicators to sustainable development: a novel approach using fuzzy set theory. Agric. Ecosyst. Environ. **86**, 173–185 (2001)
15. McGlade, J.M.: A diversity based fuzzy systems approach to ecosystem health assessment. Aquat. Ecosyst. Health Manage. **6**, 205–216 (2003)
16. Prato, T.: A fuzzy logic approach for evaluating ecosystem sustainability. Ecol. Model. **187**, 361–368 (2005)
17. Kulkarni, A.D.: Neural-fuzzy models for multispectral image analysis. Appl. Intell. **8**, 173–187 (1998)
18. Wang, Z., Zhang, B., Zhang, S., Li, X., Liu, D., Song, K., Li, J., Li, F., Duan, H.: Changes of land use and of ecosystem service values in Sanjiang Plain, Northeast China. Environ. Monit. Assess. **112**, 69–91 (2006)
19. Yang, X.: Remote sensing and GIS applications for estuarine ecosystem analysis: an overview. Int. J. Remote Sens. **26**, 5347–5356 (2005)
20. Bydekerke, L., Van Ranst, E., Vanmechelen, L., Groenemans, R.: Land suitability assessment for cherimoya in southern Ecuador using expert knowledge and GIS. Agric. Ecosyst. Environ. **69**, 89–98 (1998)
21. Ceballos-Silva, A., Lopez-Blanco, J.: Delineation of suitable areas for crops using a Multi-Criteria Evaluation approach and land use/cover mapping: a case study in Central Mexico. Agric. Syst. **77**, 117–136 (2003)

22. Kalogirou, S.: Expert systems and GIS: an application of land suitability evaluation. Comput. Environ. Urban Syst. **26**, 89–112 (2002)
23. Tong, C., Wu, J., Yong, S-p, Yang, J., Yong, W.: A landscape-scale assessment of steppe degradation in the Xilin River Basin, Inner Mongolia, China. J. Arid Environ. **59**, 133–149 (2004)

Framework of an Ecotourism Early Warning System: What can we Learn for Hainan, China?

Li-qin Zhang, Jiang-feng Li and Sheng-Quan Ma

Abstract Early warning systems (EWSs) are important for risk management. This chapter develops a framework of an ecotourism EWS which includes hazard risk, carrying capacity of service facilities and tourism sites, and tourism development impact assessment, monitoring and warning. Tourism impact is an emerging research spot. In EWSs, negative tourism impacts should be emphasized, including environmental, economic, and social impacts. After the framework is constructed and analyzed, cases based on local characteristics are selected and the concentrated points for ecotourism early warning are analyzed.

Keywords Early warning · Ecotourism · Early warning framework · Tourism impact · Hainan

1 Introduction

Early warning system (EWS) is a key element for risk reduction and adaptation strategies to extreme events. Traditional EWSs focus mainly on hazard detection and immediate warning and evacuation processes in order to save lives in the context of an extreme event. Modern EWSs can be applied to societal well-being as well as to the well-being of individuals and the environment. According to different division base, EWSs can be classified as the following types [1]: formal

L. Zhang · J. Li
Faculty of Public Administration, China University of GeoSciences, Wuhan 430070, China

L. Zhang (✉)
Department of Geography, University of Ottawa, Ottawa, Canada
e-mail: lzhan143@uottawa.ca

S.-Q. Ma
School of Information and Technology, Hainan Normal University, Haikou, China

B.-Y. Cao et al. (eds.), *Ecosystem Assessment and Fuzzy Systems Management*,
Advances in Intelligent Systems and Computing 254, DOI: 10.1007/978-3-319-03449-2_3,

and informal EWSs, political and apolitical EWSs, national and indigenous EWSs, objectively based and subjectively based EWSs, qualitative and quantitative EWSs, and so on. There are four components of EWSs: sources, channels, contents, and receivers [2]. Sources refer to the entity responsible for initiating hazard communication with the public such as government authorities and media figures. Channels are the communications medium used to transmit hazard information such as television and the Internet. Contents include the information of hazard, location, time, and guidance, based on the assessment of the existence and seriousness of a hazard and transmit information of guidance to the public for protecting themselves and adapting to the hazardous environment. Receivers remind disaster warning designers that they must consider the characteristics of the at-risk populations, which may influence the effects of the guidance seriously. Early warning is a part of risk management systems, which are a developing cycle of "hazardous event–response–recovery–mitigation–preparation–warning–adaptation and remediation" [3] (Fig. 1).

From detectability and visibility, risk can be classified into visible and invisible types. Visible risks are normally directly perceptible, or perceptible with the help of science, or virtual risks. Invisible risks generally have characteristics of being hidden from the senses, time-hidden (time latent), or scale-hidden (scale latent). Most underwater risks are hidden from the senses. Risks influenced by the ecosystem are always time-hidden. And risks in water flow normally occupy a large area and are scale-hidden [4].

Tourism management emphasizes risks prevention and related responsibilities guidance, for different stakeholders [5]. Uncertainty is one of the challenges for risk and risk tolerance simulation and forecast, especially for tourism management planning [6–8]. How to integrate individual's economic choice with coastal risk management [9], and how an EWS effects environment [10], are hot spots for research.

The objectives of ecotourism EWSs are threefold. The first objective is to reduce impacts on the environment and enhance interaction with nature. The secondary objective is to help tourists behave responsibly and enhance tourist satisfaction. The third objective is to improve the tourism impact on the local economy, to enhance community involvement and interaction with local people, and to improve local people's quality of life.

2 Ecotourism Early Warning Framework

To build an EWS, the framework should be studied and set up in advance. For sustainable ecotourism development, the complete EWS should include not only hazard forecast, simulation, and warning, but also ecotourism capacity monitoring and warning, with tourism impacts simulation and warning. Hazard forecast,

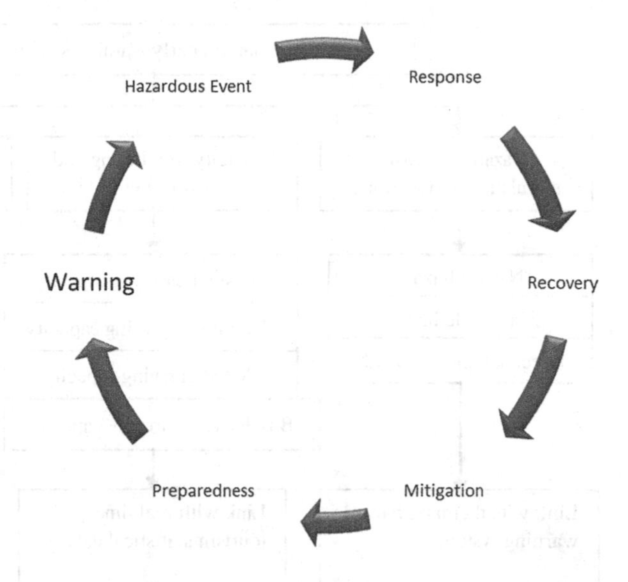

Fig. 1 Warning and risk management system

simulation, and warning are the fundamental contents of risk assessment and warning systems. It could be timely warning, mid-term warning, or long-term warning. Capacity monitoring and warning should provide receivers periodical information, i.e., hourly, daily, weekly, seasonal, or yearly. System impacts are applied to simulate the impacts of tourism development, forecast it and provide warning for negative outcomes.

Ecotourism EWSs have complex contents. Hazard subsystems include natural hazards and extreme weather such as tropical cyclones or coastal flooding; man-made hazards such as transportation accidents, fuel/resource shortages, air/water pollution/contamination, health/disease outbreaks and illness, criminal activity; and technological hazards, for example IT systems failure, telecommunications, product defect or contamination, which could be linked with the thematic hazard warning system [11]. Capacity subsystem refers to physical, economic, social, and biophysical carrying capacities, which could be linked with a real-time tourism statistical database. Tourism impact subsystem includes environmental, economic, social, and comprehensive impact simulation and warning, in which modeling and forecasts for medium- and long-term impacts are emphasized (Fig. 2).

In the framework, hazard risk monitoring is connected with the preexisting thematic hazard monitoring system. Carrying capacity monitoring should use the tourism statistical database and analyze relationships between tourism hotel, tourism sites, and tourist demands in order to find a balance. Tourism impact assessment is the most important content for research, as well as for long-term risk monitoring and warning.

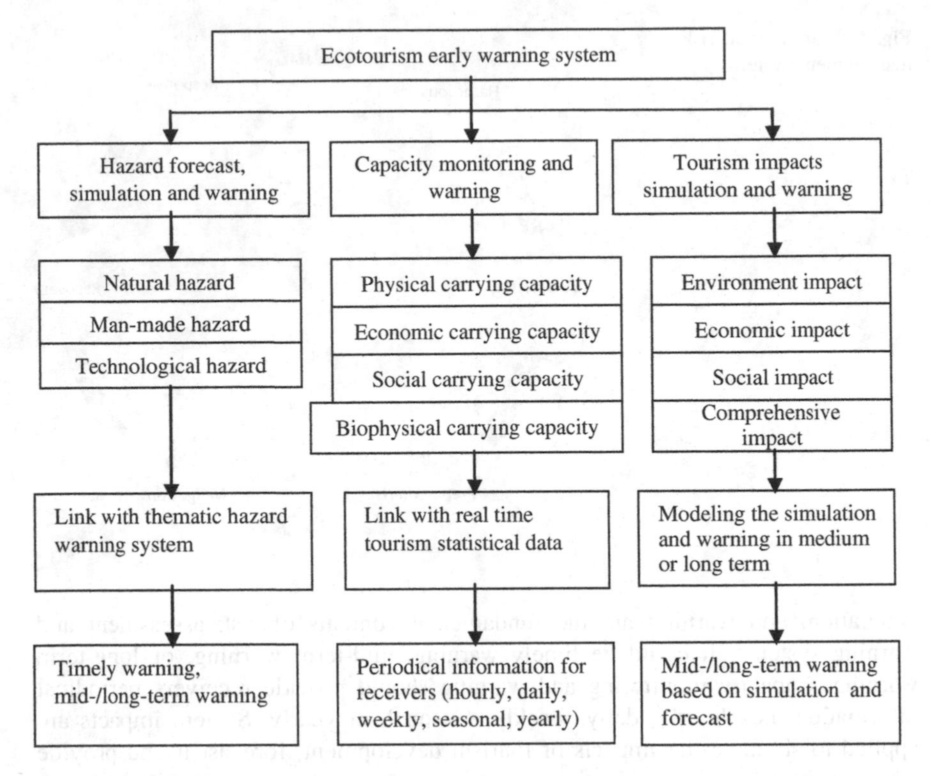

Fig. 2 The framework of ecotourism early warning system

3 Tourism Impact Assessments

3.1 Impacts of Ecotourism

Impact of tourism development is the key issue that should be studied in ecotourism warning system, since it concerns the long-term development sustainability of ecotourism itself. Ecotourism development generates positive and negative impacts on the economic, environment, and social community. Balance between different aspects is an important base for its sustainability. Ecotourism early warning focuses on negative impacts assessment, since negative aspects cannot be deducted by positive ones, even though in political decision-making processes, they normally are deducted. Potential negative environment ecotourism impacts are contained in many aspects [12], shown in Fig. 3.

From the perspectives of tourists' and indigenous communities' angles, the positive impact is the opportunity for leisure, enhancement of knowledge, improvements to indigenous communities' infrastructure, economic benefits, and the conservation of natural and cultural resources for communities. Tourists are negatively impacted when there is a disconnection between their expectations and

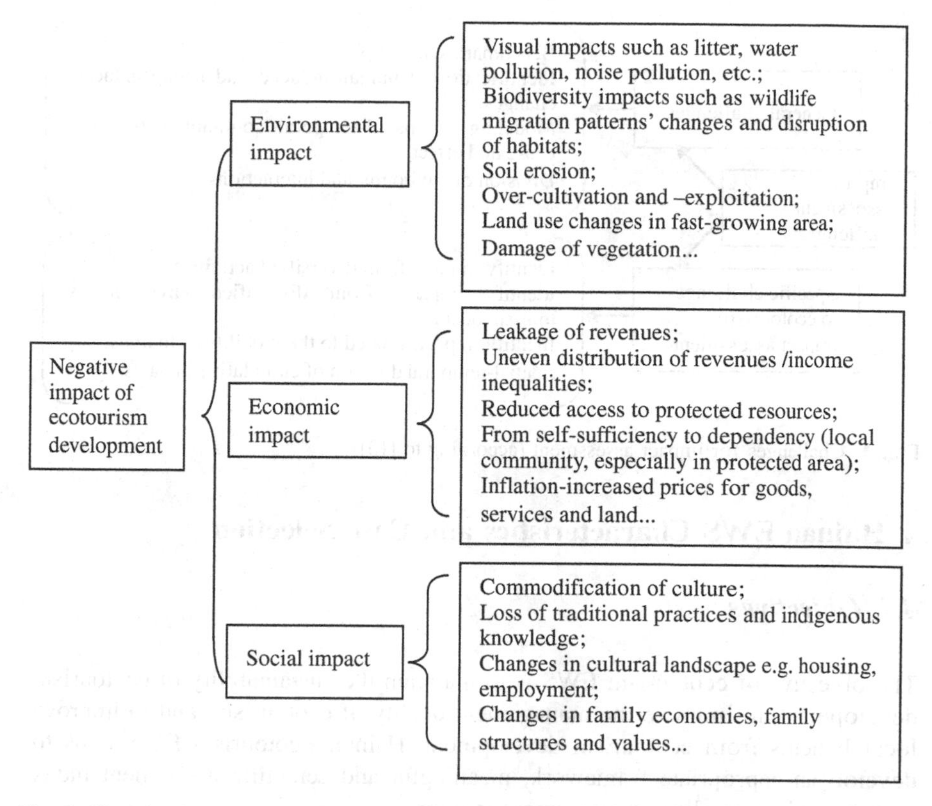

Fig. 3 Negative impact of ecotourism development (according to [12])

their actual experiences. Negative impacts for indigenous communities include over-straining infrastructure, socioeconomic unrest, and damage to the cultural traditions and the natural environment.

3.2 Challenges for Tourism Impact Assessments

Impact assessment is the most important and fundamental part for tourism EWS, for which challenges exist [13] (Fig. 4).

For all forms of impact assessment, there are still difficulties of establishing a benchmark against which to measure change; distinguishing human-induced changes from natural changes; modeling of the relationships between cause and effect; identification of the complexity of environmental interactions, and so on.

Specifically for ecotourism development impact assessment, challenges exist in identifying impacts from the diversified tourism and management activities and from the diversified environments in which tourism activities occur, as well as measuring impacts from the mobility of tourists, and dividing the impacts from accumulations.

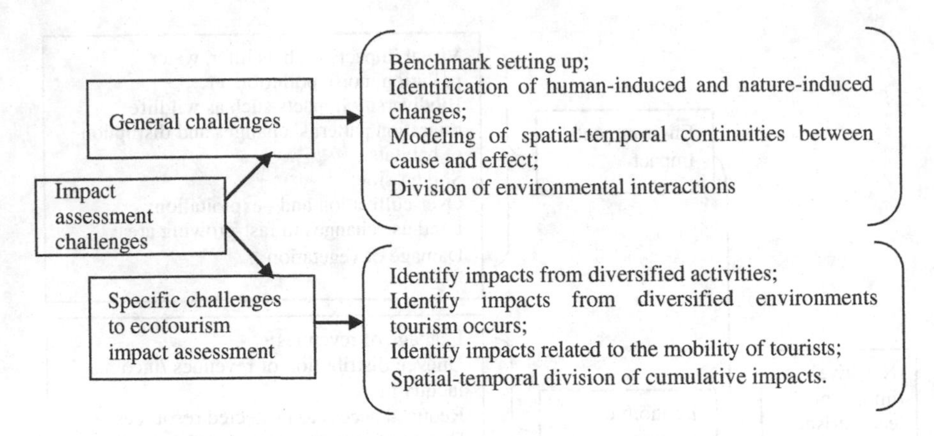

Fig. 4 Challenges for impact assessment (according to [13])

4 Hainan EWS Characteristics and Case Selection

4.1 Objectives

The objective of ecotourism EWS is to maintain the sustainability of ecotourism development, to improve the management quality of ecotourism, and to improve local benefits from ecotourism development. Hainan Ecotourism EWS aims to develop an appropriate framework, meaningful and scientific assessment index system, and a visualized management system.

The framework of this ecotourism EWS illustrates that we should focus on the main characteristics of a given study area and the negative impact assessment of ecotourism development.

Hainan is an international ecotourism island with high-quality coastal resources, high-quality forest and biodiversity, minority populations and an ocean/marine climate.

To assess ecotourism impacts on the environment, economy, and society, the focus should be on coastal regions, natural forestry regions, and indigenous communities. According to the characteristics, Sanya, Ledong, and Wuzhishan are the most appropriate cases for developing an ecotourism EWS (Table 1; Fig. 5).

4.2 Sanya

Sanya is the southernmost city on the Chinese island, with a population of 685,408 (according to the 2010 census), living in an area of 1,919.58 km^2. The city is renowned for its tropical climate and has emerged as the most popular Chinese "Sun, Sand and Sea" tourist destination. International resort hotel chains and new

Table 1 Case and characteristics of ecotourism EWS in Hainan, China

Case study area	Characteristics
Sanya	Urban, coastal (fine sandy), and developed tourism
Wuzhishan	Terrestrial, hilly/mountainous, nature-based resources, tropical forest, and developed/developing tourism
Ledong	Rural, coastal (fine sandy), minority, and developing tourism

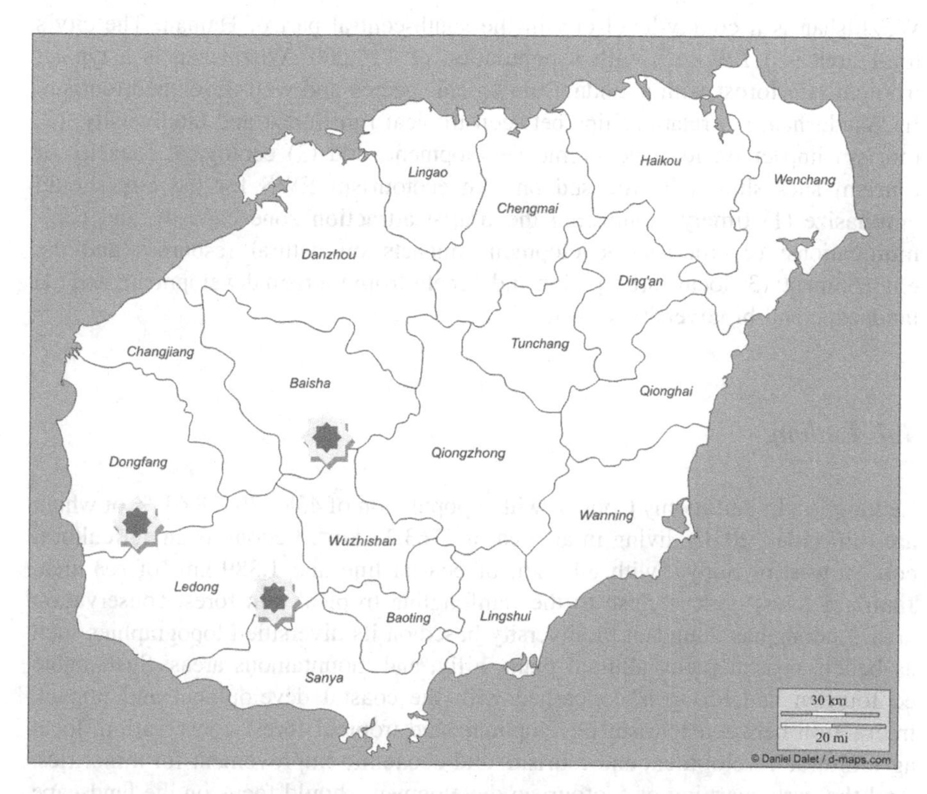

Fig. 5 Case location map in Hainan (Sanya, Wuzhishan, and Ledong) (The base map is from http://d-maps.com/carte.php?num_car=21212&lang=en)

venues for resorts have been developed in the most popular sites such as Dadonghai and Yalong Bay. Prosperous tourism development also pushed city development with tourism, recreational, and real estate development, which expanded the physical space of the city and changed the structure of the society. In Sanya, the following themes for ecotourism development should be examined: (1) coastal security from landscape/physical and biodiversity security; (2) tourism economic impact related to urban fringe equity and the interactive role between tourism development and macro economy; and (3) the capacity of tourism sites and facilities. An EWS for Sanya's ecotourism development should focus on: (1)

urban landscape change and warning the landscape security, especially for coastal area and urban fringe; (2) timely monitoring for tourism facility capacity; and (3) local people livelihood benefit from tourism industry (economic equity).

4.3 Wuzhishan

Wuzhishan is a county-level city in the south-central part of Hainan. The city's total area is 1,129 km^2, with a population of 115,000. Wuzhishan is a typical tropical rain forest with abundant biological species and well-developed tourism. In Wuzhishan, (1) relationships between tropical rain forest and biodiversity, (2) tourism impact on local economic development, and (3) ecological capacity of tourism sites should be focused on. An ecotourism EWS for the city should emphasize (1) timely monitoring the tourist attraction zone capacity and communications; (2) tourism development impacts on natural resources and the environment; (3) local participation and benefit from tourism development; and (4) landscape and biodiversity security.

4.4 Ledong

Ledong is a Li Autonomy County, with a population of 458,876 (38.61 % of which are minorities, 2010), living in an area of 2763.22 km^2. Ledong is an agricultural county west of Sanya, with 84.3 km of coastal line and 1,389 km^2 of sea area. Jianfeng forest park is close to the Jianfengling tropical rain forest conservation area. Ledong has abundant biodiversity based on its diversified topographies such as beach, coastal plain, alluvial plain, hilly, and mountainous areas. Sustainable ecotourism in Ledong is concerned with the coastal development and impact, interaction between tourism development and tropical forest conservation, local agricultural development and tourism, and economic improvement for minorities. And the early warning of ecotourism development should focus on the landscape and biodiversity security based on tourism resources development plan, the impact of drought, agricultural, local participation on the tourism sustainability, and the impact of tourism development on the local economy.

5 Conclusion and Discussion

Early warning is the most important element for risk management systems. Ecotourism EWSs include hazard risk, carrying capacity, and tourism impact monitoring, simulation and warning. Hazard risk monitoring can be linked with preexisting thematic warning system. Capacity monitoring should focus on tourism facility and resource carrying capacity, respectively. And impact assessment and simulation is the most important content for research and for long-term warning system. The impact assessment should focus on the negative aspects, not the balance between negative and positive, even though most of the political decisions are based on a balance. The impacts are for environment, economy, and society, aiming on the special characteristics within study areas. For Hainan, Sanya—the most famous southernmost coastal tourism city, Wuzhishan—the northern neighboring county to Sanya with high-quality tropical rain forest and local communities, and Ledong Li autonomous county with the expansion pressure for coastal tourism development based on the special west neighboring location to Sanya are selected as the typical cases for Ledong ecotourism EWS building.

Even though there remain lots of challenges for ecotourism impact assessment and the long-term real-time monitoring, for example, the benchmark for the changes, the relationships between cause and effect, the division between accumulate and diversified-driving-forces-induced changes, the framework and general methodology are available for further probing.

Acknowledgments Thanks to the support by the Fundamental Research Funds for the Central Universities, China University of Geosciences (Wuhan), and the project of Construction of an Early Warning Information Platform/Module based on Ecotourism Assessment and Management (EAM)—Its Application in the Building of Hainan International Tourism Island of China (2012DFA11270).

References

1. Glantz, M.H.: Early warning systems: do's and don'ts. Report of workshop 20–23 Oct. 2003, Shanghai China. Usable Sci 8 (2003). www.esig.ucar.edu/warning/
2. Grasso, V.F.: Early warning systems: state-of-art analysis and future directions. UNEP report (2007)
3. Samarajiva, R., Knight-John, M., Anderson, P.S., Zainudeen, A.: National early warning system: Sri Lanka. A participatory concept paper for the design of an effective all-hazard public warning system. Version 2.1. 2005. LIRNEasia
4. Yamashita, H.: Making invisible risks visible: education, environmental risk information and coastal development. Ocean Coast. Manag. **52**, 327–335 (2009)
5. Peattie, S., Clarke, P., Peattie, K.: Risk and responsibility in tourism: promoting sun-safety. Toursim Manage. **26**, 399–408 (2005)
6. Quintal, V.A., Lee, J.A., Soutar, G.N.: Risk, uncertainty and the theory of planned behavior: a tourism example. Tourism Manage. **31**, 797–805 (2010)
7. Williams, A.M., Balaz, V.: Tourism, risk tolerance and competences: travel organization and tourism hazards. Tourism Manage. **35**, 209–221 (2013)
8. Oroian, M., Gheres, M.: Developing a risk management model in travel agencies activity: an empirical analysis. Tourism Manage. **33**, 1598–1603 (2012)
9. Filatova, T., Mulder, J.P.M., Veen, A.: Coastal risk management: how to motivate individual economic decision to lower flood risk? Ocean Coast. Manag. **54**, 164–172 (2011)
10. Lenton, T.M.: What early warning system are there for environmental shocks? Environ. Sci. Policy **27**, S60–S75 (2012)

11. UNEP (United Nation Environment Program). Disaster risk management for coastal tourism destinations responding to climate change: a practical guide for decision makers. (2008)
12. Matthews, E.J.: Ecotourism: are current practices delivering desired outcomes? A comparative case study analysis. Major paper submitted to the Faculty of the Virginia Polytechnic Institute and State University in partial fulfillment of the requirements for the degree of Master of Urban and Regional Planning in Urban Affairs and Planning (2002)
13. Wall, G.: Ecotourism: change, impacts and opportunities. In: The Ecotourism Equation: Measuring the Impacts, Bulletin Series 99. Yale School of Forestry and Environmental Studies, New Haven, pp. 108–117 (1996)

Study on Eco-Safety Early-Warning and Assessment Index System of Hainan Province

Sheng-Quan Ma, Hu-hua Cao, Jian-xin Wang and Song Lin

Abstract To have a comprehensive evaluation on the safety of regional ecology, research is required within the fields of eco-safety and early warning on all relevant elements that affect the ecological environment of a given region. This should be complemented by research on influential elements based on regional characteristics and then followed by the construction of an eco-safety early-warning assessment index system. This article analyzes research findings of related ecological assessments and eco-safety problems of Hainan province in particular. With a focus on economic construction and sustainabe development, this analysis reflects on economic, social, environmental, technological, and institutional factors and proposes an eco-safety early-warning assessment index system which suits Hainan's ecological situation. This index system synthesizes Hainan's natural resources, ecological situation, degree of pollution, and economic development and consists of five first-class indicators (low-carbon society, land resource safety, water resource safety, air resource safety, and biological species safety), ten second-class indicators (economy, society, environmental resources, technology, low carbon, system, arable land safety, forest safety, land pollution, etc.) and 90 third-class indicators.

Keywords Eco-safety · Ecological assessment · Safety early-warning · Index system · Hainan Province

S.-Q. Ma (✉) · J. Wang · S. Lin
School of Information and Technology, Hainan Normal University, Haikou 571158, China
e-mail: mashengquan@163.com

H. Cao
Department of Geography, University of Ottawa, Ottawa, Canada
e-mail: caohuhua@uottawa.ca

B.-Y. Cao et al. (eds.), *Ecosystem Assessment and Fuzzy Systems Management*, Advances in Intelligent Systems and Computing 254, DOI: 10.1007/978-3-319-03449-2_4,

1 Introduction

With the strategy of 'Develop[ing] Hainan [as an] International Tourism Island' being implemented, Hainan will certainly increase its pace of urbanization and industrialization, drawing an abundance of tourists and leading to negative impacts on the local ecology. How to rationally develop and use natural resources; reduce the pressure of human activities on the local environment; explore the potential of available resources; ensure ecological safety; and promote sustainable local economic development have become key research foci in local development processes. Going deeply on current Hainan eco-safety status, assessing eco-safety are particularly important to guarantee the sustainable development of Hainan's economy. Eco-safety refers to the level of danger present in a system composed of human, economic, social, and environmental elements. The sources of danger are excessive consumption of natural resources and environmental degradation caused by human economic activities. The objective of the eco-safety assessment is to develop a strategy for sustainable development between humans and the environment.

2 Ecological Resource Survey of Hainan Province [1, 6]

Hainan became a province in 1988, but has a history spanning more than 2,000 years since its establishment as an administrative division in 110 BC. Now, it contains three prefecture-level cities (Hainan, Sanya, Sansha), six county-level cities, and 10 counties. In 2011, the total population was 9,078,200, among them mainly Han people, followed by Li, Miao, Zhuang, Hui, and other ethnic minorities. The Hainan minorities have long histories, profound cultural heritages, and diversified folk cultures.

Hainan province is located in the southernmost point of China between 3°20′–20°18′ north latitude and 107°50′–119°10′ east longitude. In the north, there is Qiongzhou Strait facing Leizhou Peninsula; to the West, there is Beibu Gulf opposite Vietnam. It is only about 220 nautical miles from the city of Haikou to Haiphong, in Vietnam. It faces Taiwan across the South China Sea on the east, neighbors the Philippines, Brunei, and Malaysia on the Southeast and South in the South China Sea. Hainan Island circled by the sea and neighboring other countries on three sides is the South gate of China.

The administrative region of Hainan province consists of Hainan Island, Xisha Islands, the Zhongsha Islands, the Nansha Islands, and their respective maritime spaces. The land area of the whole province covers 35,400 km^2. Hainan Island, occupying more than 34,000 km^2, is the second largest island in China next to Taiwan, and its sea territory, covering an area of around 2 million km^2, is about two-thirds of the sea space that China governs. This makes Hainan a tropical island province with the smallest land area and largest sea area in China.

The mountainous and hilly land in Hainan, concentrated in the central south of the island, takes up 38.7 % of the total area. Around the mountainous and hilly lands, there are platforms and terraces which occupy 49.5 % of the total area. The roundabouts account for 11.2 % of the total area. The coasts are formed from marine abrasion of the volcano basalt platforms, small harbors or accumulative landforms evolved from drowned valleys, and terraces surrounded by sand bank. The coast ecology is characterized by tropical mangrove and coral reef coasts.

The Xisha, Nansha, and Zhongsha Islands are relatively low and flat with an elevation of 4–5 m generally. The vrakhonisis of the Xisha Islands is the highest with an elevation of 15.9 m or so.

The mountain ranges in Hainan Island are mostly 500–800 m high. The Wuzhi, Yinggeling, Bawangling, Diaoluoshan, and Limuling Mountains are more than 1,500 m above sea level, and they are divided into three mountain ranges: Wuzhi, Yinggoling, and Yajiadaling from east to west. Wuzhi Mountain Range, the highest mountain in Hainan, is in the middle of the Island with its largest peak reaching 1,867 m above sea level. Yinggoling Mountain Range lies northwest of Wuzhi Mountain, with its largest peak being 1,811 m above sea level, while Yajialing, located in the west of the Island, has a main peak that reaches 1,519.1 m above sea level.

Hainan Island has a much higher elevation in the center of the island, with its largest rivers located in the central mountainous area, forming a radiation-like river system. There are 154 rivers flowing from Hainan into the sea, 38 of which have more than 100 km^2 of water-collecting area. The Nandujiang, Changhua, and Wanquan Rivers are the three biggest rivers in Hainan, with water-collecting areas exceeding 3,000 km^2, and a drainage area occupying 47 % of Hainan. Nandujiang River originates from Nanfeng Mountain of Baisha and passes through the middle and north part of Hainan, with a length of 331 km and a drainage area of 7,176 km^2. Changhuajiang River originates from Qiongzhong and traverses the middle and west part of Hainan, with a length of 230 km and a drainage area of 5,070 km^2. There are two branches in the upper reach of the Wanquan River, one in the north and the other in the south, both originating from Qiongzhong with its main stream being 163 km long, and with a total catchment area of 3,683 km^2.

Hainan Island is a tropical oceanic monsoon climate, warm and hot all year round with abundant rainfall, dry and wet seasons, windy weather, and frequent tropical storms and typhoons. It has an annual average temperature of 22.8–25.8 °C, with an average temperature during the hottest month (July and August) 25–28 °C, that of the coldest month (January and February) between 16 and 24 °C, and with an average extreme low temperature of above 5 °C, thus with no severe heat in summer and no severe cold in Winter. There is abundant rainfall in most areas of Hainan, with an annual average rainfall of over 1,640 mm. It is wet in the east and dry in the west, while the rainy area in the east-central mountainous area has an annual rainfall of about 2,000–2,400 mm. There is less rain in the west, with an annual average rainfall of about 1,000–1,200 mm. The rain distributes unevenly in the year with a dry season from November–May, and a wet season in May–October, with rainfall in the latter accounting for 70–90 % of that of the full year.

Xisha, Nansha, Zhongsha Islands have tropical marine climates. With long summers and no winters, the annual average temperature of these places is 26.5 °C. The highest temperature is in August, with an average temperature of 29.5 °C, and with the lowest temperature in February, with an average temperature of 22.9 °C.

Hainan province is the largest tropical area in China, not only is the natural condition superior, but the natural resources are very rich. Thanks to superior light, heat, water, and other natural resources, plants grow much more rapidly than those in other areas of China. Farm lands can be cultivated all year round, and many crops can be harvested 2–3 times a year. Hainan's land is suitable for cultivation, with the exception of steep slope lands that are more than 800 m above sea level, and sandy areas being short of water and dry that both suit only forest planting. Potential land productivity in Hainan is high, not just for general agricultural products, forest products, and animal husbandry, but also for tropical crops, rare herbs and medicines that are of high economic value.

Hainan has abundant rainfall and large river runoff. There are multiple large reservoirs in Hainan, the best of which are the Songtao, Nanfu, Changmao, and Shilu reservoirs.

The annual precipitation of Hainan is 59.6 billion cubic meters, yielding water resources of 29.7 billion cubic meters. Its fresh water resources can meet lifestyle, industrial, and agricultural demands.

Hainan is the major tropical crop producing area in China. At present, the tropical crops of high economic value in Hainan are rubber, coconut, areca-nut, black pepper, coffee, sisal hemp, citronella, cashew nut, and cocoa. Known as a 'natural greenhouse', 'tropic orchard', and 'four-season garden', Hainan is the major province developing tropical agriculture in China.

There are more than 2,000 different species of trees and shrubs in Hainan, among them, 800 of high economic value and 20 of which are subject to national protection policies. Among 4,000 plant species on the island, 3,100 are medicinal, including four famous South medicines areca-nut, alpinia oxyphylla, fructus amomi, and Radix Morindae Officinalis. Hainan accounts for 99.9 % of the total output for these medicinal plants in China.

Among animal resources in Hainan, 561 are terrestrial vertebrates and 102 are subject to national protection. There is a variety of rare animals in Hainan, among them the Eastern black crested gibbon (*Nomascus nasutus*) being one of the four anthropoids in the world and known as 'the treasure of China'; and Eld's deer (cervus eldii) being praised as a 'rare treasure'. Other animals include red deer, macaque, black bear, and the clouded leopard.

There are eight national level natural reserves covering 22,900 ha and 21 provincial level natural reserves. The national level natural reserves are Datian Natural Reserve (founded in 1986 and home to the aforementioned Eld's deer), Dongzhai Port Mangrove reserve (founded in 1980 and home to a mangrove forest), Sanya Coral Reef reserve (founded in 1990 for its coral reef), Bawangling reserve (founded in 1988, contains Nomascus hainnanus), Dazhou Island reserve (founded in 1990 for its esculent swift), Jianfengling reserve (founded in 2002), Tongguling reserve (founded in 2003), and Wuzhishan Mountain reserve (founded in 2003).

Hainan, possessing a wealth of offshore fishing grounds, a variety of species and a long catching season, is the ideal place to develop a tropical marine fishery in China. There are more than 800 aquatic products in Hainan, 600 of which are fish, 40 of which are of high economic value with muraenesox, large yellow croaker, trichiurus haumela, dogfishes, and garrupa being the main species. Marine mollusks include cuttlefish, sleeve fish, sea cucumber, and sandworms. At present, more than 20 kinds of fish, shrimp, shellfish, algae that are of high economic value are cultivated in shallow marine areas, including garrupa, abalone, knob prawn, China lobster, Pinctada maxima jameson, Madai pearl shell, and offshore oyster. Special fresh water aquaculture resources include soft-shelled turtle, frog, otter, mink, and giant salamander. The temperature of the offshore sea water is moderate, with more than 3,000 marine organisms growing in it, including more than 1,000 fish species, more than 200 algae species and more than 100 kinds of coral reef. The diverse tropical fish and coral reefs provide a beautiful view in diving tours.

Hainan is also rich in ore resources. Ninety kinds of minerals have been discovered, accounting for 55 % of all discovered minerals in China. Hainan has explored 67 kinds of minerals with proved reserves, of which 41 have been entered into national reserves. There are energy (ilmenite, zirconite, monohydralite) and nonmetal (sapphire, red diamonds, crystal, silica sand) mineral resources.

The South China Sea is one of four marine oil storage areas in the world. Shilu iron deposits account for 71 % of total national high-grade iron ore reserves. Titanium, silica, sapphire, and chemical fertilizer limestone reserves are most valuable in China, while oil shale and granite reserves are among the top in the whole nation.

Hainan Island, having a pleasant climate that remains green through the year, with an attractive tropical landscape and unique oceanic islands, has great potential for developing tourism. According to a general survey, there are a total of 241 exploitable natural and cultural resources in Hainan categorized into 11 types as follows[1, 2]:

1. 38 sandy beaches including Yalong Bay, Dadong Sea, Ends of the Sea, Gaolong Bay, Shimei Bay, etc;
2. 28 mountains including Wuzhishan Mountain, Jianfengling Mountain, Qizhiling Mountain, Dongshanling Mountain, Tongguling Mountain, etc;
3. 18 rare stone and bizarre caves including Maogongshan Mountain, Jigongshan Mountain, Emperor Cave, Pen Dropping Cave, etc;
4. 19 rivers and lakes including Wanquan River, Songtao reservoir, Nanli Lake, etc;
5. 11 waterfalls including Fengguoshan Waterfall, Baihualing Waterfall, Taipingshan Waterfall, etc;

6. 38 hot springs including Xinglong Hot Spring, Guantang Hot Spring, Nantian Hot Spring, Lanyang Hot Spring, Qixianling Hot Spring, etc;
7. 18 wildlife sightseeing spots including Datian cervus eldii, Bawangling nomascus hainnanus, Nanwan Monkey Island, Dongzhai Harbor mangrove forest, Eastern Suburb coconut forest, etc;
8. 13 islands including Wuzhizhou Island, Dazhou Island, Wild Boar Island, etc;
9. 25 historic buildings including the five Saints Temple, Qiongtai Ancient Academy, Dongpo Ancient Academy and eight ancient tombs including Hairui Tomb, Qiujun Tomb, Zhang Yuesong Tomb and Zhaoding cenotaph, etc;
10. modern historical sites including the ancestral home of Soong family, Feng Baiju Former residence; former site of Hainan Column headquarter, etc; and
11. revolutionary martyrs monuments including Hainan Revolutionary martyrs Monument, the memorial statue of the Red Detachment of Women; Statue of General Feng Baiju and the Memorial Pavilion, Tomb of Martyr Li Shuoxun, Jinniuling Martyrs Cemetery, and the Memorial Hall of Baisha Uprising.

3 Building Hainan Eco-Safety Early-Warning and Assessment Index System

The comprehensive assessment of Hainan eco-safety involves natural resources, the eco-environment, the degree of environmental pollution and economic development. Statistical data being were difficult to obtain due to the prefectures being too big while villages and towns were too small. Additionally, most statistical data are using the county as the basic statistical unit, as this facilitates the collection of data and limits mistakes. The county is in the middle of the administration divisions of prefecture and township and is the basic implementation unit of all eco-environmental protection policies. Thus, to some extent, the eco-environment situation of a certain county reflects or represents the eco-safety of the region. So, the basic units of Hainan eco-safety early-warning assessment are the 19 existing counties (cities), which are Haikou city, Sanya city, Sansha city, Wuzhishan city, Wenchang city, Qionghai city, Wanning city, Danzhou city, Dongfang city, Ding'an county, Tunchang county, Chengmai county, Lingao county, Baisha Li Autonomous county, Changjiang Li Autonomous county, Ledong Li Autonomous county, Lingshui Li Autonomous county, Baoting Li and Miao Autonomous county, Qiongzhou Li, and Miao Autonomous county. According to the above discussion on devising Hainan eco-safety early-warning and assessment index system [4–8], the index system is as follows (Table 1).

Table 1 Hainan eco-safety comprehensive assessment index system

First level	Second level	Third level	Unit
Comprehensive assessment indicators of low-carbon society	Economic indicators	GDP/GDP per capita	Yuan
		Population growth rate	%
		Population density	Person/km^2
		Proportion of growth value of tertiary industry accounting for GDP	%
		GDP growth rate	%
		GDP/green GDP per capita	Yuan
		Contribution rate of technical progress on GDP	%
		Three industrial structures	
		Growth rate of domestic investment	%
		Proportion of foreign investment accounting for fixed investments of all society	%
		Net income of rural residents	Yuan
		Disposable income of urban residents	Yuan
		Fiscal revenue per capita	Yuan
		Employment rate	%
	Social indicators	Engel coefficient	
		Urbanization ratio	%
		Illiteracy rate of population aged over 15 years old	%
		Natural growth rate of population	%
		Average life expectancy	Years old
		Gross dependency ratio	%
		Social insurance coverage rate	%
	Resource environment indicators	GDP output–input ration of main mineral products	%
		Proportion of low carbon or new energy accounting for total energy resources	%
		Disposal rate of household garbage	%
		Urban air quality	%
		Comprehensive disposing rate of three industrial wastes	%
		Proportion of environmental protection investment accounting for GDP	%
		Investment strength of ecological construction	

Table 1 (coninued)

First level	Second level	Third level	Unit
		Investment strength of pollution management	
		Generalization rate of environment education	
	Technology indicators	Energy consumption per GDP	
		Carbon dioxide emission per GDP	
		Renewable energy and new energy technology	
		Carbon dioxide catchment and burial technology	
		Resource cyclic utilization rate	%
		New energy elasticity coefficient	
		Proportion of R&D investment accounting for GDP	%
		Proportion of environmental protection investment accounting for GDP	%
	Low-carbon indicators	Proportion of buildings with low carbon energy consumption	
		Utilization rate of heat insulation building materials	
		Rate of approval of CDM projects accounting for that of the world	
		Development degree of carbon finance market	
		Gross carbon dioxide emission load	Ton
		Carbon dioxide emission load per capita	Ton
		Production value of low-carbon products	Yuan
		Development degree of low-carbon agriculture	
		Recognizing degree of low-carbon awareness	
		Publicity degree of low carbon conception	
	Institutional indicators	Mechanism and policy of eco-safety system	
		Improvement degree of ecological early-warning mechanism	
		Institutional norm of eco-safety	

Table 1 (coninued)

First level	Second level	Third level	Unit
		Construction effect of eco-safety system	
		Publicity degree of government affairs	
		Administrative supervision	
		Social responsibility assessment of the enterprises	
Safety of territorial resources	Safety of arable land	Total arable area	10,000 ha
		Arable land per capita	Hectare
		Quality index of arable land	%
		Harvest-guarantee rate of farmland during droughts and floods	%
	Safety of forest	Forest coverage rate	%
		Decrease rate of forest coverage	%
		Ecological forest area ratio	%
	Land pollution	Land pollution rate	%
		Load of three industrial wastes per unit area land	Ton/km^2
		Load of fertilizer, pesticides, agricultural film per unit area arable land	Ton/km^2
	Others	Water and soil erosion	%
		Bearing rate of population	%
		Land reserves	%
		Soil gleization	%
		Land impoverishment rate	%
		Coordination degree of water and soil	%
		Urban green space per 10,000 persons	km^2
Safety of water resource		Total water resources per capita	Cubic meter
		Fresh water resources per capita	Cubic meter
		Industrial wastewater discharge	100 million ton
		Load of industrial wastewater per unit water resource	Ton/cubic meter
		Surface water quality index	%
		Proportion of safe drinking water obtainable in urban area	%
		Proportion of safe drinking water obtainable in rural area	%
		Annual fresh water extraction accounting for total water resources	%
		Proportion of irrigable land accounting for farmland	%

(continued)

Table 1 (continued)

First level	Second level	Third level	Unit
Atmosphere safety		Sulfur dioxide emission load	10,000 ton
		Sulfur dioxide emission load per capita	Ton
		Industrial tailpipe emission	10,000 ton
		Air quality index	%
		Proportion of power generation coming from mineral fuels	%
Safety of species		Proportion of endangered mammalian and bird species	%
		Proportion of endangered higher plant species	%
		Proportion of nation-level natural reserve area accounting for national land area	%

4 Conclusion

This article discusses the frame of Hainan eco-safety early-warning assessment index system from comprehensive evaluation indicators of low-carbon society, safety of territorial resources, safety of water resources, of atmospheric and biological species, using the construction principle of eco-safety index system. It should be emphasized that building an eco-safety early-warning assessment index system must center on economic construction, target sustainable development, and integrate representative indicators of economy, society, resources environment, technology, low carbon and institutions so as to build a feasible eco-safety early-warning index system and provide a scientific basis for the sustainable development of Hainan [5–8].

Acknowledgments This work is supported by International Science & Technology Cooperation Program of China (2012DFA11270) and the Hainan International Cooperation Key Project (GJXM201105).

References

1. Wang, P., Song, J.H.: Hainan tourism geography. Hainan Publishing, China (2013)
2. Tian, J.: Research of the Ecological Security in Hainan Province. Central South University, Wuhan (2013)
3. Zhao, Y.F.: Evaluation index system study of tourism environment in Inner Mongolia. Yunnan Geogr. Environ. Res. **23**(3), 80–84 (2011)
4. Tian, T.: Ecological Risk Analysis and Evaluation of Huangshan Scenic Area. Anhui Normal University, Hefei, Anhui (2010)
5. Shang, T.C., Zhao, L.M.: A study on ecological risk analysis and ecotourism system management. J. South China Agric. Univ. **2**(2), 72–74 (2003)
6. Hainan Yearbook Editing, Hainan Year Book. Hainan Yearbook Publishing, China (2010)
7. Wu, Q.S., Yang, X.B., Han, S.M.: Study on the relationship between tourism development and ecological environment in Hainan. Nat. Sci. J. Hainan Univ. **18**(2), 159–164 (2000)
8. Yuan, F.Y.: Analyses on Tourism Development of Hainan. Tianjin University, Tianjin (2011)

Analysis of Economic Spatial Structure Evolvement and Characteristic in Qinghai Province

Ying-yi Cao

Abstract Qinghai province was an economically depressed area of China at the beginning of 1950s. From 1990s to now, the GDP has been keeping the growth rapidly. The chapter analyses the evolution of economic spatial structure in Qinghai province based on the software of ArcView and all the county data. And then, analyses the regional differences of industrial development and studies the reasons that cause the spatial difference of the industrialization factors. Results show that since 1995, the economic spatial structure in Qinghai province is agglomeration economy; the resource distribution is the key factor that influenced the spatial economic structure in Qinghai province.

Keywords Economic spatial structure · Evolvement · Characteristic

1 Introduction

Different scholars have different statements of the concept of the regional economic spatial structure. Dadao Liu believes that spatial structure indicates the degrees and forms of spatial agglomeration caused by the interaction of social economic entities. It is the relative location and the distribution for the development factors in a certain definite area, which reflects position characteristics of economic activity and correlations of regional space. Chencai holds that regional economic spatial structure is the combination relationship of human economic activities in a certain area, a sum of the relationships, which include center, periphery, and network. Juxin Zeng thinks "Spatial economic structure" is an existing

Y. Cao (✉)
School of Economics, Northwest University for Nationalities, Lanzhou 730030, China
e-mail: aabb_72@sina.com

B.-Y. Cao et al. (eds.), *Ecosystem Assessment and Fuzzy Systems Management*,
Advances in Intelligent Systems and Computing 254, DOI: 10.1007/978-3-319-03449-2_5,

form and objective entity of economic phenomenon and economic variables characterized by the characteristic of location, shape, scale, and interaction in a certain definite area. It reflects the location connection and spatial organization form of economic concept regarding geographical space as carriers. It is a sum of both spatial connection and interaction in composition of substantive factors. It embodies the nature and interrelation of things. It is a spatial organization form of human economic activity in a certain area. Yulin Lu believes that regional economic spatial structure is an organization form of human economic activity in a certain area. It contains firstly, spatial differentiation and organization relationship of economic center problem, which takes resources development and human economic activity as load. Secondly, a rating scale relationship that compose spatial entity. Thirdly, a movement form of factor existing in each spatial entity. Fan [1] thinks that it includes two connotations: spatial agglomeration form of economic activity and economic object in a certain area, and relationship of nature among economic spatial entities, which emphasizes the interaction and the relationship. Regional economic spatial structure is an important carrier and representation of regional economic development, which promotes regional economy development, makes regional economic spatial structure optimization and upgrading. At the same time, the level and degree of regional economic development can be showed by the level and optimization degree of regional economic spatial structure, and the specific index of regional economic development reflects the pattern and evolution process of regional economic spatial structure.

Owing to the fact that evolution of regional economic spatial structure has timeliness and spatiality; the economic geography method can be used to study regional economics. This chapter shows express characteristics in regional economy in geographical spatial mode using ArcView software, with through graphics thinking, observes and analyzes the evolution of the economic spatial structure in Qinghai province by directly using a great deal of data and software drawing. The study is based on a region as an element of spatial point, and an assumption that spatial units are homogeneous regions. Through the research on the process of economic development since 1995 in Qinghai province and economic spatial structural changes in each region, this chapter analyzes the spatial changes and characteristics of Qinghai economy.

Firstly, we choose per capita GDP as evaluation index of economic structure. Because per capita GDP can reflect the economic growth situation, fully reflect the extent of the regional differences, and be achieved data from official statistics. Secondly, we choose the total amount of second-industry product to GDP as evaluation index of Qinghai industrial distribution.

This chapter selects data of per capita GDP of counties (cities) in Qinghai province from 1995 to 2010. In order to observe the evolution of Qinghai economic spatial structure for 15 years, it divided time to two parts (1995–2002, 2002–2010), the mean data is 2002s. Because the administrative divisions of Qinghai changed from 1995s, the chapter uses the new administrative divisions established in 2010, it regards each county (city) as an independent area. The data comes from the Qinghai statistical yearbook.

In the ArcView, this chapter combines the corresponding properties (per capita GDP, the second-industry production/GDP) with graphics, studies the economic spatial structure and industrial distribution pattern, and analyzes the economic development situation in Qinghai province, then analyzes the characteristics of economic spatial structure in Qinghai province.

2 The Analysis of Evolution of Economic Spatial Structure in Qinghai Province

The process describes the situation of object in t_0, t_1,..., t, namely the time structure of object. The process of spatial evolution focuses on dynamic process of time and spatial coupling when factors assemble. The process of regional economic spatial structure evolution refers to the changes of spatial pattern and connection of regional economic activities on time axis [2].

2.1 Analysis of Qinghai Economic Spatial Structural Evolvement

Using 3 years per GDP of each county (city) in 1995, 2002, and 2010, the chapter divides whole counties (cities) into four categories those are well-developed, developed, under-developed, and backward region, in terms of the share of real GDP per capita from each county (city) in real GDP per capita of Qinghai province as follows.

Well-development region: >150 %;
Development region: 100 ~ 150 %;
Under-development region: 50 ~ 100 %;
Backward region: <50 %.

According to Table 1 and date from Qinghai statistical yearbooks, in 1995, under-developed regions constitute the province's basic economic spatial pattern, there are seven well-developed regions, eight developed regions, thirteen under-developed regions, and eight backward regions. Each kind of region's proportion is 19.4, 22.2, 36.1, and 22.2 %, and the highest proportion is the under-developed region, the province's economic development level is different in whole space. Furthermore, it has four well-developed regions, four developed regions, eighteen under-developed regions, and seventeen backward regions. The proportion is 9.3, 9.3, 41.9, and 39.5 %, respectively. Compare to 1995s, the proportion of under-developed and backward regions rose in 2002, economically developed areas

Table 1 Classification of Qinghai's regional economic development

Region's type	1995	2002	2010
Well-developed	[3332.96, 8886.25]	[10709.18, 53080.78]	[46274.46, 187120.7]
Developed	[2221.98, 3332.96)	[7139.457, 10709.18)	[30849.64, 46274.46])
Under-developed	[1110.99, 2221.98)	[3569.728, 7139.457)	[15424.82, 30849.64)
Backward	[212.53, 1110.99)	[2078.099, 3569.728)	[3905.471, 15424.82)

Notes Each pair of numbers represents the range of real GDP per capita, for instant: [3332.96, 8886.25) means that real GDP per capita of a region is more than 3332.96 RMB and less than 8886.25 RMB

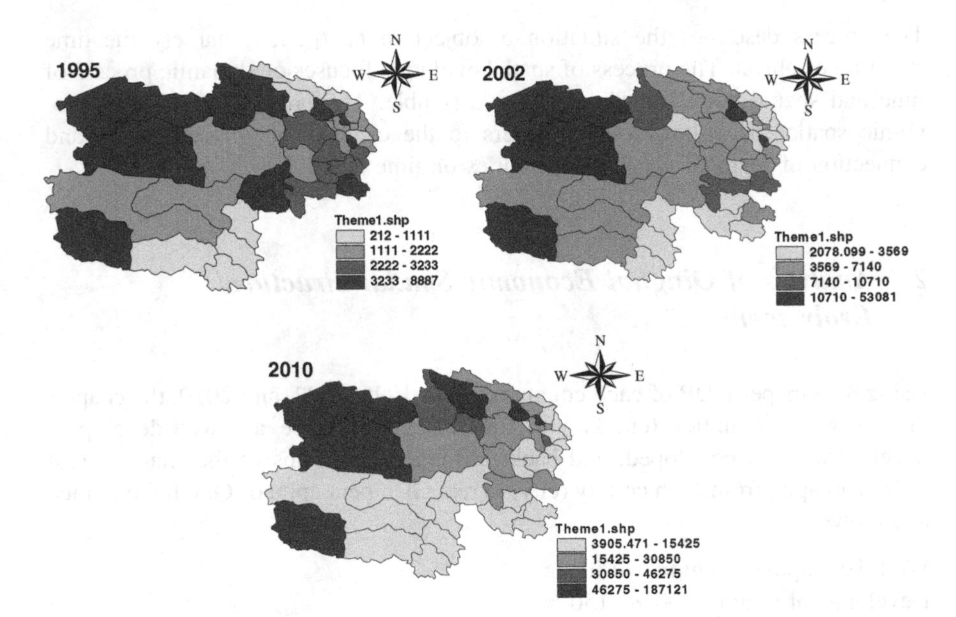

Fig. 1 Evolution of economic spatial structure in Qinghai (1995–2010)

concentrated in the Haixi Zhou. In 2010, well-developed regions in Qinghai are six, four developed regions, under-developed regions are eleven, and twenty-two counties are backward regions. The proportion is 13.95, 9.3, 25.6, and 51.2 %, respectively, Haixi Zhou still concentrated most well-developed regions (Fig. 1).

Since the 1990s, the highest economic agglomeration areas of Qinghai mainly distributed in Haixi Zhou, it shows that the economic gap between Haixi Zhou and other regions has been intensified. Backward areas concentrated in Yushu, Guoluo, Huangnan, and Hainan Zhou. In currently, Qinghai economic space structure has already formed noticeable polarization.

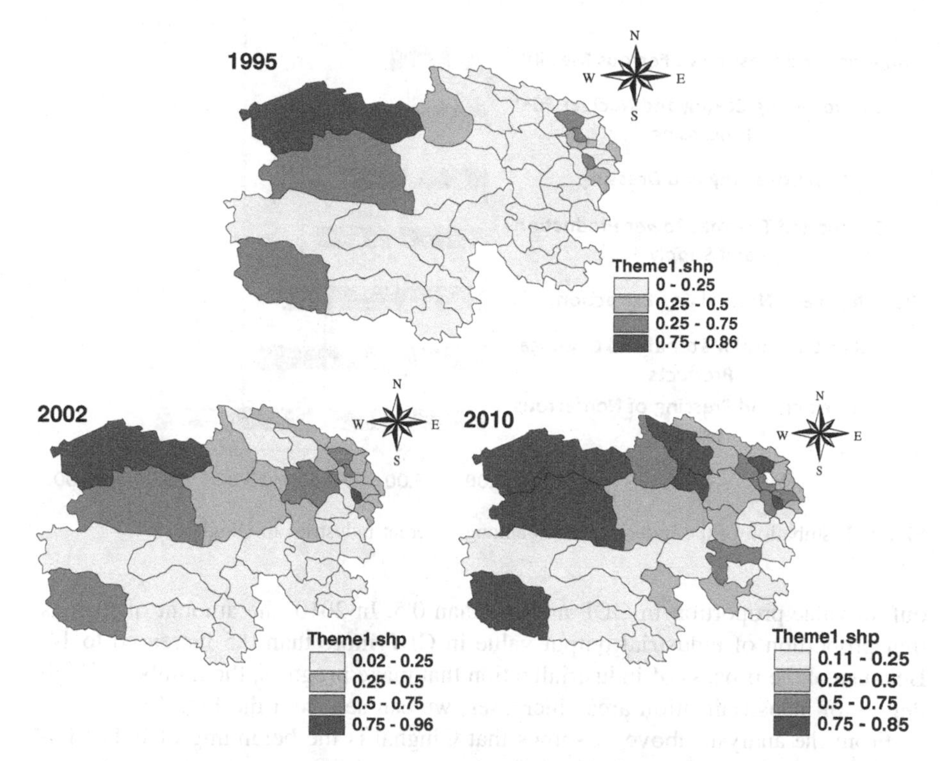

Fig. 2 Evolution of industrialization spatial pattern in Qinghai (1995–2010)

2.2 Evolution of Industry Distribution in Qinghai

Based on the proportion of industrial output value in gross domestic product (GDP) of each counties, the chapter measures the degree of regional industrialization and analyses industrial spatial distribution and evolution using the data from 1995 to 2010. Figure 2 shows that the proportion of the county's industrial output value in GDP has increased year by year, and the second industry is the main driver of regional economic growth.

Since the 1990s, the industrialization degree (overall) in Qinghai was lower. In 1995, the proportion of industrial output value in GDP is less than 0.5 in addition to Datong, Golmud, Xining, and Jianzha counties. It shows that the region that has rich resources also has high degree of industrialization, as you can see from the Fig. 2, those regions are distributed in dark color. In 2002, there are seven regions in which proportion of industrial output value in GDP are more than 0.5, those are Gonghe, golmud, Xining, Datong, Jianzha county, and regions that is directly under the Central Government (Dachai, cold lake, and Fan Ya committee), especially Fan Ya committee industrial GDP accounts for the proportion of output value as high as 0.96, but there is a big part of the counties (cities) that industry

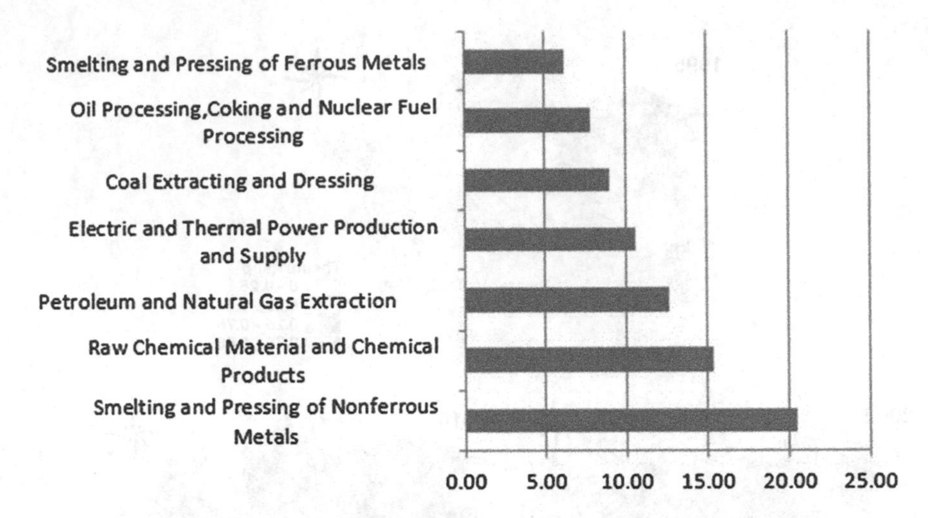

Fig. 3 Distribution of the industry values among different industries in Qinghai (%)

output value proportion in GDP are less than 0.5. In 2010, the amount of regions that proportion of industrial output value in GDP more than 0.5 increased to 14. Because of the process of industrialization that make progress, the number of high degree of industrialization areas increases, we can see from the Fig. 2.

From the analysis above, it shows that Qinghai is the beginning of industrialization. At the beginning of industrialization, mineral resources development is the initial driving force of the regional development that can directly increase revenue, promote regional economic growth, attract outside investment, improve transportation, health care, public service such as infrastructure, and create regional development condition. It also can be formed by forward and backward link of industrial cluster. The importance of all kinds of mineral resources and economic meaning change as the regional economic development change, this will lead to a change in the regional leading industries, shift of economic growth. Thus, mineral resources and animal husbandry resources endowment and its spatial difference distribution are determinants of Qinghai economic spatial structure formation and evolution.

From Fig. 3, we can see, in 2010, industrial added value accounted for the top seven sectors are nonferrous metal smelting and rolling processing industry, chemical raw materials, and chemical products manufacturing, oil and gas industry, electric power, thermal, and production and supply industry, coal mining, and washing industry, oil processing and coking and nuclear fuel processing industry, ferrous metal smelting and rolling processing industry among Qinghai industries. These seven industries mining industrial added value accounted for 82.23 % of the total industrial output. Ferrous metals, nonferrous metals, and other natural resources spatial distribution and those development of Qinghai industrial has played an important role in Qinghai industrial economic growth.

3 Conclusion

All this analysis show that Qinghai's economic development is at the initial stage of industrialization. The characteristics of spatial pattern mainly display in: (1) a powerful center—periphery period. At this period, low level economic space equilibrium is broken, the spatial structure is in the unstable state, a large number of entrepreneurs, intellectuals, labors migrate from the periphery to the center, the regional economic space structure evolutes as the center—periphery, unbalanced social and economic development; and (2) The industry, which is given priority to with resource development is developing rapidly. Industries gross fast in the regions having rich mineral resources, convenient transportation, and those regions become the regional economic center. In terms of a large number of floating people are gather in these economic centers, relative services make progress, and gradually those centers become urban or industrial and mining city. At this period, a number of closed centers may be developed into a city, slow development or decline. Qinghai economic spatial structure evolution characteristics from 1995 to 2010 as follows:

Firstly, the degree difference of economic development increase greater, the mineral resources development is the main motivation of Qinghai economic growth. The spatial pattern of Qinghai economic development level and the spatial differences of industrialization degree of industry are related to the spatial distribution of natural resources in Qinghai. The spatial distribution of mineral resources in Qinghai and its development play an important role in economic development, the spatial distribution of main mineral resources and open degree determines the spatial difference of economic development in Qinghai.

Secondly, multi-economic growth points obviously spring out in Qinghai. The well-developed regions are gather in Haixi, more and more developed counties located in Haidou and Haibei have became new economic growth points, economic elements were distributed in the different configuration space points where attract the agglomeration of production factors.

Thirdly, these regions that the animal husbandry are the leading industry are backward regions. Qinghai has large meadow area, where we should reasonably use natural resources in these areas, develop animal husbandry industry, promote economic development, increase farmer's income, narrow the development gap among regions, and balance development.

At last, in terms of the population density of counties and towns in Qinghai is low, the market size is small, consumption level and consumption structure are low too. The town distribution is more dispersed in Qinghai, against the development of the first and third industry. From the analysis results, we can see that the higher the per capita GDP value, the lower the population density in well-developed regions.

References

1. Fan, X.: Economic analysis of the spatial structure evolution of Henan province. Geogr. Inf. Sci. 70–73 (2005)
2. Bureau, C.S.: Qinghai Statistical Yearbook. China statistic press (1980–2010)

Emergy Footprint Analysis of Gannan Tibet an Autonomous Prefecture Ecological Economic Systems

Hua Liu, Yu-mei Wei and Xin Jin

Abstract Emergy footprint theory combines emergy analysis with conventional ecological footprint theory. In this paper, using emergy footprint theory, we calculated some emergy indices to evaluate and analyze the ecological economic system of Gannan Tibetan Autonomous Prefecture in Gansu province. In addition, we offer advices on how to improve its sustainable development.

Keywords Ecological footprint · Emergy footprint · Gannan Tibetan Autonomous Prefecture · Sustainable development

1 Introduction

In recent years, China has found itself confronted with a series of contradictions among natural resources, environment, and the economy. Issues such as population growth, resources depression, and environment deterioration have made sustainability a critical issue [1]. To make sustainability a reality, we must measure where we are now and how much further we can go. In the new methods of valuation, measurement of sustainability has gone from qualitative analysis to quantitative analysis. Ecosystems provide a wide variety of valuable goods and services [2]. Goods and services, in turn, must be quantified and measured on a common scale. Quantifying the value of ecosystem services has become an important vehicle for assuring social recognition and acceptance of the public management of ecosystems [3, 4].

H. Liu · X. Jin
School of Mathematics and Computer Science, Northwest University for Nationalities, Lanzhou 730124, China
e-mail: 7783360@qq.com

Y. Wei (✉)
Experimental Center, Northwest University for Nationalities, Lanzhou 730124, China
e-mail: 649118046@qq.com

B.-Y. Cao et al. (eds.), *Ecosystem Assessment and Fuzzy Systems Management*, Advances in Intelligent Systems and Computing 254, DOI: 10.1007/978-3-319-03449-2_6,

Gansu is the habitat of several minority nationalities. These are two autonomous prefectures (Gannan Tibetan Autonomous Prefecture and Linxia Hui Autonomous Prefecture) and five autonomous counties (Subei Mongolian Autonomous County, Aksay Kazak Autonomous County, Sunan Yugur Autonomous County, Tianzhu Tibetan Autonomous County, and Zhangjiachuan Hui Autonomous County). The ecological environment aggravation of minority nationality regions has seriously hindered the development of Gansu in recent years. Gannan Tibetan Autonomous Prefecture is located in the southern part of Gansu Province (100°46′–104°44′E, 33°06′–36°10′N), with a total land area of 40,201 km^2 and population 680,800. The location is at the eastern margin of the Qinghai–Tibet Plateau, in the upper reaches of Yangtze River and the Yellow River. With the Loess Plateau and Minshan Mountain forming a complex boundary, Gannan exhibits a diverse ecosystem. Gannan Tibetan Autonomous Prefecture experienced economic rapid development after 2000, but on the other hand, its natural ecosystem has deteriorated. This unreasonable economic growth increased the likelihood of ecological disasters in the region. The sustainability of Gannan's ecological economic systems directly affects the ecological security of the minority nationality regions in Gansu Province. In short, quantitative analysis and evaluation of Gannan's ecological economic systems are conducive to the sustainable development of Gansu Province as well as the Yellow River and the Yangtze River Basin.

The aim of this chapter is to demonstrate a new sustainable development index for emergy footprint and emergy capacity from calculations combining the emergy analyses with the ecological footprint model and the concept of basic sustainability.

The chapter will be structured in three parts:

1. introduction to ecological footprint and emergy theory,
2. introduction to emergy footprint,
3. calculations to emergy capacity and emergy footprint in the Gannan region.

2 Introduction to Ecological Footprint and Emergy Theory

2.1 The Ecological Footprint Methodology

Ecological footprint was developed by Wackernagel and Rees in 1996. Ecological footprint model is a biophysical assessment device to quantitatively estimate a region's sustainability. The Ecological footprint for a particular population is defined as the total 'area of productive land and water ecosystems required to produce the resources that the population consumes and assimilate the wastes that the population produces, wherever on Earth that land and water may be located' [5, 6]. Ecological footprint calculations are based on two simple assumptions; first, that we can keep track of most of the resources we use and many of the wastes we generate; second, that most of these resources and waste flows can be converted to a corresponding biological productive area.

Ecological capacity is defined the as the locally available carrying capacity. Ecological footprint methodology uses a common measurement unit to express ecological footprint and ecological capacity in terms of a biological productive area with the global average productivity, utilizing the 'equivalence factor' and 'yield factor.' So the areas are expressed in standardized 'global hectares' [7]. Therefore, ecological footprint and ecological capacity become directly comparable with each other across the globe. As the ecological footprint and ecological capacity are both measured in the same units, they can be compared directly.

Ecological footprint is estimated by the Eq. (1):

$$\mathrm{EF} = N \times \mathrm{ef} = N \times \sum_{i=1}^{n} (aa_i \times r_i) = N \times \sum_{i=1}^{n} \left(\frac{c_i}{p_i} \times r_i \right) \tag{1}$$

Ecological capacity is estimated by the Eq. (2):

$$\mathrm{EC} = N \times \mathrm{ec} = N \times \sum_{j=1}^{n} (a_j \times r_j \times y_j) \tag{2}$$

EF is the total ecological footprint; N is population size; ef is average per capita footprint; i are the kinds of natural resources or consumption items (for example, energy, food, or forest products production and consumption); a_j is the corresponding areas of No. j resources or consumption items per capita; c_i is the amount or production of No. i resource per capita; p_i is average annual productivity or yield of No. i resource; r_j is equivalence factor; y_j is yield factor. In the analysis of ecological footprint, six main categories of ecologically productive area are distinguished: crop land, pasture, forest, water area, built-up, and energy land.

If the ecological footprint of a region is larger than the ecological capacity, the region runs an ecological deficit. If the ecological capacity of a region is larger than ecological footprint, the region runs an ecological remainder.

2.2 Emergy Theory

Natural systems and economic systems are all tied to energy flow. Traditional energy analysis, however, has been criticized in many aspects. One is that various energy types, including materials and services, cannot be compared or totaled only by energy quantity. Emergy theory is a new method to evaluate natural capital and ecosystem services. Emergy analysis has been developed over the past 25 years by Odum [8]. Emergy (spelled with an 'm') is defined as the energy of one type required in transformations to generate a flow and storage. In this account, solar emergy is used. Solar emergy of a flow or storage is the solar energy required to generate that flow or storage. Its units are solar emergy emjoules (abbreviation: sej). The total emergy of an item can be expressed as Eq. (3):

$$\text{emergy} = \text{available energy of item} \times \text{transformity} \tag{3}$$

The transformity is defined as the amount of emergy of one type required directly and indirectly to generate a unit of energy of another type. It is the emergy per unit energy in units of emjoules per Joule that constitutes the ratio of emergy to available energy. The units of transformity are solar emjoules/Joule, abbreviated sej/J or solar emjoules/g (sej/g).

Emergy measures both the work of nature and that of humans in generating products and services, as a science-based evaluation system that represents both natural values and economic values with a simple, universal unit. As the products and services are both measured in the same units, they can be compared directly.

3 Emergy Footprint and Calculation of Gannan

3.1 Emergy Footprint and Emergy Capacity

Emergy footprint has been developed by Zhao et al. [9]. Emergy footprint methodology is a new method of ecological footprint calculation, based on the emergy analysis. The translation of human demand of natural resources and the supply of natural services into understandable and quantifiable concepts are the main objective of this new method. First, amounts of human consumption corresponding to six categories of ecological productive areas and amounts of natural supply are calculated. Next, these amounts are translated into common unit emergy through the emergy analysis. Finally, in this new method, the emergy footprint and emergy capacity is derived by dividing the emergy amounts by the Emergy Density. Emergy density is the emergy amount per unit time of a region; the following two Eqs. (4) and (5) are used to calculate the emergy density:

$$P_e = \frac{\text{total emergy of the earth}}{\text{areas of the earth}} = \frac{1.583 \times 10^{25}\ \text{sej}}{5.1 \times 10^{10}\ \text{hm}^2} = 3.104 \times 10^{14}\ \left(\text{sej hm}^{-2}\right) \quad (4)$$

$$P_g = \frac{\text{total emergy of Gannan}}{\text{areas of Gannan}} = \frac{6.99 \times 10^{21}\ \text{sej}}{4.0201 \times 10^{6}\ \text{hm}^2} = 1.74 \times 10^{15}\ \left(\text{sej hm}^{-2}\right) \quad (5)$$

P_e is the earth emergy density. The total emergy amount 1.583×10^{25} sej of the earth in 1 year is taken from [10]. The total emergy amount of the earth is the sum of the emergy of solar insolation, deep earth heat, and tidal energy.

P_g is the emergy density of Gannan. In calculation of the total emergy of Gannan, five kinds of renewable resources emergy are considered: sun, wind, chemical energy in rain, geo-potential energy in rain, and earth cycle energy. As shown in Table 1, the maximum item of emergy amount is regarded as the total emergy of Gannan to avoid the duplicate calculation.

Table 1 Calculations for emergy capacity in the Gannan (2010)

Item	Raw data (J)	Transformity (sej/J)	Total emergy (sej)	Emergy per cap (sej/cap)	Emergy capacity per cap (hm^2/cap)
Population of Gannan: 680,800; earth emergy density P_e = 3.104E+1014					
Sun	2.25E+20	1	2.25E+20	3.30E+14	1.07E+00
Wind	5.68E+18	6.23E+02	3.54E+21	5.20E+15	1.68E+01
Rain geo-potential	7.86E+17	8.89E+03	6.99E+21	1.03E+16	3.31E+01
Rain chemical	1.32E+17	1.54E+04	2.03E+21	2.99E+15	9.63E+00
Earth cycle	4.50E+16	2.90E+04	1.31E+21	1.92E+15	6.18E+00
Total emergy of Gannan (the maximum item)			6.99E+21		
Emergy capacity in the Gannan					3.31E+01

Emergy footprint of Gannan is estimated by the Eq. (6):

$$\mathrm{EF}' = N \times \mathrm{ef}' = N \times \sum_{i=1}^{n} a_i = N \times \sum_{i=1}^{n} \frac{c_i'}{p_g} \tag{6}$$

EF′ is the total emergy footprint; N is population size; ef′ is average per capita emergy footprint; i are the kinds of natural resources or consumption items (for example, energy, food, or forest products production and consumption); a_i is the corresponding emergy footprint of No. i resources or consumption items per capita; c_i' is the emergy amount or production of No. i resource per capita (sej); P_g is the emergy density of Gannan.

Emergy capacity of Gannan is estimated by the Eq. (7):

$$ec' = \frac{e}{P_e} \tag{7}$$

where ec′ is the emergy capacity per capita; e is the renewable resources of emergy amount per capita (sej) in Gannan; P_e is the earth emergy density.

The concept of ecological budget is defined as the sum of emergy capacity minus emergy footprint. If the ecological budget is negative, it is often interpreted as an ecological 'overshoot.' That is, 'ecological deficit' in which human consumption exceeds the carrying capacity in a given region, meaning the region is unsustainable. In a reverse situation where the ecological budget is positive and human consumption is within the carrying capacity, the state is called an 'ecological surplus,' meaning the region is sustainable.

3.2 Calculation of Gannan

To calculate the emergy footprints and the emergy capacity in Gannan Tibetan Autonomous Prefecture for 2001–2010, the method of emergy footprint was used. Table 1 shows the emergy capacity of Gannan in 2010. In order to avoid duplicate

Table 2 Calculations for emergy footprint in the Gannan (2010)

Item	Raw data (J)	Transformity (sej/J)	Total emergy (sej)	Emergy per cap (sej/cap)	Emergy footprint per cap (hm^2/cap)	Land types
Emergy footprint per capita			2.48E+22	3.65E+16	2.10E+01	
Biological resources			5.24E+21	7.69E+15	4.42E+00	
Wheat	1.38E+15	6.80E+04	9.38E+19	1.38E+14	7.93E−02	Arable land
Cereal	8.15E+12	3.59E+04	2.93E+17	4.30E+11	2.47E−04	Arable land
Beans	4.35E+12	3.59E+04	1.56E+17	2.29E+11	1.32E−04	Arable land
Tubers	9.70E+12	3.59E+04	3.48E+17	5.12E+11	2.94E−04	Arable land
Corn	5.35E+10	3.59E+04	1.92E+15	2.82E+09	4.57E−07	Arable land
Vegetables	3.15E+13	5.81E+04	1.83E+18	2.69E+12	1.55E−03	Arable land
Chinese medicine	3.89E+14	2.00E+05	7.78E+19	1.14E+14	6.57E−02	Arable land
Oil-bearing crops	7.82E+14	6.90E+05	5.40E+20	7.93E+14	4.56E−01	Arable land
Fruits	2.86E+13	5.30E+04	1.52E+18	2.23E+12	1.28E−03	Forest
Forestry	1.84E+13	2.00E+05	3.68E+18	5.41E+12	3.11E−03	Forest
Meats	4.75E+14	3.17E+06	1.51E+21	2.21E+15	1.27E+00	Pasture
Milks	1.67E+15	1.70E+06	2.84E+21	4.17E+15	2.40E+00	Pasture
Wools	3.92E+13	4.40E+06	1.72E+20	2.53E+14	1.46E−01	Pasture
Fishery	4.75E+11	2.00E+06	9.50E+17	1.40E+12	8.03E−04	Water area
Energy resources			1.96E+22	2.88E+16	1.66E+01	
Electric	1.01E+17	1.59E+05	1.61E+22	2.36E+16	1.36E+01	Water area
Coal	8.92E+16	3.98E+04	3.55E+21	5.21E+15	3.00E+00	Fossil land

calculation, the maximum item of emergy amount is regarded as the total available emergy. This amount is divided by the amount of population in the region being measured, equaling the amount of e in Eq. (7) the emergy supply of natural resources per capita. And then, the amount of e is divided by the earth emergy density P_e. We get the emergy capacity per capita (ec'). The calculation method for 2001–2009 is same as 2010.

The emergy capacity (ec') of Gannan was 33.1 hm^2/cap. As proposed by [11], at least 12 % of the earth's carrying capacity is available for biodiversity protection. Emergy capacity is reduced by 12 % for biodiversity protection. With 12 % set aside for biodiversity protection, the emergy capacity of Gannan dropped from 33.1 hm^2/cap down to 29.1 hm^2/cap.

Tables 2 and 3 show the emergy footprint of Gannan in 2010. The actual consumption amounts of two kinds of natural resources (biological resources and energy resources) are calculated, respectively, and these amounts are translated into the common units' emergy. These emergy amounts are divided by the population to get the c_i' in Eq. (6). The amount of c_i' is divided by the region emergy density of Gannan

Table 3 Ecological footprints summary in the Gannan (2010)

Emergy footprint per capita		Emergy capacity per capita	
Land types	Total (hm^2/cap)	Category	Total (hm^2/cap)
Arable land	6.03E−01	Renewable resources	3.31E+01
Forest	4.39E−03		
Pasture	3.82E+00		
Water	8.03E−04	12 % for biodiversity	3.97E+00
Built-up area	1.36E+01		
Fossil energy	3.00E+00		
Total	2.10E+01	Total	2.91E+01

Table 4 Ecological footprints and emergy capacity in the Gannan for 2001–2010

Year	Emergy footprint per capita (sej/cap)	Emergy capacity per capita (sej/cap)	Ecological budget
2001	18.0	29.72	11.72
2002	17.9	29.45	11.55
2003	18.2	29.45	11.25
2004	16.7	29.05	12.35
2005	18.6	29.20	10.60
2006	18.8	28.78	9.98
2007	19.0	28.64	9.64
2008	20.2	29.17	8.97
2009	20.4	29.16	8.76
2010	21.0	29.14	8.14

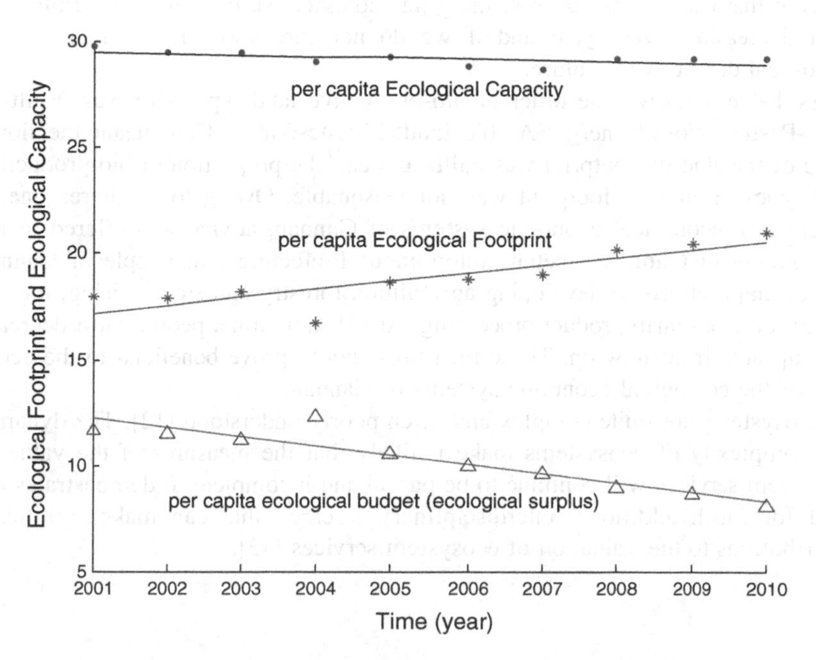

Fig. 1 Ecological footprints and emergy capacity in the Gannan for 2001–2010

P_g to get the footprint, and all the footprints are added together to get the ecological footprint per capita. The calculation method for 2001–2009 is same as 2010.

As Table 2 and Table 3 show the ecological emergy footprint of Gannan in 2010 was 21.0 hm^2/cap. The emergy capacity of Gannan was 29.1 hm^2/cap. We can draw this conclusion: the emergy capacity of Gannan is larger than the emergy footprint. Consequently, the ecological surplus was 8.131 hm^2/cap. It meant the region is sustainable in 2010.

The ecological footprint of Gannan was calculated using existing data for long periods of time: 2001–2010. Table 4 and Fig. 1 analyze development trend of Gannan's per capita ecological footprint and ecological capacity and ecological budget 2001–2010. As Fig. 1 shows, the ecological footprints of Gannan increases and the emergy capacity and ecological budget (ecological surplus) of Gannan decrease.

4 Conclusion

The aim of this paper is aim to demonstrate a new sustainable development index for emergy footprint and emergy capacity from calculations combining the emergy analyses with the ecological footprint model and the concept of basic sustainability. The per capita ecological footprint and per capita emergy capacity of Gannan from 2001 to 2010 was calculated. Results showed that the per capita emergy capacity in Gannan decreases and the ecological budget (ecological surplus) dropped from 11.72 hm^2 in 2001 to 8.14 hm^2, whereas the per capita emergy footprint increased from 18 to 21 hm^2 during the same time period. The results indicate that Gannan has been running an ecological surplus, but the surplus has been decreasing every year and if we do not care about it, Gannan will be ecological deficit in the future.

As Table 3 shows, the order of bio-productive land types size was: Built-up area>Pasture>Fossil energy>Arable land>Forest>Water. That meant the lion's share of the emergy footprint was built-up area. The proportion of bio-productive land types in emergy footprint was not reasonable. Owing to the unreasonable structure of ecological economic systems of Gannan, advice was offered to the government of Gannan Tibetan Autonomous Prefecture that people of Gannan devote major efforts to developing agriculture, forestry and stock raising, touring agriculture, and farm product processing. At the same time, people must decrease built-up area from now on. These measures should prove beneficial to the economy of the ecological economic systems of Gannan.

Ecosystems are quite complex and often poorly understood [12]. The dynamic and complexity of ecosystems make it likely that the measures of the value of ecosystem services will continue to be partial and incomplete. It demonstrates the need for much additional interdisciplinary research that can make significant contributions to the valuation of ecosystem services [13].

Acknowledgments Thanks to the support by National Natural Science Foundation of China (No. 31260098), Natural Science Foundation of Gansu Province (No. 1208RJYA037) and Program for Scientific Research Innovation Team of Northwest University for Nationalities.

References

1. Zhao, S., Li, Z.Z., Li, W.L.: A modified method of ecological footprint calculation and its application. Ecol. Model. **185**, 65–75 (2005)
2. Costanza, R, d'Arge, R., de Groot, R.S., Faber, S., Grasso, M., Hannon, B., Limburg, K., Naeem, S., O'Neill, R.V., Paruelo, J., Raskin, R.G., Sutton, P., van der Belt, M.: The value of the world's ecosystem services and natural capital. Nature **387**, 253–260 (1997)
3. Daily, G.C.: Introduction: what are ecosystem services. In: Daily, G.C. (ed.) Nature's Services: Societal Dependence on Natural Ecosystems, pp. 1–10. Island Press, Washington, DC (1997)
4. Wilson, M.A., Carpenter, S.: Economic valuation of freshwater ecosystem services in the United States: 1971–1997. Ecol. Appl. **9**(3), 772–783 (1999)
5. Rees, W.E.: Eco-footprint analysis: merits and brickbats. Ecol. Econ. **32**(3), 371–374 (2000)
6. Wackernagel, M., Onisto, L., Bello, P., Linares, A.C., Falfan, I.S.L., Garcia, J.M., Guerrero, A.I.S., Guerrero, M.G.S.: National natural capital accounting with the ecological footprint concept. Ecol. Econ. **29**, 375–390 (1999)
7. Wackernagel, M, Schulz, N.B, Deumling, D., et al.: Tracking the ecological overshoot of the human economy. Proc. National Acad. Sci. **99**(14), 9266–9271 (2002)
8. Odum, H.T.: Emergy evaluation. In: Ulgiati, S., Brown, M.T., Giampietro, M., Herendeen, R.A., Mayumi, K. (eds.) Advances in Energy Studies. Energy Flows in Ecology and Economy, pp. 99–111. MUSIS, Roma (1998)
9. Zhao, S., Li, Z.Z, Li, W.L.: A modified method of ecological footprint calculation and its application. Ecol. Model. **185**, 65–75 (2005)
10. Odum, H.T., Brown, M.T., Williams, S.B.: Handbook of Emergy Evaluations Folios 1–4. Center for Environmental Policy. University of Florida, Gainesville FL (2000)
11. WCED, World Commission on Environment and Development, Our Common Future. Oxford University Press, Oxford (1987)
12. Liu, H., Li, Z.Z., Gao, M., Dai, H.W., Liu, Z.G.: Dynamic complexities in a host-parasitoid model with Allee effect for the host and parasitoid aggregation. Ecol. Complexity **6**(3), 337–345 (2009)
13. Zhao, S., Wu, C.W., Hong, H.S., Wang, W.P.: The theoretical model for quantifying the value of ecosystem services. J. Biotechnol. **136S**, S766–S767 (2008)

The Analysis of Water Resource Ecological Carrying Capacity of Hainan International Island

Jun-kuo Cao, Jie Zhang and Sheng-Quan Ma

Abstract The ecological carrying capacity of water resources is a comprehensive concept of natural science and social science, which includes ecological society and economic society dual attributes. Since Canadian scholar Wackernagel has proposed the model of ecological footprint in 1992, it has been a popular evaluation method of ecological society sustainable development. And in the same time, it also provides a new way to analyze the capacity of regional water resources. In this paper, we used this model to assess the sustainable status of Hainan Province water resources from the year 2003 to 2009. For computing convenience and accuracy, we referred to the average water productivity modulus released annually by Department of Hainan Water Resource instead of the average water production in the calculation process. Experimental data show that we conducted deeply analysis of the current situation of Hainan water resources, which will provide scientific data for the sustainable use of water resources for international tourism island developing.

Keywords Water resources · Ecological carrying capacity · Ecological footprint model

1 Introduction

Water is the source of life, and it is indispensable to human survival and social and economic development resources. What's more, it plays an irreplaceable role in the maintenance of the ecological environment and the process of social

J. Cao · S.-Q. Ma
Department of Computer Science and Technology, Hainan Normal University, Haikou, China

J. Zhang (✉)
Department of Information and Technology, Qiongtai Teachers Colleague, Haikou, China
e-mail: zjk716@sina.com

B.-Y. Cao et al. (eds.), *Ecosystem Assessment and Fuzzy Systems Management*, Advances in Intelligent Systems and Computing 254, DOI: 10.1007/978-3-319-03449-2_7,

production. The carrying capacity of water resources refers to the size that can support human life, social production, and the ecological water use, involving population, resources, ecological environment, socioeconomic, and other systems.

With the rapid development of global warming and the social economy, water ecosystem issues become increasingly complex, coupled with the existing water quantity and quality issues, forcing people urgently to understand the bearing capacity of the water ecological system related to the human beings from a higher perspective. Obviously, the already concept about the carrying capacity of water resources and water environment cannot entirely meet the needs of reality. Even it has become the restricting factors of the development of human society. Faced with the grim situation of water shortage, the sustainable use of water resources has become a research focus of attention. Facing the severe situation of water resources and the sustainable utilization of water resources has become the research hot spot. In 1992, William [1] put forward the ecological footprint theory and his student Wackernagel developed and perfected its calculation principle and method, which later widely used by the scholars and institutions around the world [2, 3]. This theory was introduced into China by Xu et al. [4] and other scholars in 1994 and soon became the new research hot spot. In 2008, Huang et al. [5] labeled water as the seventh kind of land type into the index of ecological footprint calculation and establish the ecological footprint model, which has already been widely used in the development of regional water resources utilization and the evaluation of potential sustainable use [6–8].

2 General Situation of Water Resources in Hainan

Hainan Island is China's second largest island, which is currently in the initial stage of economic development and construction of large scale. At the same time, it also attracts the widespread concern both at home and abroad. They concern whether Hainan can maintain the ability of the extraordinary development of the economy and ensure the reasonable protection and sustainable use of environmental resources. At present, from the statistics in recent years (see Table 1), the average total water resources in Hainan Province is about 32 billion m^3 and the per capita possession of water resources 3,900 m^3 that is higher than the national level 2,200 m^3 while is still only 44 % of the world average. Therefore, because of the unique island topography, rainfall, uneven spatial and temporal distribution of water reserves, drought and floods occur almost every year along with the outstanding problems of regional and engineering water shortage.

Rainfall in Hainan Island is mainly composed of the typhoon and the southwest monsoon, and the average annual rainfall is 1,639 mm. Because the water flow is mainly from the east, in the central and eastern regions rainfall is about 2,000–2,400 mm. Slightly less in the northeast, its annual rainfall is about 1,500–2000 mm. The west to northwest is less, usually under 1,500 mm, and the least area is even less than 1,000 mm. Like the distribution of rain, the island's

Table 1 Consumption distribution of Hainan water resources from the year 1998 to 2009 (One hundred million M^3)

Year	Total water quantity	Total consumption		Domestic consumption		Industrial consumption		Agricultural consumption	
		Total water use	Proportion of total use (%)	Domestic water	Proportion of total use (%)	Industrial water	Proportion of total use (%)	Agricultural water	Proportion of total use (%)
1998	246.63	46.82	18.99	4.70	1.91	3.69	1.49	38.44	15.58
1999	337.51	45.34	13.43	4.74	1.40	3.74	1.11	36.86	10.92
2000	458.14	44.02	9.61	4.88	1.06	3.71	0.81	35.44	7.73
2001	464.15	43.55	9.38	4.91	1.06	3.73	0.80	34.91	7.52
2002	333.12	44.08	13.23	4.73	1.42	3.59	1.08	35.76	10.73
2003	291.80	46.31	15.87	4.00	1.37	4.06	1.39	36.25	12.42
2004	171.14	45.77	26.74	4.07	2.38	3.04	1.78	37.68	22.02
2005	307.29	44.04	14.33	4.18	1.36	3.18	1.03	35.47	11.54
2006	227.59	46.46	20.41	4.20	1.85	3.78	1.66	37.07	16.29
2006	283.52	46.69	16.47	4.28	1.51	4.67	1.65	36.17	12.76
2008	419.10	46.89	11.19	4.33	1.03	4.90	1.17	35.95	8.58
2009	480.70	44.64	9.29	4.44	0.92	4.07	0.85	34.34	7.14
Average	335.06	45.38	14.91	4.45	1.44	3.85	1.24	36.19	11.94

water reserves are also uneven from the perspective of spatial and temporal distribution. First of all, there exists a big difference in the flow of some big rivers within a year and between years. Secondly, the difference of the flow of the river and the different river estuaries, sea result in the uneven distribution of water resources in coastal zone, such as the non-estuarine areas is short of water, while the estuary areas are rich in water. Three major rivers of water exports account for 53 % of the surface water resources.

In addition, with the rapid development of the special social and economic construction, another adverse consequence of rapid development of industry and city water resources is causing water pollution, so as to make the (light water) to the increasingly prominent contradiction between supply and demand of water resources in Hainan. This conflict has become one of the main restricting factors for Hainan economy development [1]. Therefore, researching on the various water problems in Hainan Island and exploring the right way to solve the water crisis is a very urgent task.

3 Evaluation Model on the Carrying Capacity of Water Resources

3.1 The Ecological Footprint Model of Water Resources

Ecological footprint theory is used to measure regional sustainable development status by estimating the maintenance of natural resources for human consumption and the ecological productive space area size needed to assimilate human waste. What's more, it can also be used to compare with regional ecological carrying capacity of a given population. Water is recognized as one of the resources. The account of water resources founded in the ecological footprint model of water resources consists with the connotation of ecological footprint. The meaning of the ecological footprint can be expressed in two aspects. First of all, it is the process of human consumption of water resources in daily life and industry. Secondly, it means the stable demand for water to maintain the natural environment. According to the characteristics of water use, water can be divided into three categories: domestic water, industrial water, and ecological environment water, among which the domestic water includes water used by both the urban residents and rural residents. Industrial water includes the primary industry water (including agriculture, forestry, and animal husbandry water), the second industrial water (including industrial water and construction water), and the tertiary industry water (including commercial and service sector water use). According to Yang Zhifeng and other experts' opinion, eco-environmental water can be divided into the internal and external water use of the river. The external water of the river includes city environmental water, and wetland replenishment water, and the internal water of the river refers to the improvement of water environment of rivers, lakes, and other water allocation.

According to the three sub-domestic water footprint, the ecological footprint model of water resources can be expressed as follows:

$$\mathrm{EF}_w = \mathrm{EF}_{\mathrm{lc}} + \mathrm{EF}_{\mathrm{pc}} + \mathrm{EF}_{\mathrm{ec}} = N \times \mathrm{ef}_w$$

$$\mathrm{EF}_{\mathrm{lc}} = N \times \mathrm{ef}_{\mathrm{lc}} = \gamma_w \times (W_l/P_w)$$

$$\mathrm{EF}_{pc} = N \times \mathrm{ef}_{pc} = \gamma_w \times (W_p/P_w)$$

$$\mathrm{EF}_{\mathrm{ec}} = N \times \mathrm{ef}_{\mathrm{ec}} = \gamma_w \times (W_e/P_w)$$

In the equation, EF_w means the ecological footprint model of water resources (hm^2). ef_w means the per capita ecological footprint model of water resources (hm^2/per capita); $\mathrm{EF}_{\mathrm{lc}}$ refers to domestic water ecological footprint. It means the demand of water for citizens during a certain period, including water used by both the urban residents and rural residents. $\mathrm{EF}_{\mathrm{pc}}$ is the industrial water, including industrial water, agricultural irrigation water, animal husbandry, and fishery water; $\mathrm{EF}_{\mathrm{pc}}$ is eco-environmental water footprint, referring to the process of the improvement of the regional ecological environment that include the ecological water use and environmental water allocation. "N" is the regional total population (per capita).γ_w is the equilibrium factor of water resources. P_w is the average production capacity of regional water resources ($\mathrm{m}^3/\mathrm{hm}^2$). W_l, W_p, W_e, respectively, refer to the consumption of domestic water, industrial water, and eco-environmental water (m^3). $\mathrm{ef}_{\mathrm{lw}}$, $\mathrm{ef}_{\mathrm{pw}}$, $\mathrm{ef}_{\mathrm{ew}}$, respectively, refer to the per capita of domestic water ecological footprint, industrial water ecological footprint, and eco-environmental water footprint.

3.2 Model of the Carrying Capacity of Water Resources

The carrying capacity of water resources can be defined as follows: The supporting capacity of regional water resources in the benign development of regional water ecosystems and sustainable economic system in a specific historical stage of development. If the rate of the development and utilization of regional water resources is higher than 30 % and 40 %, it may lead to the deterioration of ecological environment. Therefore, at least 60 % of carrying capacity of water resources should be saved to maintain the health of regional ecological system and the eco-environment balance. According to the analysis, the model of the carrying capacity of water resources can be expressed [4]:

$$\mathrm{EC}_w = N \times e_{\mathrm{cw}} = \pi \times \psi \times \gamma_w \times Q_w/P_w$$

In the equation, π is the utilization rate of resource exploitation (value 0.4). EC_w is the ecological carrying capacity of water resources (hm^2). e_{cw} is the per capita of ecological carrying capacity of water resources (hm^2/per capita). γ_{w} is the

Table 2 Water resources yield factor of Hainan province from 2003 to 2009

Year	2003	2004	2005	2006	2007	2008	2009	Average
The average modulus of water	8,540	5,010	9,000	6,660	8,300	12,270	14,070	9,121
Water resources yield factor (m^3/hm^2)	2.72	1.60	2.87	2.12	2.64	3.91	4.48	2.90

equilibrium factor of water resources. ψ is the yield factor of regional water resources. Q_w is the regional total water resources (m^3). P_w is the average production capacity of water resources in the world.

4 Analysis of the Carrying Capacity of Water Resources

4.1 Model Parameters

In the process of calculating ecological footprint and ecological carrying capacity, the parameters and their values we used are as follows: Average production capacity $P_w = 140.7$ (m^3/hm^2), Balancing factor $\gamma_w = 5.19$, and Yield factor $\psi = 3.73$. In addition, we know that the average production capacity of water resources is the runoff modulus of the average global annual production capacity. Therefore, in order to facilitate the calculation and accuracy, we will use the average runoff modulus instead of the average water production capacity when doing research on s carrying capacity of water resources in Hainan Province. The average runoff modulus during 2003–2009 can be shown in Table 2.

4.2 Per Capita Ecological Profit and Loss of Water Resource

That whether the regional water resource is in a state of ecological deficit and ecological surplus can judge whether the regional production and consumption activities are within the bearing range of the ecosystem. Thus, we can measure the sustainable utilization of regional water resources. According to this definition, the model of the carrying capacity of the water resources is

$$\mathrm{er}_d = \mathrm{ef}_w - \mathrm{ec}_w$$

when $\mathrm{er}_d < 0$, it indicates that the ecological carrying capacity of regional water is greater than the ecological footprint. That means the ecological surplus and it is in the sustainable utilization condition. When $\mathrm{er}_d > 0$, it is in the state of ecological deficit and it is in the ecological destruction state.

Table 3 is about the statistics about the ecological footprint of Hainan from 2003 to 2009. It is easy to find that the highest ecological footprint in Hainan

Table 3 Statistics about the ecological footprint of Hainan from 2003 to 2009

Year	The total population (per million people)	Domestic water footprint/ hm^2	Industrial water footprint/ hm^2	Ecological water footprint/ hm^2	Total ecological water footprint/ hm^2	The per capita of domestic water ecological footprint/ $hm^2.cap^{-1}$
2003	811	243,091	2,536,050	35,248	2,814,390	0.35
2004	818	421,623	4,310,497	9,323	4,741,443	0.58
2005	828	241,047	2,293,403	5,190	2,539,640	0.31
2006	836	327,297	3,286,221	7,014	3,620,532	0.43
2007	845	267,629	2,646,275	5,628	2,919,531	0.35
2008	854	183,152	1,796,408	3,807	1,983,367	0.23
2009	864	163,778	1,479,537	3,320	1,646,635	0.19
Average	837	263,945	2,621,199	9,938	2,895,077	0.35

Table 4 Statistics about the ecological carrying capacity and the benefit and ecological deficit of water resources in Hainan Province from 2003 to 2009

Year	2003	2004	2005	2006	2007	2008	2009	Average
$EC_w 10^6 hm^2$	26.458	26.451	26.439	26.462	26.451	26.449	26.456	26.452
ec_w/hm^2	3.262	3.234	3.193	3.165	3.130	3.097	3.062	3.163
e_{rd}	−2.912	−2.654	−2.883	−2.735	−2.78	−2.867	−2.872	−2.815

Province is 0.58 $hm^2\ cap^{-1}$, the lowest one is 0.19 $hm^2\ cap^{-1}$, and the average level is 0.35 $hm^2\ cap^{-1}$. From the perspective of development trend, due to the increasing population and improvement in the utilization of water resources in recent years, the per capita water ecological footprint decreased slightly, but remained stable on the whole. In addition, on the average distribution of the ecological footprint, we know that the industrial water use accounts for 90.5 % of total water ecological footprint and the domestic water use accounts for 9.1 %, while ecological footprint water use is only 0.3 %.

Table 4 is the statistics on the ecological carrying capacity and the benefit and ecological deficit of water resources in Hainan Province from 2003 to 2009. From the chart, we can conclude that in 2003 the per capita ecological carrying capacity of water resource reached the maximum value, which is 3.262 $hm^2\ cap^{-1}$; in 2009, it experienced the minimum value 3.062 $hm^2\ cap^{-1}$. Besides, from the point of ecological water profit and loss, the Hainan international tourism island is in an ecological surplus condition and the average water ecological profit and loss is −2.815 $hm^2\ cap^{-1}$.

Based on the analysis of data as shown in Tables 3 and 4, the current water resource in Hainan international tourism island is very rich, and it is in the sustainable development of the ecological surplus condition. What's more, the use and development of water resource has a relatively vast development space.

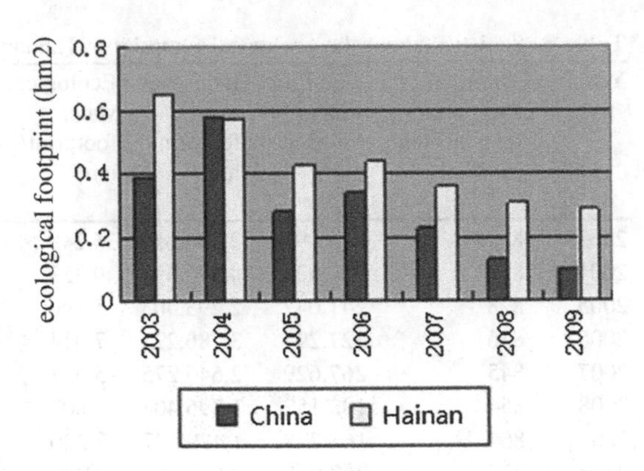

Fig. 1 Ecological footprint per ten thousand yuan GDP in Hainan from 2003 to 2009

5 Analysis of Ecological Footprint per Ten Thousand Yuan GDP

Ecological footprint per ten thousand yuan GDP means ratio of the water resources and regional ecological footprint. It can objectively measure the utilization of regional water resources and its economic growth mode. The formula is $EF_{gdp} = EF_w/GDP$

From chart 1, we can conclude that the utilization of water resources have gradually improved both in Hainan international tourism island and in the whole China. In 200, the ecological footprint per ten thousand yuan GDP of China is 0.39 hm^2, while the number is only 0.65 hm^2 in Hainan. In 2009, the ecological footprint per ten thousand yuan GDP declines to 0.10 hm^2 nationally, and the ecological footprint per ten thousand yuan GDP in Hainan is just 0.29 hm^2. Although the ecological footprint per ten thousand yuan GDP in Hainan decreased by 55.4 % from 2003 to 2009, we have to notice the gap: The rate of the utilization of water resources is only equivalent to the national level, and it accounts for 3/1 of the national level in the same period (Fig. 1).

6 Conclusion

From the perspective of the ecological footprint and carrying capacity of water resources from 2003 to 2009, it indicates that the water resource in Hainan is in the ecological surplus situation, implying that the use and development of water resource has a relatively vast development space. Judging from the development tendency of the ecological footprint of water resources in Hainan from 2003 to 2009, the efficiency of utilization of water resources in Hainan has increased. But the rate of the utilization of water resources is lower than the national level of the ecological footprint per ten thousand yuan GDP in the same period, only about a third of the country's utilization level. From distribution of the domestic water, industrial water, and ecological footprint of water resources, the industrial ecological footprint accounts for the largest share of water resources and ecological footprint of water resources shares the least of the water resources. It indicates that there exists the uneven distribution of water resource in Hainan. Measures should be taken into consideration to carry out industrial restructuring, allocate the water resources reasonably, and improve the efficiency of water use, especially to improve efficiency of the repeating utilization of industrial water and agricultural irrigation, so as to ensure the sustainable development of social economy and ecological environment in Hainan international tourism island.

Acknowledgments The authors would like to thank National Natural Science Foundation of Hainan (No. 612120), Major Scientific Projects of Haikou (No. 2012-050), International Science and Technology Cooperation Program of China (2012DFA1127), and Hainan International Cooperation Key Project (GJXM201105) for the support given to the study.

References

1. William, E.R.: Ecological footprints and appropriated carrying capacity: what urban economics leaves out? Environ. Urban. **4**, 121–130 (1992)
2. Wackenagel, M., William, E.R.: Our Ecological Footprint: Reducing Human Impact on the Earth, pp. 2–17. New Society Publishers, Gabriola Island, B. C. (1996)
3. Wackenagel, M.: An evaluation of the ecological footprint. Ecol. Econom. **31**, 315–320 (1999)
4. Xu, Z., Cheng, G., Zhang, Z., Templet, P.H., Yin, Y.: The calculation and analysis of ecological footprints, diversity and development capacity of China. J. Geog. Sci. **1**, 19–26 (2003)
5. Huang, L.N., Zhang, W.X., Jiang, C.L., Fan, X.Q.: Ecological footprint method in water resources assessment. Acta Ecologica Sinica. **28**(3), 1279–1286 (2008)
6. Zhou, W.H., Zhang, K.F., Wang, R.S.: Urban water ecological footprint analysis: a case study in Beijing. Acta Scientiae Circumstantiae **26**(9), 1524–1531 (2006)
7. Fu, C., Yao, Y., Ma, C.: Water supply and demand balance analysis and countermeasures in Haikou. J. Water Resour. Water Eng. **22**(2), 94–99 (2011)
8. Chang L.F.: dynamic analysis of water resources ecological footprint and ecological carrying capacity in Yunnan Province. Yunnan Geog. Environ. Res. **24**(5), 106–110 (2012)

Discussion on the Architecture Design of the Evaluation Management of Early-Warning Information System of Eco-Tourism

Xue-ping Zhang, Jian-xin Wang, Song Lin and Sheng-Quan Ma

Abstract Since 2005, the United Nations released the report, globally integrated assessment of ecosystem change, human welfare and strengthen the protection of ecological system of feasible countermeasures, received extensive attention and in-depth study, especially in the rapid development model and environment system and human system integration together, evaluation of eco-tourism management system to be dynamically analysis of changes in ecological tourism environment evolution in reverse trends and consequences of evolution, ecological tourism environment deterioration rate, deterioration trend is forecast, and the possible consequences of warning. The eco-tourism evaluation architecture management early-warning information system was designed; the system design principles, design goal, system structure, system function, system security, and other issues are discussed in detail; the theoretical foundation for the implementation of the development of the system is established.

Keywords Eco-tourism · Assessment management · Early warning · Information system

1 Introduction

Eco-tourism is a complex system of dissipative structure, involving a number of areas of natural and human component, its internal material flow, energy flow and information flow is extremely complex, The geographical information system (GIS) systems for research eco-tourism matter and energy in the movement and

---

X. Zhang · J. Wang · S. Lin · S.-Q. Ma (✉)
School of Information and Technology,
Hainan Normal University, Haikou 571158, China
e-mail: mashengquan@163.com

B.-Y. Cao et al. (eds.), *Ecosystem Assessment and Fuzzy Systems Management*,
Advances in Intelligent Systems and Computing 254, DOI: 10.1007/978-3-319-03449-2_8,

transformation of geospatial provide scientific space management and analyst is tools [2]. Therefore, the introduction of eco-tourism warning of GIS research, GIS and other modern geographical information technology support, thetas abolishment of a suitable ecotonal Massenet and early-warning evaluation index system and model analysis system to evaluate the overall quality of the environment Eco-tourist status and evolutionary trend, reflecting the Eco-tourie environmental threats to type, intensity and distribution in space and provide environmental quality Eco-tourism development trend sand the speed of information, Build Eco-tourism assessment management early-warning information system (Eco-tourism Assessment and Management (EAM) Early-warning Information System, short for EAMEWS).

Assessment of eco-tourism management early-warning information system's main task is forevermore evolution for eco-tourism environment trends and consequences of the evolution dynamics analysis, the rate of deterioration of eco-tourism environment, deteriorating trend forecasting, etc., and the possible consequences for alerts [3].

2 Design Principles and Objectives of EAMEWS System

2.1 Design Principles

Practical principle: System should be close to the user, according to the end-user business level application interface design system, menus, commands, concise, and friendly;

Advanced principle: System construction should give priority to the latest version of the latest technology, computer, and peripheral-equipment-based software;

Portability principle: To adapt to the ability of software and hardware system, we should establish information classification, data structure, technological process and equipment and a series of related standards, norms, rules and conventions, to ensure compatibility between systems;

Scalability principle: The construction of the system should have the foresight, and the system design to the reserved interface, application system functions, data management, system hardware, and software can be extended.

2.2 Design Objectives

1. Implementation of eco-tourism environmental background information retrieval

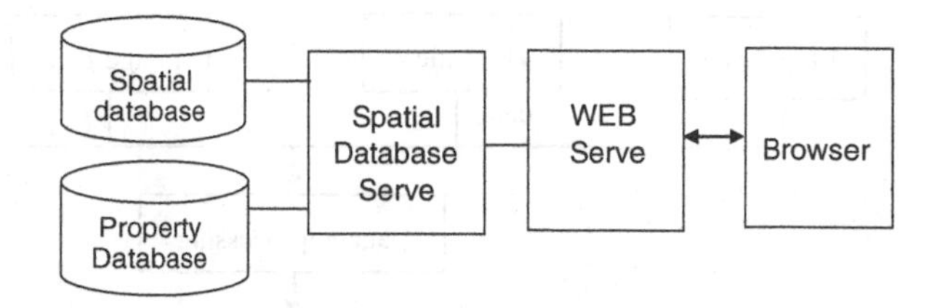

Fig. 1 System network topology frame in system

Design and establishment of standardization, standardization of basic information database, the database includes the oasis basic conditions, basic information such as the natural environment, environmental pollution, the distribution and utilization of resources, population distribution, economic development and other aspects of the implementation of ecological tourism, environmental background information query and retrieval;

2. Achieve generality in the practical basis

To further improve the system design should not only consider the experimentation area information supplement and update, function, but also consider the database design versatility, emphasize system popularization value. The system is not only a practical system, but also a typical example of the system, and the system can be extended to the development pattern and the experience of other regions and experimentation area, facing similar problems (Fig. 1).

3 System Structure and Function Design [4]

3.1 Network Structure Design of System

WebGIS is a GIS network based on Web technology standards and protocols, and it uses Web technology to expand and improve GIS. WebGIS compared with the traditional GIS, which has access to a broader range of spatial data and can be released, makes it easier for users to find the required data, graphics, and attribute data query and retrieval. Because of the implementation method provides various space models on the server, the parameters of the model through the browser receiving user input, the results will be returned, so not only can release spatial data, also can release the space model of service. In the construction of ecological tourism early-warning information system involves many departments, establishes the system structure of Browser/Server structure, based on Web has the capability of distributed computing, in favor of data updating and maintenance, improve the early-warning information, it is not only for the decision-making departments, the general public can easily browse and query of eco-tourism environment and early-warning information, greatly improve the system of social benefits (Fig. 2).

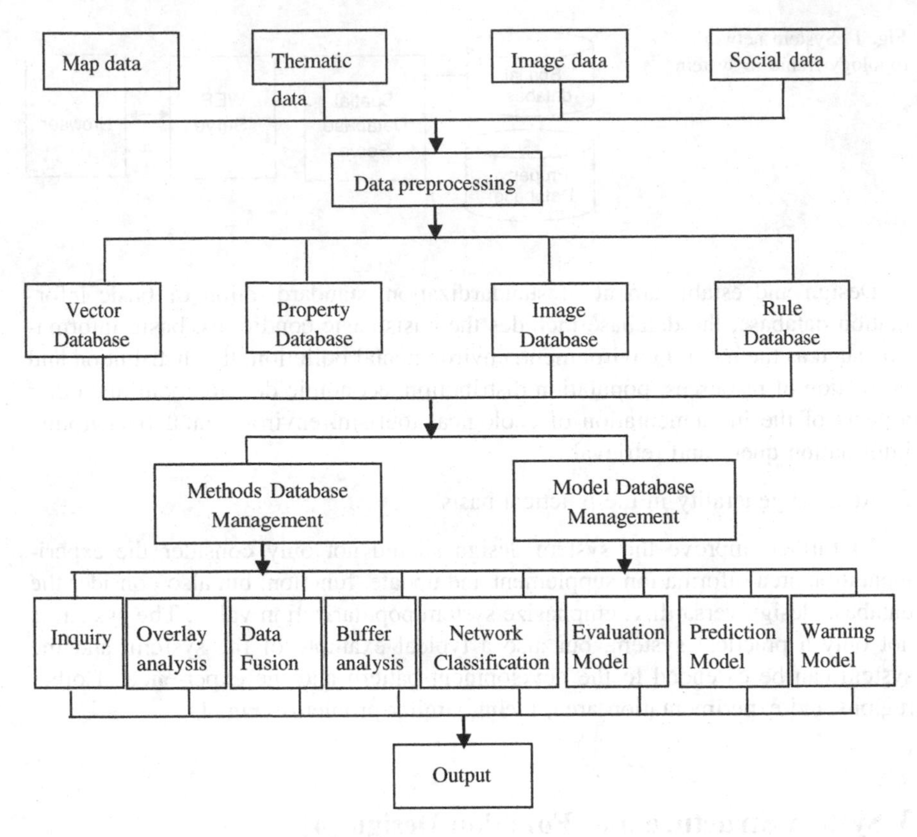

Fig. 2 General logic frame of system

Ecological tourism early-warning information system network topology structure design on the basis of data distribution in various administrative units formed characteristics, taking into account the design basis of the existing network. The whole system is divided into two parts: server systems and client systems. The server system consists of the application system, Web server, and the spatial database; client system consists of operating system, network software, and browser.

3.2 System Logic Structure Design

The function of EAMEWS by the GIS spatial analysis tools, database system and model base system consists of three parts. The database system consists of vector graphics libraries, image raster library, and attribute database; remote sensing image data mainly include the basic situation, natural environment of oasis resources condition, economic and social conditions, and various statistical data and corresponding geographical unit and also data in the form of a management file, to facilitate data editing, querying, and updating; spatial analysis tools mainly refers to the basic space GIS itself has the function analysis, comprehensive analysis, chart analysis, spatial overlay analysis and a variety of simple statistics analysis including spatial information conditions; model base system is in the GIS model library based on, using advanced programming language VB, VC++, two secondary development language to develop each function module of GIS based, and model system for regional ecological tourism warning to specific integration, including the trend model, evaluation of ecological environment quality forecast model, AHP model, gray prediction model and the fuzzy comprehensive evaluation model, analysis to support system model. Using the GIS view shows the results of analysis and thematic mapping function output model evaluation, and ultimately the formation of the data input, information management and maintenance, model evaluation and early-warning information ecological tourism output of a complete early-warning information system.

3.3 System Function Design

According to the design goal and the principle of the early-warning information system, combined with the overall design of the system structure and function results, determine the main subsystems of the system should include; basic data subsystem, spatial query and retrieval subsystem, ecological assessment of early-warning subsystem, data output subsystem and system help subsystem five main function module.

Basic data subsystem: Establishment and maintenance of the system database, including spatial database, attribute database, society, economy, population, resources, database creation, maintenance, and update. Viewing using GIS basic display function provides the visualization of data, to provide services to the early-warning decision information.

Spatial retrieval subsystem and query: This system is one of the important functions of early-warning system, including from space to space query attributes and spatial location according to the properties of the data logic query, namely the picture-text bidirectional query, and statistical analysis of query results. If the query a class object within the scope of a certain space attributes, or query in some area of objects satisfying certain conditions. Relevant maps and related attribute data are stored in the database, to facilitate the maintenance and updating of data.

Ecological evaluation of early-warning system: This is a core function of early-warning system. In GIS development language and high-level programming language support, the establishment of the ecological and environmental warning need model, forecast evaluation, current situation of ecological tourism environment trend analysis and major natural disasters, form a system of early-warning model of unified, provide the service of ecological environment early-warning information for utility area people and the government. The formulation of regional decision-making sector planning is important.

Data output subsystem: Early-warning information of the data output can be in text and tables, statistical mapping, mapping, and other forms of expression. Compared with the traditional methods, under the support of GIS data output to improve the accuracy and efficiency of the data analysis, the procedure is greatly simplified. And the use of GIS visualization function, can display multidimensional thematic elements, strengthen the early-warning information expression ability, early-warning information more convincing.

4 System Security Design

Ecological tourism early-warning information system network covers a wide range, many users, because early-warning data confidentiality requirements, which should be aimed at different user password access, according to different levels, according to the user's authentication results, corresponding user interface. Design of access permission level is divided into three levels: advanced user level, administrator level, and system and general Internet user level. The system administrator has the highest authority to access the early-warning system, which includes the early-warning information system daily maintenance and management, early-warning model construction, metadata management and managing other users of the system; system advanced user is early-warning network center and network node of each work station, to maintain and update the data to the database, you can request send warning center, the application of early-warning and forecasting model for online analysis, and can realize the distribution and update the warning information on the Internet; general customer for the majority of Internet users, can use the Internet to the client browser environment information, early-warning information retrieval and thematic map output.

Acknowledgments This work is supported by International Science and Technology Cooperation Program of China (2012DFA11270) and Hainan International Cooperation Key Project (GJXM201105)

References

1. Zhao, S.D., Zhang, Y.M.: Concepts, contents and challenges of ecosystem assessment-introduction to ecosystems and human well being: a framework for assessment. Adv. Earth Sci **19**(4), 650–657 (2004)
2. Liu, J., Li, J.: Computer system assisting the implementation of tour information prediction measure. J. Huazhong Norm. Univ. Nat. Sci. **38**(2), 245–254 (2004)
3. Li, Y.: Case-based reasoning technology in tourism planning decision. Inquiry Into Econ. Issues **10**, 115–117 (2004)

4. Yin, G., Chen, Y.: Computer-aided design and manufacturing technology. Science Press, New York (2000)
5. Low, B.T., Cheng, C.H., Motwani, J., Madan, M.S.: Expert systems in the service industry. A comprehensive survey and an application. Int. J. Comput. Appl. Technol. **10**(5–6), 289–299 (1997)

Research on Hainan Ecotourism System Early Warning

Xiu-hong Zhang, Song Lin and Mingcai Lin

Abstract This study analyzes the present situation of Hainan tourism resources; ecological tourism system early warning mechanism is put forward for sustainable development of the tourism industry and includes the tourism industry crisis warning, travel security early warning, and tourism ecological environment. This study expounds the tourism ecosystem health evaluation method based on pressure–state–response model and the capacity calculation model of tourism ecology environment early warning and discusses the requirements of Hainan ecological tourism early warning software system and the target of the software system.

Keywords Ecotourism system · Early warning · Tourism

1 Introduction

Hainan province has tropical, marine, and ecological characteristics. The conflict between tourism activities and the natural environment would be increasingly serious, with the development of the construction of Hainan international tourism island. Tourism environment protection would face many new problems and challenges. Hainan Island surrounded by seas, natural resources are relatively independent, the resources and environmental capacity is limited, sensitive, and fragile natural ecological environment. It is necessary that the development of tourism must be coordinated with environmental bearing capacity, so as to realize the sustainable development of Hainan Island.

X. Zhang · S. Lin (✉) · M. Lin
School of Information Science and Technology, Hainan Normal University,
Haikou 571158, China
e-mail: 3219957@qq.com

B.-Y. Cao et al. (eds.), *Ecosystem Assessment and Fuzzy Systems Management*,
Advances in Intelligent Systems and Computing 254, DOI: 10.1007/978-3-319-03449-2_9,

2 Hainan Tourism Resources Situation

Hainan is located at south end of China and occupies an area of 35,000 km^2. It is made up of vast coastline with natural harbors on the outside, and hills, valleys and mountains. Tropical and Seabird Island is located in the unique geographical location, it has advantage of natural and ecological environment, atmospheric environment quality is better than that of national-level standard for a long time. River and lake water quality overall in good condition, forest coverage rate reaches 53.3 %, long coastline, high quality beach cloth, blue sky blue sea. It is the sound of people health vacation paradise Ecological tourism system warning research.

2.1 Tourism Industry Crisis Early Warning

Tourism industry crisis is mainly refers to occur in other industry the negative effects of crisis spread to tourism industry, crisis damaged the lost a lot of tourists, such as war, terrorism, public health crisis, financial disturbance, natural disasters. Within the tourism industry crisis, it is to point to in operating within the scope of the tourism industry, direct threat to tourists or tourism practitioners, affects the crisis of tourism activities, such as mismanagement, rip off of tourism, tourist entertainment facilities accident.

The early warning of tourism industry crisis should research the decision-making system, information system, implementation system based on the early warning indicator system and crisis preceptor measures and strategies, etc.

2.2 The Early Warning of Travel Security

The early warning of travel security is mainly referred to high frequency in the tourism activities, relatively easy to control and influence range is relatively small, traditional travel safety and security issues.

Travel security early warning is mainly researched natural disaster early warning and forecast of the tourism destinations, tourism facilities and safety measure, tourist traffic distribution prediction based on intelligent tourism emergency response system, etc.

Early warning system can offer the functions including accident safety early warning, emergency accident treatment decisions, and interactive functions such as emergency rescue network organization based on network, geographical information system (GIS), global position system (GPS), 3D visualization, expert system (ES), tourism information system (TIS) [1, 2].

Table 1 Tourism ecological security early warning index system

Destination level	Project level	Factors level	Index level
Tourism ecological security early warning index system	Tourism ecological environment pressure warning	Pressure on tourism resources	Tourism resource utilization
			Tourism resources damage rate
		Pressure on land resources	Tourism land utilization
			Tourism demand growth in the land
		Pressure on population	Number of visitors at an annual rate
			Tourism employment demand growth
	Tourism environment quality early warning	Tourism environmental quality	Air environment quality
			Water environment quality
			Environmental noise
			Environmental health
			Environmental carrying capacity
			Quality of biological species
		Tourism ecological quality	Biological species diversity
			Ecosystem quality
			Tourism green coverage
			Tourist forest coverage
	Tourism ecological protection and control ability of early warning	Ecological protection	Ecological protection facilities
			Ecological protection measures
		Pollution abatement	Exhaust gas disposal rate
			Waste water treatment rate
			Pollution control investment intensity
		Intellectual support	Tourists quality
			Tourism practitioners quality
			Destination residents' quality

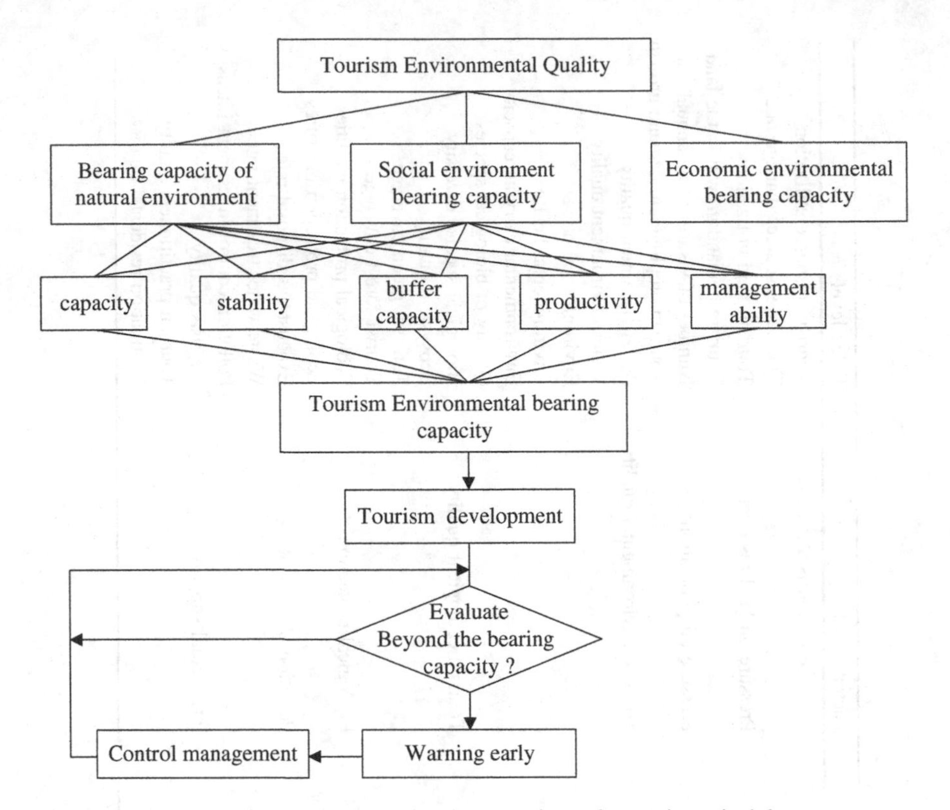

Fig. 1 Ecological tourism environment bearing capacity early warning principle

2.3 The Early Warning of Tourism Environment

2.3.1 Tourism Ecosystem Health Evaluation

Tourism ecosystem is referred to the complex large system combining of natural resources and tourism human activities. It is a nature–social–economic compound system. It has dynamic monitoring, evaluation, and optimization of tourism ecosystem health, establishes mechanism of synergy between the elements of tourism ecological system, to ensure that the natural resources of ecological integrity, stability, promotes the sustainable development of the tourism.

"Pressure–state–response" (PSR) model is widely used in tourism ecosystem health evaluation. The model framework is used to study the environmental problems and has three types of indicators, pressure index, state index, and response index [3].

Using the fuzzy evaluation method assesses ecological security of tourist destination. Tourism ecological security early warning index system is as shown in Table 1.

2.3.2 The Early Warning of Ecological Environmental Bearing Capacity

Tourism environmental bearing capacity (E, T, C, C) refers to destinations are the biggest tourist numbers at a particular time period on the premise that the worst can happen on the premise that without changes or harmful to the tourists and local residents [4]. Ecological tourism environment bearing capacity early warning principle is shown in Fig. 1.

3 Conclusion

The ecological tourism in Hainan early warning system is a composite system consisting of many factors, such as tourism, environment, resource, economy, society. It would evaluate the ecological environment of scenic area, defense anything beyond the tourism environmental bearing capacity behavior, based on sustainable development theory, and provide travel security decision-making tools for the government and tourism department.

Acknowledgments Thanks to International Science and Technology Cooperation Program of China (2012DFA11270), Hainan International Cooperation Key Project (GJXM201105). And this work is also supported by Hainan Social Development of Science and Technology Projects (SF201329), National Natural Science Foundation of Hainan (No. 612124), and Major Scientific Projects of Haikou (No. 2012-028).

References

1. Guo-ji, F.: Natural ecological carrying capacity in Hainan province. J. Nat. Resour. **123**(13), 412–421 (2008)
2. Yunpin, Y.: Research on tourist environment bearing capacity assessment and ecological security warning system of scenic spots in Tibe. J. Chonging Univ. 92–98 (2012)
3. Xin-xiang, C.: A study on ecological security alarm assessment of tourism destination. Environ. Manage. **31**(3), 30–43 (2006)
4. Chun-yu, Y.: The research on early warning system of eco-tourism environmental caring capacity. Hum. Geogr. **21**(5), 46–50 (2006.10)

The Method of Tourism Environmental Carrying Capacity

Xiu-hong Zhang, Mingcai Lin and Song Lin

Abstract This chapter introduces the connotation of tourism environmental bearing capacity, analyzes the relationship between the coordinated development of tourism economy and ecological environment resources protection, introduces the calculation modes and evaluation indexes of the tourism environmental carrying capacity in the aspects of ecological environment carrying capacity, tourism resources carrying capacity, and the carrying capacity of tourist facilities, calculates the thresholds of carrying capacity, and provides the early-warning data for the ecological tourism system.

Keywords Ecotourism system · Early warning · Tourism

1 Introduction

Tourism environmental carrying capacity refers to a system of tourist environment. It is an early-warning research that a sightseeing district can attract the biggest tourist arrivals at a particular time period in the premise of change is harmful to the tourists and local residents, such as the environmental damage of aesthetic value, ecological system damage, environmental pollution, and comfort abate. The research includes the police state of tourist environmental carrying capacity, accurate warning signal, formulation precontrol countermeasures, and provide decision basis for the tourism development strategy formulation and adjustment of economic policies.

X. Zhang · M. Lin · S. Lin (✉)
School of Information Science and Technology, Hainan Normal University,
Haikou 571158, China
e-mail: 3219957@qq.com

B.-Y. Cao et al. (eds.), *Ecosystem Assessment and Fuzzy Systems Management*,
Advances in Intelligent Systems and Computing 254, DOI: 10.1007/978-3-319-03449-2_10,

The research would help tourism destination achieve economic benefit, social benefit, and environmental benefit of coordinated development that is of great significance [1, 2, 4].

2 The Evaluation Index of Tourism Environmental Carrying Capacity

The evaluation index of tourism environmental carrying capacity mainly selects relevant factors from three aspects as natural, social, and economic, so as to evaluate tourism environmental carrying capacity, stability, productivity, buffering and control, management, and the ability of tourism sustainable development, the following four aspects.

2.1 Tourism Resources and Scenic Spots

For tourism resources, the data of type, the data of regional, the data of resources, and the data of the characteristics are included. For scenic spots, the scenic spots, scenic distribution, and scenic features are included.

2.2 Tourism Environment

For basic geographic data, terrain, water system, administrative districts, residential areas, and transportation are included. For natural environment, geology, geomorphology, hydrology, vegetation, soil, and climate are included. For natural resources, accumulated temperature, water resources, and land resources are included. For social economy, population, income, education, and ethnic composition are included.

2.3 Tourism Carrying Capacity

It describes the information on current capacity of scenic spots, such as solid waste facilities capacity, traffic capacity, space, and capacity.

2.4 Tourism Ecological Zones

It includes ecological zoning of tourist attractions and vegetation ecological zones.

3 Tourism Environmental Carrying Capacity Calculation Model

It includes ecological environmental carrying capacity, tourism resources carrying capacity, and bearing capacity of tourist facilities.

3.1 Ecological Environment Carrying Capacity

1. Ecological environmental carrying capacity of land, as in (1):

$$E_C = \sum A_j \times r_j \times y_j \quad (j = 1, 2, 3, \ldots, n) \tag{1}$$

where j represents various land use types, A_j represents all kinds of biological production area, r_j represents equilibrium factor, and y_j represents production factors.
2. Ecological environmental carrying capacity of water, as in (2):

$$\mathrm{UWECC} = W_P / P_W \tag{2}$$

where W_P represents sewage treatment and P_W represents sewage output per capita.
3. Ecological environmental carrying capacity of atmosphere, as in (3):

$$\mathrm{UAECC} = E / S \tag{3}$$

where E represents atmospheric environment quality of scenic spots and S represents the national standards for atmospheric environmental level.
4. Ecological environmental carrying capacity of solid waste, as in (4):

$$S_E = \sum_{i=1}^{n} S_i \times N_i \times R_i / B \tag{4}$$

where R_i represents the bearing capacity of the ith type dustbin, S_i represents the case of type i trash collection times per day, N_i is the number of bins I, and B represents visitors' daily waste discharge standard.
5. Ecological environmental carrying capacity of soil, as in (5):

$$\mathrm{USECC} = S / D \tag{5}$$

where S represents sample value and D represents national standards.
6. Ecological environmental carrying capacity of the life consumption, as in (5):

$$T_f = \sum_{i=1}^{n} f_i \times n \times e_{fi} \tag{6}$$

where n represents the tourists' number, f_i represents the number of tourisms, and e_{fi} represents the ecological footprint of the residence consumption in the daily life.

7. Ecological environmental carrying capacity of the transportation energy consumption, as in (5):

$$T_t = \sum_{i=1}^{n} t_i \times e_{ti} \tag{7}$$

where t_i represents the transport distance with the i kinds of the operation and e_{ti} represents the ecological footprint with the i kinds of transportation in 1 km.

8. Ecological environmental carrying capacity of residential energy consumption, as in (8):

$$T_{\mathrm{a}} = \sum_{i=1}^{n} (a_i \times n) \times e_i \tag{8}$$

where a_i represents the number of days that the tourists live the ith class hotel, e_i represents the tourists' ecological footprint in the daily life at the residence consumption, and n represents the number of the n kinds of class hotel.

9. Ecological environmental carrying capacity of Visit the energy consumption as in

$$T_s = \sum_{i=1}^{n} e_s \times n_i \tag{9}$$

where e_s represents tourist per capita energy consumption in the ith scenic spot.

3.2 Tourism Resources Carrying Capacity

The following are the four aspects of tourism resources carrying capacity:

1. Tourism space carrying capacity, as in (10):

$$Q(r) = \left[\sum_{i=1}^{n} X_i / Y_i\right] / (T/t) \tag{10}$$

where X_i represents the area of the ith tour scenic spot, t represents the average visit time per scenic spot, and Y_i represents a reasonable visit space of the ith scenic spot.

2. Water resources carrying capacity, as in (11):

$$\mathrm{WRCC} = W_S / W_C \tag{11}$$

where W_S represents tourist area water supply and W_C represents water consumption per capita.

3. Land resources carrying capacity, as in (12):

$$\text{TLCC} = \text{TLS} - \text{TLU} - \text{TLR} \quad (12)$$

where TLS represents the supply of the land used for tourism in tourist area, TLU represents the land used for tourism has been used, and TLR represents the demand for the land used for tourism.

4. Vegetation resources carrying capacity, as in (13):

$$\text{VRCC} = S/L \quad (13)$$

where S represents the area of scenic tourist and L represents the area biologically impacted per capita.

3.3 *The Carrying Capacity of Tourist Facilities*

The carrying capacity of tourist facilities mainly refers to road traffic capacity, as in (12):

$$F = \sum_{i=1}^{n} (T/t \times M \times N) \quad (12)$$

where i represents the category of the vehicle, T represents the time required for vehicles for back and forth, t represents the work time for traffic tools average daily, N represents the number of tourist vehicles that can carry, and M represents the number of available means of transport.

4 Conclusion

The purpose of tourism environmental carrying capacity is to analyze the police state of ecological tourism environmental carrying capacity, accurate warning signal, and output under the tourism environmental carrying capacity early-warning system platforms so as to develop precontrol countermeasures timely and scientifically.

Acknowledgments Thanks to International Science & Technology Cooperation Program of China (2012DFA11270) and Hainan International Cooperation Key Project (GJXM201105). And this work is also supported by Hainan Social Development of Science and Technology Projects (SF201329), National Natural Science Foundation of Hainan (No. 612124), and Major Scientific Projects of Haikou (No. 2012-028).

References

1. Xin-xiang, C.: A study on ecological security alarm assessment of tourism destination. Environ. Manage. **31**(3), 30–43 (2006)
2. Chun-yu, Y.: The research on early warning system of eco-tourism environmental carrying capacity. Human Geogr. **21**(5), 46–50 (2006)
3. Guo-ji, F.: Natural ecological carrying capacity in Hainan Province. J. Nat. Res. **123**(13), 412–421 (2008)
4. Yunpin, Y.: Research on tourist environment bearing capacity assessment and ecological security warning system of scenic spots in Tibet. J. Chongqing Univ. 92–98 (2012)

Construction of Evaluation Index System for Ecotourism City Based on Information Entropy and AHP

Yu-ping Zhou, Sheng-Quan Ma and Qing-yu Hao

Abstract The development status of ecotourism city was analyzed from the aspects of the ecotourism resources, ecological protection, urban conditions, city social factors, and international exchanges among cities, etc., and a three-level evaluation index system was constructed. The evaluation index system of eco-tourism city was screened and optimized. The weights on the criterion level were obtained through analytic hierarchy process (AHP) and the expert consult method, but weights on the indicator lever proposed to get through the information entropy method. The combination of objective and subjective empowerment was the reasonable method.

Keywords Ecotourism city · Entropy · Evaluation index system · Weight

1 Introduction

With economic development, ecotourism city has become more prosperous, and evaluation of ecotourism city will become increasingly important. Ecotourism city should be the symbiosis of tourism and city, tourism and nature, tourism and social, present and future. What it pursuits is a harmonious, economy and sustainable development. It is to use ecological theory and ideas in city tourism system, in order to make the city economy, ecology and tourism sustainable

Y. Zhou · S.-Q. Ma
College of Information Science and Technology, Hainan Normal University,
Haikou 571158 Hainan, People's Republic of China

Q. Hao (✉)
Ministry of Education Key Laboratory for Tropical Animal and Plant Ecology,
College of Life Sciences, Hainan Normal University, Haikou 571158 Hainan,
People's Republic of China
e-mail: hnhaoqy@126.com

B.-Y. Cao et al. (eds.), *Ecosystem Assessment and Fuzzy Systems Management*,
Advances in Intelligent Systems and Computing 254, DOI: 10.1007/978-3-319-03449-2_11,

development together. The term "ecotourism" was first proposed by special adviser Ceballas-Laskurain of the World Conservation Union (IUCN) in 1983. It was defined by The International Ecotourism Society in 1990 as a tourism behavior to protect the environment and improve the welfare of local residents in natural areas. It was defined by The International Ecotourism Society in 1993 as dual responsibility of tourism activities to protect the natural environment and maintain the local people's life. The connotation of ecotourism is more emphasis on the protection of the natural landscape and the sustainable development of tourism. The latest definition was to take the ecological environment with characteristics as major landscape tourism. It was also defined as a tourism way to carry out the ecological experience, ecological education, ecological awareness and get the psychosomatic good, based on the concept of sustainable development, protecting the ecological environment as a precondition, taking the development of man and nature in harmony as guidelines, relying on the natural environment and unique cultural ecosystem, and taking the ecofriendly way.

2 The Status of Ecotourism City at Home and Abroad

Domestic and international tourism is making strides forward, but the development basis of each country and the pace of progress is not the same. However, the starting point is all to take "tourism of returning to nature," "green tourism," "protecting tourism," and "sustainable development of tourism" as a main idea. At the same time, all countries in the world are carrying out ecotourism, developing ecotourism city, and a distinctive ecotourism, according to their national conditions.

Foreign developed countries with a better development of ecotourism should be the United States, Canada, Australia, and so on. Ecotourism objects of these countries had transferred from the cultural landscape and urban scenery to the natural scenery specified by Xiebeiluosi Lasikarui. It is to maintain the original nature. These natural features refer to national parks with the excellent natural ecosystems in its domestic and the good ecosystem-based virgin forest in foreign countries. It makes a lot of developing countries to become ecotourism destination, of which Caribbean and African safari became the ecotourism hot spot areas. European countries with better ecotourism cities are France, Germany, Italy, Sweden, etc. In these developed countries, traveling to nature has become a fashion in weekends and holidays. In ecotourism activities, they paid great importance to protect the natural landscape and the environment and proposed the resonant slogan of protecting the environment, such as "only footprints left, only photo taken away."

China's ecotourism was developed, mainly based on nature reserves, forest parks, and scenic spots. In 1982, China's first national forest park—Zhangjiajie National Forest Park was established, which made the good combination of tourism development and ecological environmental protection. Since then, the construction of forest parks and forest ecotourism achieved a rapid development. Although the forest tourism to be developed at this time was not an ecotourism in the strict sense, it provided a good foundation for the development of ecotourism. The ecotourism started in Hunan and Sichuan and then developed in all the country gradually. However, the ecotourism of county-level cities with a number of ecotourism resources did not enhance the tourism economy and had no chance to display specific resources, due to unwelcome tourism market, the lack of promotion and publicity [1]. At present, ecotourism areas opened to public in the domestic were mainly forest parks, scenic spots, nature reserves, etc. The ecotourism areas developed earlier and with more mature development include Shangri-La (Zhongdian), Xishuangbanna, the Changbai Mountain, the Lancang River Basin, Dinghushan, Zhaoqing of Guangdong and Xinjiang Kanas.

3 The Evaluation Index System of Ecotourism City

The rapid development of urban tourism has brought changes in the ecological environment. The evaluation of the ecotourist city often involves in the theories of ecology, economic, tourism, environment, education, and urban geography. At the same time, the degree of urban ecological civilization was also considered. Therefore, various factors, such as ecotourism resources, urban conditions, ecological protection, and sociality, will be considered to build the integrated evaluation index system of ecotourism cities. Then, the information entropy method was used to filter the many indicators.

3.1 The Principles of Constructing the Evaluation Index System for Ecotourism Cities

In recent years, there were more and more research results on the ecotourism city in domestic and foreign countries, but the evaluation index of the ecological tourist city was only studied by a minority of scholars. The studying mostly limited to conceptual, not to quantify and refine the indicators. Zhou [2] studied the standard of international scenic tourist city. Li and Bin studied the theoretical standard of building Guilin international tourist city [3]. Yan and Wang [4] summarized 27 indicators of the international tourist city, but did not do case research. Lin Qiao studied the functional criteria of international tourist city and taken Hangzhou as an example to quantify the indicators. In the paper, the methods of questionnaires and interviews were proposed to obtain data, but it was not easy to popularize due to the limitations of research methods [5].

The principles of construction of index system were as follows:

1. Concise principle. The selection of indicators should not be excessive and too detailed, and it is fine to reduce redundancy to only illustrate the problem in the case of not missing index information.
2. Scientific principles. The determination of the weight coefficient of the index system, data selection, calculation, etc., should be based on accepted scientific methods, and the evaluation has high reliability and validity.
3. Dynamic principle. The evaluation index system of ecotourism city is not immutable. It can be adjusted slightly with policies, regulations, and social environment changes. The index system must fully reflect the general characteristics of the ecotourism in all aspects. At the same time, it should also be dynamic to ensure the index system to update with the times.
4. The principle of combining qualitative and quantitative. Index systems should quantify if they can be quantifiable, but taking into account complex factors and a wide range of evaluation index system, some of the subjective evaluation need to be adopted. Therefore, it should follow the principle of combining qualitative and quantitative.

3.2 The Evaluation Index System of Ecotourism City

The integrated evaluation system was constructed by combining with the theory of ecology, sociology, economics, and sustainable development [6]. The evaluation index (Table 1) was obtained through the screening and optimization of several evaluation indexes [2, 4].

4 The Determination of Weights for the Evaluation Index System

Weights on the criterion level were obtained through analytic hierarchy process (AHP) and the expert consult method, but weights on the indicator level were obtained through information entropy. That is the combining of subjective and objective methods. The information entropy was an objective method, by which the weights were totally calculated through the data on indicator level. AHP and the expert consult method were subjective methods. Therefore, the weights on the criterion level were determined by the evaluator or experts, which can be accordance with the evaluation objective at the highest degree.

Table 1 Evaluation index of ecotourism city

Objective level	Criterion level	Labels of indicators	Indicator level
Evaluation of ecotourism city	Ecotourism resources (A1)	Tourism landscape resources	Geographical landscape resources
			The biological landscape resources
			Geographical landscape resources
		Tourism products	Tourism products richness
			Tourism product update rate
			Leisure products update rate
		Tourism services	Proportion of three-star and above hotels
			Number of service per square kilometer
			Satisfaction degree of tourism and catering
			Clearance rate of travel complaints
		Service personnel	Proportion of personnel with college and over college educational background accounted for the tourism practitioners
			Proportion of personnel with a tourism professional qualification certificate accounted for total employees
			Penetration rate of english proficiency for tourism industry employees
		Tourism achievements	Total tourism income/year
			Growth rate of total tourism revenue over the previous year
			Tourism foreign exchange earnings
			Ratio of tourism foreign exchange earnings to total income
	Ecological protection (A2)	Urban environment	City air quality index
			City noise index
			Reaching standard rate of water quality of drinking water sources
			Annual appropriate tourism days
			Public green area per capita
			Owning number of public transport vehicles per ten thousand people
		Environmental awareness	Waste disposal rate
			Environmental awareness of residents and tourists
			Times of scenic staff trained for ecology and environmental protection

(continued)

Table 1 (continued)

Objective level	Criterion level	Labels of indicators	Indicator level
	City conditions (A3)	Policy support	Ratio of government appropriate funds to scenic inputs
			Ratio of investments in environmental protection to scenic inputs
			Ratio of conservation monitoring to scenic inputs
		City public facilities	City size
			Road area per capita
			Providing rate of the public toilets per ten thousand capita
			Access service rate of international communication
	The city social conditions (A4)	The quality of residents	Coverage of resident high school education
			Friendly degree of residents
		The city social conditions	Number of police force per 10,000 person in city
			Safety and insurance of city tourism
			Number of city travel agency
	International exchange support (A5)	Marketing competition	Acceptance degree of international tourism market
			Visualization integrated index of international tourism city
			Number of English Web site of tourist class
			Number of foreign tourism enterprises
			Convenient degrees of international transportation
			Number of overseas tourism business/office
		Meeting condition	Number of large annual meeting/exhibition
			International convention and exhibition sites/total exhibition area

4.1 The Basic Theory of the Information Entropy

The information entropy in the information system is a measure of the disorder information. The greater the information entropy, the higher the degree of disorder of the information is. The utility value of the information is smaller. Conversely, the smaller the information entropy, the lower the degree of disorder of the information is, and the utility value of the information is the greater [7].

Assumed that the status of the development of tourism in a region in m years needs to be evaluated, and the evaluation index system consists of n indicators. This is a comprehensive evaluation by m samples, n indicators. The mathematical model is as follows:

The domain:

$$U = \{u_1, u_2, u_3, \ldots, u_i, \ldots u_m\} \quad (i = 1, 2, 3, \ldots, m) \tag{1}$$

Each sample (evaluation object) U_i is expressed with the data with n indices:

$$U_i = \{X_{i1}, X._{i2}, \ldots, X_{im}\} \quad (j = 1, 2, 3, \ldots, n) \tag{2}$$

Thus, initial data matrix X of the evaluation system was given:

$$X = \{X_{ij}\}_{m \times n} \tag{3}$$

On where X_{ij} represents the value in the ith samples and jth evaluation indicator. Standardization matrix of data was calculated on the basis of standardized data [8], and P and information utility values of g were obtained. The information entropy P_j in jth indicator is as follows:

$$P_j = -K \sum_{i=1}^{m} y_{ij} \ln y_{ij} \tag{4}$$

In a completely disordered information system, the order degree is zero, and its entropy is maximum, $P_i = 1$. When m samples were in a completely disordered distribution state, $y_{ij} = 1/m$, here,

$$P = -K \sum_{i=1}^{m} \frac{1}{m} \ln \frac{1}{m} = K \ln m = 1,$$

so

$$K = \frac{1}{\ln m}, \quad 0 \le P_j \le 1 \tag{5}$$

The utility value g_j of some indicator depends on the difference between the information entropy P_j of the index and 1:

$$g_j = 1 - P_j \tag{6}$$

Last, the weight in jth indicator can be obtained:

$$W_j = \frac{g_j}{\sum_{j=1}^{n} g_j} \tag{7}$$

The utility values of various indicators in each class were added, and then, the total utility values G_k ($k = 1,2,\ldots$) were obtained. Furthermore, the total utility value of all indicators was obtained, $G = \sum_{j=1}^{n} g_j$.

The weight of corresponding class is as follows:

$$W_K = \frac{G_k}{G} \tag{8}$$

4.2 The Principle of Analytic Hierarchy Process

AHP is a combination of qualitative and quantitative analysis, multifactor decision analysis methods. AHP has four basic steps: First, after the decision-making objective was determined, the factors affecting the target decision were classified and a multilayered structure constructed. Second, pairwise comparison matrix was constructed through comparing the relative importance of each factor on the same level with the same factor on the upper level. Third, the consistency of the pairwise comparison matrix was tested by the calculation. When necessary, the pairwise comparison matrix can be modified in order to make the consistency accepted. Fourth, under the premise of conforming to the consistency testing, the eigenvector corresponding to the maximum eigenvalue in the pairwise comparison matrix was calculated to determine the weight of each factor corresponding to the factor on the upper level. Then, the total ranking weight of each factor to system objective was calculated.

4.3 Determination of Weights

The weights on the criterion level were obtained based on the AHP theory and the expert consult method. Statistic and analyzed result was as follows: (A_1, A_2, A_3, A_4, A_5) = (0.4, 0.25, 0.15, 0.15, 0.05). The weights indicated that ecotourism resource (A_1) is most important to the evaluation of ecotourism city, ecotourism protection (A_2) was second important, paid higher attention, city condition (A_3) and city social condition (A_4) were the same important, and International exchange support (A_5) was less important relatively to other index. The weights on the indicator level will be obtained through the information entropy, combined with the evaluation case. In this paper, the weights on the indicator level were omitted.

5 Conclusion

The evaluation index system of ecotourism city with five criterions and 46 indicators was constructed. The screened evaluation index system covered the major aspect of the ecotourism city, which is the general evaluation index system. The major feature is to adopt the combination of subjective and objective method to get weights on the evaluation index system. It is easy to meet the total evaluation objective of evaluators to obtain weights on the criterion level by AHP and expert consult methods—subjective methods. Forty-six weights on the indicator level obtained with objective information entropy were much more reasonable.

Acknowledgments This work is supported by International Science & Technology Cooperation Program of China (2012DFA11270); Hainan International Cooperation Key Project: Construction of an Early Warning Information Platform/Module based on Ecotourism Assessment and Management (EAM)—Its Application in the Building of Hainan International Tourism Island of China (GJXM201105).

References

1. Cui, F.J.: Innovation and development of China's traditional tourist destination. China Tourism Press, Beijing (2002)
2. Zhou, L.Q.: The evaluation of the international landscape and tourist cities. Urban Plann. Rev. **23**(10), 31–34 (1999)
3. Li, Z.G., Bin, N.: Preliminary study on building the evaluation index system of a modern international tourism city—Guilin as a case study. Social Scientists **104**, 121–123 (2003)
4. Yan, Y.B., Wang, Z.: Research on the Evaluation Index System of International Tourism City. J. Hunan Finan. Econ. Coll. **23**(1), 88–91 (2007)
5. Lin, Q.: Study on the function system of international tourist city. School of Management, Zhejiang University, Zhejiang (2002)
6. Luo, M.Y.: A Study on the development of international tourist cities. J. Guilin Inst. Tourism **15**(2), 5–8 (2004)
7. Solomon, S., Weisbuch, G., de Arcangelis, L.: Social percolation models. Physica A (277), 239–247 (2000)
8. Schulze, C.: Sznajd opinion dynamics with global and local neighborhood. Int. J. Mod. Phys. C **15**(6), 867–872 (2004)

Study on Evaluation Method of Eco-Tourism City to Hainan Island

Yu-ping Zhou, Sheng-Quan Ma and Qing-yu Hao

Abstract Hainan Island has 18 city and county-level administrative regions. The evaluation and rational distribution of eco-tourism cities are significant for tourists to choose tourist destination and the future development of the international tourist of Hainan Island. The simple and easy index system was constructed and the integrated evaluation to the eco-tourism city in Hainan Island was conducted through the methods of the combination of quantitative and qualitative evaluation, the combination of single-objective, multi-objective evaluation and cluster analysis method. The evaluation results revealed that Sanya, Haikou, Wanning, Qionghai, Wenchang, Lingshui, Baoting, Wuzhishan, and Ledong were selected as top ten eco-tourism cities and counties in Hainan Island.

Keywords Eco-tourism city · Evaluation methods · Evaluation index system · Cluster analysis · Hainan Island

1 Introduction

In 2008, the Hainan provincial government released "construction action plan for Hainan international tourism island" for the first time, in which it was proposed to build Hainan Island into the world class, international tropical island vacation, and leisure travel destination with the characteristics of the internationalization of

Y. Zhou · S.-Q. Ma
College of Information Science and Technology, Hainan Normal University, Haikou 571158 Hainan, People's Republic of China

Q. Hao (✉)
Ministry of Education Key Laboratory for Tropical Animal and Plant Ecology, College of Life Sciences, Hainan Normal University, Haikou 571158 Hainan, People's Republic of China
e-mail: hnhaoqy@126.com

B.-Y. Cao et al. (eds.), *Ecosystem Assessment and Fuzzy Systems Management*, Advances in Intelligent Systems and Computing 254, DOI: 10.1007/978-3-319-03449-2_12,

tourism facilities, tourism services, tourism products, and eco-tourism environment, distinctive features, security, and foreign tourists yearning, after 20 years of efforts [1]. According to the construction action plan of Hainan International Tourism Island, Hainan will be built into "the Island of tourism and opening up, the Island of enjoy and Sun, the Island of leisure and holiday, the Island of ecological harmony, the Island of civilized service." It is the core of the construction of the international tourist city to improve the construction of Hainan eco-tourism increasingly during the process of building Hainan International Tourism Island, in addition to the software and hardware construction.

Since the 1980s, eco-tourism has been considered a critical endeavor by environmentalists, so that future generations may experience destinations relatively untouched by human intervention [2]. What is eco-tourism? It is difficult to give an authorized definition, but eco-tourism is defined as "responsible travel to natural areas that conserves the environment and improves the well-being of local people" [3]. Eco-tourism in Wikipedia free encyclopedia is defined as a form of tourism involving visiting fragile, pristine, and relatively undisturbed natural areas, intended as a low impact and often small-scale alternative to standard commercial (mass) tourism. No matter which definition, the purpose of ecotourism may be to educate the traveler, to provide funds for ecological conservation, to directly benefit the economic development and political empowerment of local communities, or to foster respect for different cultures and for human right [2]. The so-called international eco-tourism city refers to a tourist city to open natural and cultural landscapes for tourists from around the world and the one with sustainable development under the standards of no ecological environmental pollution and destruction through limiting the number of visitors [4]. Including Sansha city newly established, Hainan Province has a total of 19 cities and counties. Is it not very clear by now which cities have conditions and competitiveness of eco-tourism. The purpose of this study was to systematically evaluate and select top ten eco-tourism cities from Hainan Island, by selecting the reasonable evaluation method and constructing the evaluation index systems, in order to provide the reference for tourists to make the decision on the destination selection and also provide a useful practice for the construction of the eco-tourism city.

2 The Advantages of Hainan Island Eco-Tourism Resources

Hainan Island tourism resources are very rich, distinctive, high quality, and deeply welcomed by domestic and foreign tourists. Hainan has about 335 of natural tourism resources, in which resource numbers of physiographic landscape, biological landscapes, water scenery, and the planetarium climates accounted for 50.0, 22.1, 21.0 and 6.9 %, respectively. The data indicated that the natural tourism resources in Hainan are mainly the categories of landscape class, biological landscape, and water scenery [5]. The main advantages of eco-tourism in Hainan Island were as following:

1. The advantage in natural resources. It includes the advantages of tropical coastal scenery resource, tropical rainforest resources, tropical climate resources, tropical agricultural cash crop resources and hot spring resources, etc, such as: Yalong Bay, Shimei Bay, and Perfume Bay. Numerous national and provincial nature reserves and ecological demonstration zones—Jianfengling, Wuzhishan, Diaoluoshan tropical rainforests, and mangrove wetland system, etc.
2. The advantage in natural environments. Such as: fresh air and high density of negative oxygen ions due to the island surrounded by the sea and covered by dense vegetations, sunny weather, light industrial pollution, etc. Moreover, the average life expectancy in Hainan Province ranks first in China. Therefore, many Chinese and foreigners take Hainan Island as "rare unspoilt virgin land in the world today."
3. The advantage in geographic location. Hainan is located in the circular belt of western Pacific, only way that must be passed in the traffic between the Pacific and Indian Ocean, the crossroads between China, Japan, South Korea, and Southeast Asian countries, and ideally transit and rest place for the past travelers.
4. The advantage in policy. Hainan is the China's largest special economic zone and enjoys country's unique special policies, such as "landing visa," "offsite visa," duty-free shops, and others.
5. The advantage in ethnic customs and overseas Chinese. Hainan is a multiethnic province, in which Hainan Li and Miao ethnic customs, folk dance, ethnic costumes were very popular to tourists. Hainan overseas Chinese have more than 200 million, which is a valuable asset for Hainan to expand tourist markets [6].

3 Selection of Evaluation Methods

The selection of evaluation methods is the core of the evaluation, because the evaluation method is related to the credibility of the evaluation results. The evaluation results on the same item might be totally different if different evaluation methods were adopted, due to differences of the theoretical basis of the evaluation method, the application scope, and inherent characteristics; therefore, the selection of evaluation methods must consider the evaluated item and characteristics of evaluation methods at the same time. There are many evaluation methods to be used by now, but the analysis of hierarchy process (AHP), clustering, principal component analysis, and entropy weight methods were commonly used to evaluate the eco-tourism, such as AHP used to evaluate ecological tourism resource [7]. Other evaluation methods were also used in practice, such as: the method of principal component analysis used in tourism attraction grade assessment [8] and clustering method used to evaluate the tourism competitiveness factor of cities [9]. AHP is a subjective evaluation method, in which the evaluation weights need to be designed according to human's ideas. The entropy and principal component analysis methods are the objective evaluation method, in which the weighting was calculated objectively according to the differences of the evaluation index data. Because the empowerment method is completely dependent on the differences of evaluation index data, the result of the evaluation often exists some differences with the actual situation, so those methods can only be suited for secondary evaluation means [10]. The cluster analysis is also an assist evaluation method. It will achieve ideal effects when only combined with other evaluation methods. This paper attempts to use the methods of combining quantitative and qualitative evaluations, single-objective, multi-objective and clustering evaluation to evaluate the eco-tourism cities in Hainan Island.

4 The Construction of Evaluation Index Systems

4.1 The Principles of Constructing Index Systems for the Eco-Tourism City Evaluation

1. Rich in eco-tourism resources, high-impact index, and high quality. Such as the categories, quantities, scale, well-known and reputation, esthetic and recreational values of tourism resources.
2. A large number of tourists accepted and events held frequently.
3. Beautiful ecological environment to accord with eco-tourism requirements. Such as air and water quality, sanitation condition, vegetation covering rate, biodiversity index, tourism capacity, etc.
4. Great development potential in eco-tourism resources. Such as the development of new eco-tourism items, improvement on the quality of tourism items, and infrastructure investments.
5. Distinct characteristics in tourism resources, beautiful human environments, sound finance, etc.

4.2 The Construction of Evaluation Index System for Eco-Tourism Cities

According to the principles of constructing evaluation index system for eco-tourism cities, ten indicators were initially selected as the evaluation index system for eco-tourism cities, such as (1) number of the city's scenic spots, (2) number of A and over A level scenic spots, (3) number of golf course, (4) major events, (5) proposed projects on 12th Five-Year Plan, (6) GDP, (7) tourists accepted in hotels, (8) tourists accepted in restaurants, (9) landing typhoons, and (10) times of mud-rock flow and landslides.

4.3 Features of Evaluation Index System for Eco-Tourism Cities

The evaluation index system was simple, but covering the most important aspects possessed by an eco-tourism city. Among them, the number of tourists accepted in hotels and restaurants is an indicator of reflecting the number of visitors. The greater the number of tourists accepted in hotels and restaurants, the more abundant in tourism resources and the higher the quality of tourism resources. GDP is an indicator of the overall urban economic level. With the improvement of the GDP, eco-tourism facilities, ecological environmental conditions, quality of service, and potential development on tourism resources, etc. will be improved and enhanced. The quantities of scenic spots, A and over A level scenic spots, golf courses, and large-scale events and activities can fully reflect the scale, quality, and influence of tourism resources in a city. Landing typhoons, mud-rock flow, and landslides are the natural disaster indicators, which will have bad effects on the eco-tourism cities.

5 The Application of the Evaluation Method

Only 18 cities and counties in Hainan Island were evaluated in this chapter, in which newly established Sansha city was not included, because the tourism projects in Sansha city has not yet officially launched, and data used as the evaluation have not happened. Top ten will be selected from 18 cities and counties in Hainan Province by above-selected evaluation methods. The evaluation data were mainly collected from the local chronicles and the Hainan Provincial tourism resources Web sites. Original data for 18 cities and counties in Hainan Island were listed in Table 1.

5.1 The Single-objective Evaluation

According to the data in Table 1, evaluation results over various evaluation factors were listed in Table 2. The results showed that six counties and counties of Sanya, Wanning, Haikou, Qionghai, Danzhou, and Wenchang were kept in the top ten among 18 cities and counties in the ranking of all single-objective evaluations; therefore, these six cities and counties were evaluated as the candidates of top ten eco-tourism cities in Hainan Province. Three evaluation indicators of tourists accepted in hotels, tourists accepted in restaurants, and GDP for Baoting were outside the top ten. Two evaluation indicators of events and activities and GDP for Lingshui were outside the top ten. The indicator of the quantities of scenic spots

Table 1 Original data used as the evaluation of eco-tourism cities

Cities and counties in Hainan Island	No. of *A* and over *A* level scenic spots	No. of scenic spots	No. of proposed projects on 12th Five-Year Plan	Tourists accepted in restaurants in May 2012 (10,000 people)	Tourists accepted in hotels in May 2012 (10,000 people)	No. of landing typhoon in past 12 years	No. of mud-rock flow and landslides	Events and activities	Golf courses	GDP in the first half of 2011(RMB 10,000)
Sanya	9	16	12	56.3	69.01	5	0	19	7	1,352,526
Wanning	7	10	8	21.86	23.19	3	0	6	0	592,283
Haikou	5	11	7	43.61	71.12	1	0	74	7	3,521,093
Qionghai	4	8	5	8.53	13.86	1	1	6	2	690,115
Danzhou	3	17	5	4.77	7.52	0	0	4	2	1,733,047
Lingshui	2	3	8	7.92	9.37	1	2	1	1	314,835
Baoting	2	7	6	1.33	2.05	0	0	4	0	83,197
Dingan	1	6	8	2.23	3.45	0	0	4	0	215,752
Wuzhishan	1	5	8	1.85	2.8	0	0	0	0	79,871
Chengmai	1	2	6	2.15	4.27	0	0	5	0	632,079
Wenchang	1	8	4	5.35	10.15	10	0	2	0	752,208
Dongfang	0	4	6	1.21	3.26	0	0	2	1	536,174
Baisha	0	3	6		1.14	0	0	0	0	76,953
Ledong	0	1	5	0.7	2.34	1	0	1	0	296,649
Lingao	0	1	5	0.65	1.75	0	0	1	0	405,559
Tunchang	0	3	4	0.57	1.36	0	0	4	1	136,647
Changjiang	0	1	4	1.86	2.73	0	0	1	0	316,062
Qiongzhong	0	2	2	1.09	1.97	0	0	0	0	62,928

Table 2 Single-objective ranking

Single-objective ranking	No. of *A* and over *A* level scenic spots	No. of scenic spots including proposed projects on 12th Five-Year Plan and golf courses	Tourists accepted in restaurants in May 2012	Tourists accepted in hotels in May 2012	Events and activities	GDP in the first half of 2011
1	Sanyan	Sanya	Sanya	Haikou	Haikou	Haikou
2	Wanning	Haikou	Haikou	Sanya	Sanya	Danzhou
3	Haikou	Danzhou	Wanning	Wanning	Wanning3	Sanya
4	Qionghai	Wanning	Qionghai	Qionghai	Qionghai3	Wenchang
5	Danzhou	Qionghai	Lingshui	Wenchang	Chengmai5	Qionghai
6	Baoting6	Dingan	Wenchang	Lingshui	Danzhou6	Chengmai
7	LIngshui6	Baoting7	Danzhou	Danzhou	Dingan6	Wanning
8	Chengmai8	Wuzhishan7	Dingan	Chengmai	Baoting6	Dongfang
9	Dingan8	Lingshui9	Chengmai	Dingan	Tunchang6	Lingao
10	Wenchang8	Wenchang9	Changjiang	Dongfang	Wenchang10	Changjiang
11	Wuzhishan8	Dongfang	Wuzhishan	Wuzhishan	Dongfang10	Lingshui
12	Tunchang12	Baisha	Baoting	Changjiang	Lingshui12	Ledong
13	Dongfang12	Chengmai13	Dongfang	Ledong	Changjiang12	Dingan
14	Ledong12	Tunchang13	Qiongzhong	Baoting	Ledong12	Tunchang
15	Lingao12	Ledong15	Ledong	Qiongzhong	Lingao12	Baoting
16	Changjiang12	Lingao15	Lingao	Lingao	Wuzhishan16	Wuzhishan
17	Baisha12	Changjiang	Tunchang	Tunchang	Qiongzhong16	Baisha
18	Qiongzhong12	Qiongzhong	Baisha	Baisha	Baisha16	Qiongzhong

Note Same figures behind the city names mean the coordinate ranking for Chengmai was outside the top ten. The indicator of GDP for Dingan was outside of top ten.

Only indicator of the quantities of scenic spots for Wuzhishan was in the top ten. Other cities and counties were all ranked in outside of top ten.

5.2 Multi-Objective Ranking

Based on the same weight coefficient for each evaluation index, sequence numbers of six evaluation indexes in Table 2 for each city or county was added to get a numeric value, which was considered as the multi-objective evaluation score. Based on those scores, 18 cities and counties were ranked. Multi-objective ranking results and scores obtained were listed in Table 3. According to the multi-objective ranking scores, the sequence of top ten cities or counties were as follows from top one to top ten: Sanya, Haikou, Wanning, Qionghai, Danzhou, Wenchang, Lingshui Chengmai, Dingan, and Baoting. Among them, the top six cities and counties of Sanya, Haikou, Wanning, Qionghai, Danzhou, and Wenchang were the same results with single-objective evaluation results.

5.3 Selecting Eco-Tourism Cities According to the Layout of the Construction and Development Plan for Hainan International Tourism Island (2010–2020)

According to the plan, the land use in Hainan Province was divided into six functional areas [11]: (1) integrated industrial development area for the northern cities, including Haikou, Wenchang, Lingao, and Chengmai (2) integrated tourism development area for the eastern cities, including Wenchang, Qionghai, and Wanning (3) integrated tourism development area for the southern cities, including Sanya, Lingshui, and Ledong (4) industrial town development area for the western cities, including Danzhou, Changjiang, Dongfang, and Ledong (5) terraced agricultural comprehensive development areas around the Island, including terraced area around the Island of Haikou, Wenchang, Qionghai, Wanning, Dingan, Tunchang, Chengmai, Lingao, Danzhou, Baisha, Changjiang, Dongfang, and Ledong, etc. cities (counties) (6) central ecological protection area, including the part areas of Wuzhishan, Qiongzhong, Baoting, Danzhou, Baisha, Changjiang, Dongfang, Ledong, Sanya, and Lingshui. According to the plan, the eastern and southern cities and counties, such as Wenchang, Qionghai, Wanning, Sanya, Lingshui, and Ledong, were classified into the tourism development areas; therefore, those six cities and counties should be among the top ten eco-tourism cities. With the exceptions of Ledong, other five cities were already selected as the top ten, based on the multi-objective evaluation results.

Table 3 Multi-objective ranking

Multi-objective ranking	1	1	3	4	5	6	7	7	9	10	11	12	13	14	14	16	17	17
Cities and counties	Sanya	Haikou	Wanning	Qionghai	Danzhou	Wenchang	Lingshui	Chengmai	Dingan	Baoting	Dongfang	Wuzhishan	Changjiang	Ledong	Tunchang	Lingao	Baisha	Qiongzhong
Ranking score	10	10	22	25	30	42	49	49	50	60	64	69	73	79	79	80	93	93

5.4 Cluster Analysis

Sample clustering is clustered one by one, based on the distance between the samples. The samples with close relationship were clustered into a small-class unit first, then those with distant relationship were clustered into a large class unit until all the samples were clustered. Based on the data in Table 1, 18 cities and counties were clustered through the sample clustering analysis method, with Euclidean squared distance coefficient, the average linkage (between groups) clustering method. Clustering results were shown in the dendrogram of Fig. 1. According to Fig 1, the 18 cities and counties were clustered into three categories: (1) Haikou, (2) Sanya, Danzhou, (3) the others. Haikou, Sanya, and Danzhou in category (1) and (2) were selected as the candidates of top ten eco-tourism cities by the multi-objective evaluation method.

After excluding Haikou, Sanya, and Danzhou, re-clustering results were shown in Fig. 2. Based on the dendrogram in Fig. 2, 15 cities and counties were clustered into four categories: (1) Wanning, Chengmai, Dongfang, (2) Qionghai, Wenchang, (3) Baisha, Baoting, Wuzhishan, Qiongzhong, Tunchang, (4) Lingshui, Changjiang, Ledong, Dingan, Lingao. One to two cities in each clustering category were the top ten candidates obtained with the multi-objective evaluation method. They were Wanning and Chengmai in category (1), Qionghai and Wenchang in category (2), Baoting in category (3), and Lingshui and Dingan in category (4). The clustering results showed that top ten candidates were evenly distributed in each category. It indicated that the distribution of eco-tourism cities was reasonable and representative.

5.5 Qualitative Analysis Between Ledong and Dingan, Wuzhishan and Chengmai

Dingan and Chengmai were top ten candidates in multi-objective ranking; however, Ledong and Wuzhishan were not in the top ten. Dingan and Ledong were not close to each other in the multi-objective ranking, but the cluster analysis showed that they belong to the same category, so they were highly similar in the evaluation index. The tourism resources in Dingan were rich than those in Ledong, but features were not prominent. Tropical montane rainforest ecosystem in Ledong is typical in the world. In addition, the southern part of Ledong was classified into the integrated tourism development area in the construction development plan for Hainan International Tourism Island; therefore, Ledong will have a huge potential for development in the future. Furthermore, Ledong was an important part of the central ecological protection region in the construction development plan and Li Autonomous County. According to the above points of view, Ledong was selected into one of top ten eco-tourism city to replace Dingan.

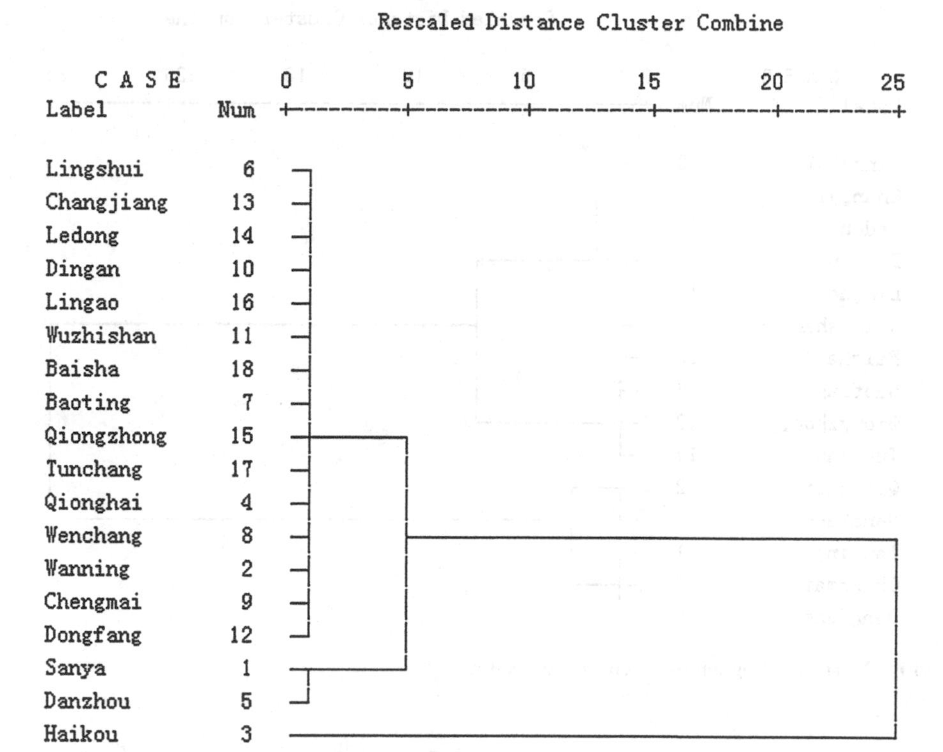

Fig. 1 The dendrogram of 18 cities and counties

Wuzhishan and Chengmai were not close either in the multi-objective evaluation ranking, but compared with Wuzhishan, tourism resources in Chengmai were not rich and Chengmai belongs to the non-tourist development areas in the plan. In contrast, Wuzhishan was classified into the central ecological protection areas, and the tropical rainforest ecosystem in Wuzhishan was very famous in China. Therefore, Wuzhishan was selected into the one of top ten eco-tourism cities to replace Chengmai.

6 Results and Discussion

Based on the combination of quantitative and qualitative evaluation methods, single-objective, multi-objective, and clustering evaluation methods, the final evaluation results of top ten eco-tourism cities in Hainan were Haikou, Sanya, Wanning, Danzhou, Qionghai, Lingshui (Li Autonomous County), Wenchang,

```
Dendrogram using Average Linkage (Between Groups)

                        Rescaled Distance Cluster Combine

    C A S E          0         5        10        15        20        25
  Label       Num    +---------+---------+---------+---------+---------+

  Lingshui      3    -+
  Changjiang   10    -+-+
  Ledong       11    -+ |
  Dingan        7    ---+---------+
  Lingao       13    ---+         |
  Wuzhishan     8    -+           +-----------------------------------+
  Baisha       15    -+           |                                   |
  Baoting       4    -+           |                                   |
  Qiongzhong   12    -+-----------+                                   |
  Tunchang     14    -+                                               |
  Qionghai      2    -+-----+                                         |
  Wenchang      5    -+     +-----------------------------------------+
  Wanning       1    -+     |
  Chengmai      6    -+-----+
  Dongfang      9    -+
```

Fig. 2 The dendrogram of 15 cities and counties

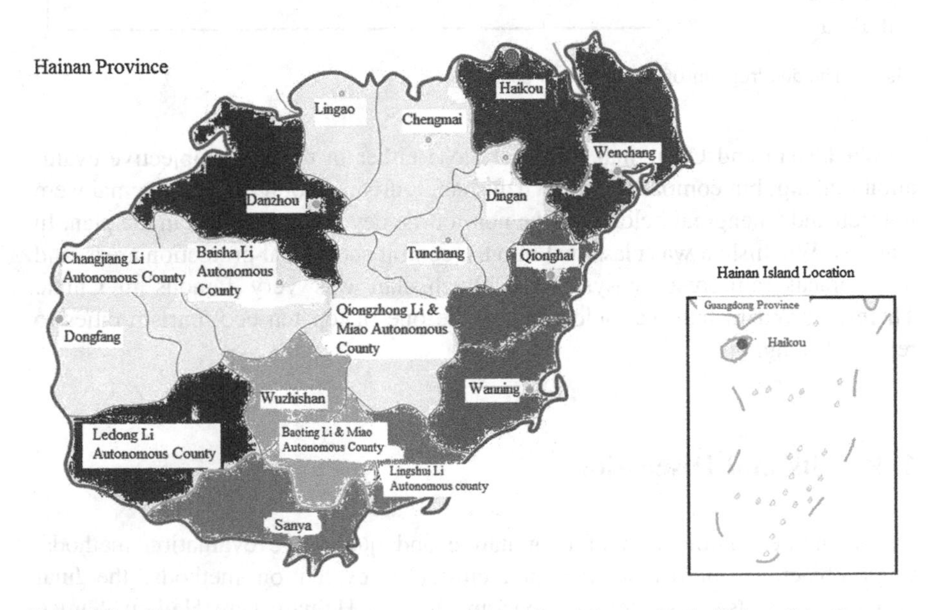

Fig. 3 The distribution map of top ten eco-tourism cities and counties in Hainan Island

Table 4 Major feature of top ten eco-tourism cities and counties

Cities and counties	No. of *A* and over *A* level scenic spots	No. of scenic spots	No. of proposed projects on 12th Five-Year Plan	Land use function	Characteristics	Human environment	Natural environment	Famous scenic spots
Haikou	5	11	7	Integrated industrial development area-northern cities	Capital city and convenient on transportation	Political, economic and cultural center; many famous historical and cultural monuments	Healthy City in the world, China Excellent Tourism City, excellent living environment and excellent garden city, national historical, and cultural famous city	1. Haikou Geopark with Rock Hill and volcanic group 2. Barbette 3. Hainan Tropical Wildlife Park 4. Holiday Beach in Haikou 5. Haikou Tropical Ocean World 6. five Ancestral 7. Ever Green Park 8. Jinniuling Park 9. Dongzhaigang 10. Qiongya former site of first congress
Sanya	9	16	12	Integrated tourist town development region-Southern cities	The most beautiful tourist destination	Important port for foreign trade	Many beautiful harbor, bay, and Island	1. Ends of the earth 2. Westerly Island 3. Yalong Bay 4. Zhujiang Nantian hot spring 5. Wuzhizhou Island 6. fish spa 7. Daxiaodongtian, 8. Luhuitou 9. Nanshan Cultural Tourism Zone 10. Dadonghai

(continued)

Table 4 (continued)

Cities and counties	No. of *A* and over *A* level scenic spots	No. of scenic spots	No. of proposed projects on 12th Five-Year Plan	Land use function	Characteristics	Human environment	Natural environment	Famous scenic spots
Wanning	7	10	8	Integrated tourism development area-eastern cities	Hometown of overseas Chinese		Tourism resources rich city	1. Xinglong Tropical Botanical Garden of the Southern medicine 2. Shimei Bay 3. Xinglong Tropical Garden 4. Wanning Dongshan Ridge 5. Shenzhou Peninsula 6. Xinglong Hot Springs 7. Sun Moon Bay Haimen 8. Xinglong Asian style garden 9. Perfume Bay 10. Coconut Island
Danzhou	3	17	5	Industrial development area-Western cities	Old shire	Yangpu Economic Development Zone, the bonded port area, "National Poetry town," "village" of Chinese couplets, the "hometown of Chinese folk art"	Excellent Tourism City	1. Yangpu Economic Development Zone 2. Baimajing monuments 3. Yunyue lake 4. tone flower water tunnel 5. Liangyuan Botanical Garden 6. Songtao Reservoir 7. Lanyang Spa 8. Dongpo Academy 9. Egret Paradise 10. Sugar Palm hut nunnery

(continued)

Table 4 (continued)

Cities and counties	No. of *A* and over *A* level scenic spots	No. of scenic spots	No. of proposed projects on 12th Five-Year Plan	Land use function	Characteristics	Human environment	Natural environment	Famous scenic spots
Qionghai	4	8	5	Integrated tourism development area-eastern cities	One of the key provincial tourist area	Boao Forum for Asia, a lot of businessmen of overseas Chinese	Ecological beauty	1. Boao Aquarium 2. Kuantang Spa 3. Baishi Ridge 4. Boao Jade Belt Beach 5. Wanquan River 6. Coconut Village and Farmhouse 7. The site of Boao Forum for Asia 8. Boao Oriental Culture Garden 9. leisure rafting in Wanquan River 10. Statue of the Red Detachment of Women in Qionghai
Lingshui	2	3	8	Integrated tourism development area-Southern cities	Feng shui treasure	Site of Soviet government, South Tyrant Manor, a street in the Qing Dynasty, Li Autonomous County	Bays, beaches, islands, coconut, virgin forests, waterfalls, hot springs, monkeys, ostriches	1. Boundaries Island 2. Nanwan Monkey Island 3. Diaoluo Mountain 4. Coconut field and Li and Miao village 5. Soviet government site

(continued)

Table 4 (continued)

Cities and counties	No. of *A* and over *A* level scenic spots	No. of scenic spots	No. of proposed projects on 12th Five-Year Plan	Land use function	Characteristics	Human environment	Natural environment	Famous scenic spots
Wenchang	1	8	4	Integrated industrial development area-northern cities	Cultural village, the hometown of overseas Chinese, coconut Township, volleyball township, the hometown of the generals, the hometown of the mother of the country	Song ancestral	Coconut Grove, harbor, natural bath	1. Coconut Grove in eastern outskirts 2. Qizhou Islands 3. Song ancestral 4. Tonggu Ridge 5. Gaolong Bay 6. Mangrove in Bamen Bay 7. Qixing Ridge 8. Yun-long Bay 9. Coconut Grand Park 10. Mulan Bay
Wuzhishan	1	5	8	Central Ecological protection area-central cities	Mountain city with the highest elevation in Hainan Island, Original Prefecture of Hainan Li and Miao Autonomous		"Natural air conditioning," "natural oxygen bar", "jade mountain," "natural animal and plant kingdom," "fairyland paradise" and "China's health city"	1. Wuzhi Mountain 2. Rafting in Wuzhishan Grand Canyon 3. Li villages 4. Taipingshan Falls 5. Baoting Tropical Botanical Garden 6. Emerald Hill 7. Maoan Miao Village 8. Qiongya Memorial Pavilion of Public School

(continued)

Table 4 (continued)

Cities and counties	No. of *A* and over *A* level scenic spots	No. of scenic spots	No. of proposed projects on 12th Five-Year Plan	Land use function	Characteristics	Human environment	Natural environment	Famous scenic spots
Baoting	2	7	6	Ecological protection area-central cities	Planting base for Southern medicine, special local product-rambutan	Li and Miao culture, Li and Miao Autonomous County	The forest coverage rate of 81.5 %, rich in natural resources and tourism resources	1. Qixianling Spa 2. Yanoda rainforest 3. Baoting Betel Palm Park 4. Maogan Xianan Stone Forest 5. Betel Palm Valley 6. Li and Miao style tourist village 7. MaoganQianlong hole
Ledong	0	1	5	Integrated tourism development area-southwest cities	White sand and blue see, first choice of vacation and leisure	Li Autonomous County	"Bananas town in China"	1 Jianfengling National Forest Park

Baoting (Li and Miao Autonomous County), Wuzhishan, and Ledong (Li Autonomous County). The distribution pattern of the ten eco-tourism cities was as following: one in the southwest, one in the western, two in the southern, two in the central, two in the eastern, and two in the northern (Fig. 3). The main characteristics for top ten eco-tourism cities were shown in Table 4.

This chapter was a preliminary evaluation to eco-tourism cities in Hainan, because the evaluation index system was relatively simple and needs to be further improved. The results indicated that only if the multi-objective evaluation method was taken as the foundation method, cluster analysis as an assistant means, and the qualitative analysis considered, the relatively satisfied evaluation results will be achieved.

Acknowledgments This work was supported by International Science and Technology Cooperation Program of China (2012DFA11270); Hainan International Cooperation Key Project: Construction of an Early Warning Information Platform/Module based on Ecotourism Assessment and Management (EAM)—Its Application in the Building of Hainan International Tourism Island of Chin (GJXM201105).

References

1. Sun, T.T.: A research on the internationalization of tourism industry in Hainan. Northeast University (2009)
2. Honey, M.: Ecotourism and sustainable development: Who owns paradise? 2nd edn, p. 33. Island Press, Washington, DC (2008). ISBN: 1-59726-125-4, 978-1597261258
3. The International Ecotourism Society: The definition [EB/OL]. http://www.ecotourism.org/what-is-ecotourism, 3 May 1990/2013
4. Sun, S.T., Wu, C.D., Zhang, H.L.: The Construction of Jixi as an International Ecological Tourist City. J. Harbin Univ. **23**(9), 51–54 (2002)
5. Fu, G.J.: Investigation, classification and evaluation of natural tourism resources in Hainan. Nat. Sci. J. Hainan Univ. **28**(1), 52–58 (2010)
6. Cao, T.L.: My opinion on the advantages of Hainan tourism resources. New Orient. **6**, 76–78 (1996)
7. Peng, L.S., Mou, R.F.: Application research into the analytic hierarchy process on ecological tourism resource assessment. Environ. Sci. Manage. **31**(3), 177–180 (2006)
8. Chu, D.P., Zheng, Y.X.: The application of the method of principal component analysis in tourism attraction grade assessment. J. Jimei Univ. (Nat. Sci.) **7**(40), 34–358 (2002)
9. Yang, S.S.: Clustering assessment to tourism competitiveness factor of cities in Provinces. Dev. Res. **6**, 42–43 (2006)
10. Hao, Q.Y., Liu, Q., Zhong, Q.X., Zhou, Y.P.: Comparison and selection of comprehensive evaluation methods for protection efficiency of coastal shelterbelts. For. Resour. Manage. **2**, 82–88 (2010)
11. Notice on the land use overall planning in Hainan Province (2006–2020) issued by Hainan Provincial People's Government, Bulletin of Hainan Provincial People's Government, 15 Feb 2010

An Analytical Study on Tourism Informatization in Hainan International Tourism Island

Wen-juan Jiang and Bin Meng

Abstract Tourism informatization plays a very important role in the development of Hainan into an international tourism island. This chapter makes an analysis of the importance of Hainan's tourism informatization. It then elaborates on the current situation of and problems in the province's tourism informatization and puts forward relevant suggestions for improving its tourism informatization.

Keywords Tourism informatization · Hainan · Hainan International Tourism Island

The project of developing Hainan Province into an international tourism island was formally upgraded to a national strategy on January 4, 2010, after the State Council issued its opinions on promoting the construction and development of Hainan International Tourism Island. Hainan, which has since then seen a surge in tourists number, found there is a much urgent need for tourism informatization.

Tourism informatization refers to the process in which electronic technology, information technology, data base technology, and internet technology are used to transform the traditional mechanism on the production, distribution, and consumption activities in the tourism industry. In this process, technological approaches are adopted to improve the operation of tourism economy and to boost the development of tourism industry by upgrading the traditional tourism industry into a modern one. Tourism informatization mainly includes the three aspects: informatization of tourism enterprises, tourism e-business, and tourism e-governance [1].

W. Jiang (✉)
School of Information Science and Technology, Hainan Normal University,
Haikou 571158, China
e-mail: jwj@hainnu.edu.cn

B. Meng
Hainan Administration of Surveying, Mapping and Geoinformation,
Haikou 570203, China
e-mail: may_jwj@163.com

B.-Y. Cao et al. (eds.), *Ecosystem Assessment and Fuzzy Systems Management*,
Advances in Intelligent Systems and Computing 254, DOI: 10.1007/978-3-319-03449-2_13,

1 Current Situation of Hainan's Tourism Informatization

1.1 Informatization of Tourism Enterprises

Presently, Hainan Province has a total of 349 travel agencies, of which, 24 are engaged in outbound trip operations, and 103 branches. Most of these travel agencies are distributed in Haikou and Sanya cities. Only the large and medium-sized travel agencies have set up their own official Web sites, while many of them simply post their company introduction, address, and contact information on some tourism-related Web sites. For those which have developed their own Web sites, what they display online are mainly the services' general introduction, tourism routes, hotels, and very simple maps.

1.2 Tourism e-governance

Hainan Provincial Commission of Tourism Development plays an indispensable role in the province's tourism informatization, as it is the government authority responsible for coordinating the development of the provincial tourism industry, formulating provincial tourism development plans and protecting the provincial tourism resources. The Web site—tourism.hainan.gov.cn, known as the official Web site of the provincial tourism development commission, functions in the following aspects:

1. Releasing news and information related to Hainan's tourism industry;
2. Offering information on tourist attractions, itineraries, accommodation, catering, and travel agencies;
3. Delivering e-government services for the public;
4. Serving as an e-business platform on which tourists can book hotels and air tickets.

The tourism development commissions in the major cities and counties in Hainan also have their own official Web sites. For instance, www.haikoutour.gov.cn for Haikou City, www.sanyatour.com for Sanya City, www.qionghai.gov.cn for Qionghai City, and www.wenchangtour.com for Wenchang City. However, what these Web sites focus on are mainly the introduction to the tourism attractions, folk customs, classical itineraries, and accommodations. A few of them can deliver e-government services and offer links to shopping malls, but most of them are only Web pages containing static information.

The provincial tourism development commission's subordinate institutions and the other tourism-related organizations in Hainan have also developed their official Web sites. The following are some of the major official Web sites established by these organizations:

1. www.hainanta.com for Hainan Provincial Tourism Association. The Web site mainly offers introduction and news on the association, as well as the policies, laws and regulations regarding the tourism industry. One of its features is that it has launched a series of questionnaires and surveys on the hot issues in the tourism industry;
2. www.hi898.com for Hainan Tourism Development Research Association. The Web site mainly offers general introduction to the association, as well as tourism news, policies, laws and regulations regarding the tourism industry. There are also some online surveys.

1.3 Tourism e-business

According to the 31st Statistical Report on Internet Development in China released by the China Internet Network Information Center (CNNIC) in January 2013, China has had a total of 242 million online shoppers and the utilization ratio of online shopping rose to 42.9 % by the end of December 2012. Compared to 2011, online shoppers rose by 48.07 million with the growth rate of 24.8 %. However, despite the rapid growth of online shopping, e-business in tourism industry has maintained a relatively slow development trend. Most travel agencies have still been processing the clients' orders in the traditional and manual way. The provincial tourism development commission, which has an e-business platform, can only offer very limited tourism products and is lagged behind in promotion and marketing efforts. As the hotel industry is concerned, only a very limited number of hotels can offer online booking and payment services.

1.4 Other Tourism Informatization Projects

Different government authorities and organizations in Hainan have been making great efforts to develop the province into an international tourism island since 2010. For instance, the Hainan Administration of Surveying, Mapping and Geoinformation, with a view to supporting the construction of Hainan as an international tourism island, has applied to develop a digital geospatial framework, which has a public version that can offer the electronic maps, image maps, and 3D maps of the whole province and its 18 cities and counties. It can also provide the panorama of some major scenic spots. In addition, a special section on Hainan's tourism is set up under the framework to include the following modules: tourism routes, food and catering, shopping centers, hotels and accommodation, scenic spots, and 360 Panorama. Its major feature lies in that the specific positions of these modules are indicated by special pictures.

2 Problems and Countermeasures in Hainan's Tourism Informatization

2.1 Tourism Administrations Playing Leading Role in Tourism Informatization

The tourism administrations should play a leading role in tourism informatization as they are the authorities guiding the development of local tourism. Generally speaking, almost all tourism development commissions at various levels have opened their own official Web sites, but most of them are static, rather than dynamic. Though the provincial tourism development commission added the panoramic views of some scenic spots and hotels onto its official Web site, most of the local tourism administrations' Web sites only contain dull and out-of-date information. It is important to enrich and update timely the Web site information to better serve tourists. In the meanwhile, efforts should be made to encourage the administrations' staff to be engaged in online office and bring into full play their enthusiasm in using the e-government platform in processing daily work.

2.2 Avoiding Repetitive Construction and Monotony in Web site Content

The tourism companies and tourism administrations have attached more importance to tourism informatization with the rapid development of information and internet technologies. The first step they take is to set up their own official Web sites. However, it has been found that most of these Web sites lack their distinct features, as they usually only include information on tourist attractions, tourism routes, accommodation, and food and catering [2].

It is important to use various technical methods to help a Web site stand out from the pack. For instance, apart from news and photos, other technical forms, such as adequate use of flash and video, can make a Web site more friendly, eye-catching, and helpful. A lot of new technologies, such as Panorama, GIS, and GPS, can also be applied.

2.3 Developing Sharing Mechanism Between Government Administrations and Tourism Web sites

Even though tourism administrations and tourism enterprises have developed many tourism Web sites, they are lack of a resources sharing mechanism. This has, to a great extent, affected the efficiency of tourism-related operations and businesses.

Therefore, it is important for the tourism administrations to plan the informatization process and contents in a scientific and sound way, so that the whole province's tourism informatization may advance smoothly and avoid redundant work and waste of resources. Only in this way, can the relevant departments and enterprises have more time to think about what the distinctive features they have in tourism informatization are and how to increase their work efficiency and operation benefits.

2.4 Applying GIS Tech into Tourism Industry

The GIS and computer technologies can be applied into the tourism industry to help develop the traveling geographic information system (TGIS), which is able to offer tourists a full range of tourism information in a more direct way [3]. It is a trendy in modern tourism informatization to develop a user-friendly TGIS with versatile functions [4].

2.5 Cultivating Professionals in Tourism Informatization

The tourism informatization professionals should be inter-disciplinary talent personnel who are well-versed in management, tourism, and information technology. However, there have been a very limited number of such professionals so far, which somehow obstructs Hainan's tourism informatization.

To solve this problem, the higher education institutions can set up tourism informatization programs to cultivate the relevant talent personnel, and various training organizations can also open more relevant training programs. It is also important that relevant units provide their employees with training and study opportunities in this aspect to enhance their professional knowledge and capabilities. To cultivate high-quality talent for tourism informatization is one of the prerequisites for fueling tourism informatization.

3 Conclusion

In line with the national strategy, Hainan is scheduled to have been basically developed into a world-class recreation and vacation Mecca by 2020. It will become an open, green, and harmonious island by that time. To achieve this goal, it is important to boost tourism informatization, so as to overcome the obstacles encountered by the tourism industry in its upgrading process. To sum up, tourism informatization will help the traditional tourism industry to make significant adaptations and develop in a modern way.

Acknowledgments Thanks to the support by National Natural Science Foundation of Hainan (No. 612122, 612126). And this work is supported by International Science and Technology Cooperation Program of China (2012DFA1127); Hainan International Cooperation Key Project (GJXM201105).

References

1. Chen, S., Feng, X.G.: Preliminary research on information construction of city tourism, a case study of Hangzhou city. East China Econ. Manage. **19**(3), 8–11 (2005)
2. Guo, J.C., Li, Z., Zhu, J.F., Ma, Y.C.: To build Hainan International tourism island informatization must go ahead. J. Qiongzhou Univ. **16**(5), 1–4 (2009)
3. Li, X.Y., Jiang, W.B.: Research and analysis of the development of digital Hainan regional tourism geography information system. Mod. Comput. **11**, 73–76 (2012)
4. Zhang, B.H., Yan, Y.F.: A review on the domestic and overseas tourism information research. Geogr. Geo-Information Sci. **28**(5), 95–99 (2012)

Part II
Intelligent Algorithm, Fuzzy Optimization and Engineering Application

Land Use and Land Cover Mapping Using Fuzzy Logic

Ruibo Han

Abstract The problem of mixed pixels has resulted in the low accuracy of land cover classification in urban areas. Conventional methods for the classification of multispectral remote sensing imagery, such as parallelepiped, minimum distance between means, and maximum likelihood, only utilize spectral information and consequently have limited success in classifying urban multispectral images. Fuzzy classification techniques allow pixels to have a probability of belonging (membership) in more than one class, and through this membership, the imprecise nature of the data is better represented. In this chapter, a fuzzy logic-based hierarchical classification method that incorporates both spectral and spatial information is then developed. Meanwhile, the maximum likelihood classifier (MLC) and knowledge-based expert system are also applied to provide comparison references. This fuzzy expert system is proved to produce a substantial increase in the classification accuracy of urban land cover maps in comparison with the traditional maximum likelihood and expert system classification approach.

Keywords Fuzzy logic · Hierarchical classification method · Fuzzy expert system · NDVI · NDBI · PCA

1 Introduction

Urban and economic development requires an up-to-date information system to deal with the numerous issues associated with the increasing rate of urbanization. Access to products that can supply timely and accurate information is essential for government and private agencies to make effective decisions regarding the urban

---

R. Han (✉)
Science and Technology Branch, Agriculture and Agri-Food Canada, Ottawa, Ontario K1A 0C6, Canada
e-mail: ruibo.han@agr.gc.cahan@live.ca

B.-Y. Cao et al. (eds.), *Ecosystem Assessment and Fuzzy Systems Management*, Advances in Intelligent Systems and Computing 254, DOI: 10.1007/978-3-319-03449-2_14,

environment [1]. Land use mapping, which provides knowledge about land use and land cover, has become increasingly important in overcoming the problems of haphazard and uncontrolled sprawl, deteriorating environmental quality, loss of agricultural lands, destruction of important wetlands, etc. [2]. With the development of remote sensing and geospatial technologies, satellite imagery has been widely used as an effective and timely source for the generation of digital images of land use and land cover maps, which can be easily integrated into existing GIS databases and utilized to aid officials in planning and decision-making processes. Applications for urban land cover maps include, but not limited to, land use change detection, environmental planning and assessment, infrastructure planning and administration, and ecosystem management [3–5].

Remotely sensed medium-resolution imageries (spatial resolution of 10–100 m) and multispectral imageries (e.g., Landsat and ASTER) have typically been used to identify urban areas or to distinguish between residential, industrial, and commercial zones in an urban landscape [6–10]. Accurate image classification results are a prerequisite for further applications; however, producing a satisfactory classification image from medium-resolution imagery is not a straightforward task. Factors contributing to this difficulty include the availability of suitable images for the desired time, ancillary and ground reference data, as well as the variables and algorithms selected to run the classification [11]. First, it has been proven that a combination of ancillary data and spectral features can significantly improve the classification performance [12–14]. Out of the variety of available ancillary data, which includes texture, context, and base maps like transportation networks, the digital elevation model (DEM) has been proven to be the most efficient for improving classification accuracy in an urban context [11, 12, 15, 16].

Multiple spectral imageries provide various dimensions of information about land features; however, high input dimension also causes the critical problem of space and time consumption in processing procedures, such as land cover classification [15]. Therefore, reducing the dimensions of data, without losing much information, would increase the overall effectiveness of remote sensing classification. Several methods have been developed to transform multispectral bands into band ratios and vegetation indices, such as Normalized Difference Vegetation Index (NDVI), Tasseled Cap Transformation, and principal component analysis (PCA) [17]. In contrast to other transformations that use predefined linear equations on a number of selected channels, PCA employs a statistical procedure to convert all applicable data channels into several principal output channels. This procedure uses the interband correlation and decorrelation of multiband imagery to minimize data redundancy and correlation between bands.

Since it is usually not possible to control the availability of desired images, classification algorithms are often resorted in order to improve classification accuracy. Conventional methods for the classification of multispectral remote sensing imagery, such as parallelepiped, minimum distance from means, and maximum likelihood, only utilize spectral information and consequently have limited success in classifying urban multispectral images [1, 2, 18]. In addition, urban landscapes are usually composed of features with similar spectral signatures, and this makes mixed pixel a common problem, especially for medium-resolution imagery. However, conventional classification methods are based on crisp classifications, i.e., each pixel can only be classified as one class in the classification process. Therefore, the problem of mixed pixels has resulted in the low accuracy of land cover classification in urban areas [11].

Fuzzy logic, also synonymous with the theory of fuzzy sets, is a theory that relates to the classification of objects with imprecise boundaries. Due to fuzzy logic's ability to resolve ambiguity, it is able to process an input space to an output space through a mechanism of if–then inference rules. Fuzzy logic techniques in the form of approximate reasoning offer powerful reasoning capabilities for decision support and expert systems [19]. Fuzzy classification techniques allow pixels to have a probability of belonging (membership) in more than one class, and through this membership, the imprecise nature of the data is better represented.

In this chapter, DEM is treated as ancillary data, and several transformations including NDVI, Normalized Difference Built-Up Index (NDBI), and PCA are employed to provide a better represented dataset with distinctive characteristics. A fuzzy logic-based hierarchical classification method that incorporates both spectral and spatial information is then developed. Meanwhile, the maximum likelihood classifier (MLC) and knowledge-based expert system are also applied to provide comparison references. In conclusion, this fuzzy expert system produces a substantial increase in the classification accuracy of urban land cover maps in comparison with the traditional maximum likelihood and expert system classification approach.

2 Methodology

This work applies an expert system approach to a portion of urban land cover of Beijing, China (See Fig. 1). An area that is comprised of a variety of land cover types and terrains is selected so that the classification system can be tested in a general context. Located in the northwestern districts of Haidian and Changping of Beijing, the study area occupies approximately 1,859 km^2. The major body in the area is a flat landscape with urban buildup and agricultural land that is surrounded by mountainous terrain to the north and west. Urban and suburban development of the study region has proceeded at a fairly rapid rate with the widespread conversion of adjacent undeveloped regions and agricultural lands into residential and commercial areas. These districts have been ranked as the fastest growing regions in the Beijing since 1990. The primary motivation for this work is to conduct historical land cover classification and to monitor and plan future land cover change based on the remotely sensed imageries.

Fig. 1 Location of study area and true-color 3D image from Landsat TM and DEM

2.1 Data Processing and Atmospheric Correction

The sequence of steps shown in Fig. 2 was carried out to preprocess Landsat imagery and use this imagery to derive land cover features and to simulate urban growth in future research.

2.1.1 Convert DN to At-sensor Spectral Radiance (Q_{cal}-to-L_λ)

Calculation of at-sensor spectral radiance is the fundamental step in converting image data from multiple sensors and platforms into a tangible and accessible radiometric scale [6]. The pixel values acquired in the L1T product are stored as an 8-bit digital number (DN, quantized calibrated pixel value ranging from 0 to 255 for Landsat), which is rendered from absolute radiance (units: $\mathrm{Wm^{-2}\,sr^{-1}\,\mu m^{-1}}$). It is usually required that the digital numbers are converted back to at-sensor spectral radiance in order to minimize changes in the instrument radiometric calibration. Then, they are converted to top-of-atmosphere (TOA) reflectance to minimize remote sensing variations introduced by variations in the Sun–Earth distance, the solar geometry, and exoatmospheric solar irradiance arising from spectral band differences [20]. The conversion to radiance is important for the application of Landsat data for large spatial and temporal coverage. The conversion of the radiance sensed at the Landsat reflective to reflectance (unitless) provides information that has physical meaning and significantly provides data that can be used to derive biophysical products [21].

The following equation is used to convert DN back to radiance units:

$$L_\lambda = G_{\mathrm{RESCALE}} \times Q_{\mathrm{CAL}} + B_{\mathrm{RESCALE}}$$

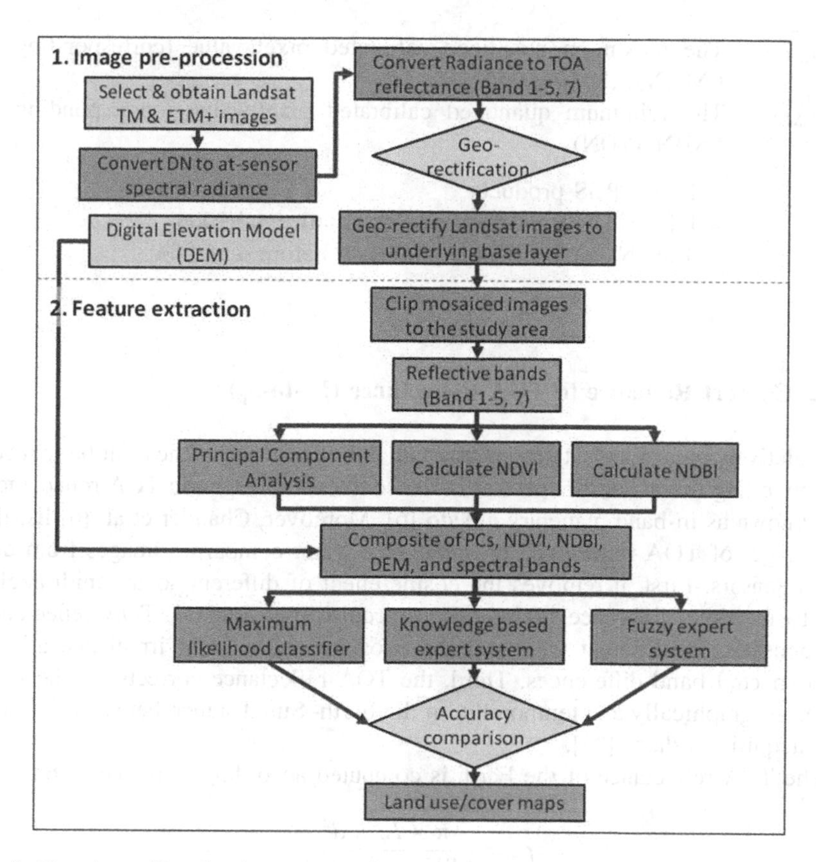

Fig. 2 Flowchart of Landsat image processing

which is also expressed as

$$L_\lambda = ((\mathrm{LMAX}_\lambda - \mathrm{LMIN}_\lambda)/(Q_{\mathrm{CALMAX}} - Q_{\mathrm{CALMIN}})) \times (Q_{\mathrm{CAL}} - Q_{\mathrm{CALMIN}}) + \mathrm{LMIN}_\lambda$$

where:

- L_λ Spectral Radiance at the sensor's aperture ($\mathrm{Wm}^{-2}\ \mathrm{sr}^{-1}\ \mu\mathrm{m}^{-1}$)
- G_{RESCALE} Rescaled gain (the data product "gain" contained in the Level 1 product header or ancillary data record) ($\mathrm{Wm}^{-2}\ \mathrm{sr}^{-1}\ \mu\mathrm{m}^{-1}\ \mathrm{DN}^{-1}$)
- B_{RESCALE} Rescaled bias (the data product "offset" contained in the Level 1 product header or ancillary data record) ($\mathrm{Wm}^{-2}\ \mathrm{sr}^{-1}\ \mu\mathrm{m}^{-1}\ \mathrm{DN}^{-1}$)
- Q_{CAL} The quantized calibrated pixel value (DN)
- LMAX_λ The spectral radiance that is scaled to Q_{CALMAX} ($\mathrm{Wm}^{-2}\ \mathrm{sr}^{-1}\ \mu\mathrm{m}^{-1}$)), available from Chander et al. [6]
- LMIN_λ The spectral radiance that is scaled to Q_{CALMIN} ($\mathrm{Wm}^{-2}\ \mathrm{sr}^{-1}\ \mu\mathrm{m}^{-1}$), available from Chander et al. [6]

Q_{CALMAX} The maximum quantized calibrated pixel value (corresponding to $LMAX_\lambda$) (DN) = 255

Q_{CALMIN} The minimum quantized calibrated pixel value (corresponding to $LMIN_\lambda$) (DN)

= 1 for LPGS products
= 1 for NLAPS products processed after 4/4/2004
= 0 for NLAPS products processed before 4/5/2004

2.1.2 Convert Radiance to TOA Reflectance (L_λ-to-ρ_p)

For relatively clear Landsat scenes, the variability between scenes can be reduced by converting the at-sensor spectral radiance to exoatmospheric TOA reflectance, also known as in-band planetary albedo [6]. Moreover, Chander et al. [6] list the advantages of TOA reflectance over radiance when comparing images from different sensors. First, it removes the cosine effect of different solar zenith angles due to the time difference between data acquisitions. Second, TOA reflectance compensates for different values of the exoatmospheric solar irradiance arising from spectral band differences. Third, the TOA reflectance corrects for the variation, geographically and temporally, in the Earth–Sun distance between different data acquisition dates [22].

The TOA reflectance of the Earth is computed according to the equation

$$\rho_p = \frac{\pi \times L_\lambda \times d^2}{\mathrm{ESUN}_\lambda \times \cos \theta_S}$$

where

ρ_p Planetary TOA reflectance (unitless)

L_λ Spectral radiance at the sensor's aperture ($\mathrm{Wm}^{-2}\ \mathrm{sr}^{-1}\ \mu\mathrm{m}^{-1}$)

d Earth–Sun distance in astronomical units for day of the year (DOY), available at http://landsathandbook.gsfc.nasa.gov/handbook/excel_files/d.xls or from Chander et al. [6]

ESUN_λ Mean solar exoatmospheric irradiances ($\mathrm{Wm}^{-2}\ \mathrm{sr}^{-1}\ \mu\mathrm{m}^{-1}$), available from Chander et al. [6]

θ_S Solar zenith angle (degree)

= 90-solar elevation angle (degree), and the solar elevation angle is typically stored in the Level 1 product header file.

Note that the TOA reflectance values usually range from 0 to 1, so the TOA reflectance values for all six reflective bands (bands 1–5, 7) were stored as signed 8-bit integers after being scaled by 255.

2.1.3 Georectify Landsat Images to Underlying Base Layer

Although Landsat L1T products have been orthorectified using GCPs and DEM to attain absolute geodetic accuracy, there may still be significant discrepancies between the multiple-date images and the underlying GIS base layers in certain areas in some of the landscapes. In the instance of such errors, control points are used to geometrically rectify the misaligned scenes with the underlying base layer. Geometric ratification should result in the two images having a root mean square error (RMSE) of $\leq$0.5 pixel.

2.1.4 Calculate NDVI

The process of automatically mapping land covers has been streamlined by the positive results of certain efforts such as the use of indices. One of the most commonly used indices is the Normalized Difference Vegetation Index (NDVI). NDVI is an indicator of the density and photosynthetic activity of living vegetation [17, 23]. It successfully uses the unique shape of the reflectance curve of vegetation to its advantage and consequently has been used globally for mapping vegetation from multiple sources of data such as the Advanced Very High Resolution Radiometer (AVHRR) data [24].

This index is a function of the relative reflectance in the red and near-infrared bands and is calculated according to the following formula:

$$\mathrm{NDVI} = (\mathrm{NIR} - \mathrm{RED})/(\mathrm{NIR} + \mathrm{RED})$$

where

NIR Reflectance in the near-infrared band

RED Reflectance in the red band.

The red and NIR images of TOA reflectance are obtained and used to calculate an NDVI value for each pixel. The NDVI equation produces values in the range of -1.0 to 1.0, where vegetated areas will typically have values greater than zero and negative values indicate non-vegetated surface features such as water, barren land, ice, snow, or clouds. In order to maximize the range of values and provide numbers that are appropriate to display in an 8-bit image, the NDVI value must be scaled. This scaling converts a number between -1.0 and 1.0 into a pixel value that is appropriate for a gray tone display. One example of scaling an NDVI value for display is the following equation:

$$\text{Scaled NDVI} = \mathrm{Int}(100 \times \mathrm{NDVI}) + 100$$

Thus, using this equation, the NDVI computed value is scaled to the range of 0–200, where computed -1.0 equals 0, computed 0 equals 100, and computed 1.0 equals 200. As a result, NDVI values less than 100 now represent clouds, snow, water, and other non-vegetative surfaces and values equal to, or greater than, 100 represent vegetative surfaces.

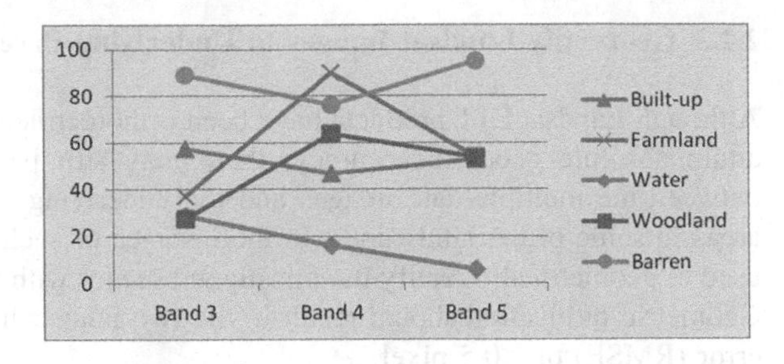

Fig. 3 Land cover's spectral profiles of bands 2, 3, and 4 of Landsat TM

2.1.5 Calculate NDBI

Figure 3 illustrates that the spectral disparity of the represented areas is most distinguishable in bands 3, 4, and 5. NDVI, as an index of vegetation, is derived from bands 3 and 4. In addition to NDVI, Normalized Difference Built-Up Index (NDBI) can be devised from bands 4 and 5 to map the urban buildup. Built-up areas and barren land experience a drastic increment in their reflectance from band 4 to band 5, while vegetation has a slightly larger or smaller DN value on band 5 than on band 4 (see Fig. 3). This pace of increment significantly exceeds that of the other covers. The minimum and maximum DNs in band 4 are much smaller than those in band 5 for the same cover. Therefore, the standardized differentiation of these two bands (as in the equation below) will result in close to 0 for woodland and farmland pixels, negative for water bodies, but positive values for built-up pixels, enabling the latter to be separated from the remaining covers [24].

$$\text{NDBI} = (\text{TM5} - \text{TM4})/(\text{TM5} + \text{TM4})$$
$$\text{Scaled NDBI} = \text{Int}(100 \times \text{NDBI}) + 100$$

As a result, the scaled NDBI values of less than 100 now represent vegetation and water, while values greater than 100 represent built-up and barren areas.

2.1.6 Principal Component Analysis

Remotely sensed data are highly correlated between the adjacent spectral bands. For instance, the visible bands 1, 2, and 3 in Landsat TM/ETM+ are highly correlated, as are bands 5 and 7 [11, 18]. The global analysis conducted by Small [25] indicates that >98 % of ETM+ image spectral variance can be represented within a three-dimensional spectral mixing space and that >90 % of the variance can be described with a two-dimensional projection of the mixing space. Principal component analysis (PCA) has been proven to be of value in reducing the dimensionality of multispectral datasets [11, 25, 26]. After the principal component rotation minimizes the correlations among dimensions, the resulting principal component bands (PCs) represent orthogonal components of diminishing variance

[25]. Classification of these principal components and of the different images was found to yield generally higher accuracies than other methods [27].

In practice, PCA is used to project reflective bands of Landsat TM and ETM+ into a new coordinate set of bands that have no correlation. The result of PCA resembles that of the Kauth–Thomas transformation (K–T). The K–T components of brightness, greenness, and wetness are clearly related to the biophysical properties of land cover and therefore have a useable interpretive value. The first principal component (PC1) is a sum of different bands that highlights the overall brightness. The PC2 is a difference between the sum of bands 4 and 5 and the sum of visible bands and band 7, which enhances vegetation information; and PC3 is the difference between short-wave infrared bands (TM5 and TM7), and the sum of visible bands and the near-infrared band [11]. The PC1 contains the largest percentage of the data variance, the second PC has the next largest data variance, and so on. However, when it comes to the classification of images, PC3 usually conveys useful information for distinguishing between specific bands based on their particular characteristics.

The original bands of TM data can be set aside since the first three PCs account for most of the variance in the dataset, and the following image enhancement and classification can be performed based solely on these three PCs [18]. This greatly reduces the amount of space and time necessary for classifying remotely sensed data.

Use of other components may be necessary in specific cases when the fourth component contains information on crops and forests, and the fifth contains information about roads [18]. However, it is necessary to remove the noise in these components before using them.

2.1.7 Create Final Composite of PCs, NDVI, NDBI, and DEM

The next step is to create composite data from the results of previous steps: principal components, NDVI, as well as temperature, and/or apply the NO DATA masks to create final raster dataset with the maximum available data coverage. This final dataset is ready to be classified to extract land features.

2.2 Classification

2.2.1 Supervised Maximum Likelihood Classification

Multispectral classification involves sorting pixels into individual classes based on their reflective or transformed values. An example of a classified image is a land cover map that displays features like urban sprawl, vegetation, bare land, water, etc. Supervised classification refers to a wide variety of feature extraction approaches. However, it is traditionally used to identify the use of specific decision

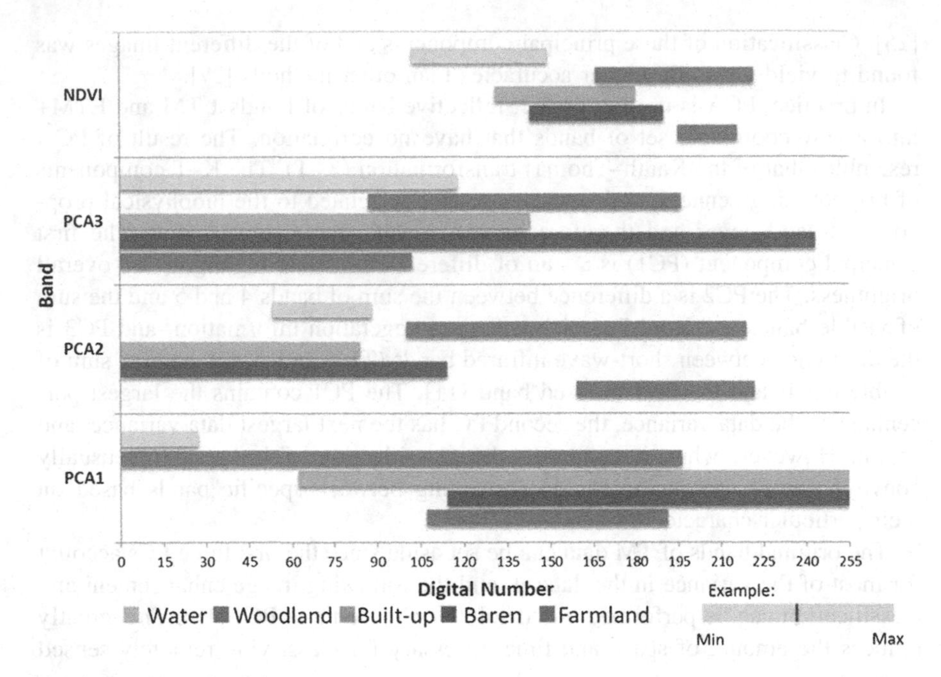

Fig. 4 Statistics of training areas

rules and classifiers such as maximum likelihood, minimum distance and Mahalonobis distance.

A supervised classification of the Landsat data was performed using the algorithm of maximum likelihood classifier (MLC), and all of the composite bands except DEM were used. Since this research is urbanization related and conducted at the regional scale, the definition of land cover classes was adopted from Anderson et al.'s [2] Land Use And Land Cover Classification System. Based on the land use and land cover situations, five types of land cover classes were distinguished in the classification scheme: built-up land (BL), wood land (WL), farm land (FL), water (WA), and barren land (BL).

Training areas for the land cover classes were selected with the assistance of ancillary geological, land use, and field data. Both the Jeffries-Matusita and transformed divergence separability were measured to ensure that the selected training areas for the land cover types are statistically separate. The statistics of training areas were calculated for each class and are presented in Fig. 4.

2.2.2 Knowledge-Based Expert System

One of the major disadvantages of these techniques is that they are all per-pixel-based classifiers. Each pixel is treated in isolation to determine which feature or class to assign it to. There is usually no incorporation of ancillary data such as context, shape, and proximity, which the human visual interpretation system takes for granted when interpreting what it sees. Expert knowledge-based classification, however, overcomes these limitations by setting up a decision tree of conditions and defining variables based on raster imagery, vector layers, or spatial models.

The application of expert systems or knowledge-based algorithms in the remote sensing classification falls into two categories: (1) the expert system is used to improve the accuracy after the conventional pixel-based classification method, such as the maximum likelihood classifier (MLC), was applied. It is also referred to as post-classification refinement, in which a low-level constituent information gets abstracted into a set of high-level informational classes. In such applications, ancillary data, including band ratios, textures, land cover maps, and administrative boundaries, were usually incorporated into conventional, and afterward expert, system classifications [7, 10, 14]. (2) In the other applications, expert systems were used directly to conduct the classification: Rule sets or decision trees were designed firstly based on expert knowledge or training experiments; a hierarchical system of IF–THEN rules were subsequently applied to extract the results [16, 28]. Modern methods like neural networks and fuzzy logics were usually used in combination with the expert system in order to obtain improved results [15].

Essentially, an expert classification system is a hierarchy of rules, or a decision tree, that sets down the conditions under which a group of low-level constituent information gets abstracted into a set of high-level informational classes. A rule is a conditional statement, or list of conditional statements, about the variable's data values and/or attributes that determine an informational component or hypotheses. Multiple rules and hypotheses can be linked together into a hierarchy that ultimately describes a final set of target informational classes or terminal hypotheses. Based on the knowledge of NDVI and NDBI, the multiband image can be segmented in a hierarchy of input and output. To achieve a higher accuracy in classifying the land cover, principal component and DEM are also incorporated into the decision system (see Fig. 5).

2.2.3 Fuzzy Logic Classifier

Fuzzy Expert System

Membership function is the mathematical function that defines the degree of an element's membership in a fuzzy set and therefore is the core of fuzzy logic applications [29]. In terms of the application of fuzzy logic in image classification, two ways to define membership functions have been widely adopted:

(a) Use predefined membership function based on a priori knowledge or data exploration, such as results from supervised classification, and the fuzzy logic inference rules are constructed and tested through the simulation of classification procedure.

Fig. 5 Multicriteria decision tree of the expert system

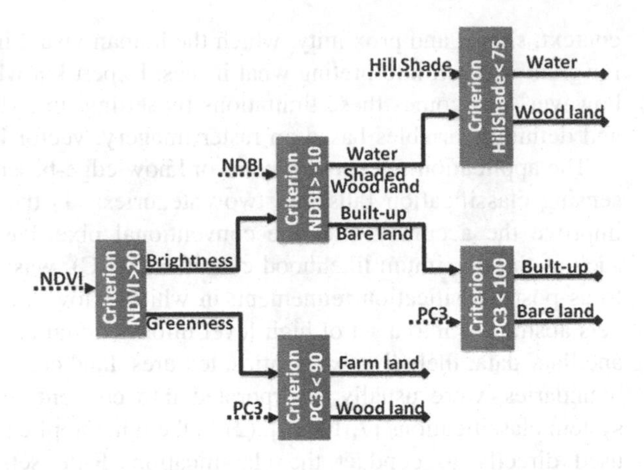

(b) Adaptive neuro-fuzzy inference system (ANFIS) is a neuro-adaptive learning technique developed in MATLAB. As opposed to choosing the parameters for a given membership function arbitrarily, these parameters can be purposefully determined to customize the membership functions to account for a greater number of variations in the data [30]. There are two methods that ANFIS learning employs for updating membership function parameters: backpropagation for all parameters (a steepest descent method) and a hybrid method consisting of backpropagation for the parameters associated with the input membership functions and least squares estimation for the parameters associated with the output membership functions.

The fuzzy inference system (FIS) contains both the model structure, which specifies such items as the number of rules in the FIS, the number of membership functions for each input, etc., and the parameters, which denote the shapes of membership functions.

Because the model structure used for ANFIS is fixed, it is prone to overfit the data on which is it trained, especially for a large number of training epochs. The checking data, which are used to cross-validate the fuzzy inference model, are important for learning tasks for which the input number is large, and/or the data itself are noisy. When the checking data option is used with ANFIS, the model parameters that correspond to the minimum checking error are returned at each training epoch. The FIS membership function parameters computed when both training and checking data are loaded are associated with the training epoch that has a minimum checking error [30].

Consequently, the training error decreases, at least locally, throughout the learning process. Therefore, the more the initial membership functions resemble the optimal ones, the easier it will be for the model parameter training to converge. Human expertise about the target system being modeled may contribute to the setup of these initial membership function parameters in the FIS structure.

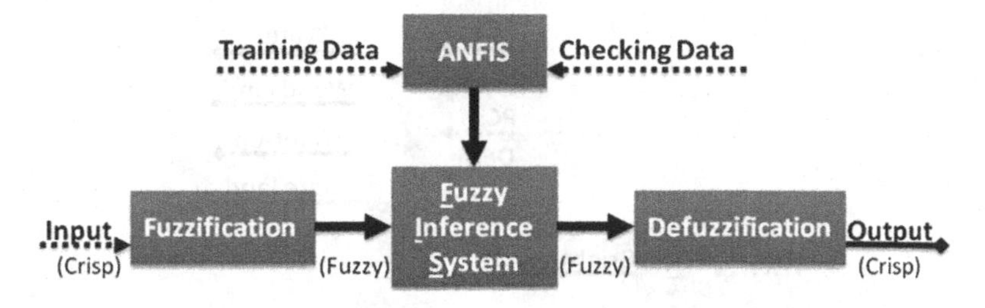

Fig. 6 Definition of a fuzzy logic system

Figure 6 represents a definition of the fuzzy expert system or fuzzy control system. The first operation is fuzzification, which is the mapping from a crisp point to a fuzzy set. The second operation is inferencing, which is the evaluation of the fuzzy rules in the form of IF–THEN. The last operation is defuzzification, which maps the fuzzy output of the expert system into a crisp value.

Hierarchical Structure

In Sect. 2.2.1, we have defined the five land cover classes to be separated. Normally, it is difficult to achieve a good performance with only two or three input bands, especially for the similar classes since there is not much differentiation between their spectral bands. To discriminate the similar classes, we have introduced non-spectral bands, PC bands, NDVI, NDBI, and DEM. For the fuzzy classifiers discussed above, it must be assumed that an input band is partitioned by m membership functions; then, n input bands will consist of $n^m \times 5$ rules for the classification so that the classification complexity increases exponentially with the rising number of input bands. In this study, each input is partitioned using 3 membership functions, and the number of rules needed would be $3^6 \times 5 = 3{,}645$, and the time required for the classification would be unreasonable. A hierarchical structure is proposed to solve the problem (see Fig. 7).

The hierarchical structure simplifies multiclass one-level classification to gradual levels of classification. In the first level, the land cover classes are separated into two groups based on the characteristics of NDVI and NDBI. Each group is then further divided into subgroups in the next level by the additional band inputs. For the group classification, additional input bands are selected based on observations of the signature data and consulting with geographical experts.

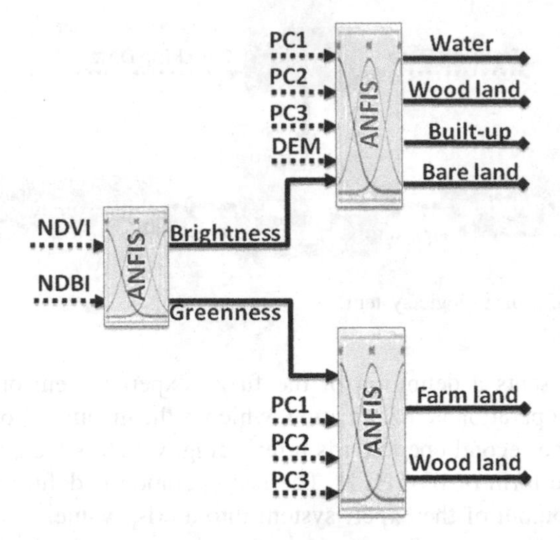

Fig. 7 Hierarchical structure of the fuzzy expert system

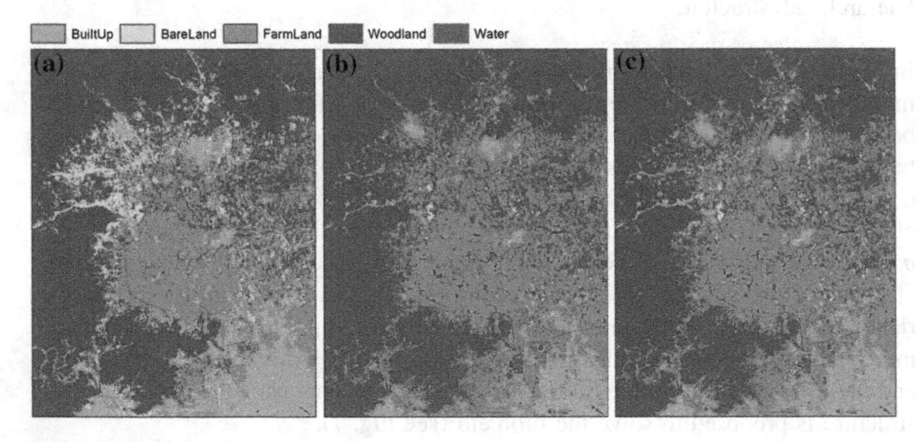

Fig. 8 Land cover classification

3 Result and Discussion

Figure 8 presents the results of the land cover classification of the study area by applying three different classifiers. The results of the accuracy assessment for maximum likelihood classification are presented in Table 1.

Inspection of Table 1 indicates that the maximum likelihood classifier has an overall accuracy of 77.39 %, lowest among these three classifiers.

Table 1 Error matrix of the classification from Landsat TM data

Reference data

	Classification	Built up	Water	Woodland	Farmland	Bareland	Row total
Maximum likelihood classification	Built up	66	12	1	7	6	92
	Water		46	1			47
	Woodland	1	1	53	9		64
	Farmland				30		30
	Bareland	13		6	7	24	50
	Column total	80	59	61	53	30	283
	Overall accuracy = 77.39 %, K_{hat} = 71.23 %						
Knowledge-based expert system	Built up	48	1				49
	Water	1	47	2			50
	Woodland	7	1	57	4	2	71
	Farmland	5	5	1	48		59
	Bareland	19	5	1	1	28	54
	Column total	80	59	61	53	30	283
	Overall accuracy = 80.57 %, K_{hat} = 75.73 %						
Fuzzy expert system	Built up	59	1	1	1	1	63
	Water	1	52	2			55
	Woodland	6	1	57	5	2	71
	Farmland	3	2		46	1	52
	Bareland	11	3	1	1	28	42
	Column total	80	59	61	53	30	283
	Overall accuracy = 84.70 %, K_{hat} = 80.66 %						

4 Conclusion

Land cover classification of urban areas has been problematic because of the significant subpixel mixing that results from the heterogeneity and small spatial size of the surficial materials [31]. This problem is exacerbated when the discrimination between multiple classes is necessary. Recently, however, significant improvements have been achieved in the accuracy of land cover classification in urban areas through the implementation of a variety of sophisticated approaches including the following: (1) the use of neural networks [32, 33]; (2) fuzzy classification [1, 15, 34]; and (3) image texture analysis [12, 13, 35]. Instead of using the multispectral bands as the main input for classification, this study utilizes the principal component bands, NDVI, NDBI, and the DEM as the main components of the classification system. Three kinds of classification systems, namely the supervised maximum likelihood classifier, the knowledge-based expert system, and the fuzzy expert system, are implemented separately in order to examine the variances in accuracy of their respective products. The results presented here demonstrate the usefulness of satellite imagery for urban land cover mapping and some of the shortcomings of conventional classification techniques, such as maximum likelihood. It was found that the maximum likelihood classification of high-resolution multispectral imagery over urban areas produced a significant amount of misclassification errors between spectrally similar classes such as road and building classes.

For the supervised classification system, NDVI and especially NDBI enable built-up areas to be mapped at a higher degree of accuracy and objectivity in comparison with supervised classification systems that are based on multispectral bands. Furthermore, due to the absence of training samples from the mapping, the same results can be derived regardless of a human analyst's subjectivity or how many times the mapping is repeated. This redundancy of a human analyst considerably accelerates the mapping process since it can be accomplished by the direct subtractions of original spectral bands. Also, NDBI does not require complex mathematical computation; the arithmetic manipulation of TM bands and the recoding of intermediate images is sufficient. Thus, it is concluded that the proposed NDBI is much more effective and advantageous in mapping built-up areas than the maximum likelihood method, and consequently, it can be used as a valuable alternative for quickly mapping urban land [24].

In the knowledge-based expert system, the arbitrary selection of training areas or signatures, which generates the confusion error to compromise the final result, is avoided. Although the building of If–Then rules at the first step requires expertise and abundant knowledge about the spectral signature of different land covers, it is relatively efficient and more accurate than the traditional supervised classification system. Additionally, once the system is constructed, it can be easily applied to images of a similar geophysical composition or images of the same location at different times.

Finally, the hierarchical fuzzy classification method proves to be the most accurate method in classifying the data by utilizing both spectral and spatial information. The classification accuracies of the fuzzy classifier were approximately 7 % greater than the maximum likelihood system in terms of overall accuracy and 9 % greater in terms of K_{hat} accuracy. Accordingly, there were significant decreases in the number of misclassifications between spectrally similar classes. Further work is needed to improve the performance of the fuzzy classifier in dense urban areas and to produce more detailed urban land cover maps by identifying features such as roads, business areas, and residential areas. We believe an image segmentation approach combined with morphological feature operators may be used to further improve upon the results presented here.

Acknowledgments Thanks to the support by International Science and Technology Cooperation Program of China (2012DFA1127); Hainan International Cooperation Key Project (GJXM201105).

References

1. Shackelford, A.K., Davis, C.H.: A hierarchical fuzzy classification approach for high-resolution multispectral data over urban areas. Geosci. Remote Sens. IEEE Trans. **41**, 1920–1932 (2003)
2. Anderson, J.R., Hardy, E.E., Roach, J.T., Witmer, R.E.: A Land Use and Land Cover Classification System for Use with Remote Sensor Data. United States Government Printing Office, Washington (1976)
3. Alberti, M., Waddell, P.: An integrated urban development and ecological simulation model. Integr. Assess. **1**, 215–227 (2000)
4. Civco, D.L., Hurd, J.D., Wilson, E.H., Song, M., Zhang, Z.: A comparison of land use and land cover change detection methods. ASPRS-ACSM annual conference and FIG XXII congress (2002)
5. Herold, M., Couclelis, H., Clarke, K.C.: The role of spatial metrics in the analysis and modeling of urban land use change. Comput. Environ. Urban Syst. **29**, 369–399 (2005)
6. Chander, G., Markham, B.L., Helder, D.L.: Summary of current radiometric calibration coefficients for Landsat MSS, TM, ETM+, and EO-1 ALI sensors. Remote Sens. Environ. **113**, 893–903 (2009)
7. Kahya, O., Bayram, B., Reis, S.: Land cover classification with an expert system approach using Landsat ETM imagery: a case study of Trabzon. Environ. Monit. Assess. **160**, 431–438 (2010)
8. Lo, C.P.: The application of geospatial technology to urban morphological research. Urban Morphol. **11**, 81–90 (2007)
9. Xie, Y., Fang, C., Lin, G.C.S., Gong, H., Qiao, B.: Tempo-spatial patterns of land use changes and urban development in globalizing China: a study of Beijing. Sensors **7**, 2881–2906 (2007)
10. Wentz, E.A., Nelson, D., Rahman, A., Stefanov, W.L., Roy, S.S.: Expert system classification of urban land use/cover for Delhi, India. Int. J. Remote Sens. **29**, 4405–4427 (2008)
11. Lu, D., Weng, Q.: Urban classification using full spectral information of Landsat ETM Imagery in Marion County, Indiana. Photogram. Eng. Remote Sens. **71**, 1275–1284 (2005)
12. Zhang, Y., Chen, L., Yu, B.: Multi-scale texture analysis for urban land use/cover classification using high spatial resolution satellite data. Geoinformatics 2007: Remotely Sensed Data and Information Nanjing, China, (2007)
13. Coburn, C., Roberts, A.: A multiscale texture analysis procedure for improved forest stand classification. Int. J. Remote Sens. **25**, 4287–4308 (2004)
14. Stefanov, W.L., Ramsey, M.S., Christensen, P.R.: Monitoring urban land cover change: an expert system approach to land cover classification of semiarid to arid urban centers. Remote Sens. Environ. **77**, 173–185 (2001)
15. Wang, Y., Jamshidi, M., Neville, P., Bales, C., Morain, S.: Hierarchical fuzzy classification of remote sensing data. In: Nikravesh, M., Kacprzyk, J., Zadeh, L.A. (eds.) Forging New Frontiers: Fuzzy Pioneers I, pp. 333–350. Springer, Berlin (2007)
16. Wang, Y., Jamshidi, M.: Multispectral landsat image classification using fuzzy expert systems. World Automation Congress, Seville (2004)
17. Bannari, A., Morin, D., Bonn, F., Huete, A.R.: A review of vegetation indices. Remote Sens. Rev. **13**, 95–120 (1995)
18. Jensen, J.R.: Introductory Digital Image Processing: A Remote Sensing Perspective, 3rd edn. Prentice Hall, New Jersey (2005)
19. Kulkarni, A.D.: Neural-fuzzy models for multispectral image analysis. Appl. Intell. **8**, 173–187 (1998)
20. Chander, G., Chengquan, H., Limin, Y., Homer, C., Larson, C.: Developing consistent landsat data sets for large area applications: the MRLC 2001 protocol. Geosci. Remote Sens. Lett. IEEE **6**, 777–781 (2009)

21. Justice, C.O., Townshend, J.R.G., Vermote, E.F., Masuoka, E., Wolfe, R.E., Saleous, N., Roy, D.P., Morisette, J.T.: An overview of MODIS land data processing and product status. Remote Sens. Environ. **83**, 3–15 (2002)
22. Bannari, A., Morin, D., Bénié, G.B., Bonn, F.J.: A theoretical review of different mathematical models of geometric corrections applied to remote sensing images. Remote Sens. Rev. **13**, 27–47 (1995)
23. Bannari, A.A., Omari, K., Teillet, P.M., Fedosejevs, G.: Multisensor and multiscale survey and characterization for radiometric spatial uniformity and temporal stability of Railroad Valley Playa (Nevada) test site used for optical sensor calibration. Barcelona, Spain (2004)
24. Zha, Y., Gao, J., Ni, S.: Use of normalized difference built-up index in automatically mapping urban areas from TM imagery. Int. J. Remote Sens. **24**, 583–594 (2003)
25. Small, C.: The Landsat ETM+ spectral mixing space. Remote Sens. Environ. **93**, 1–17 (2004)
26. Roy, D.P., Ju, J., Kline, K., Scaramuzza, P.L., Kovalskyy, V., Hansen, M., Loveland, T.R., Vermote, E., Zhang, C.: Web-enabled landsat data (WELD): Landsat ETM+ composited mosaics of the conterminous United States. Remote Sens. Environ. **114**, 35–49 (2010)
27. Collins, J.B., Woodcock, C.E.: An assessment of several linear change detection techniques for mapping forest mortality using multitemporal landsat TM data. Remote Sens. Environ. **56**, 66–77 (1996)
28. Lawrence, R., Wright, A.: Rule-based classification systems using classification and regression tree (CART) analysis. Photogram. Eng. Remote Sens. **67**, 1137–1142 (2001)
29. Zadeh, L.A.: Fuzzy sets. Inf. Control **8**, 338–353 (1965)
30. MathWorks: Fuzzy Logic Toolbox™ 2 User's Guide. Natick, MA, The MathWorks, Inc. (2010)
31. Ridd, M.: Exploring a V-I-S(vegetaton-impervious surface-soil) model for urban ecosystem analysis through remote sensing: comparative anatomy for cities. Int. J. Remote sens. (Print) **16**, 2165–2185 (1995)
32. Stathakis, D., Vasilakos, A.: Satellite image classification using granular neural networks. Int. J. Remote Sens. **27**, 3991–4003 (2006)
33. Tzeng, Y.C., Chen, K.S.: A fuzzy neural network to SAR image classification. Geosci. Remote Sens. IEEE Trans. **36**, 301–307 (1998)
34. Wu, W., Gao, G.: An application of neuro-fuzzy system in remote sensing image classification. In: Computer Science and Software Engineering, 2008 International Conference on (2008)
35. Anys, H., Bannari, A., He, D., Morin, D.: Texture analysis for the mapping of urban areas using airborne MEIS-II images. In: The First International Airborne Remote Sensing Conference and Exhibition, Strasbourg (1994)

Global Stability of Two Linear Non-autonomous Takagi–Sugeno Fuzzy Systems

Min Zhou, Sheng-gang Li and Xiao-fei Yang

Abstract In our present chapter, we investigate two linear non-autonomous Takagi–Sugeno fuzzy dynamical systems. By constructing appropriate Lyapunov function and using analytical techniques, we obtain global stability conditions for two linear non-autonomous Takagi–Sugeno fuzzy systems, which is helpful for fuzzy control system design.

Keywords Stability conditions · Non-autonomous · Fuzzy systems

1 Introduction

In the real word, due to many natural or man-made factors, many evolutionary processes in natural are affected by uncertain and irregular changes. In the viewpoint of mathematics, such uncertain and irregular factors could be characterized by fuzzy rules. For this reason, studies on Takagi–Sugeno fuzzy systems have aroused widespread interest of mathematic researchers. Many important and interesting results on the fuzzy control and the dynamical behaviors for such systems could be seen in [1–5].

In [1], Bong-Jae Rhee gave the global ability condition for a class of the continuous-time TS fuzzy system, which can be described by the following fuzzy rules:

R_i : If $x_1(t)$ is $M_1^{\alpha_{i1}}, \ldots,$ and $x_n(t)$ is $M_n^{\alpha_{in}}$ then

M. Zhou (✉) · S. Li
College of Mathematics and Information Science, Shaanxi Normal University, Xi'an 710062, China
e-mail: zhouminjy@126.com

X. Yang
School of Science, Xi'an Polytechnic University, Xi'an 710048, China

B.-Y. Cao et al. (eds.), *Ecosystem Assessment and Fuzzy Systems Management*, Advances in Intelligent Systems and Computing 254, DOI: 10.1007/978-3-319-03449-2_15,

$$x'(t) = A_i x(t) + B_i u(t), \quad i = 1, 2, \ldots, r \tag{1}$$

where α_{ij} specifies x_j-based fuzzy set which is used in the ith fuzzy rule and $1 \leq \alpha_{ij} \leq r_j$ for any i and j. Let $\omega_j^{\alpha_i^j}(x_j(t))$ be the membership function of $\mu_j^{\alpha_i^j}$. Then the normalized membership function is

$$\mu_j^{\alpha_i^j}(x_j(t)) = \frac{\omega_j^{\alpha_i^j}(x_j(t))}{\sum_{\alpha_{ij=1}}^{r_j} \omega_j^{\alpha_i^j}(x_j(t))},$$

so

$$0 \leq \mu_j^{\alpha_i^j}(x_j(t)) \leq 1, \quad \sum_{\alpha_{ij=1}}^{r_j} \mu_j^{\alpha_i^j}(x_j(t)) = 1.$$

Then, the membership function $h_i(x(t))$ of the ith fuzzy rule becomes

$$h_i(x(t)) = \prod_{j=1}^{n} \mu_j^{\alpha_{ij}}(x_j(t)).$$

Therefore, we have

$$0 \leq h_i(x(t)) \leq 1, \quad \text{and} \quad \sum_{i=1}^{r} h_i(x(t)) = 1.$$

For simplicity, we omit t in $x(t)$ and $h_i(x(t))$; then, (1) can be rewritten as

$$x' = \Phi(x)x + \Psi(x)u,$$

where $\Phi(x) = \sum_{i=1}^{r} h_i(x)A_i(t)$, $\Psi(x) = \sum_{i=1}^{r} h_i(x)B_i(t)$.

Inspired by Rhee and Won [1], in this chapter, we first consider a non-autonomous TS fuzzy system described by the following fuzzy rules:

$$R_i : \text{If } x_1 \text{ is } M_1^{\alpha_{i1}}, \text{and} \ldots x_n \quad \text{is } M_1^{\alpha_{in}}, \tag{2}$$

then $x' = A_i(t)x, \quad i = 1, 2, \ldots, r$. The fuzzy system can be rewritten as

$$x'(t) = A(x,t)x(t), \tag{3}$$

where

$$A(x,t) = \sum_{i=1}^{r} h_i(x)A_i(t),$$

$$B(x,t) = \sum_{i=1}^{r} h_i(x)B_i(t),$$

here

$$A_i(t) = \begin{pmatrix} a^i_{11}(t) & a^i_{12q}(t) & \dots & a^i_{1n}(t) \\ a^i_{21}(t) & a^i_{22}(t) & \dots & a^i_{2n}(t) \\ \vdots & \vdots & \vdots & \vdots \\ a^i_{n1}(t) & a^i_{n2}(t) & \dots & a^i_{nn}(t) \end{pmatrix},$$

$$B_i(t) = \begin{pmatrix} b^i_{11}(t) & b^i_{12q}(t) & \dots & b^i_{1n}(t) \\ b^i_{21}(t) & b^i_{22}(t) & \dots & b^i_{2n}(t) \\ \vdots & \vdots & \vdots & \vdots \\ b^i_{n1}(t) & b^i_{n2}(t) & \dots & b^i_{nn}(t) \end{pmatrix}$$

and $h_i(x)$ be the membership function of the ith fuzzy rule.

Furthermore, we consider the following non-autonomous TS fuzzy system based on the above fuzzy rules (2).

$$x'(t) = A(x,t)x(t) + B(x,t)u(t). \tag{4}$$

2 Global Stability Analysis

In this section, we first give the stability condition for system (3).

Theorem 2.1 *If the following conditions hold,*

1.

$$\sum_{i=1}^{\gamma} h_i(x)a^i_{kk}(t) \le \sigma_k, \quad \sigma_k < 0 \quad \text{and} \quad \left|\sum_{i=1}^{\gamma} h_i(x)a^i_{kj}(t)\right| \le c_{kj}, \tag{2.1}$$

2. *there exist positive numbers* $\varepsilon_1, \varepsilon_2, \ldots, \varepsilon_k$ *such that real symmetric matrix* $S = (s_{kj})_{n\times n}$ *is negative defined, where*

$$s_{kj} = \begin{cases} -2\varepsilon_k\sigma_k & -k = j \\ \varepsilon_k c_{kj} + \varepsilon_j c_{jk} & k \ne j \end{cases}. \tag{2.2}$$

Then, system (3) *is globally stable.*

Proof Since

$$h_i(x)A_i(t) = \begin{pmatrix} h_i(x)a^i_{11}(t) & h_i(x)a^i_{12}(t) & \dots & h_i(x)a^i_{1n}(t) \\ h_i(x)a^i_{21}(t) & h_i(x)a^i_{22}(t) & \dots & h_i(x)a^i_{2n}(t) \\ \vdots & \vdots & \vdots & \vdots \\ h_i(x)a^i_{n1}(t) & h_i(x)a^i_{n2}(t) & \dots & h_i(x)a^i_{nn}(t) \end{pmatrix},$$

we have

$$A(x) = \sum_{i=1}^{r} h_i(x)A_i(t)$$
$$= \begin{pmatrix} \sum_{i=1}^{r} h_i(x)a_{11}^i(t) & \sum_{i=1}^{r} h_i(x)a_{12}^i(t) & \dots & \sum_{i=1}^{r} h_i(x)a_{1n}^i(t) \\ \sum_{i=1}^{r} h_i(x)a_{21}^i(t) & \sum_{i=1}^{r} h_i(x)a_{22}^i(t) & \dots & \sum_{i=1}^{r} h_i(x)a_{2n}^i(t) \\ \vdots & \vdots & \vdots & \vdots \\ \sum_{i=1}^{r} h_i(x)a_{n1}^i(t) & \sum_{i=1}^{r} h_i(x)a_{n2}^i(t) & \dots & \sum_{i=1}^{r} h_i(x)a_{nn}^i(t) \end{pmatrix}$$

Similarly, we can easily obtain

$$B(x) = \sum_{i=1}^{r} h_i(x)B_i(t) = \begin{pmatrix} \sum_{i=1}^{r} h_i(x)b_{11}^i(t) & \sum_{i=1}^{r} h_i(x)b_{12}^i(t) & \dots & \sum_{i=1}^{r} h_i(x)b_{1n}^i(t) \\ \sum_{i=1}^{r} h_i(x)b_{21}^i(t) & \sum_{i=1}^{r} h_i(x)b_{22}^i(t) & \dots & \sum_{i=1}^{r} h_i(x)b_{2n}^i(t) \\ \vdots & \vdots & \vdots & \vdots \\ \sum_{i=1}^{r} h_i(x)b_{n1}^i(t) & \sum_{i=1}^{r} h_i(x)b_{n2}^i(t) & \dots & \sum_{i=1}^{r} h_i(x)b_{nn}^i(t) \end{pmatrix},$$

and then, $x'(t) = A(x,t)x(t)$ can be rewritten as

$$\begin{pmatrix} x_1' \\ x_2' \\ \vdots \\ x_n' \end{pmatrix} = \begin{pmatrix} \sum_{i=1}^{r} h_i(x)a_{11}^i(t) & \sum_{i=1}^{r} h_i(x)a_{11}^i(t) & \dots & \sum_{i=1}^{r} h_i(x)a_{11}^i(t) \\ \sum_{i=1}^{r} h_i(x)a_{21}^i(t) & \sum_{i=1}^{r} h_i(x)a_{22}^i(t) & \dots & \sum_{i=1}^{r} h_i(x)a_{2n}^i(t) \\ \vdots & \vdots & \vdots & \vdots \\ \sum_{i=1}^{r} h_i(x)a_{n1}^i(t) & \sum_{i=1}^{r} h_i(x)a_{n2}^i(t) & \dots & \sum_{i=1}^{r} h_i(x)a_{nn}^i(t) \end{pmatrix} \begin{pmatrix} x_1 \\ x_2 \\ \vdots \\ x_n \end{pmatrix}.$$

Then, we can acquire the solution of (2.1) as follows:

$$x_k' = \left(\sum_{i=1}^{r} h_i(x)a_{kk}^i(t) \right) x_k + \sum_{j=1, j \neq k}^{n} \left[\left(\sum_{i=1}^{r} h_i(x)a_{kj}^i(t) \right) x_j \right] \quad (k = 1, 2, \dots, n),$$

Take the Lyapunov function of kth system as

$$V_k(t, x_k) = x_k^2 \quad (k = 1, 2, \dots, n);$$

then, we have

$$\frac{\mathrm{d}V_k(t, x_k)}{\mathrm{d}t} = 2x_k x_k' = 2\left(\sum_{i=1}^{r} h_i(x)a_{kk}^i(t) \right) x_k^2 + 2 \sum_{j=1, j \neq k}^{n} \left[\left(\sum_{i=1}^{r} h_i(x)a_{kj}^i(t) \right) x_k x_j \right].$$

Furthermore, we take the Lyapunov function of system (3) as

$$v(t,x) = \sum_{k=1}^{n} \varepsilon_k v_k(t, x_k),$$

where $\varepsilon_k > 0$, $(k = 1, 2, \ldots, n)$, $v(t,x)$ is obvious positive defined and infinite function with lower bound, and therefore, the derivative of $v(t,x)$ along the solution of system (3) can be expressed as

$$\begin{aligned}
\frac{\mathrm{d}v(t,x)}{\mathrm{d}t} &= \sum_{k=1}^{n} \varepsilon_k v'_k(t, x_k) \\
&= \sum_{k=1}^{n} \left[2\varepsilon_k \left(\sum_{i=1}^{\gamma} h_i(x) a^i_{kk}(t) \right) x_k^2 + 2\varepsilon_k \sum_{j=1, j\neq k}^{n} \left[\left(\sum_{i=1}^{\gamma} h_i(x) a^i_{kk}(t) \right) x_k x_j \right] \right] \\
&= \sum_{k=1}^{n} 2\varepsilon_k \left(\sum_{i=1}^{\gamma} h_i(x) a^i_{kk}(t) \right) x_k^2 + \sum_{k=1}^{n} \sum_{j=1, j\neq k}^{n} 2\varepsilon_k \left(\sum_{i=1}^{\gamma} h_i(x) a^i_{kj}(t) \right) x_k x_j \\
&\leq \sum_{k=1}^{n} (2\varepsilon_k \sigma_k) x_k^2 + \sum_{k=1}^{n} \sum_{j=1, j\neq k}^{n} 2\varepsilon_k \left| \sum_{i=1}^{\gamma} h_i(x) a^i_{kj}(t) \right| x_k x_j \\
&\leq \sum_{k=1}^{n} (2\varepsilon_k \sigma_k) x_k^2 + \sum_{k=1}^{n} \sum_{j=1, j\neq k}^{n} 2\varepsilon_k c_{kj} |x_k| |x_j| \\
&= W^T S W,
\end{aligned}$$

where

$$s_{kj} = \begin{cases} 2\varepsilon_k \sigma_k, & k = j \\ \varepsilon_k c_{kj} + \varepsilon_k c_{jk}, & k \neq j, \end{cases}$$

so considering the conditions (2.1) and (2.2), system (3) is globally stable.

Next, we consider the global stability condition for system (4).

Theorem 2.2 *If the following conditions hold,*

1. $h_i(x) a^i_{ij}(t) \in C([t_0, +\infty))$, $(t_0 > 0)$, $\frac{\partial f_k(t,x)}{\partial x}$ *is continuous,* $\sum_{i=1}^{r} h_i(x) a^i_{ij}(t) + \frac{\partial f_i(t,x)}{\partial x} \leq -\sigma_i \quad (\sigma_i > 0)$, *and*

$$\left| \sum_{i=1}^{r} h_i(x) a^i_{ij}(t) + f'_{i(\mu x_j)}(t,x) \right| \leq C_{ij} \quad (i \neq j, i, j = 1, 2, \ldots, n). \tag{2.3}$$

2. *There exist positive numbers* $\varepsilon_1, \varepsilon_2, \ldots, \varepsilon_k$ *such that real symmetric matrix* $S = (s_{kj})_{n\times n}$ *is negative defined, where*

$$s_{kj} = \begin{cases} -2\varepsilon_k \sigma_k & k = j \\ \varepsilon_k c_{kj} + \varepsilon_j c_{jk} & k \neq j \end{cases} \tag{2.4}$$

Then, system (1.4) *is globally stable.*

Proof $x'(t) = A(x,t)x + B(x,t)u(t)$ can be written as

$$\begin{pmatrix} x_1' \\ x_2' \\ \vdots \\ x_n' \end{pmatrix} = \begin{pmatrix} \sum_{i=1}^{\gamma} h_i(x)a_{11}^i(t) & \sum_{i=1}^{r} h_i(x)a_{12}^i(t) & \dots & \sum_{i=1}^{\gamma} h_i(x)a_{1n}^i(t) \\ \sum_{i=1}^{\gamma} h_i(x)a_{21}^i(t) & \sum_{i=1}^{\gamma} h_i(x)a_{22}^i(t) & \dots & \sum_{i=1}^{\gamma} h_i(x)a_{2n}^i(t) \\ \vdots & \vdots & \vdots & \vdots \\ \sum_{i=1}^{\gamma} h_i(x)a_{n1}^i(t) & \sum_{i=1}^{\gamma} h_i(x)a_{n2}^i(t) & \dots & \sum_{i=1}^{\gamma} h_i(x)a_{nn}^i(t) \end{pmatrix} \begin{pmatrix} x_1 \\ x_2 \\ \vdots \\ x_n \end{pmatrix}$$
$$+ \begin{pmatrix} \sum_{i=1}^{\gamma} h_i(x)b_{11}^i(t) & \sum_{i=1}^{\gamma} h_i(x)b_{12}^i(t) & \dots & \sum_{i=1}^{\gamma} h_i(x)b_{1n}^i(t) \\ \sum_{i=1}^{\gamma} h_i(x)b_{21}^i(t) & \sum_{i=1}^{\gamma} h_i(x)b_{22}^i(t) & \dots & \sum_{i=1}^{\gamma} h_i(x)b_{2n}^i(t) \\ \vdots & \vdots & \vdots & \vdots \\ \sum_{i=1}^{\gamma} h_i(x)b_{n1}^i(t) & \sum_{i=1}^{\gamma} h_i(x)b_{n2}^i(t) & \dots & \sum_{i=1}^{\gamma} h_i(x)b_{nn}^i(t) \end{pmatrix} \begin{pmatrix} u_1 \\ u_2 \\ \vdots \\ u_n \end{pmatrix}.$$

Note $f_k(t,x) = \sum_{j=1}^{n} \left(\sum_{i=1}^{\gamma} h_i(x)b_{kj}^i(t) \right) u_j(t)$, $(k = 1, 2, \dots, n)$, and $f_k(t, 0) = 0$ is obvious, and then, we have

$$\begin{pmatrix} x_1' \\ x_2' \\ \vdots \\ x_n' \end{pmatrix} = \begin{pmatrix} \sum_{i=1}^{\gamma} h_i(x)a_{11}^i(t) & \sum_{i=1}^{\gamma} h_i(x)a_{12}^i(t) & \dots & \sum_{i=1}^{\gamma} h_i(x)a_{1n}^i(t) \\ \sum_{i=1}^{\gamma} h_i(x)a_{21}^i(t) & \sum_{i=1}^{\gamma} h_i(x)a_{22}^i(t) & \dots & \sum_{i=1}^{\gamma} h_i(x)a_{2n}^i(t) \\ \vdots & \vdots & \vdots & \vdots \\ \sum_{i=1}^{\gamma} h_i(x)a_{n1}^i(t) & \sum_{i=1}^{\gamma} h_i(x)a_{n2}^i(t) & \dots & \sum_{i=1}^{\gamma} h_i(x)a_{nn}^i(t) \end{pmatrix} \begin{pmatrix} x_1 \\ x_2 \\ \vdots \\ x_n \end{pmatrix}$$
$$+ \begin{pmatrix} f_1(t,x) \\ f_2(t,x) \\ \vdots \\ f_n(t,x) \end{pmatrix}.$$

So the above equation can be decomposed as

$$x_k' = \left(\sum_{i=1}^{\gamma} h_i(x)a_{kk}^i(t) \right) x_k + \sum_{j=1, j\neq k}^{n} \left[\left(\sum_{i=1}^{\gamma} h_i(x)a_{kj}^i(t) \right) x_j \right] + f_k(t,x),$$
$$(k = 1, 2, \dots, n).$$

Take the Lyapunov function of kth system as

$$v_k(t, x_k) = x_k^2 \quad (k = 1, 2, \dots, n),$$

and then, we have

$$\begin{aligned}\frac{\mathrm{d}v_k(t,x_k)}{\mathrm{d}t} &= 2x_kx_k' \\ &= 2\left(\sum_{i=1}^{\gamma} h_i(x)a_{kk}^i\right)x_k^2 + 2\sum_{j=1,j\neq k}^{n}\left[\left(\sum_{i=1}^{\gamma} h_i(x)a_{kj}^i(t)\right)x_kx_j\right] + 2x_kf_k(t,x)\end{aligned}$$

since

$$\begin{aligned}\frac{\mathrm{d}f_k(t,\mu_x)}{\mathrm{d}\mu} &= \frac{\mathrm{d}}{\mathrm{d}\mu}f_k(t,\mu x_1,\mu x_2,\ldots,\mu x_n) \\ &= f'_{k,(\mu x_1)}(t,\mu x)x_1 + \cdots + f'_{k,(\mu x_n)}(t,\mu x)x_n \\ &= f'_{k,(\mu x_k)}(t,\mu x)x_k + \sum_{j=1,j\neq k}^{n} f'_{k,(\mu x_j)}(t,\mu x)x_j.\end{aligned}$$

Considering $f_k(t,0)=0$, we can easily obtain

$$f_k(t,x) = \int_0^1\left[f'_{k,(\mu x_k)}(t,\mu x)x_k + \sum_{j=1,j\neq k}^{n} f'_{k,(\mu x_j)}(t,\mu x)x_j\right]\mathrm{d}\mu, \tag{5}$$

then by mean value theorem of integrals, we have

$$\text{Equation (5)} = f'_{k,(\mu x_k)}(t,\theta x)x_k + \sum_{j=1,j\neq k}^{n} f'_{k,(\mu x_j)}(t,\theta x)x_j \quad 0<\theta<1.$$

So we can acquire

$$\begin{aligned}\frac{\mathrm{d}v_k}{\mathrm{d}t} &= 2\left(\sum_{i=1}^{\gamma} h_i(x)a_{kk}^i(t)\right)x_k^2 + 2x_k\left[f'_{k,(\mu k_k)}(t,\theta x)x_k + \sum_{j=1,j\neq k}^{n} f'_{k,(_j)}(t,\theta x)x_j\right] \\ &\quad + 2\sum_{j=1,j\neq k}^{n}\left[\left(\sum_{i=1}^{\gamma} h_i(x)a_{kj}^i(t)\right)x_kx_j\right] \\ &= 2\left[\sum_{i=1}^{\gamma} h_i(x)a_{kk}^i(t) + f'_{k,(\mu x_k)}(t,)\right]x_k^2 \\ &\quad + 2\left[\sum_{j=1,j}^{n} f'_{k,(\mu x_j)}(t,\theta x) + \sum_{j=1,j\neq k}^{n}\left(\sum_{i=1}^{\gamma} h_i(x)a_{kj}^i(t)\right)\right]x_kx_j.\end{aligned}$$

Furthermore, we take the Lyapunov function of system (4) as

$$v(t,x) = \sum_{k=1}^{n} \varepsilon_k v_k(t,x_k),$$

where $\varepsilon_k > 0, (k = 1, 2, \ldots, n)$, $v(t, x)$ is obvious positive-defined infinite function with lower bound, and therefore, the derivative of $v(t, x)$ along the solution of system (4) can be expressed as

$$\begin{aligned}
\frac{\mathrm{d}v(t,x)}{\mathrm{d}t} &= \sum_{k=1}^{n} \varepsilon_k v_k(t, x_k) \\
&= \sum_{k=1}^{n} \left\{ 2\varepsilon_k \left[\sum_{i=1}^{\gamma} h_i(x) a_{kk}^i(t) + f'_{k,(\mu x_k)}(t, \theta x) \right] x_k^2 \right. \\
&\quad \left. + 2\varepsilon_k \left[\sum_{j=1, j\neq k}^{n} f'_{k(\mu x_j)}(t, \theta x) + \sum_{j=1, j\neq k}^{n} \left(\sum_{i=1}^{\gamma} h_j(x) a_{kj}^i(t) \right) \right] x_k x_j \right\} \\
&= \sum_{k=1}^{n} 2\varepsilon_k \left[\sum_{i=1}^{\gamma} h_i(x) a_{kk}^i(t) + f'_{k(_k)}(t, \mu x) \right] x_k^2 \\
&\quad + \sum_{k=1}^{n} \left\{ \sum_{j=1, j\neq k}^{n} 2\varepsilon_k \left[f'_{k(\mu x_j)}(t, \mu x) + \sum_{i=1}^{\gamma} h_i(x) a_{kj}^i(t) \right] x_j x_k \right\} \\
&\leq \sum_{k=1}^{n} 2\varepsilon_k \left[\sum_{i=1}^{\gamma} h_i(x) a_{kk}^i(t) + f'_{k(\mu x_k)}(t, \mu x) \right] x_k^2 \\
&\quad + \sum_{k=1}^{n} \left\{ \sum_{j=1, j\neq k}^{n} 2\varepsilon_k \left| f'_{k,(\mu x_j)}(t, \mu x) + \sum_{i=1}^{\gamma} h_i(x) a_{kj}^i(t) \right| |x_i||x_j| \right\} \\
&= W^{\mathrm{T}} S W,
\end{aligned}$$

where $s_{kj} = \begin{cases} -2\varepsilon_k \sigma_k & k = j \\ \varepsilon_k c_{kj} + \varepsilon_j c_{jk} & k \neq j \end{cases}$ and $W = (|x_1|, |x_2|, \ldots, |x_n|)^{\mathrm{T}}$.

So by the conditions (2.3) and (2.4), system (4) is globally stable.

Acknowledgments The authors thank the National Natural Science Foundation of China (No. 11071151) and the Special Fund of Shaanxi Provincial Education Department (No. 2013JK0568) for the support. And this work is also supported by International Science and Technology Cooperation Program of China (2012DFA1127) and Hainan International Cooperation Key Project (GJXM201105).

References

1. Rhee, B.J., Won, S.: A new fuzzy Lyapunov function approach for a Takagi-Sugeno fuzzy control system design. Fuzzy Sets Syst. **157**, 1211–1228 (2006)
2. Biglarbegin, M., Melek, W.W.: On the stability of interval type-2 TSK fuzzy logic control systems. IEEE Trans. Fuzzy Syst. Man. Cybern. B Cybern. **40**(3), 798–818 (2010)
3. Phu, N.D., Dung, L.Q.: On the stability and controllability of fuzzy control set differential equations. Int. J. Reliab. Saf. **5**, 320–335 (2011)

4. Feng, G.: A survey on analysis and design of model-based fuzzy control system. IEEE Trans. Fuzzy. Syst. **14**(5), 676–697 (2006)
5. Ding, Z., Kandel, A.: On the controllability of fuzzy dynamical systems. J. Fuzzy Math. **18**(2), 295–306 (2000)

A New Measurement of Similarity About Rough Vague Set

Jian-xin Wang, Sheng-Quan Ma and Zhi-qing Zhao

Abstract This chapter discussed the concept of rough set and rough Vague (RV) set, the way of representation of knowledge. And also blend rough set with Vague set, described the concept of RV set, gave related concept of rough Vague (RV) value and a new method of measurement of similarity, studied the related property and the method of measurement of similarity for the RV sets.

Keywords Rough sets · Vague sets · Rough Vague sets · Similarity measure

1 Introduction

The membership function of fuzzy set assigns a number, which is between zero and one for each object as the degree of membership, and it not only includes the proof that the element belongs to the set, but also includes the proof that the element does not belong to the set. For overcoming the insufficiency of the information by the single value description, Zadeh [1] led to go into the interval-valued fuzzy set in 1975, used [0, 1] of inside closed sub-interval to represent an element how belongs to a set, it descends to carry to order a necessity of means the object belongs to, the top end point the possibility that means the object belongs to. In 1986, Atanassov [2] considered the fuzzy set from a different angle of generalization, he adopted two number to depict a element belonging to the fuzzy set, leading to go into belong to a degree with belonged to one degree concept not, the Atanassov call that definition from here of set for the intuitionistic fuzzy set.

J. Wang · S.-Q. Ma (✉) · Z. Zhao
School of Information and Technology, Hainan Normal University, Haikou, China
e-mail: mashengquan@163.com

J. Wang
e-mail: 81665146@qq.com

B.-Y. Cao et al. (eds.), *Ecosystem Assessment and Fuzzy Systems Management*, Advances in Intelligent Systems and Computing 254, DOI: 10.1007/978-3-319-03449-2_16,

In 1989, the Atanassov and Gargov [3] point out the interval-valued fuzzy set to intuitionistic fuzzy set, which is the fuzzy set expansion in of two equivalent generalizations. In 1993, the Gau and Buehrer [4] pass "the vote model" to set to carry on understanding to release to the Vague sets. Speak from the essence, the interval-valued fuzzy set, the intuitionistic fuzzy set and the Vague set to have no hypostatic differentiation (see [5]).

At the computer science and its application realm, especially in the artificial intelligence (AI), date mining and knowledge discovery in database (KDD), the theories of rough set have important of physically application, rough set collectively now to description of the thing connection and the whole characteristic, provide important tool to the inside contact of the research thing. The Vague set provides a kind of knowledge to representation of new tool, it is fresh to give people clearly to can know to the thing the degree and scope mean, can carry on a good description to the thing attribute from the form and the top of the contents. Both the rough set theories and Vague set theories study the uncertainty problem in information system, rough set the point of departure that theories solve problem to lie in the knowledge undistinguished in the information system, but the Vague set the fuzzy that theories then is fix attention on in the concept content and person to the concept know of not precision. However, in many situations, the concept is not only misty, and cannot distinguish, cause people's understanding to the concept also impossibly and completely accurate with overall. According to this because of, need to gather rough theories and the Vague theories carry on blend to make up they are alone it is each while handling an actual problem from of shortage.

In the application study of rough set and Vague set theories, measurement of similarity is an important problem, and it is the application realm of foundation of fuzzy gather, pattern recognition, approximate reasoning, etc. This text owing to this from, studies the problem of similarity measure of RV sets, gives a kind of new measurement method for rough Vague sets.

2 The Concept of Rough Set and Vague Set

Let U be a universe of discourse. $X \subseteq U$ be an object space, and a Vague set V defined on $X \subseteq U$ can use a true membership function $t_V(x)$ and a false membership function $f_V(x)$ to mean. $t_V(x)$ is the membership degree's bounded to the below from the that the proof of support lead, then $f_V(x)$ is from the that opposed proof lead of the negation membership degree's bounded to the below, $t_V(x)$ and $f_V(x)$ establish a contact between a real number on [0, 1] to each one point in X. i.e., $t_V : X \to [0,1]$; $f_V : X \to [0,1]$. $\forall x \in X$.

The membership degree of V is denoted by

$$V(x) = [t_V(x), 1 - f_V(x)],$$

where

$$0 \leq t_V(x) + f_V(x) \leq 1.$$

So that, we can noted by $V = \{(x, t_V(x), f_V(x))|x \in X\}$. $[t_V(x), 1 - f_V(x)]$ be called a Vague value of the point at x in V. Noted by $x_V = [t_V(x), 1 - f_V(x)]$.

Let $U = \{x_1, x_2, \ldots, x_n\}$ be a limited set, R be an equivalent relation on U, U/R are all the equivalent classes on U; $[x]_R$ mean containment of x of R equivalent class, $\forall x \in U$, $\forall X \subseteq U$.

$$\underline{R}(X) = \{x \in U|[x]_R \subseteq X\},$$
$$\overline{R}(X) = \{x \in U|[x]_R \cap X \neq \emptyset\}.$$

$\underline{R}(X)$ and $\overline{R}(X)$ are called low approximate and up approximate. When low approximate and up approximate are not equal, X is a rough set for R. Use $(\underline{R}, \overline{R})$ to represent the rough set in brief; $\mathrm{Bn}_R(X) = \overline{R}(X) - \underline{R}(X)$ is called R boundary of X; $\mathrm{POS}_R(X) = \underline{R}(X)$ is called R positive area of X, namely the positive area of rough set; $\mathrm{Neg}_R(X) = U - \underline{R}(X)$ is called the negative area of X.

3 Rough Vague Sets and Rough Vague Value

Definition 3.1 Let U be a universe of discourse, R be an equivalent relation, and V be a Vague set. The rough Vague sets (RV) are constituted by R with V (the RV sets) definition as follows:

$$\underline{\mathrm{Rt}}(V) = \inf\{t_V(x)|x \in [x]_R\}; \quad \overline{\mathrm{Rt}}(V) = \sup\{t_V(x)|x \in [x]_R\};$$
$$\underline{\mathrm{Rf}}(V) = \sup\{f_V(x)|x \in [x]_R\}; \quad \overline{\mathrm{Rf}}(V) = \inf\{f_V(x)|x \in [x]_R\},$$

where $\underline{\mathrm{Rt}}$, $\overline{\mathrm{Rt}}$ is the least and the biggest value of the true membership degree at same equivalent class and $\underline{\mathrm{Rf}}$, $\overline{\mathrm{Rf}}$ is the least and the biggest value of the false membership degree at the same equivalent class. Up and down approximate Vague set mean respectively $\overline{V} = [\overline{\mathrm{Rt}}(V), 1 - \overline{\mathrm{Rf}}(V)]$ $\underline{V} = [\underline{\mathrm{Rt}}(V), 1 - \underline{\mathrm{Rf}}(V)]$, then $V = (\underline{V}, \overline{V})$ be called a rough Vague set.

Definition 3.2 For $X \subseteq U$, $V = (\underline{V}, \overline{V})$ be a rough Vague set of R with V constitute, $\forall x \in X$, Record,

$$\overline{V}(x) = [\overline{\mathrm{Rt}}(x), 1 - \overline{\mathrm{Rf}}(x)], \quad \underline{V}(x) = [\underline{\mathrm{Rt}}(x), 1 - \underline{Rf}(x)],$$

$\langle[\underline{\mathrm{Rt}}(x), 1 - \underline{\mathrm{Rf}}(x)], [\overline{\mathrm{Rt}}(x), 1 - \overline{\mathrm{Rf}}(x)]\rangle$ be called the rough Vague (RV) value of $V = (\underline{V}, \overline{V})$, briefly for x, i.e., $x = \langle[\underline{\mathrm{Rt}}(x), 1 - \underline{\mathrm{Rf}}(x)], [\overline{\mathrm{Rt}}(x), 1 - \overline{\mathrm{Rf}}(x)]\rangle$.

Definition 3.3 For two rough Vague values

$$x = \langle [\underline{\mathrm{Rt}}(x), 1 - \underline{\mathrm{Rf}}(x)],\ [\overline{\mathrm{Rt}}(x), 1 - \overline{\mathrm{Rf}}(x)] \rangle$$

and

$$y = \langle [\underline{\mathrm{Rt}}(y), 1 - \underline{\mathrm{Rf}}(y)],\ [\overline{\mathrm{Rt}}(y), 1 - \overline{\mathrm{Rf}}(y)] \rangle,$$

x with y be called equal, if and only if $\underline{\mathrm{Rt}}(x) = \underline{\mathrm{Rt}}(y)$, $\underline{\mathrm{Rf}}(x) = \underline{\mathrm{Rf}}(y)$, and $\overline{\mathrm{Rt}}(x) = \overline{\mathrm{Rt}}(y)$, $\overline{\mathrm{Rf}}(x) = \overline{\mathrm{Rf}}(y)$.

Definition 3.4 The complement of $x = \langle [\underline{\mathrm{Rt}}(x), 1 - \underline{\mathrm{Rf}}(x)],\ [\overline{\mathrm{Rt}}(x), 1 - \overline{\mathrm{Rf}}(x)] \rangle$ is denoted by x^c, if $\underline{\mathrm{Rt}}(x^c) = \underline{\mathrm{Rf}}(x)$, $\underline{\mathrm{Rf}}(x^c) = \underline{\mathrm{Rt}}(x)$, $\overline{\mathrm{Rt}}(x^c) = \overline{\mathrm{Rf}}(x)$, $\overline{\mathrm{Rf}}(x^c) = \overline{\mathrm{Rt}}(x)$.

The complement of the rough Vague set $V = (\underline{V}, \overline{V})$ is denoted by $V^c = (\underline{V}^c, \overline{V}^c)$.

Definition 3.5 $(\underline{V}, \overline{V}) \subseteq (\underline{W}, \overline{W})$ if and only if $\underline{V} \subseteq \underline{W}$ and $\overline{V} \subseteq \overline{W}$, i.e.,

$$\underline{\mathrm{Rt}}_V(x) \le \underline{\mathrm{Rt}}_W(x),\ \underline{\mathrm{Rf}}_V(x) \ge \underline{\mathrm{Rf}}_W(x) \quad \text{and} \quad \overline{\mathrm{Rt}}_V(x) \le \overline{\mathrm{Rt}}_W(x),\ \overline{\mathrm{Rf}}_V(x) \ge \overline{\mathrm{Rf}}_W(x).$$

4 Similarity Measurement of Rough Vague Value

In order to study a rough Vague value, we have to first analyze the situation of Vague sets. In the similarity measurement of Vague sets, people often adopt similarity measurement of the Vague value method.

For a Vague value $x = [t(x), 1 - f(x)]$, people use $S(x) = t(x) - f(x)$ record a cent for x, obviously $S(x) \in [-1, 1]$; we have $\phi(x) = \frac{1}{2}\{1 - f(x) + t(x)\}$ the middle point of x, and obviously $|\phi(x) - \phi(y)| = \frac{1}{2}|S(x) - S(y)|$; with $\pi(x) = 1 - f(x) - t(x)$ as the length (the interval length) of x, use $\pi(x)$ to explain Vague's set V an unknown degree. Correspondently, its known degree can use $K(x) = 1 - \pi(x) = f(x) + t(x)$ as described and use the degree (affirmation) that its reflection may support.

We know the characteristics of an interval has four important parameters generally, is the left(right) point, the interval length and the middle point. See from the vote model, the Vague value reflected the information of three parts, namely "approve number," "opposed number," and "abstain number." Therefore, we want to measure the similarity of two Vague values and should consider this information and approve a tendency information. For this, the consideration of our comprehensive these aspects, can give the definition to similarity of two Vague values as follows.

Definition 4.1 Let $x = [t(x), 1 - f(x)]$ and $y = [t(y), 1 - f(y)]$ be two Vague values. Then,

$$M(x, y) = 1 - a|t(x) - t(y)| - b|f(x) - f(y)| - c|\pi(x) - \pi(y)| - d|\phi(x) - \phi(y)|$$

is a kind of measurement of the Vague value x with y, where $a, b, c, d \geq 0$ and $a + b + c + d = 1$.

So two Vague sets V and W defined finite universe of discourse X, and its measurement of similarity can be given as below:

Definition 4.2 Let $X = \{x_1, x_2, \ldots, x_n\}$ finite universe of discourse and V and W be two Vague sets. Then,

$$M(V, W) = \frac{1}{n}\sum_{i=1}^{n}\{1 - a|t_V(x_i) - t_W(x_i)| - b|f_V(x_i) - f_W(x_i)| \\ - c|\pi_V(x_i) - \pi_W(x_i)| - d|\phi_V(x_i) - \phi_W(x_i)|\}$$

is a kind of measurement of the Vague sets V and W, where $a, b, c, d \geq 0$ and $a + b + c + d = 1$.

According to the above discussion, for $X \subseteq U$, a rough Vague $V = (\underline{V}, \overline{V})$ is constituted by R with V set, and its RV value $x = \langle[\underline{\mathrm{Rt}}(x), 1 - \underline{\mathrm{Rf}}(x)], [\overline{\mathrm{Rt}}(x), 1 - \overline{\mathrm{Rf}}(x)]\rangle$ can be written as

$$\underline{S}(x) = \underline{\mathrm{Rt}}(x) - \underline{\mathrm{Rf}}(x), \quad \overline{S}(x) = \overline{\mathrm{Rt}}(x) - \overline{\mathrm{Rf}}(x), \quad S(x) = \alpha_1\underline{S}(x) + \alpha_2\overline{S}(x),$$

where $0 \leq \alpha_1, \alpha_2 \leq 1$ and $\alpha_1 + \alpha_2 = 1$.

Then, $S(x)$ can record a cent of a RV value x. Obviously, $S(x) \in [-1, 1]$.

Concerning the middle point of x, we record

$$\underline{\phi}(x) = \frac{1}{2}[1 - \underline{\mathrm{Rf}}(x) + \underline{\mathrm{Rt}}(x)], \quad \overline{\phi}(x) = \frac{1}{2}\left[1 - \overline{\mathrm{Rf}}(x) + \overline{\mathrm{Rt}(x)}\right],$$
$$\phi(x) = \beta_1\underline{\phi}(x) + \beta_2\overline{\phi}(x)$$

where $0 \leq \beta_1, \beta_2 \leq 1$ and $\beta + \beta_2 = 1$. Record $\underline{\pi}(x) = 1 - \underline{\mathrm{Rf}}(x) - \underline{\mathrm{Rt}}(x)$, $\overline{\pi}(x) = 1 - \overline{\mathrm{Rf}}(x) - \overline{\mathrm{Rt}}(x)$, $\pi(x) = \gamma_1\underline{\pi}(x) + \gamma_2\overline{\pi}(x)$.

$$\underline{K}(x) = \underline{\mathrm{Rt}}(x) + \underline{\mathrm{Rf}}(x), \quad \overline{K}(x) = \overline{\mathrm{Rt}}(x) + \overline{\mathrm{Rf}}(x), \quad K(x) = \gamma_1\underline{K}(x) + \gamma_2\overline{K}(x),$$

where $0 \leq \gamma_1, \gamma_2 \leq 1$ and $\gamma_1 + \gamma_2 = 1$, then, $\pi(x)$ can be considered the unknown degree of rough value x and $K(x)$ is x known degree of x.

Therefore, we can give a following definition to two rough Vague values:

Definition 4.3 Let two rough Vague values $x = \langle[\underline{\mathrm{Rt}}(x), 1 - \underline{\mathrm{Rf}}(x)], [\overline{\mathrm{Rt}}(x), 1 - \overline{\mathrm{Rf}}(x)]\rangle$ and $y = \langle[\underline{\mathrm{Rt}}(y), 1 - \underline{\mathrm{Rf}}(y)], [\overline{\mathrm{Rt}}(y), 1 - \overline{\mathrm{Rf}}(y)]\rangle$,

$$M(x,y) = 1 - \frac{1}{2}\{a|S(x) - S(y)| + b[|K(x) - K(y)|| + |\phi(x) - \phi(y)|] + c[|\underline{\mathrm{Rt}}(x) - \underline{\mathrm{Rt}}(y)|| + |\underline{\mathrm{Rf}}(x) - \underline{\mathrm{Rf}}(y)|] + d[|\overline{\mathrm{Rt}}(x) - \overline{\mathrm{Rt}}(y)| + |\overline{\mathrm{Rf}}(x) - \overline{\mathrm{Rf}}(y)|]\}.$$

Then, $M(x,y)$ is a kind of similarity measurement for RV value x and y, where $a,b,c,d \geq 0$ and $a+b+c+d=1$, and it reflects the tendency of various information.

So that, we can get a following property by the above definition:

Theorem 4.1

1. $M(x,y) \in [0,1]$;
2. $M(x,y) = M(y,x)$;
3. $M(x^c,y^c) = M(x,y)$;
4. *If* $x = y$, *then* $M(x,y) = 1$.
5. *When* $x = \langle[0,0],[0,0]\rangle$, $y = \langle[1,1],[1,1]\rangle$ *or* $y = \langle[0,0],[0,0]\rangle$ $x = \langle[1,1],[1,1]\rangle$, $M(x,y) = 0$.

The (2), (4), (5) in the theorems establish obviously. The one that is underneath proves (1), (3) only.

Proof (1) For arbitrarily RV value x, we can know according to the top definition: $S(x) \in [-1,1]$, then $|S(x) - S(y)|| \in [0,2]$; $\underline{\mathrm{Rt}}(x) \in [0,1], \underline{\mathrm{Rf}}(x) \in [0,1]$, $\overline{\mathrm{Rt}}(x) \in [0,1], \overline{\mathrm{Rf}}(x) \in [0,1]$, hence $|K(x) - K(y)| \in [0,1], |\phi(x) - \phi(y)| \in [0,1]$, $|\underline{\mathrm{Rt}}(x) - \underline{\mathrm{Rt}}(y)| \in [0,1], |\,|\overline{\mathrm{Rt}}(x) - \overline{\mathrm{Rt}}(y)| \in [0,1]|$,

$$|\underline{\mathrm{Rf}}(x) - \underline{\mathrm{Rf}}(y)| \in [0,1],\ |\overline{\mathrm{Rf}}(x) - \overline{\mathrm{Rf}}(y)| \in [0,1], \text{ and } a+b+c+d=1.$$

Thus, $M(x,y) \in [0,1]$.

Proof (3) For arbitrarily RV value x, since

$$\underline{\mathrm{Rt}}(x^c) = \underline{\mathrm{Rf}}(x), \quad \underline{\mathrm{Rf}}(x^c) = \underline{\mathrm{Rt}}(x), \quad \text{and} \quad \overline{\mathrm{Rt}}(x^c) = \overline{\mathrm{Rf}}(x), \overline{\mathrm{Rt}}(x^c) = \overline{\mathrm{Rf}}(x).$$

So we have $S(x^c) = -S(x)$, $|S(x^c) - S(y^c)| = |S(x) - S(y)|$;

$$|\underline{\mathrm{Rt}}(x^c) - \underline{\mathrm{Rt}}(y^c)| = |\underline{\mathrm{Rf}}(x) - \underline{\mathrm{Rf}}(y)|,\ |\underline{\mathrm{Rf}}(x^c) - \underline{\mathrm{Rf}}(y^c)| = |\underline{\mathrm{Rt}}(x) - \underline{\mathrm{Rt}}(y)|;$$
$$|\overline{\mathrm{Rt}}(x^c) - \overline{\mathrm{Rt}}(y^c)| = |\overline{\mathrm{Rf}}(x) - \overline{\mathrm{Rf}}(y)|,\ |\overline{\mathrm{Rf}}(x^c) - \overline{\mathrm{Rf}}(y^c)| = |\overline{\mathrm{Rt}}(x) - \overline{\mathrm{Rt}}(y)|.$$

Hence, $M(x^c,y^c) = M(x,y)$.

5 Similarity Measurement of Rough Vague Sets

According to the above discussion, let $X = \{x_1, x_2, \cdots, x_n\}$. For $\forall x_i \in X$, two RV sets V with W are given by

$$V = (\underline{V}, \overline{V}) = \langle [\underline{\mathrm{Rt}}_V(x_i), 1 - \underline{\mathrm{Rf}}_V(x_i)], [\overline{\mathrm{Rt}}_V(x_i), 1 - \overline{\mathrm{Rf}}_V(x_i)] \rangle;$$
$$W = (\underline{W}, \overline{W}) = \langle [\underline{\mathrm{Rt}}_W(x_i), 1 - \underline{\mathrm{Rf}}_W(x_i)], [\overline{\mathrm{Rt}}_W(x_i), 1 - \overline{\mathrm{Rf}}_W(x_i)] \rangle.$$

Then, the similarity measurement of $V = (\underline{V}, \overline{V})$ with $W = (\underline{W}, \overline{W})$ can be defined as follows.

Definition 5.1 Let $X = \{x_1, x_2, \ldots, x_n\}$ be finite. The above-mentioned one gives two rough Vague sets, and its similarity measurement can be given as follows:

Let

$$M(V, W) = \frac{1}{n} \sum_{i=1}^{n} M[V(x_i), W(x_i)]$$
$$= \frac{1}{n} \sum_{i=1}^{n} \left\{ 1 - \frac{1}{2} \left[aS_{VW}(x_i) + 2bK_{VW}(x_i) + c\underline{R}_{VW}(x_i) + d\overline{R}_{VW}(x_i) \right] \right\};$$

where $S_{VW}(x_i) = |S_V(x_i) - S_W(x_i)|; K_{VW}(x_i) = |K_V(x_i) - K_W(x_i)|;$

$$\underline{R}_{VW}(x_i) = |\underline{\mathrm{Rt}}_V(x_i) - \underline{\mathrm{Rt}}_W(x_i)| + |\underline{\mathrm{Rf}}_V(x_i) - \underline{\mathrm{Rf}}_W(x_i)|;$$
$$\overline{R}_{VW}(x_i) = |\overline{Rt}_V(x_i) - \overline{Rt}_W(x_i)| + |\overline{Rf}_V(x_i) - \overline{Rf}_W(x_i)|; \quad i = 1, 2, \ldots, n,$$

$$a, b, c, d \geq 0, \quad \text{and} \quad a + b + c + d = 1.$$

Then, $M(V, W)$ is a kind of the similarity measurement of rough Vague set V with W.

Defining from here can know, $M(V, W)$ having a following property:

Theorem 5.1 *According to Definition 5.1, the similarity measurement $M(V, W)$ of rough Vague sets V and W has the following property:*

1. $M(V, W) \in [0, 1]$;
2. $M(V, W) = M(W, V)$;
3. $M(V^c, W^c) = M(V, W)$;
4. *When* $V = (\underline{V}, \overline{V}) = W = (\underline{W}, \overline{W})$, *then* $M(V, W) = 1$.

Proof The one that is underneath proves (1) only.

In fact, since $S_{VW}(x_i) = |S_V(x_i) - S_W(x_i)| \in [0, 2]; K_{VW}(x_i) = |K_V(x_i) - K_W(x_i) | \in [0, 1];$

$$\underline{R}_{VW}(x_i) = |\underline{\mathrm{Rt}}_V(x_i) - \underline{\mathrm{Rt}}_W(x_i)| + |\underline{\mathrm{Rf}}_V(x_i) - \underline{\mathrm{Rf}}_W(x_i)| \in [0, 2];$$

$\overline{R}_{VW}(x_i) = |\overline{\mathrm{Rt}}_V(x_i) - \overline{\mathrm{Rt}}_W(x_i)| + |\overline{\mathrm{Rf}}_V(x_i) - \overline{\mathrm{Rf}}_W(x_i)| \in [0, 2];$ and $a + b + c + d = 1.$

So

$0 \leq aS_{VW}(x_i) + 2bK_{VW}(x_i) + c\underline{R}_{VW}(x_i) + d\overline{R}_{VW}(x_i) \leq 2a + 2b + 2c + 2d = 2.$

Hence,
$0 \leq 1 - \frac{1}{2}\left[aS_{VW}(x_i) + 2bK_{VW}(x_i) + c\underline{R}_{VW}(x_i) + d\overline{R}_{VW}(x_i)\right] \leq 1, \quad i = 1, 2, \ldots, n.$
Therefore, $0 \leq M(V, W) \leq 1$.

6 Conclusion

RV sets theories is the mathematics tool of a kind of new processing uncertainty information, very suitable for processing since have the knowledge (concept) that cannot distinguish and have fuzzy. This text synthesizes various circumstance that should consider, giving the new method of the similarity measurement for rough Vague sets, provided the theories foundation for the applied realm that the rough Vague sets.

Acknowledgments This work is supported by International Science & Technology Cooperation Program of China (2012DFA11270) and Hainan International Cooperation Key Project (GJXM201105).

References

1. Zadeh, L.A.: The concept of a linguistic and its application to approximate reasoning. Inf. Sci. **8**(3), 199–219 (1975)
2. Atanassov, K.T.: Intuitionistic fuzzy sets. Fuzzy Sets Syst. **20**(1), 87–96 (1986)
3. Atanassov, K., Gargov, G.: Interval valued intuitionistic fuzzy sets. Fuzzy Sets Syst. **31**(3), 343–349 (1989)
4. Gau, W.L., Buehrer, D.J.: Vague sets. IEEE Trans. SMC **23**(2), 610–614 (1993)
5. Deschrijver, G., Kerre, E.E.: On the relationship between some extensions of fuzzy set theory. Fuzzy Sets Syst. **133**(2), 227–235 (2003)

Agricultural Product Traceability with Services Resource Provisioning

Bin Wen, Zi-qiang Luo, Chang-qing Liu and Song Lin

Abstract Faced with the key issues for domain-oriented and requirements-dominated services-based software system, active provisioning of services resource will be focused. Segments of requirements knowledge-driven active personalized customization in consumer-centric service-oriented architecture (SOA) will be systematically investigated with service-oriented requirements model and distributed requirements elicitation. In this chapter, regarding the fault of traditional bar code traceability, we present a service-oriented approach for agricultural product traceability based on QR two-dimensional code and data aggregation with services resource provisioning. Service-based architecture for product traceability with ATOM protocol will be chosen to meet the heterogeneous multi-client and lightweight requirements. Active and adaptive production for services provisioning will be investigated, and its application in the domain of agricultural product traceability will also be designed. Empirical application and analysis for Hainan agricultural product traceability have proved the feasibility and effectiveness of the proposed approach.

Keywords Services computing · Product traceability · Web service

1 Introduction

Traceability of agricultural product refers to retaining their traceability in circulation and will provide with relevant information on traceability for the agencies or consumers in case of need. When agricultural products are found hazardous, it can

---

B. Wen · Z. Luo · S. Lin
School of Information Science and Technology, Hainan Normal University, Haikou 571158, China

C. Liu (✉)
Library, Hainan Normal University, Haikou 571158, China
e-mail: changqing@hainnu.edu.cn

B.-Y. Cao et al. (eds.), *Ecosystem Assessment and Fuzzy Systems Management*,
Advances in Intelligent Systems and Computing 254, DOI: 10.1007/978-3-319-03449-2_17,

be recalled from the market in a timely manner to avoid entering the market. This is achieved by information means such as bar code (multi-dimensional), electronic tags, and identification.

Hainan is China's largest tropical fruit producing area and winter-season vegetable base. National tropical modern agricultural base is an important development support of international tourism island, which will make Hainan become the country's winter vegetable base, tropical fruit base, southern seed breeding base, fisheries export base, and natural rubber base. Currently, the world increasingly pays high degree of attention to food quality and safety. The USA, EU, and other developed countries and regions have enacted strict food market access standards. Many parts of the country are constantly introduced to a number of local mandatory standards to limit the polluted agricultural products entering the market. Meanwhile, controllable quality and safety traceability will be implemented to improve the market competitiveness for sales of agricultural products.

Services resource represents a set of well-defined and interoperable Web software component. The basic physical unit will be understood as Web service, namely autonomous, open, self-describing, and implementation-independent networked component [1]. Services resources provisioning means the production style, the provisioning methods, and sources structure of services resource.

The importance of services resources provisioning is mainly reflected as follows: (1) Services resource that is adequate to meet the personalized needs is material basis of services aggregation and service-based software production and (2) effective services resource provisioning approaches should be widely used to ensure smooth implementation and efficient completion of service-based software production.

The importance of services resource provisioning has become the consensus of industrialized service-oriented software development and applied to software production. And it has been partially supported by mainstream tools.[1]

The starting point of current study assumed that services resource is sufficient, but actual development stage is not the case. Services resource that meets users' needs often does not exist or cannot be run satisfactorily. As described in the literature [2], "If no services are available for some parts, the application developer can register them in the service broker directory and wait until the needed services are available." Thus, it leads to a serious case with difficulty, complexity, and low availability of services composition.

At the same time, due to the dynamic nature of Web services and error-prone services resource provisioning environment, a variety of exceptions in services resources provisioning can occur during the execution of composing services [3]. Exceptions refer to service failure (fault), network errors, and abnormal events that are subject to resources or requirement changes. Problems are caused by lack of exception handling mechanism, such as poor performance, waste of resources,

[1] http://www.microsoft.com/whdc/driver/kernel/kb-drv.aspx; http://www.ibm.com/developerworks/.

non-optimized service provider, or even the failure of process execution. Services resource provisioning mechanism must be able to take actively, i.e., they should be able to adapt to the runtime exceptions.

2 Motivation

Tracking and tracing agricultural products can greatly improve the efficiency of agricultural products' traceability and the quality of protection through information technology, in which traceability information mainly focuses on service-oriented agricultural information network digital sampling, compression, digital anti-counterfeiting technology, and rendering technology concerns online on-demand material for traceability of agricultural products. These have a lot of underlying legacy software resources. In addition, for the existing agricultural information tracing method, information representation and content will often demonstrate independence.

Therefore, how to make coding a trace record can be achieved by manual identification and information processing automatically? Traceability information can be queried either directly through the browser or mobile phones and other reading devices to meet the network users' (agricultural producers and consumers) immediate and effective access for information and application services in the support of services computing. The above methods and techniques will be considered in this study as the starting point and distinguished foothold for other studies. Services computing as a new software production technology for network era has been recognized by the industry. But for the ultimate goal of computing, namely on-demand service, it has not been able to achieve or rarely involves by far. It also is not effectively applied to actual work, such as Hainan agricultural areas of traceability.

Producing software as manufacture has been the dream that software engineers pursued. Recalling the history of the evolution of software development approaches, regardless of machine-oriented, process-oriented, object-oriented, or component-oriented, they are forward along the road to industrial large-scale production. Ford's assembly line is a typical representative of major industrialized production mode, through the standardization system, to achieve large-scale production capability. In accordance with the "building blocks" to produce, assemble, and substitute the products, a large-scale, rapid, and low-cost production style can be achieved. With the amazing production capacity released from large-scale production, today, we are facing a world that has abundant material supply. In order to cope with competition of such rich world in a buyer's market, the manufacturer's production must be user-centric and further lead to market segmentation; customized production has become a key competitive factor. The final fusion with the users also come in—the users' needs to become part of design and production, more flexible, convenient, and economical service for the customer.

Previously we have designed conceptual model with large-scale customization for networked software [4], and this model also assumes that services resource is rich enough, in which the shaded part refers to individual needs of stakeholders. If personalized services resource does not exist and also cannot be achieved through value-added services composition, then it will relate to explore the issue of customized production which captured individual needs of service based on component resources. How to associate the back-end services resource with individual needs fragment of production model? How to embody on-demand customization requirements of service component resources and on-the-fly production? These issues are directions and concerns of this chapter focusing on active customized provisioning of services resource.

Customized services refer to supplying suitable and satisfactory value of experience according to their own needs of consumers. Service-based software development is distinguished into mass customization and personalized services resource customization. Mass customization combines the advantages of two production modes of custom manufacturing and mass production to meet individual needs of customers. At the same time, it retains lower production costs and shorter delivery time. But personalized customization of services resource aims at capturing unfound individual needs to customized production for abounding service component library resources and improving services resource provisioning style from service provider-centric mode to consumer requirements-driven production. The two styles complement each other to improve service quality of experience for stakeholders together.

Regarding the traditional service-oriented architecture (SOA) [5], services resource provisioning strategy is based on service providers' priority. Service providers produce services resource and publish these services' description information in service registries. Service requesters (consumers) can establish dynamic binding association with real services by querying service registries. Traditional SOA does not furnish the consumers with publishing-related information functions for services production. Consumers can only retrieve, find, and match provided services resource of service providers in the registry according to the needs of application, and the process is essentially a passive selection.

Institute of Scientific Technical Information of Chinese Academy of Tropical Agricultural sciences has developed quality security traceability platform for Hainan tropical agricultural product. But, considering user needs with service-oriented perspective to meet, service-based methods and techniques have not been effectively addressed and reasonably applied for Hainan agricultural traceability. An urgent need is to use the latest services computing and requirements engineering methods for conducting interdisciplinary and innovative research in depth to promote their levels of networked service.

This chapter is a research about agricultural product traceability system with services resource provisioning and focuses on services resource personalized active customized production approach.

3 Customization Model for Services Resource

Definition 1 (*Customization of Services Resource*)

Given service requesters set $R = \{R_1, R_2, \ldots, R_n\}$ and service providers set $P = \{P_1, P_2, \ldots, P_m\}$, for customization requirements CReq_i from $R_i \in R$ $(1 \le i \le n)$, if there exists $P_j \in P$ $(1 \le j \le m)$ which can complete services resource production and publishing within the specified time parameters T, we call service provider P_j in response to services resource customization of service requester R_i.

Definition 2 (*Customized Matching Degree*)

For customized matching degree of services resource, $C_{\text{match}} = \text{Semantic}$ $Match_{P \to R} * e^{-\frac{|t-T|}{T}}$, where semantic matching degree $\text{SemanticMatch}_{P \to R}$ denotes semantic matching degree between service semantics of provider's customization production and requirements semantic of requester, t represents customization completion time of provider, and T is the required customization time of requester.

Definition 3 (*Custom Preferred Provider*)

Given the current services resource customization matrix RP_{Matrix}, custom requesters and providers pair off and implement assignment operating. We call the provider the related requester's custom preferred provider if there exists a provider which corresponds to arrive at the sum of the maximum matching result.

The mathematical models for solving custom preferred provider of services resource are described as follows:

$$\max C = \sum_{i=1}^{n} \sum_{j=1}^{m} RP_{ij} * x_{ij} \tag{1}$$

$$s.t. \begin{cases} x_{ij} = 0, 1; & i = 1, 2, \ldots, n,\ j = 1, 2, \ldots, m \\ \sum_{i=1}^{n} x_{ij} = 1; & j = 1, 2, \ldots, m \\ \sum_{j=1}^{m} x_{ij} = 1; & i = 1, 2, \ldots, n \end{cases} . \tag{2}$$

This is similar to combinatorial optimization assignment problem, with reference to the Hungarian algorithm [6]. Its related conditions of the original algorithm are improved, such as the solved total minimum are transformed to maximum (the custom matching degree, higher the better). Meanwhile, tag matrix is designed. The improved method is more efficient than original algorithm, easy to operate and implement. The time complexity of Custom Preferred Provider selection algorithm is $O(n^3)$.

For the services resource custom provider selection, if the total custom matching degree is not required to reach a maximum just based on past selection chance to predict next selection, it is very suitable for the use of random Markov prediction method. A comprehensive forecast of the event, to be able to point out not only the incident with all possible outcomes, but also the probability of each outcome, must be given with current state of the probability to predict the best choice. This could be based on historical statistical data to build a custom provider state transition probability matrix T_{RP}, in which rows of the matrix correspond to the requester and columns correspond to the custom provider. P_{ij} denotes selection probability between requester i and custom j in current state.

Definition 4 (*Prediction Selection of Custom Provider*)

Given custom providers state transition probability matrix T_{RP} and the completed service customization selection state, the next service customization selection state can be found according to Markov process without aftereffect and Bayesian conditional probability formula, in which the provider corresponding to maximum probability for each requester is the service custom provider prediction selection.

Custom provider prediction model under the random Markov chain is described below.

$$\pi_j(k) = \sum_{i=1}^{n} \pi_j(k-1)P_{ij} \quad (j = 1, 2, \cdots, n) \tag{3}$$

where the state probability $\pi_j(k)$ indicates the probability of custom provider j in the moment (period) k under the known initial state ($k = 0$) to pass state transfer with k times, and $\sum_{j=1}^{n} \pi_j(k) = 1$.

4 Active Customized Production for Services Resource Provisioning

Requirements sign ontology (RSO) [7] is semantic description of overall business process including requirements-induced services resource (components) and control structure, and it also denotes knowledge about business workflow to guide services resource identifying and complete services aggregation.

For unable matched services, requirements-driven active customization will be used to complete custom produce for service providers during services aggregating. The main difference is to be able to identify (business) process and services resource between service-oriented requirements engineering and the traditional requirements engineering. We have designed the metadescription of networked software (see [7]), in which stakeholders are composed of participants playing different roles whose required goals will be achieved by the business process. As a software agent, services resource is the constituent element of business process.

```
1 INPUT: QR 2-dimension code(XML)
2 OUTPUT: Atom file for QR 2-dimension code

 1: Intialize Atom file header;
 2: form feed title,ID etc.
 3: while (queue)<>NIL do
 4:     get a QR code segment from queue; //queue length decrease
 5:     create a entry header; //<entry>
 6:     create title part from ID of QR code segment;
 7:     create ID; //GUID
 8:     create content header;//Type="application/xml"
 9:       begin
10:         create name part from ID;
11:         create address from production address;
12:         create time from production time;
13:         create Dateofpackaging from packaging time;
14:         create service site from RESTful address;
15:       end
16:     form updated part;
17:     create summary part from comment
18: end while
19: if (entry part)<>NIL then
20:     write a Atom file;
21:     return the file;
22: end if
23: return
```

Algorithm 1: QRcode2Atom

On the one hand, the framework will guide how to organize domain ontology to create asset model; on the other hand, domain ontology depicts concepts and its semantic associations to support semantics for asset modeling that is characterized by ontology description (OWL or OWL-S format). It will facilitate the semantic reasoning in the SODR asset repository.

4.1 Atom-Based Information Publishing

Experimental platform mainly adopts Apache Abdera[2] to generate Atom message and publishing protocol. Semantics in RSO are expressed in an OWL-S format. We can list the corresponding relationships between every item of entry in Atom format and service semantics in RSO.

The transformation algorithm from semantic description of QR two-dimensional code (XML) to Atom is shown in Algorithm 1.

[2] http://incubator.apache.org/abdera/.

4.2 QR Two-Dimensional Code

Automatic identification technology refers to non-keyboard data input technology, like programmable logic controller (PLC), computer, or other microprocessor apparatus, including sound image recognition, bar code, radio frequency identification (RFID), biometric identification, and magnetic stripe card technology, in which bar code and RFID technology are used widely in automatic identification.

Traditional traceability systems utilize one-dimensional bar code to identify agricultural products. Through the bar code technology applied in agricultural products, the bar code is classified as product information and vendor information and then, respectively, query product information and vendor information based on product code and manufacturer code. Bar codes have input speed, high reliability, large amount of collected information, and economic advantages and are easy to produce, so bar code technology is suitable for use as an information carrier for traceability systems of agricultural supply chain. For example, in March 2010, Institute of Scientific Technical Information of Chinese Academy of Tropical Agricultural sciences has established quality security traceability platform for Hainan tropical agricultural product.

However, due to the one-dimensional bar code encoding limits, accommodation of bar code information is extremely limited. Meanwhile, fault tolerance of bar code is poor, they can no longer be properly read after defacement, destruction, or deformation. The most important thing is that code can be identified only through a specific reader and then check the relevant product information from database. It is passive and very troublesome. Therefore, agricultural traceability system based on one-dimensional bar code has not solved the problem for the safety of agricultural products. We are eager to provide an agricultural product traceability system with the facility of easy reading and quickly getting the products information.

To facilitate the derived application of products traceability, RESTful Web services address will be designed for two-dimension agricultural metadata code. Through the client program, corresponding HTTP request (GetPutPostDelete) is sent to the target URI. Client applications can process the traceability information (mainly XML format) to achieve the purpose of further application (Fig. 1).

4.3 Agricultural Product Traceability Service Design with Services Resource Active Provisioning

Based on the Abdera library, thinfeeder,[3] an open source RSS reader, can be revised to display Atom and subscribe the services or traceability information of agricultural product coded by QR two-dimension code. Also, adding notify

[3] http://thinfeeder.sourceforge.net/.

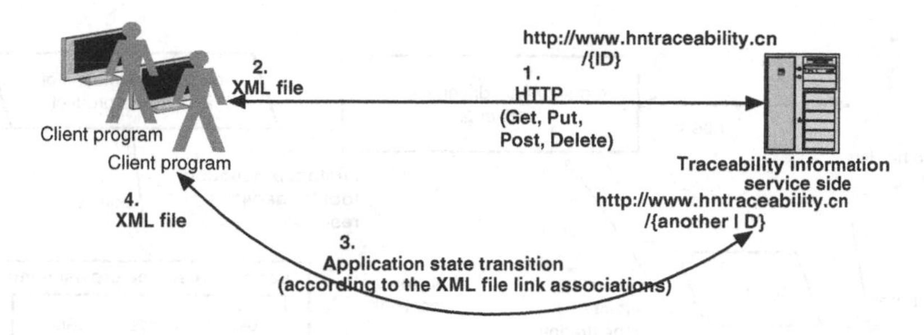

Fig. 1 RESTful Web service process for Agricultural Traceability

function to thinfeeder, aggregator can get customized services data (WSDL or OWL-S file location, service endpoint URI, etc.) to facilitate rerunning aggregation flow after completing the feedbacks of customized service queue.

Prototype of user requirements-driven services engineering is developed as a series of requirements-conducted services platform that has realized the main features, where requirements asset modeling for domain, service registration, requirements capture, and analysis modules has launched a number of versions. The main part of prototype adopts language written in Java and PHP using Apache as a Web server and MySQL database as a background. The whole system is based on the J2EE specification to achieve distributing operations with multi-user network environment and also solve the issue of heterogeneity. The platform uses a distributed SOA as a background system architecture, and the whole system is divided into three parts, namely requirements semantic acquisition module, service aggregation module, and services resource customized modules. Among them, requirements semantic export and service operation governance are packaged as RESTful interface to provide the modules of service aggregation and services resource customization for calling.

Regarding the experiment carrier for services resource provisioning, we select the Hainan agricultural product online traceability service system as application field. In a service-oriented agricultural traceability domain, stakeholders facing the domain user have rich and diverse individual needs, which led to a variety of customization requirements. Meanwhile, numerous legacy software about information processing of agricultural products exists to meet empirical demands of the project with active services resource provisioning.

How better to achieve real-time and on-demand digital service for Hainan online agricultural product traceability system under the Internet with customization-based resource provisioning ecological chain extension? The issue is the focus for empirical carrier research.

The preliminary overall application architecture is shown in Fig. 2. Specific functional requirements of services resource active provisioning can be acquired through full communication with stakeholders. Preliminary functional requirements of CASE tool are shown in Table 1.

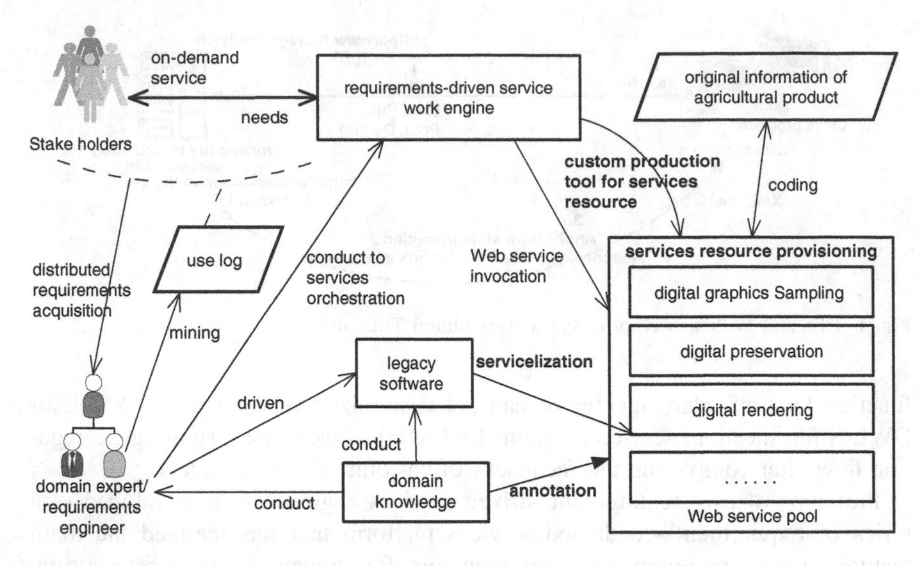

Fig. 2 Architecture for agricultural traceability service system

Table 1 Functional requirements of CASE tool for services resource active provisioning

Resource activities	Provisioning functional requirements of CASE tool
1. Requirements acquisition	1.1 Requirements model import
	1.2 Participate to get consistent and complete requirements
2. Custom manufacturing of services resources	2.1 Customization requirements slicing (splitting)
	2.2 Custom description information ConvertOWL-S or WSDL
	2.3 Customized information push and provider selection
	2.4 Registration and feedback of on-the-fly custom information
	2.5 Custom management
3. Legacy software servicelization	3.1 Legacy software signature analysis for specific code (such as PHP)
	3.2 Legacy software slicing (module dependency diagram)
	3.3 Wrapper and produce WSDL description
	3.4 ServiceS resource information publish (register)
4. Common features	Domain knowledge annotation; services resources query

Figure 3 shows agricultural traceability service system with B/S structure, and we have developed an app program of cell phone that is application for agricultural product traceability service based on Android platform, which can generate and parse QR two-dimensional code for agricultural product traceability and complete online inquiries combined with server to achieve the mobile client-side access service simultaneously.

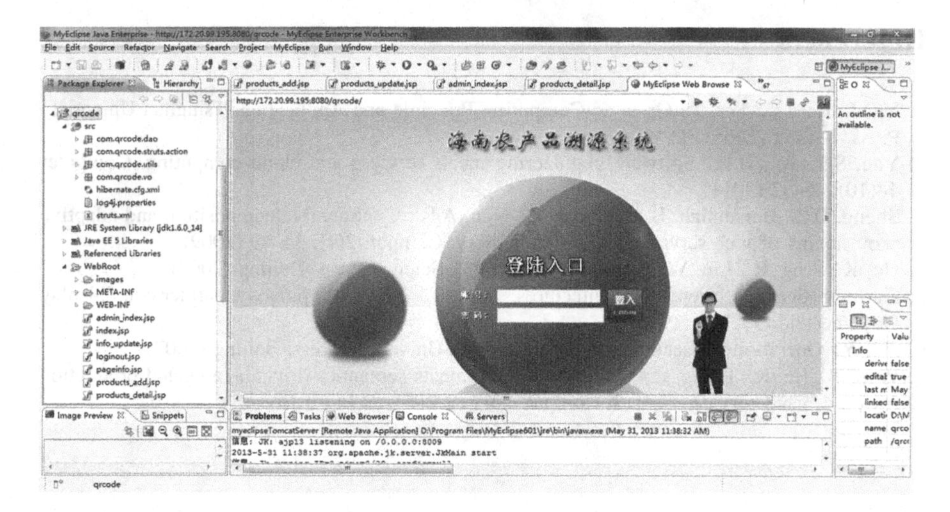

Fig. 3 Agricultural traceability service with B/S

5 Conclusions

Service-oriented approaches are gaining more and more attention since they claim to provide new and flexible ways of supporting the activities in an organization. The chapter has established software architecture with personalized active custom for services resource that will select unsatisfactory requirements fragment of customers to dynamically generate custom needs in the running of the services composition. By means of custom preferred provider selection approach designated to push the customization requirements, service providers complete the services resource production and customized products evaluation just in time, and the process will effectively prevent failure of services aggregation and services resource provisioning exceptions from occurring. Service computing architecture of traceability with Atom protocol will be chosen to meet the heterogeneous multi-client and lightweight requirements. Active and adaptive production for services information provisioning will be explored, and its application architecture in the domain of agricultural product traceability will also be designed.

Acknowledgments This research has been supported by the Natural Science Foundation of Hainan Province (No. 613162) and Higher School Scientific Research Project of Hainan Province (No. Hjkj2013-17).

References

1. Yu, J., Han, Y.: Service-Oriented Computing-Principle and Application. Tsinghua University Press, Beijing (2006)
2. Yau, S.S., An, H.G.: Software engineering meets services and cloud computing. Computer **44**(10), 46–52 (2011)
3. Sheng, Q.Z., Benatallah, B., Maamar, Z., Ngu, A.H.: Configurable composition and adaptive provisioning of web services. IEEE Trans. Serv. Comput. **2**(1), 34–49 (2009)
4. He, K., Peng, R., Liu, W.: Networked Software. Science Press, Beijing (2008)
5. Gartner: Service oriented architectures: part 1 and 2. http://www.gartner.com/DisplayDocument?id=302868 (1996)
6. Hu, Y.: Operations Research, 3rd edn. Tsinghua University Press, Beijing (2007)
7. Wen, B., He, K., Liang, P., Wang, J.: Requirements semantics-driven aggregated production for on-demand service. Chin. J. Comput. **33**(11), 2163–2176 (2010)

Raster-Based Parallel Multiplicatively Weighted Voronoi Diagrams Algorithm with MapReduce

Ming Xu, Han Cao and Chang-ying Wang

Abstract While Voronoi diagram has been used in many fields, most vector-based methods of generating Voronoi diagrams focus mostly on point features, but they have difficulties in handling generators like lines or areas, which can be easily generated by raster-based methods, however, with substantial calculation cost. For the sake of integrating Voronoi diagram models with Web GIS, which inevitably encounters generators like lines and areas, we present a parallel algorithm with MapReduce for generating raster-based multiplicatively weighted Voronoi diagrams. The experiments and case studies show that the algorithm significantly improves the efficiency of generating Voronoi diagrams on large-scale raster data with potential use in urban public green space planning and optimal path planning.

Keywords Weighted Voronoi diagram · Parallel algorithm · Hadoop · MapReduce

1 Introduction

In mathematics, a weighted Voronoi diagram (WVD) in n dimensions is a Voronoi diagram for which the Voronoi cells are defined in terms of a distance defined by some common metrics modified by weights assigned to generator points. The multiplicatively weighted Voronoi diagram (MWVD) is defined when the distance between points is multiplied by positive weights [1].

M. Xu · H. Cao (✉)
Department of Computer Science, Shaanxi Normal University, Xian 710062, China
e-mail: caohan@snnu.edu.cn

C. Wang
Department of Computer Science, Fujian Agriculture and Forestry University, Fuzhou 350002, China

B.-Y. Cao et al. (eds.), *Ecosystem Assessment and Fuzzy Systems Management*, Advances in Intelligent Systems and Computing 254, DOI: 10.1007/978-3-319-03449-2_18,

The dominance region defined by subsequent Voronoi regions is generalized by the type of generator, such as points, lines, and polygons, various weights, plane constraints, and metric space used. Many researchers have focused on issues of WVD generation.

Hu et al. [2] used map algebra method to solve the problem of generating MWVDs for point, line, and polygon features on raster data. Wang et al. [3] presented a raster-based algorithm supported by the ArcInfo, which is capable to generate Voronoi diagrams for points, lines, and polygons in two-dimensional space. On the basis of Ref. [4], Dong [5, 6] provided a discrete algorithm for generating line segment WVD. However, the algorithm did not consider more complex generators such as line features with an arbitrary shape. Wu and Luo [7] proposed an algorithm based on cellular automata for constructing Voronoi diagram of complex entities in grid space. Dong [8] noted applications of MWVDs for points, polylines, and polygons using ArcGIS. Fan [9] presented a vector-based algorithm for generating WVDs, which, however, is also not suitable for complex spatial objects. Gong et al. [10] developed a vector-based algorithm to generate and update MWVDs for points, polylines, and polygons in C# and present several examples. However, like all other vector-based methods, when dealing with complex generators such as lines and areas, the algorithm needs to first decompose them into simpler elements, which may violate spatial integrity. Very recently, many researchers have used graphics hardware to improve the performance of computing Voronoi diagram. Rong et al. [11] presented a GPU-assisted Voronoi diagram algorithm for computing centroidal Voronoi tessellation (CVT). Xu et al. [12] proposed a raster-based algorithm to construct WVDs with GPU, which is capable to generate discrete WVDs in real time. Besides, Afsin et al. [13] proposed an approach of generating spatial index, Voronoi diagram, and efficient processing of a wide range of geospatial queries.

Although theoretical and computational aspects of WVDs have been extensively discussed, there are still some problems especially in case of complex generators. There are three major issues of generating WVDs: (1) Vector-based algorithms handle so many factors in calculation and storage, that it is difficult to generate WVDs for complex generators directly. (2) Raster-based approaches involve judging and computing distances for each grid, so it costs large amount of calculation and has to tolerate precision limitation. (3) Since sequential algorithms, especially raster-based approaches are inefficient to handle large-scale massive map data, parallel and high-efficiency methods for complex spatial features based on cloud computing are reasonable. Hence, considering types of generators and various weights, we present a raster-based parallel approach with Hadoop to generate MWVDs for polygons.

The remainder of this chapter is organized as follows. In Sect. 2, we first provide a review of relevant concepts of Voronoi diagrams, which build the foundation of proposed method in the chapter. In Sect. 3, raster-based parallel algorithm with pseudo-code is discussed. Section 4 presents experimental results to verify the performance and scalability of proposed approach. Potential applications of proposed approach are discussed in Sect. 5. Finally, Sect. 6 gives conclusions of the chapter and directions for future work.

2 Background: Voronoi Diagram

Voronoi diagram divides a space into disjoint polygons where the nearest neighbor of any generator inside a polygon is the generator of the polygon. In this section, we review the relevant concepts of the Voronoi diagrams [14].

2.1 Ordinary Voronoi Diagram

Consider a set of limited number of points, called generator points, in the Euclidean plane (in general, generators can be any type of spatial object). Every location in the plane can be assigned to the closest generator(s) with a certain distance metric. The set of locations assigned to each generator forms Voronoi polygon of that generator. The set of Voronoi polygons associated with all the generators is called the Voronoi diagram. The Voronoi polygon and Voronoi diagram can be formally defined as follows: Assume a set of generators $G = \{g_1, g_2, \ldots, g_n\}$, where $2 < n < \infty$ and $g_i \neq g_j$ for $i \neq j$, $i, j \in I_n = \{1, \ldots, n\}$. For each generator, the set of Voronoi polygon given by

$$V(g_i) = \bigcap_{j \neq i} \{p | d(p, g_i) < d(p, g_j)\} \tag{1}$$

where $d(p, g_i)$ specifies the minimum distance between p and g_i and is called the Voronoi diagram generated by G.

2.2 Weighted Voronoi Diagram

In many applications, not only the location but also the weight (or importance) and the spatial extent of a site should be taken into account. The influence of different generators on the surrounding is different, so the ordinary Voronoi diagram always cannot meet the needs of general spatial analysis. We need to improve and generalize the approach. WVDs can be divided into two types: MWVDs [15] and additively WVDs. For the former, the distance between points is multiplied by positive weights. For the latter, positive weights are subtracted from the distances between points. Based on the above concept, now we add the weight value of distance to ordinary Voronoi diagram.

Let $g_i \in G$ be an element with positive weight ω_i. The weight distance $d_\omega(g, g_i)$ between p and g_i is $d_\omega(p, g_i) = d(p, g_i)/\omega_i$. Then, the dominance of generators, called $V_\omega(g_i)$, can be represented by

$$V_\omega(g_i) = \bigcap_{j \neq i} \{p | d_\omega(p, g_i) < d_\omega(p, g_j)\}. \tag{2}$$

2.3 Weighted Voronoi Diagram for Polygons

In general, generators can be any type of spatial object, such as points, lines, and polygons. WVD for polygons [9, 16] is an important generalization of the ordinary Voronoi diagram in two sides of generator and weight. Most methods of computing Voronoi diagrams have some difficult in handling complex generators (not points). Many scholars have discussed the Voronoi diagram for lines or for polygons. Next, we focus on the WVD for polygons.

Assume a set of polygons $P = \{p_1, p_2, \cdots, p_n\}$, $d(p, p_i)$ specifies the minimum distance between p and p_i, and then, the Voronoi Diagram generated by P can be represented by

$$V_{\omega}(P_i) = \bigcap_{j \neq i} \left\{ p \left| \frac{d(p, p_i)}{\omega_i} < \frac{d(p, p_j)}{\omega_j} \right. \right\}. \quad (3)$$

3 Raster-Based MWVDs Parallel Algorithm with MapReduce

The general idea of the raster-based method [17] is to calculating raster distance of points and obtaining neighbor relationship between generators based on the idea of distance transformation. After map rasterization, all spatial features are translated into grid points. In the raster metric space, points, lines, and polygons are processed at spatial raster. All diagrams can be represented by a discrete grid lattice of size $N \times N$, which gives N^2 points in the plane. A unique identifier was placed in each grid cell. But there have some difference between three situations. Our approach focus on finding the raster point sets of polygons. Taking polygon feature as an example, our method increases the process of edge points extraction which is based on traditional Voronoi diagrams generation for points.

3.1 Raster-Based Weighted Voronoi Diagram Generation

After rasterization of primitive data, complex generators, such as polygons, can translate into lots of grid cells. Then, it makes huge computational burden and increases the time, especially if a large number of features are involved for generating Voronoi diagrams. While calculating the distance, we only need to get the minimum distance between polygons and other points. On the other hand, assume area features without internal voids, let a raster point specifies any grid cell except generators, the minimum distance can be the distance between the point and

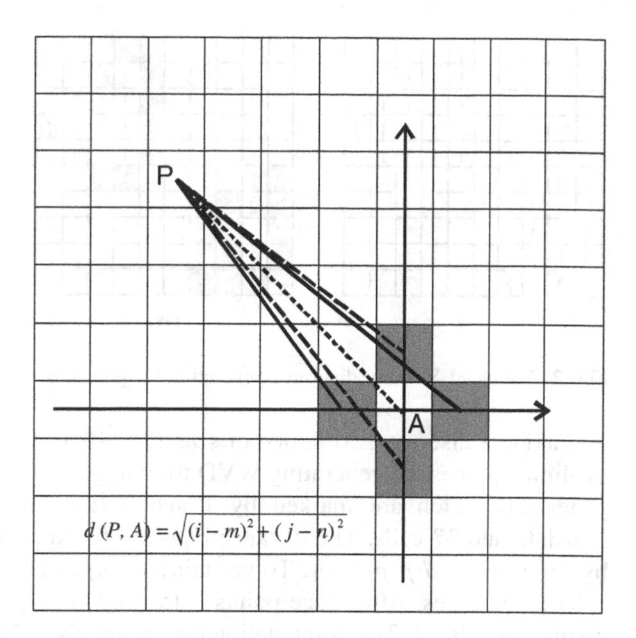

Fig. 1 Distance calculation from point to polygons

generator edge. Therefore, we need to get a simplified treatment of polygons to shorten the calculation time.

Edge points extraction of polygon namely judge whether a raster point is a boundary point or not. Our proposed approach uses a simple rule of criteria for determining pixels based on judging its four-neighbor points. If there are blank cells, the raster point can be a boundary point, otherwise delete it (Fig. 1). In the raster space, point $P(m, n)$ is any other blank cell and $A(i, j)$ is one raster point; the shadow part is its four-neighbor. If its four-neighbor is all not blank cell, so point A is not a boundary point, then the distance between $P(m, n)$ and $A(i, j)$ cannot be the minimum distance between the five distances. So it is unnecessary to compute the distance between internal points (like A) and blank cells. We just need compute the distance between boundary points and blank points.

The steps of the method are as follows:

Step 1. Rasterization of primitive data. For every area generator, use a limited number of raster points to approach its region. And other regions are translated into blank cells.

Step 2. Edge points extraction of polygon. Traverse all raster points of generators. Judge its four-neighbor points are blank cells or not. If it is, mark the point as a boundary point and record it.

Step 3. Computing the weight distance and deciding each grid's character. Calculate the weighted distance between every blank cell and set of boundary points. Label the character of boundary point of nearest distance until values of all blank cells are determined.

Step 4. Output the final matrix.

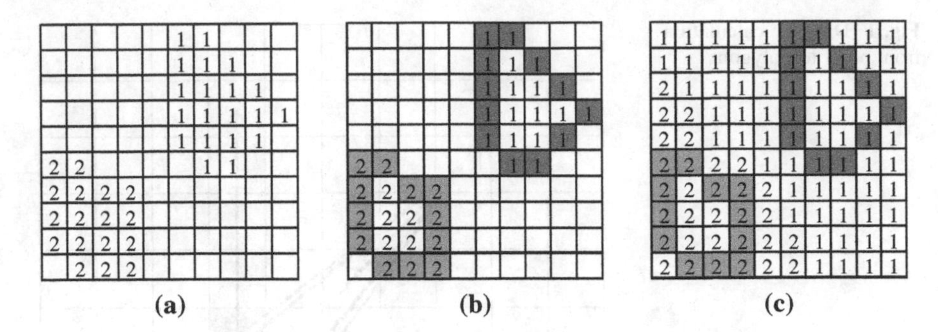

Fig. 2 Weighted Voronoi diagrams generation for polygons

Taking a discrete grid lattice of size 10×10 as an example, Fig. 2 shows the evolution process of generating WVD for polygons. There are two area features as generators which are marked by 1 and 2 (Fig. 2a). After rasterization, they translate into 37 cells. The distance between P and polygon 1 can be represented by $D_1(p,p_1) = d(p,p_1)/\omega_1$. To get $\min\{D_1(p,p_1), D_2(p,p_2)\}$, there would be 37 calculation times. After edge points extraction of polygon (Fig. 2b), the times of computation is 24. The computation cost is greatly reduced by using this method. Let $\omega_1 : \omega_2 = 3 : 2$.

3.2 MapReduce Parallelization

MapReduce programming model is simply represented in two functions, namely a map function and a reduce function. The MapReduce job processes a key/value pair to generate a set of intermediate key/value pairs in the map function, while merges all intermediate values associated with the same intermediate key [18]. The two functions are written by the user. It makes programmers design parallel and distributed applications easily. In our Hadoop implementation, the generation of WVD is implemented in one MapReduce job (Tables 1 and 2).

The task of map function is to determine the minimum distance between blank cells and generators and deciding each grid's character. We choose two files as input data, one is the original raster data file, another records boundary points. The map function takes as input a $<i, recordLine>$ pair in which i is the line number of the original matrix. Every *recordLine* deposit a row of grid cells. The map function must read record file for getting the location of generators in the original matrix. Finally, it outputs the $<i, newLine>$ pair as intermediate output. The output value deposits a row of grid's character.

In reduce function, the intermediate results are merged, sorted, and summed to output the final matrix. It takes as input a $<i, recordLine>$ pair. The output of the function is the same as the input. Because the MapReduce model has ranked the record lines following keys in map function, so reduce function only need to output the final data.

Table 1 Algorithm: VoronoiMapper (key, value)

Input: $<key, value>$ pair, where key is the line number, and value is the content of a row of original matrix.
Output: $<key', value'>$ pair, where key is the line number, and value is the content of a row of grid's character.

Map(Text Key , Text value){
4 Read the original dataset, where the point is expressed by a triple group-$p[i](i, j, value)$ with row, column and value.
5 Get the record file of boundary points, every point is represented by $obj[i](ID, i, j, w)$, which has four attributes: ID, row, column, and weight.
6 for i=0 to row number
7 **min_distance**= $d_\omega(p[m], obj[0])$;
8 for j=0 to column number
9 calculate the weight distance $d_\omega(p[i], obj[j])$;
10 if($d_\omega(p[i], obj[j])$<**min_distance**) then
11 do **min_distance**= $d_\omega(p[i], obj[j])$;
12 $p[i].value = obj[j].ID$;
13 record the row $obj[j].i$ and column $obj[j].j$;
14 else ($d_\omega(p[i], obj[j])$==**min_distance**)
15 then $p[i].value = obj[j'].ID$;($obj[j']$ is expressed as the generator which has smaller coordinate)
16 end if
17 end for
18 output $(key', value')$;}

Table 2 Algorithm: VoronoiReducer(key, value)

Input: $<key, value>$ pair, where key is the line number, and value is a string of the intermediate results.
Output: $<key', value'>$ pair, where key is a null string, and value is a string of the intermediate results.

Reduce(Writable key , Iterator<Text> value){
19 Initiate string key' as a null string.
20 output $(key', value')$;}

3.3 Analysis

In practice, the MapReduce implementation of the method runs on multiple machines in parallel. When discussing the method complexity, we consider the fact that it runs in parallel and discuss the parallel method complexity. In the MapReduce job, the computational complexity of the associated reduction is as follows:

$$(m \times n - g_k) \times g_k \times O/\text{nodes}$$

where $m \times n$ is the total number of grid cells, k is the number of generators, g_k is the total number of boundary points after edge points extraction of polygon, nodes is the number of the Hadoop nodes, and O is the computational complexity of weight distance computing.

4 Results and Discussion

In this section, we evaluate the performance impact of algorithm implementation and not its accuracy on our cluster system. The core idea of Apache Hadoop [19] is the MapReduce programming model. Our experimental hardware consists of nine nodes cluster: one namenode and eight datanode. Each node in the cluster is equipped with four quad-core 3.10 GHZ Intel Core(TM) i5-2400 processors, 4 GB of memory and 500 G of disk, runs Fedora15, and is connected with fast Ethernet. In this chapter, all experiments described are obtained using Hadoop version 0.20.2 and Java 1.7.0.04, while the data are stored with two replicas per block in HDFS.

4.1 Single Machine Environment Versus Hadoop Pseudo-Distributed Environment

When there are same data scales and same hardware configuration environments, we compare the time of generating WVDs under single machine environment and Hadoop pseudo-distributed environment. From the experimental results, the algorithm running in the single machine environment needs less time when the data scale is small. But when the scale of data increases to a certain extent, it reports out of memory and cannot complete calculation tasks, where the tasks can be treated successfully under Hadoop pseudo-distributed environment.

In our analysis, the control between nodes and task schedule take most part of resources when there is a small scale of data. So the time of calculation tasks is longer. When the scale of data increases, the single machine environment cannot meet the demand of computing because of many reasons such as the growth of memory resource consumption. However, the Hadoop platform can easily handle large datasets (Table 3).

4.2 Experiment Analysis for Cluster System

In the following experiments, we choose three gigabit-scale datasets as original datum: DS1, DS2, and DS3. The datasets are described in detail in Table 4.

Table 3 Contrast of execution time between single and parallel systems

Times	File size/MB	Record lines	Serial τ_1/s	Hadoop τ_2/s
1	2	1,750	1.716	22
2	6	5,000	4.805	26
3	24	20,000	18.548	43
4	45	37,450	Out of memory	61
5	90	75,000	Out of memory	73

Table 4 Experimental datasets

Datasets	Size of original data/ MB	Record lines/ 10^5	Polygons	Numbers of boundary points	Data blocks
DS1	1,116.4	9.3	3	5,786	18
DS2	2,233.6	18.6	3	5,786	35
DS3	4,468.4	37.2	3	5,786	70

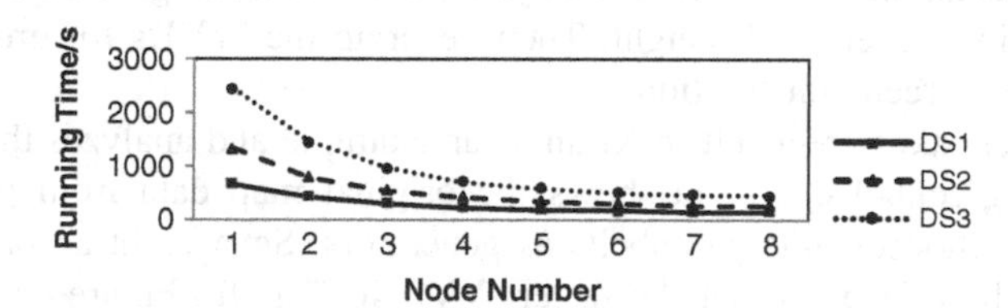

Fig. 3 Running time

In light of three datasets, kill the certain quantity of datanodes every time and the running time of each experiment is shown in Fig. 3. We can see that the time decreased with the growth of node numbers. Increasing the number of nodes can significantly improve the processing ability of the cluster when the data scale is the same. The running speed is similar to linear growth with the increase of data nodes. It shows that the speed of generating WVDs for polygons is increased markedly on Hadoop distributed environment.

5 Case Study

In the former section, we verify the performance efficiency of parallel algorithm. Now we make a trial to use our approach in some practical application. Because of the limit of experiment condition, we only choose some part of maps as original map data. Our original datasets are the raster data are obtained by map rasterization using ArcGIS software.

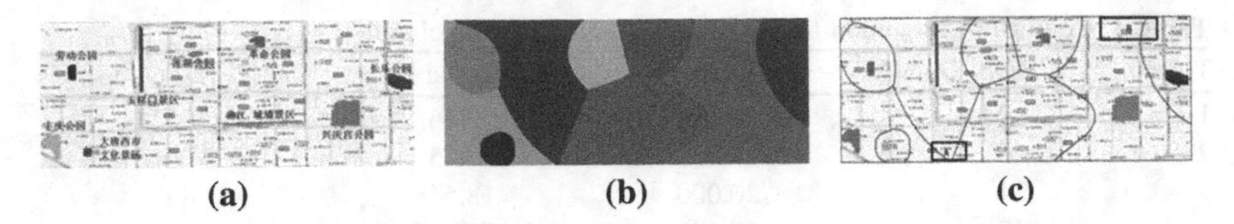

Fig. 4 WVDs for polygons of greenbelts

5.1 Case One: Green Space Planning

According to the green space planning in Xi'an city, the city will vigorously develop the green space system to become a national garden city. The city greenbelt element can appear as green space blocks in various shapes and size. Park, shelter-forest, and water conservation district are strip or area features. The affection of greenbelts on the environment depends on varied factors such as area, size, tree species, and purification ability. Now we abstract greenbelts into polygons and assign generators' weight. Then generate the WVDs for greenbelts for the decision of green construction.

Our chapter takes greenbelt in Xi'an as an example and analyzes the sphere of influence of greenbelts. Fig. 4a shows the original map data from part area of Xi'an city. It chooses nine greenbelts as generators. Set weight according to the size of greenbelts, we can get the raster WVD in Fig. 4b. Figure 4c is the final vector diagram. It marks off the sphere of influence for every greenbelt. Regions A and B are located at the interface of several greenbelts and in a position of the edge. They are weakly influenced by the greenbelts. So we suggest increasing greenbelt in these regions.

5.2 Case Two: Optimal Path Planning Problem

Given some obstacles in space or in a plane, the retraction method for motion planning uses the Voronoi diagram to determine whether there exists an optimal path from an initial posit onto a final position. If obstacle can be approximately regarded as particle, the safest path can follow Voronoi edges. If obstacle cannot be approximately expressed by particle, the expansion of Voronoi diagram can be employed (Generators are lines, polygons, or polyhedrons). On the other hand, for different obstacles, they may have different criticality. So their weights are different. For example, when obstacles are contaminated zones or danger, the path need to be away from them. So, their weights should be smaller than safe obstacles. Our approximate fast algorithm computes the WVDs for polygons to get the Voronoi edges, which are the final optimal path. Figure 5a gives distribution of some obstacles with different weights. Then, we will get the optimal path by following the Voronoi edges (Fig. 5b).

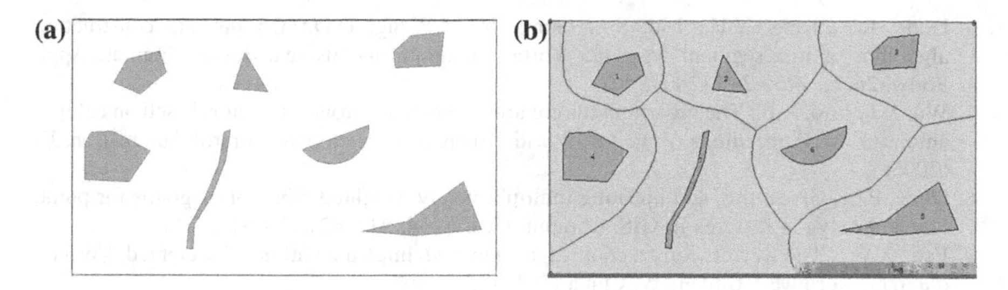

Fig. 5 Optimal path generation of obstacles

6 Conclusion and Future Work

The chapter presents a parallel approach for generating weighted raster-based Voronoi diagrams on Hadoop platform. The presented approach is based on traditional distance transformation and MapReduce model. Considering type and weight of generators, distance computation is simplified by extracting edge points of polygon. The experiments show that the approach significantly improves the performance with a linear scale-up in response to the increase in nodes. Application cases show that the approach is successfully applied in urban green space planning. Some further work can also be expected with the presented method: selecting reasonable weights for different cases and applying the algorithm to applications such as data collection in sensor networks, emergency modeling, and Voronoi-based geospatial query.

Acknowledgments The authors would like to thank the National Natural Science Foundation of China (No. 41271387), the tourism constructing study soft science project of Xian city (No. SF1228-3), and the academician innovation project of Shaanxi Normal University (No. 999521) for the support given to the study.

References

1. Lauro, C.G., Antonio, G.N., Novaes, J.E., Souza, D.C., João, C.S.: A multiplicatively-weighted Voronoi diagram approach to logistics districting. Comput. Oper. Res. **33**, 93–114 (2006)
2. Hu, P., You, L., Hu, H., et al.: Voronoi diagram. In: An Introduction to Map Algebra. Surveying and Mapping Press, China (2008)
3. Wang, X.S., Li, Q., Guo, Q.S., Wu, H.H., Fu, F.Y.: The generalization and construction of Voronoi diagram and its application on delimitating city's affected coverage. J Central China Normal Univ. (Nat. Sci.), **36**(1), 107–111 (2002)
4. Zhang, Y.H.: Voronoi diagram of weighted segments. Chin. J. Comput. **18**(11), 822–829 (1995)
5. Dong, R.: The discrete construction algorithm of line segment weighted Voronoi diagram. Hebei Normal University, Hebei (2006)

6. Dong, R., Zhang, Y.H., Liu, S.J., Gong, X.Y., Wang, D.D.: On discrete construction algorithm of line segment weighted Voronoi diagram and its realization. Comput. Appl. Softw. **26**(7), 245−247 (2009)
7. Wu, X.J., Luo, X.F.: The algorithm for creating weighted Voronoi diagrams based on cellular automata. In: Proceedings of the 6th World Congress on Intelligent Control Automation, EI (2006)
8. Dong, P.L.: Generating and updating multiplicatively weighted Voronoi diagrams for point, line and polygon features in GIS. Comput. Geosci. **34**, 411–421 (2008)
9. Fan, X.W.: The vector construction algorithm and implementation of weighted Voronoi diagram. Northwest University, China (2011)
10. Gong, Y.X., Li, G.C., Tian, Y., et al.: A vector-based algorithm to generate and update multiplicatively weighted Voronoi diagrams for points, polylines, and polygons. Comput. Geosci. **42**, 118–125 (2012)
11. Rong, G., Liu, Y., Wang, W., Yin, X., Gu, X., Guo, X.: GPU-assisted computation of centroidal Voronoi tessellation. IEEE Trans. Vis. Comput. Graph. **17**(3), 345–356 (2011)
12. Xu, Z.H., Kong, D.H., Xiao, X.F.: GPU-based weighted Voronoi diagram computing. J. Syst. Simul. (20), 29−32 (2008)
13. Afsin, A., Ugur, D., Farnoush, B.K., Cyrus, S.: Voronoi-based geospatial query processing with MapReduce. In: Proceedings of the IEEE Second International Conference on Cloud Computing Technology and Science, 9–16 (2010)
14. Kei, K., Kokichi, S.: Crystal Voronoi diagram and its applications. Future Gener. Comput. Syst. **18**, 681–692 (2002)
15. Barry, B., South, R.: Modeling retail trade areas using higher-order, multiplicatively weighted Voronoi diagrams. J. Retail. **73**(4), 519–536 (1997)
16. Atsuyuki, O., Atsuo, S.: Locational optimization problems solved through Voronoi diagrams. Eur. J. Oper. Res. **98**, 445–456 (1997)
17. Ickjai, L., Christopher, T.B., Kyungmi, L.: Map segmentation for geospatial data mining through generalized higher-order Voronoi diagrams with sequential scan algorithms. Expert Syst. Appl. **39**, 11135–11148 (2012)
18. He, Q., Shang, T.F., Zhuang, F.Z., Shi, Z.Z.: Parallel extreme learning machine for regression based on MapReduce. Neurocomputing **102**, 52–58 (2013)
19. Zhang, J.B., Li, T.R., Ruan, D., Gao, Z.Z., Zhao, C.B.: A parallel method for computing rough set approximations. Inf. Sci. **194**, 209–223 (2012)

Design of Evaluation System of Residents' Tourism Quality of Tourism Area Based on Fuzzy Evaluation

Ruiqiong Zhou, Song Lin and Mingcai Lin

Abstract Residents' tourism quality of tourism area has important influence on the local tourism development. The chapter established a set evaluation system based on the fuzzy evaluation to assess the quality of residents' Tourism. Firstly, the direct and indirect factors of tourism area residents' tourism are classified, and then the main factors and their relative weights according to the Delphi method are determined, and finally a systematic evaluation index system is formed. Key of this system is that can be sensitive to reflect the tourism system-state index by scientific and rational selection. In the process of determining the index value, objective and comprehensive considerations of various factors are used.

Keywords Tourism quality · Fuzzy evaluation · Index

1 Introduction

Tourism is a modern service industry and new industry in the world, all countries attach great importance to the development of the tourism industry. China's tourism industry has developed rapidly and has become the economic and social development to a new stage in key areas of new consumption of residents. According to the World Tourism Organization predicts, in 2020, China will become the world's largest tourist destination and the fourth largest exporter of tourists. As a major strategic plan, the state in January 4, 2010, the State Council issued the "opinions" of the State Council on accelerating the construction of Hainan international tourism island, Hainan International Tourism Island construction officially on the right track, countries in the future will initially build

R. Zhou · S. Lin (✉) · M. Lin
Center of Network, Hainan Normal University, Haikou 571158, China
e-mail: 3219957@qq.com

B.-Y. Cao et al. (eds.), *Ecosystem Assessment and Fuzzy Systems Management*, Advances in Intelligent Systems and Computing 254, DOI: 10.1007/978-3-319-03449-2_19,

Hainan as world-class resort island. It not only brings the opportunity to the development of Hainan, also on the tourism industry, people's demand for tourism in Hainan, the better development of tourism has brought new demands and challenges, one of which is as the host of Hainan residents should be how to become new citizen in tourist island, becomes an important issue.

The present research directions about tourism development is rapider, mostly about tourists satisfaction, and research on bearing capacity and environmental capacity, tourism ecological environment and conditions, leisure tourism, tourists are also common quality, but as a tourist and local residents, in many respects, its quality directly and indirectly affect the development of tourism, tourist satisfaction, but related research is rarely.

The tourism quality of local residents is important for sustainable and healthy development of tourism. This chapter attempts to establish a set of scientific evaluation system for the evaluation of tourism quality of residents in destination. Firstly, established a scientific evaluation index system, reflect the tourism in the development of abnormal phenomenon in operation, and feedback to the tourism warning subsystem. Key of this system is that can be sensitive to reflect the tourism system-state index selection of scientific and rational, to establish a set of index system. In the process of determining the index value, fair and objective and comprehensive rules had been considered in many factors. In addition, due to the rapid pace of economic development, to update the index value in a timely manner, to ensure its value. Fuzzy comprehensive evaluation method used in this chapter has been applied in many fields though, but there is little research in the evaluation of tourism quality of residents.

2 Related Concepts

2.1 Tourism Quality

Quality is the physiological concept, refers to the anatomical characteristics of the congenital physiological, mainly refers to the characteristics and sensory organs and organ motion nervous system, the brain and, in addition, also has been used to refer to people of innate and acquired through training, shaping, characteristics of exercise and physical and personality.

We believe that quality has a variety of meanings, including one's cultural level; healthy body; and the family genetic in their own inertia thinking ability and insight into the ability of things, management ability, and IQ, EQ level and occupation skill level attained embodied integrated. Man's quality includes natural quality, psychological quality, and social quality, the quality of people once form a relatively stable intrinsic characteristics.

As a tourist area residents, one is living in tourism area, need as a living master basic quality; faced with all kinds of tourists and its various behavior. On the other hand, they need to face with tourists in harmony quality; third is to adapt to the local new living environment, including in the more open environment, many shoppers environment, prices with more market-oriented environment, traffic environment, and the stranger growing environment.

American psychologist McClelland in 1973 put forward the Iceberg Model of Competence, in his model quality has 21 general quality elements, these 21 qualities which are divided into six specific quality group, the following specific quality six quality group and the: (1) Management group, including teamwork, training, monitoring ability, leadership ability; (2) Cognitive group, including deductive thinking, inductive thinking, professional knowledge and skills; the self-concept, including confidence; (3) Self-concept group, including confidence; (4) The influence group, including influence, relationship building; (5) Target and action group, including the achievement orientation, initiative, information collection; (6) Help and services group, including interpersonal understanding, customer service. From the tourist area residents, last five groups' qualities are more important, especially the help and service family. This not only has the basic characteristics of individuals, such as personal and cultural level, personal morality, behavior civilization, law-abiding degree, and constitute the whole social quality local part, such as the formation of social security, clean, civilized etiquette environment quality, such as the quality of the cultural quality, knowledge of history, care for the environment, love hometown, hospitality, and other social. The present study investigated from the point of view, the cultural quality of the residents, moral quality, aesthetic quality, social quality four aspects. The establishment of population quality and its evaluation model of tourism area is helpful for us to analyze the Hainan as the ideal level of tourism island, differences in quality status and good level of residents, to better guide the residential quality training and improve the relevant work, from the residents quality promote the tourism island become tourists yearning, in Hainan happy travel.

2.2 Introduction to the Fuzzy Evaluation Method

Fuzzy evaluation method is a kind of important method to establish the system of scientific index system, widely used in the field of social research. According to the principle of fuzzy evaluation, element has the following several aspects:

1. Target. Evaluation of the system goal is to guide and the purpose of the whole system, it is subordinate to and in the service of the goals of the organization. In this study, need to understand the residents of the whole tourism quality, the formation of a quantitative index.
2. Object. Evaluation object is the tourism destination resident, who is the host.
3. Index. Evaluation index refers to the characteristics and behavior of the residents of tourism areas, the formation of a structured system. Concerned about

the evaluation system is the relationship between local residents and to establish a good tourism atmosphere.

4. Standard. A criterion level of quality classification of tourism.
5. The analysis results. Evaluated according to certain standards draw on the tourism destination residents' qualities conclusion, formated report of evaluation.

Below, we discuss fuzzy evaluation method. Set $U = \{u_1, u_2,\ldots, u_n\}$ is the evaluation index set, u_i is the index such as the cultural quality, and $V = \{v_1, v_2,\ldots, v_m\}$ is the evaluation grade set, such as five grades in this article with the set {very satisfied, satisfied, indifferent, not satisfied, not very satisfied}. Firstly, evaluation index set u_i $(i = 1, 2,\ldots, n)$ single index in U, Single index evaluation in the assessment grade v_j $(j = 1, 2,\ldots, m)$, get the $r_i = (r_{i1}, r_{i2},\ldots, r_{im})$ value evaluation index. Subject to $\sum_{j=1}^{m} r_{ij} = 1$ evaluation of n index value consists of a fuzzy evaluation matrix $R(r_{ij})_{nm}$. Where r_{ij} is the membership degree evaluation index u_i about the evaluation grade of v_j.

Multiple index evaluation also needs a evaluation index importance (weights) vector. Set the weights to $W = \{w_1, w_2,\ldots, w_m\}$ and w_i is the important degree evaluation index u_i. In this chapter, Delphi method is used to determine the evaluation index importance vector W.

To make accurate evaluation on the tourism area residents, evaluation tools (mathematical model) is the choice of a critical step. The evaluation index and some are "clearly state either this or that," there is also a lot of "also" transition state (or fuzzy state), has the character of gradual change, and the relationship between the index attribute most of the nonlinear relationship, the relationship between the various indexes cannot generally be used to accurately describe the quantitative relationship. Plus the fuzziness of subjective arbitrariness and on understanding, with the absolute (or explicit) comments to evaluation is evaluation object, the evaluation result is not accurate. Fuzzy evaluation model can quantitatively assess the clarity on these indexes. The multi-level fuzzy comprehensive evaluation is used in this chapter, the mathematical models are as follows:

U, V, R, W meaning is as described above, and to establish membership function, derived membership degree. Establish fuzzy mapping $f_i{:}u_i \rightarrow F(V)$, The fuzzy evaluation matrix R_i from the membership of factor set to get. Taking (u_i, V, R_i) as the original model, weighting factors are given in u_i, $w_i = \{w_{i1}, w_{i2}\ldots, w_{ini}\}$, (the weight of each factor and to 1). The comprehensive evaluation results can be obtained as follows:

$$S_i = w_i \odot R_i \in F(V)(i = 1, 2, \ldots, n). \quad (1)$$

Normalization, S_i labeled $\check{S}$. Consider the second layer factors set $U = \{u_1, u_2,\ldots, u_n\}$, taking $\check{S}$ as the single factor evaluation factors of u_i. The establishment of fuzzy mapping: $f{:}U \rightarrow F(V)$, $u_i \rightarrow F(u_i) = S_i'$. Get two layer evaluation matrix $R = (S_1', S_2', \Lambda, S_n')^T$. Taking (U, V, R) as the original model, weighting factors are given in U, $W = \{w_1, w_2, \ldots, w_m\}$, (the weight of each factor and to 1). The comprehensive evaluation results can be obtained also as follows:

$$S = W \odot R \in F(V) \tag{2}$$

More layers of evaluation and so on can obtain results evaluation. The results of comprehensive evaluation in general normalized, normalized evaluation, according to the principle of maximum degree of membership reached a final evaluation results.

In this model, comprehensive evaluation of various factors on the first minimum level, layer upon layer sequentially to review, the top until, and get the final evaluation results, namely, fuzzy comprehensive evaluation is the level to start by the end of the evaluation.

Talk about fuzzy operator selection. The "$\odot$" in formula (1), (2) is referred to as the fuzzy operator symbols. There are six fuzzy operators often be used. Choosing a different operator will form different evaluation model. Membership in order to highlight the main factors and the single factor evaluation, to retain all the information of single factor evaluation, to obtain a more accurate evaluation results, this chapter adopts $M(\cdot, \oplus)$ operator.

3 Determining the Factors and Weights

3.1 Determining the Evaluation Index

This chapter considers that there are four first-level index of tourism quality of tourism area, such as cultural quality, moral quality, aesthetic quality, social quality, covers the basic characteristics of personal quality, good moral character and aesthetic ability, interpersonal, and social integration. Indicators of two levels and connotation of corresponding are as follows (Table 1).

3.2 Determining the Weights of Factors

Here, this chapter adopts Delphi determined in consultation to determine the weight distribution of the factors. Delphi consulting method had used the knowledge and experience of experts, but also reduces the effect of authority. Took the anonymous way to seek expert advice, repeated feedback correction, finally get the comprehensive views of experts. Set $U = \{u_1, u_2, \ldots, u_n\}$ as the evaluation index set. The index weight vector is set to $W = \{w_1, w_2, \ldots, w_n\}$. The basic steps of Delphi consulting as follows:

Table 1 Residents' tourism quality classification and connotation of tourism area

First-level index	Second-level index and its connotation
Cultural quality	Level of education: including the level of education
	Understanding of the place of residence of the natural environment: including landscape, transportation, and other knowledge
	Understanding of the humanistic environment of residential area: including local customs, language, cultural habits, understanding
Moral quality	Sincere quality: including the sense of responsibility
	Integrity, positive attitude: including the enthusiasm, humility, tolerance, patience, initiative
	Civilized habits: including the etiquette of civilization, hygiene, care for the environment
Aesthetic quality	Ability to find the good things: including curiosity, observation ability
	Ability of beauty appreciation: including the ideal, appreciation
Social quality	Open field of vision: including the understanding of others, understanding of foreign cultural
	Ability to communicate with visitors: care about others, including appropriate communication mode
	The ability to work well with others: including the mutual adaptation, support
	The ability to respond to emergencies: including the rational, self-control, in accordance with the law

Table 2 Evaluation index importance ranking K member of the value evaluation

Evaluation index	u_1	u_2	…	u_n
According to the order of importance value	H_{1-k}	H_{2-k}	…	H_{n-k}

1. Organization of the expert consultation table. Let the experts to determine the importance of each evaluation index ranking u_i $(i = 1, 2,\ldots, n)$ value. Every judgement will had the importance of each index provides a sort of value H_i where $H_1, H_2,\ldots, H_n$ is a permutation of $1, 2,\ldots, n$. In each evaluation that the u_i index is the most important, the H_i value is 1; u_j index the least important, its H_j value is n. The K evaluation members evaluation index importance sequence index u_i given values as H_{i-k}. (see Table 2).
2. Generation of priority score matrix.

The index of importance K set up the first evaluation members offer value priority score matrix generated by H_{i-k} is

$$W(K) = \begin{pmatrix} W_{11}^{(K)} & W_{12}^{(K)} & \cdots & W_{1n}^{(K)} \\ W_{21}^{(K)} & W_{22}^{(K)} & \cdots & W_{2n}^{(K)} \\ \cdots & & & \\ W_{n1}^{(K)} & W_{n^2}^{(K)} & \cdots & W_{nn}^{(K)} \end{pmatrix}$$

where

$$W_{ij}^{(K)} = \begin{cases} 1, & \text{when } H_{i-k} < H_{j-k} \\ 0, & \text{other} \end{cases}$$

e members to participate in the evaluation of the total *m* of human. Accumulated all participate in the deliberations of the member's value, i.e., $W_{ij} \sum_{k=1}^{m} W_{ij}^{(K)} (i = 1, 2,\ldots, n, j = 1, 2,\ldots, n)$. The following priority score matrix was organized.

$$W = \begin{bmatrix} W_{11} & W_{12} & \cdots & W_{1n} \\ W_{21} & W_{22} & \cdots & W_{2n} \\ & \cdots & & \\ W_{n1} & W_{n2} & \cdots & W_{nn} \end{bmatrix}$$

3. Set $S_i = \sum_{j=1}^{n} W_{ij}$, $S_{\max} = \max\{S_1, S_2,\ldots, S_n\}$, $S_{\min} = \min\{S_1, S_2,\ldots, S_n\}$, obviously with the $S_{\max}$ corresponding to the highest degree of evaluation index, evaluation index and corresponding to the lowest degree of important $S_{\min}$.
4. Calculation of differential d. Take $a_{\max}$, $a_{\min} \in [0, 1]$ and $a_{\max} > a_{\min}$ For example, $a_{\max} = 1$, $a_{\min} = 0.1$, So

$$\mathrm{d} = (S_{\max} - S_{\min})/(a_{\max} - a_{\min})$$

5. Calculation of evaluation index importance coefficient w_i.

$$\text{Set } c_i = (S_i - S_{\min})/d + 0.1\ (i = 1, 2,\ldots, n)$$

So obtained vector $C = (c_1, c_2,\ldots, c_m)$. The *C* normalization, Finally get the evaluation index set importance vector: $W = \{w_1, w_2,\ldots, w_n\}$.

The weight of 4 first class indexes and two-level index is shown in Table 3.

3.3 Calculation of the Membership Degree of Evaluation Index

Determination of membership function properly is the key of fuzzy comprehensive evaluation; it can get a grade evaluation accurate membership. Because the level of evaluation index system in this chapter is qualitative indicators, can not use the membership functions to describe the change rule, it therefore, in order to get more accurate and objective result, we compute an evaluation index membership degree is obtained by comprehensive scoring by experts.

According to the classification and evaluation of the index set $V = \{v_1, v_2,\ldots, v_m\}$, such as five grades in this article with the set {very satisfied, satisfied, indifferent, not satisfied, not very satisfied with the}. The expert group for each factor score in the five grade evaluation set, The five levels of each factor score and 1. Score for each

Table 3 Residents' tourism quality evaluation index system of tourism area

First-level index	Index weight	Second-level factors	Index weight
Cultural quality	0.4	Level of education	0.2
		Understanding of the place of residence of the natural environment	0.4
		Understanding of the humanistic environment of residential area	0.4
Moral quality	0.2	Sincere quality	0.4
		Integrity, positive attitude	0.3
		Civilized habits	0.3
Aesthetic quality	0.1	Ability to find the good things	0.5
		The ability of beauty appreciation	0.5
Social quality	0.3	Open field of vision	0.2
		Ability to communicate with visitors	0.3
		The ability to work well with others	0.3
		The ability to respond to emergencies	0.2

factor level geometry and the average value of the expert group evaluation. So, the evaluation grade of each factor of geometric average membership is composed of the factors. For example $(8, 10) \in v_1$, $(6, 8] \in v_2$, $(4, 6] \in v_3$, $(2, 4] \in v_4$, $(0, 2] \in v_5$.

4 Conclusion

The healthy development of tourism destination resident's quality affects directly the tourist experience and tourism. This paper probes into the residents' tourism quality problems in tourism, cultural quality, moral quality, aesthetic quality, social quality four aspects and 13 s level index to establish a set of evaluation index system. Fuzzy evaluation method is used in the evaluation process. The evaluation index weights are determined by Delphi method. Through the establishment of the corresponding evaluation index system, it can help to promote tourism to improve the work, enhance the quality of tourism.

Acknowledgments This work is supported by International Science and Technology Cooperation Program of China (2012DFA11270); Hainan International Cooperation Key Project (GJXM201105). And this work is also supported by Hainan Social Development of Science and Technology Projects (SF201329), National Natural Science Foundation of Hainan (No. 612124, 612126), and Major Scientific Projects of Haikou (No. 2012-028).

References

1. Lian, Y., Wang, X.: Research and application of customer satisfaction evaluation. Tourism Tribune, **5**, 006 (2004)
2. Dong, G., Yang, F.: Study on the evaluation system of tourist satisfaction tourism scenic spot. Tourism Tribune **1**, 27–30 (2005)
3. Wang Q.: Geographical research model of tourist satisfaction index about tourism environment. Geogr. Res. **1** (2006)
4. Zhang, X., Zhu, Z.: Study on the tourism environmental capacity of Jiuzhaigou scenic spot. Tourism Tribune **9** (2007)
5. Li, Z., Bu, X., Wang, B.: Study on ecological carrying capacity of tourism environment. J. Hunan Univ. (Soc. Sci.) **4** (2008)
6. Liu, T.: Study on the features of leisure Yunnan folk tourism. Acad. Explor. **1** (2009)
7. Rao, H.: Xu Xiake and the National Tourism quality research. J. Central South Univ. Forest. Technol. (Soc. Sci.) **5** (2012)
8. Yang, L.: Principle and Application of Fuzzy Mathematics. South China University of Technology Press, Guangzhou (2000)

Qualitative Evaluation of Software Reliability Considering Many Uncertain Factors

Peng Cao, Guo-chun Tang, Yu Zhang and Zi-qiang Luo

Abstract There are many uncertainty factors affecting software reliability in software development. This chapter proposed a modified Delphi hierarchy process based on cloud model to determine the weights of the factors affecting software reliability, and then, we can obtain software reliability qualitative rules using standard weighted association rule mining algorithm. Thus, software reliability qualitative evaluation can be achieved through uncertainty reasoning based on cloud model.

Keywords Software reliability · Uncertainty · Cloud model · Qualitative evaluation

1 Introduction

Traditional software reliability models are often founded on failure data from the software testing or the actual running phase, ignoring the influence of various uncertainty factors in the software development process. If there is very few data

P. Cao
Department of Mathematics and Physics, Qiongtai Teachers College, Haikou, China
e-mail: lanyuan97@163.com

G. Tang
Department of Information Technology, Qiongtai Teachers College, Haikou, China
e-mail: tangguochun@163.com

Y. Zhang · Z. Luo (✉)
College of Information Science and Technology, Hainan Normal University, Haikou 571158, China
e-mail: luo_letian@163.com

Y. Zhang
e-mail: 344248003@qq.com

B.-Y. Cao et al. (eds.), *Ecosystem Assessment and Fuzzy Systems Management*, Advances in Intelligent Systems and Computing 254, DOI: 10.1007/978-3-319-03449-2_20,

or no information related to software failure, we can indirectly evaluate software reliability by means of investigating some uncertain factors which determine the level of software reliability in the software development process. Abundance of uncertain information for software reliability engineering requires a formal symbol system to portray. Concepts and the overall characteristics of things are perceived through language, not by precise calculation of some values. There is a very good relationship between the uncertainty inherent in natural language and fuzzy prevalent in the objective world. Cloud model [1] can be used to describe language atoms and then just to reflect this relationship.

For comprehensive analysis of the relationship between several uncertain factors affecting software reliability and software reliability level, we will first determine the weighs of the factors, and then, we can obtain software reliability qualitative rules using standard weighted association rule mining algorithm [2]. Thus, software reliability qualitative evaluation can be achieved through uncertainty reasoning based on cloud model.

2 Uncertain Factors Affecting Software Reliability

Software development is a very complex dynamic process including people, development tools, and application background. Researchers [3, 4] realized that the implied useful knowledge, which can be dug out from the various factors affecting the software reliability, software application features, and software failure data in the software development process, may be used for qualitative and quantitative software reliability assessment process. Thus, more credible and accurate software evaluation results can be obtained. Because software reliability is subjected to common effect of numerous uncertainties factors, we need to analyze comprehensively such factors to assess. The degree of the factors impacting on the software reliability need to be determined by means of engineering experience and subjective judgment, inevitably bearing subjectivity and uncertainty. Cloud model theory is suitable for processing the information with uncertain attributes.

According to the survey of existing research and analysis, there are 32 uncertainty factors affecting software reliability during the entire software development process. These factors constitute a qualitative evaluation index system for software reliability. Here mainly introduce the six most important factors.

1. Program Complexity, abbreviated as PC: the more complex the software is, in the development and maintenance process, the more resources are consumed, and the greater possibility of errors are introduced in the design. The program size, function modules themselves, and the connections between modules are the main factors determining software complexity.
2. Programmer Skills, abbreviated as PS: programmer skills mainly depend on the programmer's level, number of developed software products, the scale of the

developed software, and the experience of developing software. In general, the programmer skills directly affect software reliability. Due to personal programming experience, work stress, and personal qualities, programmers often make errors in the program modules.

3. Testing Efforts, abbreviated as TEF: the number of test cases, test workload, and test time reflect the testing efforts.
4. Testing Coverage, abbreviated as TC: the proportion of source code covered by test cases.
5. Testing Environment, abbreviated as TE: in order to find more software bugs in the testing phase, testing environment should be as close as possible to the actual operating environment.
6. Frequency of Specification Change, abbreviated as FSC: because users have new requirements, or developers do not properly understand the user's needs, specification needs to add or correct during the period of software development. This will certainly affect the subsequent software development. The more the number of changes or the later change occurs, the greater the impact on reliability.

Because the factors affecting software reliability, which are taken from the software development process at all stages, are closely related to software reliability levels, can truly reflect the software reliability from different angles. But various factors have primary and secondary points, so you need to assigning weights to highlight the main factors.

3 Delphi Hierarchy Process Based on Cloud Model

Delphi hierarchy process based on cloud model [5, 6], DHP for short, synthesizing a number of experts' opinion on many factors affecting software reliability, cleaning up and selecting factors, and assigning weights can implement visualized heuristic optimization *n* of the complex data objects.

By using DHP method, we can select factors affecting software reliability and calculate initial weights with Delphi method, and then, using the analytic hierarchy process, known as AHP, we can obtain hierarchical factors and determine the final weights.

DHP is briefly introduced as follows: First, the issues to be assessed and the necessary background material in the form of communication may be sent to experts, and then, the replies should be synthesized and concluded based on cloud model. Once again back to the experts for further advice, so repeatedly 2–4 times, until we get satisfactory results. Second, the relative weights of each level can be calculated, and the combined weights of the elements can also be calculated. Finally, the total weight can be calculated.

4 The Mining Approach of Software Reliability Qualitative Rules

4.1 Establishment and Transformation of Related Databases

For some factors affecting software reliability, we cannot get the real data, but with qualitative description language, such as "programmer skills are ordinary" and "testing method is good". These language values can be represented appropriately by the cloud model in the domain of [0, 1], which can be converted to a numeric representation through a forward cloud generator. Thus, by some typical case studies, we can get numeric databases about factors and software reliability, as shown in Table 1.

Obviously, using raw numeric data for data mining, it will be hard to find useful rules. Therefore, firstly, adopt cloud transformation method for discretization of continuous attributes, raise the basic concepts to the appropriate concept hierarchy, and then make a soft partition to the original data set by maximum determination principle according to the gained concept sets. The six most important factors using qualitative language values may be constructed with evaluation concept sets including five grades, such as follows:

PC {very low, low, medium, high, very high};
PS {very poor, poor, average, good, very good};
TEF {very few, few, average, many, a great many};
TC {very small, small, medium, large, very large};
TE {not very close, not close, medium, close, very close};
FSC {very few, few, average, many, a great many}.

Other factors affecting software reliability can be similarly constructed. Of course, we can get more coarse or fine division according to the specific situation.

Table 2 shows the soft partition for original attribute values based on concept sets. Because of the uncertainties of soft partition based on cloud model, the percentage of the value attribute table is slightly different in different times, and then, we can calculate the mean result for multiple soft partitions. Clearly, the amount of data is substantially reduced, and the intrinsic correlation between attributes is also highlighted after the soft partition. According to the requirement of the association rule mining algorithm, the category type attribute Table 2 can be converted to a Boolean database as shown in Table 3.

Table 1 Databases about factors and software reliability

PC	PS	TEF	TC	TE	FSC	…	Software reliability
0.01	0.96	0.98	0.99	0.999	0.02	…	0.99
0.2	0.99	0.97	0.98	0.997	0.03	…	0.78
…	…	…	…	…	…	…	…

Table 2 Soft partition for the attribute values table

PC	PS	TEF	TC	TE	FSC	…	Software reliability	Count/ %
Very low	Very good	A great many	Very large	Very close	Very few	…	Very high	4
Low	Very good	Many	Large	Close	Very few	…	High	9
…	…	…	…	…	…	…	…	…

Table 3 Boolean database from the converted category type attribute Table 2

PC	Very low	Low	Medium	High	Very high
	1	0	0	0	0
	0	1	0	0	0
	…	…	…	…	…
PS	Very poor	Poor	Average	Good	Very good
	0	0	0	0	1
	0	0	0	0	1
	…	…	…	…	…
Software reliability	Very low	Low	Medium	High	Very high
	0	0	0	0	1
	0	0	0	1	0
	…	…	…	…	…

4.2 Standard Weighted Association Rules Mining Based on Cloud Model

When the weighs of the factors affecting software reliability are gained, for a given level of soft support threshold and soft confidence threshold, we can get a standard weighted association rules list with "software reliability" as rule consequent using standard weighted association rules algorithm mining based on cloud model.

If the conjunction of "PC," "PS," and "TEF" is the rule antecedent, "software reliability" as the rule consequent, the following six qualitative rules are assumed to be produced on the above conceptual granularity:

If PC is "very low," PS is "very good," and TEF is "a great many," then software reliability is "very high";

If PC is "low," PS is "very good," and TEF is "a great many," then software reliability is "high";

If PC is "medium," PS is "very good," and TEF is "a great many," then software reliability is "high";

If PC is "medium," PS is "good," and TEF is "many," then software reliability is "medium";

If PC is "high," PS is "average," and TEF is "a great many," then software reliability is "low";

If PC is “very high,” PS is “average,” and TEF is “many,” then software reliability is “very low.”

These rules reflect the relationships between qualitative concepts such as software complexity, programmer skills, testing efforts, and software reliability (also available for other software reliability indexes).

We might as well set [0, 1] as the three factors’ domain U, cloud representations of the given five qualitative concepts in the domain of software complexity are as follows:

$$C_{A_{11}} = C(0,\ 0.08,\ 0.0008),\ C_{A_{12}} = C(0.25,\ 0.08,\ 0.0008),$$
$$C_{A_{13}} = C(0.5,\ 0.08,\ 0.0008),\ C_{A_{14}} = C(0.75,\ 0.08,\ 0.0008),$$
$$C_{A_{15}} = C(1,\ 0.08,\ 0.0008).$$

Cloud representations of the given five qualitative concepts in the domain of PS are as follows:

$$C_{A_{21}} = \begin{cases} 1 & x \in [0, 0.6] \\ C(0.6,\ 0.03,\ 0.0003) & \text{else} \end{cases},$$
$$C_{A_{22}} = C(0.7,\ 0.03,\ 0.0003),\ C_{A_{23}} = C(0.8,\ 0.03,\ 0.0003),$$
$$C_{A_{24}} = C(0.9,\ 0.03,\ 0.0003),\ C_{A_{25}} = C(1,\ 0.03,\ 0.0003).$$

Cloud representations of the given five qualitative concepts in the domain of TEF are as follows:

$$C_{A_{31}} = \begin{cases} 1 & x \in [0, 0.9] \\ C(0.9,\ 0.03,\ 0.0003) & \text{else} \end{cases},$$
$$C_{A_{32}} = C(0.99,\ 0.003,\ 3 \times 10^{-5}),\ C_{A_{33}} = C(0.999,\ 3 \times 10^{-4},\ 3 \times 10^{-6}),$$
$$C_{A_{34}} = C(0.9999,\ 3 \times 10^{-5},\ 3 \times 10^{-7}),\ C_{A_{35}} = C(1,\ 3 \times 10^{-5},\ 3 \times 10^{-7}),$$

where A_{11}, A_{21}, and A_{31} are descending half-clouds, and A_{15}, A_{25}, and A_{35} are ascending half-clouds.

5 Software Reliability Qualitative Evaluation

The summary steps of software reliability qualitative evaluation method are as follows:

1. clean up, select, and assign weights for the 32 factors affecting software reliability, using DHP method based on cloud model;
2. generate some association rules about the factors and software reliability, using standard weighted association rule mining method based on cloud model;

3. give a new determined input, activate the corresponding rules based on these qualitative rules, and then generate uncertainty output from the rule generator based on cloud model;

Based on multi-conditions and multi-rules reasoning mechanism, firstly, activate these six qualitative rules using the input values through quantitative analysis; secondly, synthesize the above generated clouds using geometric cloud technology; and finally, the expected value Ex of the generated geometric cloud C(Ex, En, He) can be set as the quantitative assessment of software reliability. Reference to the software reliability evaluation conception set, the quantitative evaluation of the software cloud reliability is actually a virtual cloud in software reliability domain [0, +1]. With the input of expectation Ex, X condition cloud generators may individually output the certain degrees corresponding to the different evaluation concepts. Finally, according to maximum determination principle for membership concept may judge the software reliability.

Acknowledgments This work is supported by Grants of Higher School Scientific Research Project of Hainan Province (Nos. Hjkj2012-14, Hjkj2013-17), National Natural Science Foundation of Hainan Province (Nos. 613161, 613162), International Science & Technology Cooperation Program of China (No. 2012DFA11270), Hainan International Cooperation Key Project (No. GJXM201105), and Qiongtai Teachers College Scientific Research Project (No. qtky 201111).

References

1. Li, D., Du, Yi.: Artificial Intelligence with Uncertainty. National Defense Industry Press, Beijing (2005)
2. Du, Y.: Research and application of association rules in data mining. PhD Thesis for PLA University of Science and Technology, Nanjing (2000)
3. Zhang, X., Hoang, P.: An analysis of factors affecting software reliability. J. Syst. Softw. **50**, 43–56 (2000)
4. Wang, T., Li, M.: A fuzzy comprehensive evaluation model for software reliability. Comput. Eng. Appl. **38**(20), 23–26 (2002)
5. Wang, S.: Data field and cloud model based spatial data mining and knowledge discovery. PhD Thesis for Wuhan University, Wuhan (2002)
6. Li, D., Wang, S., Li, D.: Spatial Data Mining Theories and Applications. Science Press, Beijing (2006)

Quantum-Behaved Particle Swarm Optimization with Diversity-Maintained

Hai-xia Long and Shu-lei Wu

Abstract Quantum-behaved particle swarm optimization (QPSO) algorithm is a global-convergence-guaranteed algorithm, which outperforms original PSO in search ability but has fewer parameters to control. But QPSO algorithm is to be easily trapped into local optima as a result of the rapid decline in diversity. So this paper describes diversity-maintained into QPSO (QPSO-DM) to enhance the diversity of particle swarm and then improve the search ability of QPSO. The experiment results on benchmark functions show that QPSO-DM has stronger global search ability than QPSO and standard PSO.

Keywords Diversity · Quantum-behaved particle swarm optimization · Diversity-maintained · Benchmark function

1 Introduction

Particle swarm optimization (PSO) is a kind of stochastic optimization algorithms proposed by Kennedy and Eberhart [1] that can be easily implemented and is computationally inexpensive. The core of PSO is based on an analogy of the social behavior of flocks of birds when they search for food. PSO has been proved to be an efficient approach for many continuous global optimization problems. However, as demonstrated by Van Den Bergh [2], PSO is not a global-convergence-guaranteed algorithm because the particle is restricted to a finite sampling space for each of the iterations. This restriction weakens the global search ability of the algorithm and may lead to premature convergence in many cases.

H. Long (✉) · S. Wu
Department of Computer Science and Technology, Hainan Normal University,
Haikou 571158, China
e-mail: 64169486@qq.com

B.-Y. Cao et al. (eds.), *Ecosystem Assessment and Fuzzy Systems Management*,
Advances in Intelligent Systems and Computing 254, DOI: 10.1007/978-3-319-03449-2_21,

Several authors developed strategies to improve on PSO. Clerc [3] suggested a PSO variant in which the velocity to the best point found by the swarm is replaced by the velocity to the current best point of the swarm, although he does not test this variant. Clerc [4] and Zhang et al. [5] dynamically change the size of the swarm according to the performance of the algorithm. Eberhart and Shi [6], He et al. [7] adopted strategies based on dynamically modifying the value of the PSO parameter called *inertia weight*. Various other solutions have been proposed for preventing premature convergence: objective functions that change over time [8]; noisy evaluation of the function objective [9]; repulsion to keep particles away from the optimum [10]; dispersion between particles that are too close to one another [11]; reduction in the attraction of the swarm center to prevent the particles clustering too tightly in one region of the search space [12]; hybrids with other metaheuristic such as genetic algorithms [13]; ant colony optimization [14], etc. An up-to-date overview of the PSO is introduced in [15].

Recently, a new variant of PSO, called quantum-behaved particle swarm optimization (QPSO) [16, 17], which is inspired by quantum mechanics and particle swarm optimization model. QPSO has only the position vector without velocity, so it is simpler than standard particle swarm optimization algorithm. Furthermore, several benchmark test functions show that QPSO performs better than standard particle swarm optimization algorithm. Although the QPSO algorithm is a promising algorithm for the optimization problems, like other evolutionary algorithm, QPSO also confronts the problem of premature convergence and decreases the diversity in the latter period of the search. Therefore, a lot of revised QPSO algorithms have been proposed since the QPSO had emerged. In Sun et al. [18], the mechanism of probability distribution was proposed to make the swarm more efficient in global search. Simulated annealing is further adopted to effectively employ both the ability to jump out of the local minima in simulated annealing and the capability of searching the global optimum in QPSO algorithm [19]. Mutation operator with Gaussian probability distribution was introduced to enhance the performance of QPSO in Coelho [20]. Immune operator based on the immune memory and vaccination was introduced into QPSO to increase the convergent speed using the characteristic of the problem to guide the search process [21].

In this chapter, QPSO with diversity-maintained (QPSO-DM) is introduced. This strategy is to prevent the diversity of particle swarm declining in the search of later stage.

The rest of the chapter is organized as follows. In Sect. 2, the principle of the PSO is introduced. The concept of QPSO is presented in Sect. 3, and the QPSO with diversity-maintained is proposed in Sect. 4. Section 5 gives the numerical results on some benchmark functions and discussion. Some concluding remarks and future work are presented in the last section.

2 PSO Algorithm

In the original PSO with M individuals, each individual is treated as an infinitesimal particle in the D-dimensional space, with the position vector and velocity vector of particle i, $X_i(t) = (X_{i1}(t), X_{i2}(t), \ldots, X_{iD}(t))$, and $V_i(t) = (V_{i1}(t), V_{i2}(t), \ldots, V_{iD}(t))$. The particle moves according to the following equations:

$$V_{ij}(t+1) = V_{ij}(t) + c_1 \cdot r_1 \cdot (P_{ij}(t) - X_{ij}(t)) + c_2 \cdot r_2 \cdot (P_{gj}(t) - X_{ij}(t)) \quad (1)$$

$$X_{ij}(t+1) = X_{ij}(t) + V_{ij}(t+1) \quad (2)$$

for $i = 1, 2, \ldots M; j = 1, 2 \ldots, D$. The parameters c_1 and c_2 are called the acceleration coefficients. Vector $P_i = (P_{i1}, P_{i2}, \ldots, P_{iD})$ known as the *personal best position* is the best previous position (the position giving the best fitness value so far) of particle i; vector $P_g = (P_{g1}, P_{g2}, \ldots, P_{gD})$ is the position of the best particle among all the particles and is known as the *global best position*. The parameters r_1 and r_2 are two random numbers distributed uniformly in (0, 1), that is, $r_1, r_2 \sim U(0, 1)$. Generally, the value of V_{ij} is restricted in the interval $[-V_{\max}, V_{\max}]$.

Many revised versions of PSO algorithm are proposed to improve the performance since its origin in 1995. Two most important improvements are the version with an Inertia Weight [22] and a Constriction Factor [23]. In the inertia-weighted PSO, the velocity is updated by using

$$V_{ij}(t+1) = w \cdot V_{ij}(t) + c_1 \cdot r_1(P_{ij}(t) - X_{ij}(t)) + c_2 \cdot r_2 \cdot (P_{gj} - X_{ij}(t)) \quad (3)$$

while in the Constriction Factor model, the velocity is calculated by using

$$V_{ij}(t+1) = K \cdot [V_{ij}(t) + c_1 \cdot r_2 \cdot (P_{ij}(t) - X_{ij}(t)) + c_2 \cdot r_2 \cdot (P_{gj} - X_{ij}(t))] \quad (4)$$

where

$$k = \frac{2}{\left|2 - \varphi - \sqrt{\varphi^2 - 4\phi}\right|} \qquad \varphi = c_1 + c_2, \quad \varphi > 4 \quad (5)$$

The inertia-weighted PSO was introduced by Shi and Eberhart [6] and is known as the standard PSO.

3 QPSO Algorithm

Trajectory analyses in Clerc and Kennedy [24] demonstrated the fact that convergence of PSO algorithm may be achieved if each particle converges to its local attractor $p_i = (p_{i1}, p_{i2}, \ldots p_{iD})$ with coordinates

$$p_{ij}(t) = (c_1 r_1 P_{ij}(t) + c_2 r_2 P_{gj}(t))/(c_1 r_1 + c_2 r_2), \text{or} \, p_{ij}(t) \\ = \varphi \cdot P_{ij}(t) + (1 - \varphi) \cdot P_{gj}(t) \quad (6)$$

where $\varphi = c_1 r_1/(c_1 r_1 + c_2 r_2)$. It can be seen that the local attractor is a stochastic attractor of particle i that lies in a hyper-rectangle with P_i and P_g being two ends of its diagonal. We introduce the concepts of QPSO as follows.

Assume that each individual particle moves in the search space with a δpotential on each dimension, of which the center is the point p_{ij}. For simplicity, we consider a particle in one-dimensional space, with point p the center of potential. Solving Schrödinger equation of one-dimensional δ potential well, we can get the probability distribution function $D(x) = e^{-2|p-x|/L}$. Using Monte Carlo method, we obtain

$$x = p \pm \frac{L}{2} \ln(1/u), \quad u \sim U(0, 1) \quad (7)$$

The above is the fundamental iterative equation of QPSO.

In Sun et al. [17], a global point called Mainstream Thought or Mean Best Position of the population is introduced into PSO. The mean best position, denoted as C, is defined as the mean of the personal best positions among all particles. That is

$$C(t) = (C_1(t), C_2(t), \ldots, C_D(t)) \\ = \left(\frac{1}{M} \sum_{i=1}^{M} P_{i1}(t), \ \frac{1}{M} \sum_{i=1}^{M} P_{i2}(t), \ \ldots, \ \frac{1}{M} \sum_{i=1}^{M} P_{iD}(t) \right) \quad (8)$$

where M is the population size and P_i is the personal best position of particle i. Then, the value of L is evaluated by $L = 2\alpha \cdot |C_j(t) - X_{ij}(t)|$, and the position is updated by

$$X_{ij}(t+1) = p_{ij}(t) \pm \alpha \cdot |C_j(t) - X_{ij}(t)| \cdot \ln(1/u) \quad (9)$$

where parameter α is called Contraction–Expansion (CE) Coefficient, which can be tuned to control the convergence speed of the algorithms. Generally, we always call the PSO with Eq. (9) quantum-behaved particle swarm optimization (QPSO). In most cases, α decrease linearly from can be controlled to α_0 to $\alpha_1 (\alpha_0 < \alpha_1)$.We outline the procedure of the QPSO algorithm as follows:

Procedure of the QPSO algorithm:

Step 1: Initialize the population;
Step 2: Computer the personal position and global best position;
Step 3: Computer the mean best position C;
Step 4: Properly select the value of α;
Step 5: Update the particle position according to Eq. (9);
Step 6: While the termination condition is not met, return to Step 2;
Step 7: Output the results.

4 QPSO-DM Algorithm

QPSO is a promising optimization problem solver that outperforms PSO in many real application areas. First of all, the introduced exponential distribution of positions makes QPSO global convergent. In the QPSO algorithm in the initial stage of search, as the particle swarm initialization, its diversity is relatively high. In the subsequent search process, due to the gradual convergence of the particle, the diversity of the population continues to decline. As the result, the ability of local search ability is continuously enhanced, and the global convergence ability is continuously weakened. In early and middle search, reducing the diversity of particle swarm optimization for contraction efficiency improvement is necessary; however, in late stage of search, because the particles are gathered in a relatively small range, particle swarm diversity is very low, the global search ability becomes very weak, the ability for a large range of search has been very small, and the phenomenon of premature will occur in this algorithm.

To overcome this shortcoming, we introduce diversity-maintained into QPSO.

The population diversity of the QPSO-DM is denoted as diversity($p$best) and is measured by average Euclidean distance from the particle's personal best position to the mean best position, namely

$$\text{diversisty}(p\text{best}) = \frac{1}{M \cdot |A|} \sum_{i=1}^{M} \sqrt{\sum_{j=1}^{D} (\text{pbest}_{i,j} - p\overline{\text{best}}_j)} \quad (10)$$

where M is the population of the particle, $|A|$ is the length of longest the diagonal in the search pace, and D is the dimension of the problem. Hence, we may guide the search of the particles with the diversity measures when the algorithm is running.

In the QPSO-DC algorithm, only low-bound d_{low} is set for diversity($p$best) to prevent the diversity from constantly decreasing. The procedure of the algorithm is as follows. After initialization, the algorithm is running in convergence mode. In process of convergence, the convergence mode is realized by Contraction–Expansion (CE) Coefficient α. On the course of evolution, if the diversity measure diversity($p$best) of the swarm drops to below the low-bound d_{low}, the particle swarm turns to be in explosion mode in which the particles are controlled to explode to increase the diversity until it is larger than d_{low}.

5 Experiment Results and Discussion

To test the performance of the QPSO with diversity-maintained (QPSO-DM), six widely known benchmark functions listed in Table 1 are tested for comparison with standard PSO (SPSO), QPSO. These functions are all minimization problems with minimum objective function values as zeros. The initial range of the

Table 1 Expression of the five tested benchmark functions

	Function expression	Search domain
Sphere	$f_1(X) = \sum_{i=1}^{n} x_i^2$	$-100 \le x_i \le 100$
Rosenbrock	$f_2(X) = \sum_{i=1}^{n-1} (100 \cdot (x_{i+1} - x_i^2)^2 + (x_i - 1)^2)$	$-100 \le x_i \le 100$
Rastrigrin	$f_3(X) = \sum_{i=1}^{n} (x_i^2 - 10 \cdot \cos(2\pi x_i) + 10)$	$-10 \le x_i \le 10$
Greiwank	$f_4(X) = \frac{1}{4000} \sum_{i=1}^{n} x_i^2 - \prod_{i=1}^{n} \cos\left(\frac{x_i}{\sqrt{i}}\right) + 1$	$-600 \le x_i \le 600$
Ackley	$f_5 = 20 + e - 20e^{-\frac{1}{5}\sqrt{\frac{1}{n}\sum_{i=1}^{n} x_i^2}} - e^{\frac{1}{n}\sum_{i=1}^{n} \cos(2\pi x_i)}$	$-30 \le x_i \le 30$
Schwefel	$f_6 = 418.9829n - \sum_{i=1}^{n} (x_i \sin\sqrt{\lvert x_i \rvert})$	$-500 \le x_i \le 500$

Table 2 The initial range of population for all the tested algorithms and V_{max} for spso

	Initial range	V_{max}
f_1	(50, 100)	100
f_3	(15, 30)	100
f_3	(2.56, 5.12)	10
f_4	(300, 600)	600
f_5	(15, 30)	30
f_6	(250, 500)	500

population listed in Table 2 is asymmetry as used in Shi and Eberhart [25]. Table 2 also lists V_{max} for SPSO. The fitness value is set as function value, and the neighborhood of a particle is the whole population.

As in Angeline [22], for each function, three different dimension sizes are tested. They are dimension sizes 10, 20, and 30. The max number of generations is set as 1,000, 1,500, and 2,000 corresponding to the dimensions 10, 20, and 30 for first six functions, respectively. The maximum generation for the last function is 2,000. In order to investigate whether the QPSO-DM algorithm is good or not, different population sizes are used for each function with different dimension. They are population sizes 20, 40, and 80. For SPSO, the acceleration coefficients are set to be $c_1 = c_2 = 2$, and the inertia weight is decreasing linearly from 0.9 to 0.4 as in Shi and Eberhart [25]. In experiments for QPSO, the value of CE Coefficient varies from 1.0 to 0.5 linearly over the running of the algorithm as in [18], while in QPSO-DM, the value of CE Coefficient is listed in Table 3. From the Table 3, we also obtain the CE cofficient of QPSO-DM decreases from 0.8–0.5 linearly. We had 50 trial runs for every instance and recorded mean best fitness and standard deviation.

The mean values and standard deviations of best fitness values for 50 runs of each function are recorded in Tables 4, 5, 6, 7, and 8.

The results show that both QPSO and QPSO-DM are superior to SPSO except on Schwefel and Shaffer's f6 function. On Sphere Function, the QPSO works better than QPSO-DM when the warm size is 40 and dimension is 10, and when the warm size is 80 and dimension is 20. Except for the above two instances, the best result is QPSO-DM. The Rosenbrock function is a monomodal function, but

Table 3 Parameter value of QPSO-dM

CE coefficient	Sphere function		Rosenbrock function		Rastrigrin function	
	Mean best	St. dev.	Mean best	St. dev.	Mean best	St. dev.
(1.0, 0.5)	1.2237e−013	1.9534e−013	75.0328	52.2408	88.5367	14.3207
(0.9, 0.5)	2.4890e−017	2.6749e−017	54.3227	69.7425	90.7490	18.2660
(0.8, 0.5)	**3.2054e−026**	1.6573e−025	**33.2065**	28.2984	78.4024	11.3524
(0.7, 0.5)	9.7934e−022	4.2890e−021	76.4626	88.0403	77.6020	9.8336
(1.0, 0.4)	4.6827e−012	2.1674e−011	39.7421	36.8124	69.7656	15.9507
(0.9, 0.4)	1.9889e−014	6.3148e−014	71.6725	89.3760	92.8337	13.6210
(0.8, 0.4)	5.9680e−015	2.5249e−014	69.5267	97.4625	83.3520	9.8675
(0.7, 0.4)	1.5027e−018	4.2109e−018	86.2217	122.5409	73.4665	17.6453
(1.0, 0.3)	1.8963e−010	4.5227e−010	56.3226	68.2468	59.2417	18.6628
(0.9, 0.3)	2.0527e−011	5.3745e−011	70.6849	98.7489	47.5948	15.0963
(0.8, 0.3)	5.6849e−014	6.1342e−013	90.3573	92.5246	61.0499	16.4421
(0.7, 0.3)	2.5864e−013	8.3346e−013	121.7752	145.4575	**40.7232**	18.3726

CE coefficient	Griewank function		Ackley function		Schwefel function	
	Mean best	St. dev.	Mean best	St. dev.	Mean best	St. dev.
(1.0, 0.5)	0.0328	0.0307	3.0209e−013	4.3354e−013	4.9953e + 003	216.7453
(0.9, 0.5)	0.0243	0.0152	3.2649e−014	8.3201e−014	4.8774e + 003	206.1504
(0.8, 0.5)	**0.0065**	0.0064	**1.1453e−014**	8.4250e−015	**2.8249e + 003**	415.4217
(0.7, 0.5)	0.0187	0.0097	3.9826e−014	5.6935e−014	4.1370e + 003	336.8420
(1.0, 0.4)	0.0651	0.0086	3.7380e−011	1.3192e−010	4.9682e + 003	209.5413
(0.9, 0.4)	0.0281	0.0170	1.5916e−012	3.8720e−012	5.1393e + 003	278.3024
(0.8, 0.4)	0.0106	0.0107	8.0437e−013	1.2495e−012	4.9504e + 003	255.6286
(0.7, 0.4)	0.0164	0.0165	0.0213	0.1634	4.5336e + 003	492.6527
(1.0, 0.3)	0.0137	0.0136	4.0969e−010	7.6405e−010	4.9972e + 003	252.7829
(0.9, 0.3)	0.0190	0.0190	7.0642e−011	1.1408e−010	5.0918e + 003	213.0963
(0.8, 0.3)	0.0151	0.0182	1.3296e−010	2.2764e−010	4.0437e + 003	324.4462
(0.7, 0.3)	0.0137	0.0126	0.1862	0.1324	3.8596e + 003	264.1462

its optimal solution lies in a narrow area that the particles are always apt to escape. The experiment results on Rosenbrock function show that the QPSO-DM outperforms the QPSO. Rastrigrin function and Griewank function are both multimodal and usually tested for comparing the global search ability of the algorithm. On Rastrigrin function, it is also shown that the QPSO-DM generated best results than QPSO. On Griewank function, QPSO-DM has better performance than QPSO and PSO algorithms. On Ackley function, QPSO-DM has best performance when the dimension is 10; except for these, the QPSO-DM has minimal value. On the Schwefel function, SPSO has the best performance in any situation. QPSO-DM is better than QPSO. Generally speaking, the QPSO-DM has better global search ability than SPSO and QPSO.

Figure 1 shows the convergence process of the three algorithms on the first four benchmark functions with dimension 30 and swarm size 40 averaged on 50 trail runs. It is shown that, although QPSO-DM converges more slowly than the QPSO

Table 4 Numerical results on sphere function

M	Dim.	Gmax	SPSO		QPSO		QPSO-DM	
			Mean best	St. Dev.	Mean best	St. dev.	Mean best	St. dev.
20	10	1000	4.6119e−021	1.0352e−020	3.1979e−043	2.2057e−042	4.9172e−045	1.4682e−045
	20	1500	9.0266e−012	3.5473e−011	1.9197e−024	7.1551e−024	1.2247e−026	2.6428e−026
	30	2000	3.9672e−008	7.0434e−008	7.3736e−015	2.1796e−014	2.4369e−016	4.0917e−016
40	10	1000	1.3178e−024	3.7737e−024	2.7600e−076	1.8954e−075	7.8025e−074	1.6432e−074
	20	1500	2.3057e−015	6.6821e−015	3.9152e−044	1.8438e−043	4.1583e−045	7.6524e−045
	30	2000	1.1286e−010	2.2994e−010	2.6424e−031	6.3855e−031	2.3208e−032	1.0527e−032
80	10	1000	1.6097e−028	7.1089e−028	2.5607e−103	6.4847e−103	2.1084e−104	1.6582e−104
	20	1500	6.3876e−018	1.4821e−017	1.4113e−068	7.4451e−068	1.3586e−068	3.7647e−068
	30	2000	3.2771e−013	7.5971e−013	6.5764e−050	3.6048e−049	1.9688e−050	1.7052e−050

Table 5 Numerical results on Rosenbrock function

M	Dim.	Gmax	SPSO		QPSO		QPSO-DM	
			Mean best	St. dev.	Mean best	St. dev.	Mean best	St. dev.
20	10	1,000	51.0633	153.7913	9.5657	16.6365	6.7435	1.6753
	20	1,500	100.2386	140.9822	82.4294	138.2429	30.2483	22.5409
	30	2,000	160.4400	214.0316	98.7948	122.5744	64.2516	48.3277
40	10	1,000	24.9641	49.5707	8.9983	17.8202	3.0418	2.4972
	20	1,500	59.8256	95.9586	40.7449	41.1751	12.4916	10.3762
	30	2,000	124.1786	269.7275	43.5582	38.0533	34.5349	21.8962
80	10	1,000	19.0259	41.6069	6.8312	0.3355	4.0263	3.2651
	20	1,500	40.2289	46.8491	33.5287	31.6415	14.8572	1.3806
	30	2,000	56.8773	57.8794	44.5946	31.6739	20.5704	1.6127

Table 6 Numerical results on rastrigrin function

M	Dim.	Gmax	SPSO		QPSO		QPSO-DM	
			Mean best	St. dev.	Mean best	St. dev.	Mean best	St. dev.
20	10	1,000	5.8310	2.5023	4.0032	2.1409	1.8261	0.3362
	20	1,500	23.3922	6.9939	15.0648	6.0725	11.0704	10.5382
	30	2,000	51.1831	12.5231	28.3027	12.5612	20.4728	3.6157
40	10	1,000	3.7812	1.4767	2.6452	1.5397	0.9504	1.3854
	20	1,500	18.5002	5.5980	11.3109	3.5995	10.3207	6.4872
	30	2,000	39.5259	10.3430	18.9279	4.8342	13.4697	5.7639
80	10	1,000	2.3890	1.1020	2.2617	1.4811	0.9650	1.0348
	20	1,500	12.8594	3.6767	8.4121	2.5798	5.7236	3.7684
	30	2,000	30.2140	7.0279	14.8574	5.0408	11.6570	4.1366
Numerical results on griewank function								
20	10	1,000	0.0920	0.0469	0.0739	0.0559	0.0425	0.0652
	20	1,500	0.0288	0.0285	0.0190	0.0208	0.0163	0.0216
	30	2,000	0.0150	0.0145	0.0075	0.0114	0.0077	0.0516
40	10	1,000	0.0873	0.0430	0.0487	0.0241	0.0468	0.0724
	20	1,500	0.0353	0.0300	0.0206	0.0197	0.0122	0.0364
	30	2,000	0.0116	0.0186	0.0079	0.0092	0.0013	0.0108
80	10	1,000	0.0658	0.0266	0.0416	0.0323	0.0388	0.0528
	20	1,500	0.0304	0.0248	0.0137	0.0135	0.0041	0.0086
	30	2,000	0.0161	0.0174	0.0071	0.0109	1.0273e−006	4.9264e−006

during the early stage of search, it may catch up with QPSO at later stage and could generate better solutions at the end of search.

From the results above in the tables and figures, it can be concluded that the QPSO-DM has better global search ability than SPSO and QPSO.

Table 7 Numerical results on ackley function

M	Dim.	Gmax	SPSO		QPSO		QPSO-DM	
			Mean best	St. dev.	Mean best	St. dev.	Mean best	St. dev.
20	10	1000	2.0489e−011	3.0775e−011	6.8985e−012	1.2600e−011	4.6281e−016	4.6027e−017
	20	1500	0.0285	0.2013	1.5270e−008	2.1060e−008	8.6631e−016	4.7815e−016
	30	2000	0.2044	0.4899	4.3113e−007	4.4188e−007	6.6520e−014	2.1354e−013
40	10	1000	2.4460e−013	5.2901e−013	2.5935e−015	5.0243e−016	1.8209e−016	3.2296e−017
	20	1500	2.6078e−008	4.0653e−008	6.3491e−013	1.3305e−012	4.6917e−016	2.2475e−016
	30	2000	3.7506e−006	6.4828e−006	5.5577e−011	8.6059e−011	5.6051e−016	7.8722e−016
80	10	1000	5.5778e−015	7.4138e−015	2.4514e−015	8.5229e−016	3.8246e−016	2.6115e−016
	20	1500	5.2979e−010	7.2683e−010	5.8620e−015	1.6446e−015	3.6504e−016	4.7535e−017
	30	2000	1.6353e−007	2.1300e−007	6.7253e−014	4.4759e−014	4.5725e−016	1.8864e−016

Table 8 Numerical results on schwefel function

M	Dim.	Gmax	SPSO		QPSO		QPSO-DM	
			Mean best	St. dev.	Mean best	St. dev.	Mean best	St. dev.
20	10	1,000	630.6798	169.9164	1.0784e + 003	342.1967	421.0562	188.3162
	20	1,500	2.0827e + 003	327.0313	4.0026e + 003	679.1461	2.8447e + 003	1.2683e + 003
	30	2,000	4.1097e + 003	481.8801	6.5075e + 003	589.9931	5.6023e + 003	1.4219e + 003
40	10	1,000	494.0236	169.5594	1.4359e + 003	246.9290	628.2261	290.3428
	20	1,500	1.8516e + 003	273.8690	4.5580e + 003	242.8220	2.1046e + 003	756.1462
	30	2,000	3.4042e + 003	384.3150	7.3380e + 003	288.9463	6.5209e + 003	741.2416
80	10	1,000	426.0796	130.5357	1.6060e + 003	132.0009	528.3466	273.1564
	20	1,500	1.5466e + 003	270.5587	4.6407e + 003	225.9431	2.5592e + 003	425.3320
	30	2,000	3.0325e + 003	419.8580	8.0583e + 003	226.3796	6.3488e + 003	560.3284

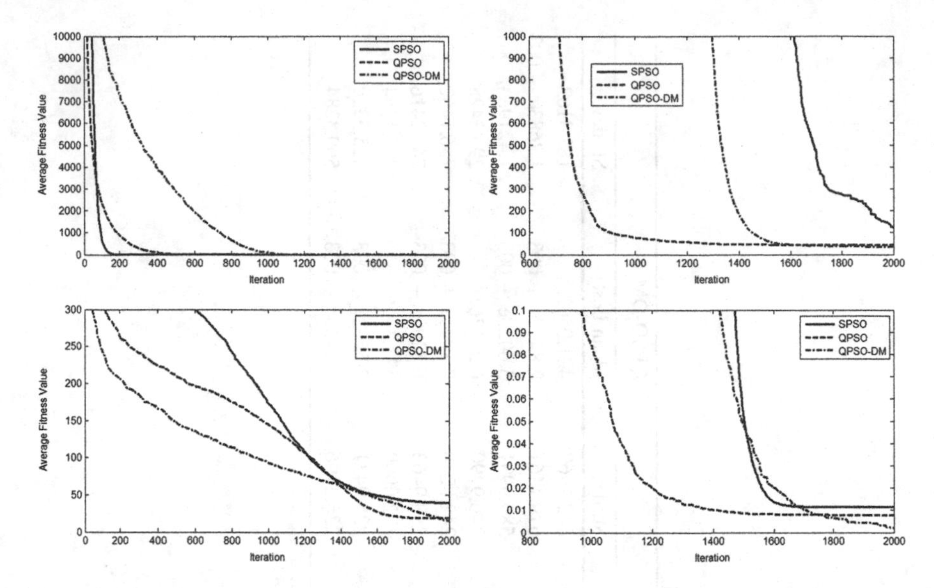

Fig. 1 Convergence process of the three algorithms on the first four benchmark functions with dimension 30 and swarm size 40 averaged on 50 trail runs

Acknowledgments This work is supported by National Natural Science Fund (No. 61163042), Higher School Scientific Research Project of Hainan Province (Hjkj2013-22), International Science and Technology Cooperation Program of China (2012DFA11270), and Hainan International Cooperation Key Project (GJXM201105).

References

1. Kennedy, J., Eberhart, R.: Particle swarm optimization. In: Proceedings IEEE International Conference Neural Networks, pp.1942–1948 (1995)
2. Van den Bergh, F.: An Analysis of Particle Swarm Optimizers. University of Pretoria, South Africa (2001)
3. Clerc, M.: Discrete particle swarm optimization illustrated by the traveling salesman problem. New optimization techniques in engineering, Berlin: Springer pp. 219–239 (2004)
4. Clerc, M.: Particle swarm optimization, ISTE, 2006
5. Zhang, W., Liu, Y., Clerc, M.: An adaptive PSO algorithm for reactive power optimization. In: 6th International Conference on Advances in Power Control, Operation and Management, Hong Kong (2003)
6. Eberhart, R.C., Shi, Y.: Comparing inertia weights and constriction factors in particle swarm optimization. In: IEEE International Conference on Evolutionary Computation, pp. 81–86 (2001)
7. He, S., Wu, Q.H., Wen, J.Y., Saunders, J.R., Paton, R.C.: A particle swarm optimizer with passive congregation. Biosystems **78**, 135–147 (2004)
8. Hu, X., Eberhart, R.C.: Tracking dynamic systems with PSO: where's the cheese?. In: Proceedings of the Workshop on Particle Swarm Optimization, Indianapolis (2001)
9. Parsopoulos, K.E., Vrahatis, M.N.: Particle swarm optimizer in noisy and continuously changing environments. Artif. Intell. Soft Comput., 289–294 (2001)

10. Parsopoulos, K.E., Vrahatis, M.N.: On the computation of all global minimizers thorough particle swarm optimization. IEEE Trans. Evol. Comput. **8**, 211–224 (2004)
11. LoZvbjerg, M., Krink, T: Extending particle swarms with self-organized criticality. In: Proceedings of the IEEE congress on evolutionary computation, pp. 1588–1593 (2002)
12. Blackwell, T., Bentley, P.J: Don't push me! Collision-avoiding swarms. In: Proceedings of the IEEE congress on evolutionary computation, pp. 1691–1696 (2002)
13. Robinson, J., Sinton, S., Rahmat-Samii, Y.: Particle swarm, genetic algorithm, and their hybrids: optimization of a profiled corrugated horn antenna. In: IEEE swarm intelligence symposium, pp. 314–317 (2002)
14. Hendtlass, T.: A combined swarm differential evolution algorithm for optimization problems. In: Lecture notes in computer science 2070, pp. 11–18.(2001)
15. Poli, R., Kennedy, J., Blackwell, T.: Particle swarm optimization. Swarm Intelligence **1**, 33–57 (2007)
16. Sun, J., Xu, W.B., Feng, B.: Particle swarm optimization with particles having quantum behavior. In: Proceedings Congress on Evolutionary Computation, pp. 325–331 (2004)
17. Sun, J., Xu, W.B., Feng, B.: A global search strategy of quantum behaved particle swarm optimization. In: Proceedings IEEE Conference on Cybernetics and Intelligent Systems, pp. 111–116 (2004)
18. Sun, J., Xu, W.B., Fang, W.: Quantum-behaved particle swarm optimization with a hybrid probability distribution. In: proceeding of 9th Pacific Rim International Conference on Artificial Intelligence (2006)
19. Liu, J., Sun, J., Xu, W.B.: Improving quantum-behaved particle swarm optimization by simulated annealing, LNAI 4203, pp. 77–83. Springer, Italy (2006)
20. Coelho, L.S.: Novel Gaussian quantum-behaved particle swarm optimizer applied to electromagnetic design. Sci, Meas Technol. **1**, 290–294 (2007)
21. Liu, J., Sun, J., Xu, W.B.: Quantum-behaved particle swarm optimization with immune memory and vaccination. In: Proceedings IEEE International conference on Granular Computing, USA, pp. 453–456 (2006)
22. Angeline, P.J.: Using selection to improve particle swarm optimization. In: Proceedings 1998 IEEE International Conference on Evolutionary Computation. Piscataway, pp. 84–89 (1998)
23. Shi, Y., Eberhart, R.C.: A modified particle swarm. In: Proceedings 1998 IEEE International Conference on Evolutionary Computation, Piscataway, pp. 69–73 (1998)
24. Clerc, M., Kennedy, J.: The particle swarm: explosion, stability, and convergence in a multidimensional complex space. IEEE Trans. Evol. Comput., Piscataway **6**, 58–73 (2002)
25. Shi, Y., Eberhart, R.: Empirical study of particle swarm optimization. In: Proceedings 1999 Congress on Evolutionary Computation, Piscataway, pp. 1945–1950 (1999)

A Delay Fractioning Approach to Global Synchronization of Delayed Complex Networks with Neutral-Type Coupling

Hongli Wu, Ya-peng Zhao, Huan-huan Mai and Zheng-xia Wang

Abstract The global issues of synchronization of complex networks with neutral-type coupling are investigated in this chapter, which is not adequately considered in existing literatures. Based on these new complex models, we derive asymptotical and exponential criterions via delay fraction approach. Numerical examples are then given to illustrate the effectiveness of our scheme and to compare with the recent proposals. We also make (some) attempts to explore the relationship between delay fraction numbers and the conservatism of our criterions.

Keywords Synchronization · Complex networks · Neutral-type coupling · Delay fractioning

1 Introduction

Synchronization is a ubiquitous and interesting phenomenon in nature. For example, how dose thousands of neurons or fireflies or crickets suddenly fall into step with one another, all firing or flashing or chirping at the same time, without

H. Wu
College of Information and Technology, Hainan Normal University, Haikou, Hainan, China

Y. Zhao
Institute of Intelligent Manufacturing and Control, Wuhan University of Technology, Wuhan 430063 Hubei, People's Republic of China

H. Mai (✉)
College of Computer Science, Chongqing University, Chongqing 400044, China
e-mail: huanhuanmai@126.com

Z. Wang
Department of Information and Computing Science, Chongqing Jiaotong University, Chongqing 400074, China

B.-Y. Cao et al. (eds.), *Ecosystem Assessment and Fuzzy Systems Management*,
Advances in Intelligent Systems and Computing 254, DOI: 10.1007/978-3-319-03449-2_22,

any leader or signal from the environment [1]? A complex network is a large set of interconnected nodes, in which each node is a fundamental unit with specific state. Recently, many attempts devoting to a better understanding of synchronization take advantage of the topology of complex networks, and they also contribute to the understanding of general emergent properties of networked systems. In fact, synchronization of complex networks has allured much attention as an interdisciplinary subject and provoked various applications in lots of fields such as neuroscience, engineering, computer science, economy, and social sciences.

It is well known that the way we connecting nodes plays an important role in the efficiency of synchronization in large networks. Li et al. studied the global synchronization of complex networks without delays [2–4]. Many researchers believed that there must be some time delays in spreading and responding due to the finite speed of transmission as well as traffic congestions, so delayed coupling should be modeled in order to simulate more realistic networks [5–10]. Moreover, it is natural and important to consider the neutral-type coupling delay in complex dynamical networks. Dai gave an example on it that when complex dynamical networks are used to model a stock transaction system, each node's state is defined as a behavior of the agent such as buying, selling, or holding. And the stock transaction system dynamically in terms of the current and historical fluctuating rate records [11]. On the other hand, the neutral-type coupling may be essential in specific applications such as secure communication. Solís-Perales founded that the derivative term under certain network topology leads to the chaotic synchronous behavior, whereas the standard coupling network reaches the equilibrium or a limit cycle [12]. To the authors' best knowledge, there are only two literatures [11, 12] that introduced the derivative term into the coupling of complex networks. By numerical simulations, [12] illustrated that the derivative terms in coupling have a significant influence on the synchronization. However, the impact of time delay is curtly neglected. While [11] considered the asymptotically synchronization of the neutral-type delay coupling complex networks by classical Lyapunov method.

In this chapter, we will introduce the synchronization of the new complex models first. Combining several techniques such as delay fraction method, the free-weighting matrices approach, Lyapunov–Krasovskii functional, and linear matrix inequality (LMI), we then study the asymptotically synchronization conditions and exponentially synchronization conditions of them. Finally, some simulations will be exercised to demonstrate the effectiveness and applicability of the proposed criterions together with some attempts to discuss the fraction number's influence on the criterions.

2 Model Description and Preliminaries

Notations: R^n denotes the n dimensional Euclidean space, and $R^{m\times n}$ is the set of all $m \times n$ real matrices. $||\cdot||$ denotes the Euclidean norm in R^n or $R^{m\times n}$. Let the Euclidian norm be $||\phi||_\tau = \sup_{-\tau\leq\theta\leq 0} ||x(\theta)||$, $||\phi^*||_\tau = \sup_{-\tau\leq\theta\leq 0} ||\dot{x}(\theta)||$ for a given continuous function. A^T denotes the transpose of matrix A, $\lambda_{\mathrm{Max}}(M)$ denotes the maximum eigenvalue of M, and $\lambda_{\mathrm{Min}}(M)$ denotes the minimum eigenvalue of M.

Consider the following continuous-time complex dynamical network with neutral-type coupling.

$$\dot{x}_i(t) = f(x_i(t), t) + c\sum_{j=1}^{N} G_{ij}\left(Ax_j(t-h) + B\dot{x}_j(t-\tau)\right), \quad i = 1,\ldots,N. \tag{1}$$

With the initial condition $x(t) = \phi(t)$, $t \in [-h_{\max}, 0]$, $h_{\max} = \max\{h, \tau\}$ where $f : R^n \to R^n$ is continuously $x_i = (x_{i1}, x_{i2}, \ldots, x_{in})^T \in R^n$ are the state variables of node i, the constant $c > 0$ is the coupling strength, h, τ are the retarded delay and the neutral delay, respectively. $A = (a_{ij}) \in R^{n\times n}$ is a constant inner-coupling matrix of the nodes about the retarded delay, and $B = (b_{ij}) \in R^{n\times n}$ regarding to the neutral one. $G = G_{ij} \in R^{N\times N}$ is the outer-coupling configuration matrix of the network, in which G_{ij} is defined by:

$$G_{ii} = -\sum_{j=1, j\neq i}^{N} G_{ij} = -\sum_{j=1, j\neq i}^{N} G_{ji} \quad i = 1,\ldots,N.$$

Suppose that the network (1) is connected in the sense that there are no isolated clusters. That is, G is an irreducible matrix.

Definition 1 The synchronized state of the entire networks is denoted by $s(t) \in R^n$ is a solution of an isolate node, namely $\dot{s}(t) = f(s(t))$, $s(t)$ may be a limit cycle, or a chaotic orbit in the phase space.

Definition 2 The dynamical network (1) is said to achieve asymptotic synchronization if

$$x_1(t) = x_2(t) = \ldots = x_N(t) = s(t) \text{ as } t \to \infty, \tag{2}$$

Definition 3 The dynamical network (1) is said to be globally exponentially synchronized if, for any solution *x*(*t*), if there exist constants $\mu > 0$ and $\alpha > 0$ such that

$$\lim_{t\to\infty} \|x_i(t) - s(t)\| \leq \mu e^{-\alpha t}\max\left\{\|\phi\|_\tau, \|\phi^*\|_\tau\right\}, \quad i = 1, 2, \ldots, n. \tag{3}$$

Lemma 1 ([5]) *Suppose that an irreducible matrix* $G = (G_{ij})_{N\times N}$ *satisfies the above conditions. Then,* 0 *is an eigenvalue of matrix* G*, associated with eigenvector* $(1, 1, \ldots, 1)^T$;

All the other eigenvalues of G *are real-valued and are strictly negative.*

Let λ_i, $i = 1, 2, \ldots, N$ be the nonzero eigenvalues of G. Lemma 1 is, without loss of generality, all the eigenvalues of G are real numbers and ordered as

$$0 = \lambda_1 \geq \lambda_2 \geq \ldots \geq \lambda_N$$

Lemma 2 ([13]) *For scalar $r > 0$, let $M \in R^{m\times m}$ be a positive semi-definite matrix and $\rho : [0, r] \to R^m$ be a vector function. If the interactions concerned are well defined, then the following inequity holds*

$$r\int_0^r \rho^T(s)M\rho(s)ds \geq \left(\int_0^r \rho(s)ds\right)^T M\left(\int_0^r \rho(s)ds\right) \tag{4}$$

Lemma 3 ([14]) *The following LMI* $\begin{bmatrix} Q(x) & S(x) \\ S^T(x) & R(x) \end{bmatrix} > 0$ *where* $Q(x) = Q^T(x)$, $R(x) = R^T(x)$, *and* $S(x)$ *depend affinely on* x, *is equivalent to* $R(x) > 0$, $Q(x) - S(x)R^{-1}(x)S^T(x) > 0$.

Lemma 4 *If the following* $(N-1) \times n$ *dimensional neutral-type delay differential equations* $\dot{w}_i(t) = J(t)w_i(t) + c\lambda_i(Aw_i(t-h) + B\dot{w}_i(t-\tau))$ $i = 2, \ldots N$ *where* $J(t)$ *is the Jacobian of* $f(x(t), t)$ *at synchronized state* $s(t)$ *are asymptotically stable about their zero solution, then the synchronization states are asymptotically stable for the complex networks.*

Lemma 5 *If the following* $(N-1) \times n$ *dimensional neutral-type delay differential equations*

$$\dot{w}_i(t) = J(t)w_i(t) + c\lambda_i(Aw_i(t-h) + B\dot{w}_i(t-\tau)) \quad i = 2, \ldots, N \tag{5}$$

where $J(t)$ is the Jacobian of $f(x(t), t)$ *at synchronized state* $s(t)$ *are exponentially stable about their zero solution, then the synchronization states are exponentially stable for the complex networks* (1).

Proof In order to investigate the stability of the synchronized states (2),
Set

$$x_i(t) = s(t) + e_i(t), \quad i = 1, 2, \ldots, N.$$

Substituting (5) into (1), we have

$$\dot{e}_i(t) = f(s(t) + e_i(t)) - f(s(t)) + c\sum_{j=1}^N G_{ij}\big(Ae_j(t-h) + B\dot{e}_j(t-\tau)\big),$$
$$i = 1, \ldots, N.$$

Since f is continuous differentiable, then we obtain

$$\dot{e}_i(t) = J(t)e_i(t) + cAe_i(t-h)G^T + cB\dot{e}_i(t-\tau)G^T,$$

there exists a nonsingular matrix, $\varphi = (\phi_1, \ldots, \phi_N) \in R^{N\times N}$ such that $G^T\Lambda = \varphi\Lambda$ with $\Lambda = \mathrm{diag}(\lambda_1, \ldots, \lambda_N)$. Using the nonsingular transform $w(t) =$

$(w_1(t),\ldots,w_N(t)) \in R^{n\times N}$, we have the following matrix equation: $\dot{w}(t) = J(t)w(t) + c(Aw(t-h) + B\dot{w}(t-\tau))\Lambda \quad i = 2,\ldots,N$.
Namely,

$$\dot{w}_i(t) = J(t)w_i(t) + c\lambda_i(Aw_i(t-h) + B\dot{w}_i(t-\tau)) \quad i = 2,\ldots,N.$$

Then, the global exponential synchronization problem of the dynamical networks (1) is equivalent to the problem of global exponential stabilization of the error dynamical system (5).

3 Main Results

Theorem 1 *The synchronous state* $s(t)$ *of complex networks with neutral-type coupling* (1) *is globally asymptotically stable if there exist matrices* O_i, $i = 1,2,3,4$ *and symmetrical positive definite matrices* P, *positive definite matrices* Z, R_i, $i = 1,2,3$, *and* $\Omega_1 \in R^{(m+4)\times(m+4)}$ *such that*

$$\Omega_1 = \left[\begin{array}{cccccc} \begin{array}{c}2PJ(t)+R_1+R_2-mZ\\ +2O_1J(t)+J^T(t)R_3J(t)\end{array} & mZ & mZ & \ldots & mZ & c\lambda_iPA2+c\lambda_iJ^T(t)R_3A+c\lambda_iO_1A+J^T(t)O_3^T \\ mZ^T & -2mZ & 0 & \cdots & 0 & 0 \\ mZ^T & 0 & -2mZ & \cdots & 0 & 0 \\ \vdots & \vdots & \vdots & \ddots & \vdots & \vdots \\ mZ^T & 0 & 0 & 0 & -2mZ & 0 \\ 0 & 0 & 0 & 0 & 0 & (c\lambda_i)^2A^TR_3A-mZ-R_1 \\ 0 & 0 & 0 & 0 & 0 & 0 \\ O_2J(t)-O_1^T & 0 & 0 & 0 & 0 & 0 \\ \begin{array}{c}c\lambda_iB^TP^T+O_4J(t)\\ +2c\lambda_iB^TR_3J(t)c\lambda_iO_1B\end{array} & 0 & 0 & 0 & 0 & 2(c\lambda_i)^2B^TR_3^TA+B^TO_3^T+O_4A \end{array}\right.$$

$$\left.\begin{array}{ccc} 0 & J^T(t)O_2^T-O_1 & 2c\lambda_iJ^T(t)R_3Bc\lambda_iO_1B+J^T(t)O_4^T+c\lambda_iPB \\ 0 & 0 & 0 \\ 0 & 0 & 0 \\ \vdots & \vdots & \vdots \\ 0 & 0 & 0 \\ 0 & 0 & 2(c\lambda_i)^2A^TR_3B+O_3B+A^TO_4^T \\ -R_2 & 0 & 0 \\ 0 & h^2Z-2O_2 & 0 \\ 0 & 0 & (c\lambda_i)^2B^TR_3B-R_3+2c\lambda_iO_4B \end{array}\right] < 0. \tag{6}$$

Proof Select a Lyapunov–Krasovskii functional as

$$V_i(w_i(t)) = V_{i1}(w_i(t)) + V_{i2}(w_i(t)) + V_{i3}(w_i(t)) + V_{i4}(w_i(t)) + V_{i5}(w_i(t)), \tag{7}$$

where

$$V_{i1}(w_i(t)) = w_i^T(t)Pw_i(t),$$

$$V_{i2}(w_i(t)) = \int_{-h}^{0} w_i^T(t+\xi)R_1w_i(t+\xi)d\xi,$$

$$V_{i3}(w_i(t)) = \int_{-\tau}^{0} w_i^T(t+\xi)R_2w_i(t+\xi)d\xi,$$

$$V_{i4}(w_i(t)) = \int_{-\tau}^{0} \dot{w}_i^T(t+\xi)R_3\dot{w}_i(t+\xi)d\xi,$$

$$\begin{aligned} V_{i5}(w_i(t)) = & h\int_{-h}^{-\frac{(m-1)h}{m}}\int_{t+\theta}^{t} \dot{w}_i^T(\xi)Z\dot{w}_i(\xi)d\xi d\theta \\ & + h\int_{-\frac{(m-1)h}{m}}^{-\frac{(m-2)h}{m}}\int_{t+\theta}^{t} \dot{w}_i^T(\xi)Z\dot{w}_i(\xi)d\xi d\theta \\ & + h\int_{-\frac{(m-2)h}{m}}^{-\frac{(m-3)h}{m}}\int_{t+\theta}^{t} \dot{w}_i^T(\xi)Z\dot{w}_i(\xi)d\xi d\theta + \cdots \\ & + h\int_{-\frac{2h}{m}}^{-\frac{h}{m}}\int_{t+\theta}^{t} \dot{w}_i^T(\xi)Z\dot{w}_i(\xi)d\xi d\theta \\ & + h\int_{-\frac{h}{m}}^{0}\int_{t+\theta}^{t} \dot{w}_i^T(\xi)Z\dot{w}_i(\xi)d\xi d\theta, \end{aligned}$$

The derivative of $V_{i1}(w(t))$ along the solution of the dynamic system (5)

$$\begin{aligned} \dot{V}_{i1}(w(t)) &= 2w_i^T(t)P\dot{w}_i(t) \\ &= 2w_i^T(t)P(J(t)w_i(t) + c\lambda_iAw_i(t-h) + c\lambda_iB\dot{w}_i(t-\tau)), \end{aligned} \tag{8}$$

$$\dot{V}_{i2}(w_i(t)) = w_i^T(t)R_1w_i(t) - w_i^T(t-h)R_1w_i(t-h) \tag{9}$$

$$\dot{V}_{i3}(w_i(t)) = w_i^T(t)R_2w_i(t) - w_i^T(t-\tau)R_2w_i(t-\tau), \tag{10}$$

$$
\begin{aligned}
\dot{V}_{i4}(w_i(t)) &= \dot{w}_i^T(t)R_3\dot{w}_i(t) - \dot{w}_i^T(t-\tau)R_3\dot{w}_i(t-\tau) \\
&= [J(t)w_i(t) + c\lambda_i A w_i(t-h) + c\lambda_i B\dot{w}_i(t-\tau)]^T \\
&\qquad R_3[J(t)w_i(t) + c\lambda_i A w_i(t-h) + c\lambda_i B\dot{w}_i(t-\tau)] \\
&\quad - \dot{w}_i^T(t-\tau)R_3\dot{w}_i(t-\tau) \\
&= w_i^T(t)J^T(t)R_3J(t)w_i(t) \\
&\quad + (c\lambda_i)^2 w_i^T(t-h)A^TR_3Aw_i(t-h) \\
&\quad + \dot{w}_i^T(t-\tau)\left((c\lambda_i)^2B^TR_3B - R_3\right)\dot{w}_i(t-\tau) \\
&\quad + 2c\lambda_i w_i^T(t)J^T(t)R_3Aw_i(t-h) \\
&\quad + 2c\lambda_i w_i^T(t)J^T(t)R_3B\dot{w}_i(t-\tau) \\
&\quad + 2(c\lambda_i)^2 w_i^T(t-h)A^TR_3B\dot{w}_i(t-\tau),
\end{aligned}
\tag{11}
$$

$$
\begin{aligned}
\dot{V}_{i5}(w_i(t)) &= h\dot{w}_i^T(t)Z\dot{w}_i(t) - h\int_{-h_M}^{-\frac{(m-1)h}{m}} \dot{w}_i^T(t+\xi)Z\dot{w}_i(t+\xi)\mathrm{d}\xi \\
&\quad - h\int_{-\frac{(m-1)h}{m}}^{-\frac{(m-2)h}{m}} \dot{w}_i^T(t+\xi)Z\dot{w}_i(t+\xi)\mathrm{d}\xi \\
&\quad - h\int_{-\frac{(m-2)h}{m}}^{-\frac{(m-3)h}{m}} \dot{w}_i^T(t+\xi)Z\dot{w}_i(t+\xi)\mathrm{d}\xi - \cdots \\
&\quad - h\int_{-\frac{2h}{m}}^{-\frac{h}{m}} \dot{w}_i^T(t+\xi)Z\dot{w}_i(t+\xi)\mathrm{d}\xi \\
&\quad - h\int_{-\frac{h}{m}}^{0} \dot{w}_i^T(t+\xi)Z\dot{w}_i(t+\xi)\mathrm{d}\xi,
\end{aligned}
$$

According to Lemma 2, we immediately get

$$
\begin{aligned}
&-h\int_{-\frac{(i+1)h}{m}}^{-\frac{ih}{m}} \dot{w}_i^T(t+\xi)Z\dot{w}_i(t+\xi)\mathrm{d}\xi \le -m\left(\int_{-\frac{(i+1)h}{m}}^{-\frac{ih}{m}} \dot{w}_i^T(t+\xi)\mathrm{d}\xi\right)Z\left(\int_{-\frac{(i+1)h}{m}}^{-\frac{ih}{m}} \dot{w}_i(t+\xi)\mathrm{d}\xi\right) \\
&\le -m\left[w_i(t-\tfrac{ih}{m}) - w_i\left(t-\tfrac{(i+1)h}{m}\right)\right]^T Z\left[w_i(t-\tfrac{ih}{m}) - w_i\left(t-\tfrac{(i+1)h}{m}\right)\right],
\end{aligned}
$$

then

$$
\begin{aligned}
\dot{V}_{i5}(w_i(t)) \leq\ & h^2\dot{w}_i^T(t)Z\dot{w}_i(t) - m\left[w_i\left(t - \frac{(m-1)h}{m}\right) - w_i(t-h)\right]^T \\
& \times Z\left[w_i\left(t - \frac{(m-1)h}{m}\right) - w_i(t-h)\right] \\
& - m\left[w_i\left(t - \frac{(m-2)h}{m}\right) - w_i\left(t - \frac{(m-1)h}{m}\right)\right]^T \\
& \times \left[w_i\left(t - \frac{(m-2)h}{m}\right) - w_i\left(t - \frac{(m-1)h}{m}\right)\right] \\
& - m\left[w_i\left(t - \frac{(m-3)h}{m}\right) - w_i\left(t - \frac{(m-2)h}{m}\right)\right]^T \\
& \times \left[w_i\left(t - \frac{(m-3)h}{m}\right) - w_i\left(t - \frac{(m-2)h}{m}\right)\right] \\
& - \cdots - m\left[w_i\left(t - \frac{h}{m}\right) - w_i\left(t - \frac{2h}{m}\right)\right]^T \\
& \times \left[w_i\left(t - \frac{h}{m}\right) - w_i\left(t - \frac{2h}{m}\right)\right] \\
& - m\left[w_i(t) - w_i\left(t - \frac{h}{m}\right)\right]^T \\
& \left[w_i(t) - w_i\left(t - \frac{h}{m}\right)\right] \\
\leq\ & h^2\dot{w}_i^T(t)Z\dot{w}_i(t) - mw_i^T(t-h)Zw_i(t-h) \\
& - 2mw_i^T\left(t - \frac{(m-1)h}{m}\right)Zw_i\left(t - \frac{(m-1)h}{m}\right) \\
& - 2mw_i^T\left(t - \frac{(m-2)h}{m}\right)Zw_i\left(t - \frac{(m-2)h}{m}\right) \\
& - 2mw_i^T\left(t - \frac{(m-3)h}{m}\right)Zw_i\left(t - \frac{(m-3)h}{m}\right) - \cdots \\
& - 2mw_i^T\left(t - \frac{2h}{m}\right)Zw_i\left(t - \frac{2h}{m}\right) \\
& - 2mw_i^T\left(t - \frac{h}{m}\right)Zw_i\left(t - \frac{h}{m}\right) - mw_i^T(t)Zw_i(t) \\
& + 2mw_i^T(t-h)Zw_i\left(t - \frac{(m-1)h}{m}\right) \\
& + 2mw_i^T\left(t - \frac{(m-1)h}{m}\right)Zw_i\left(t - \frac{(m-2)h}{m}\right) \\
& + 2mw_i^T\left(t - \frac{(m-2)h}{m}\right)Zw_i\left(t - \frac{(m-3)h}{m}\right) + \cdots \\
& + 2mw_i^T\left(t - \frac{h}{m}\right)Zw_i\left(t - \frac{2h}{m}\right) \\
& + 2mw_i^T(t)Zw_i\left(t - \frac{h}{m}\right)
\end{aligned} \tag{12}
$$

For any real matrices O_i, $i = 1,2,3,4$ with compatible dimensions

$$
\begin{aligned}
&2\left[w_i^T(t) \quad \dot{w}_i^T(t) \quad w_i^T(t-h) \quad \dot{w}_i^T(t-\tau)\right]\begin{bmatrix} O_1 \\ O_2 \\ O_3 \\ O_4 \end{bmatrix} \\
&(-\dot{w}_i(t) + J(t)w_i(t) + c\lambda_i(Aw_i(t-h) + B\dot{w}_i(t-\tau))) \\
&\quad = -2y^T(t)O_1J(t)\dot{y}(t) \\
&\qquad + 2w_i^T(t)\left(J^T(t)O_2^T - O_1\right)\dot{w}_i(t) - 2\dot{w}_i^T(t)O_2\dot{w}_i(t) \\
&\qquad + 2w_i^T(t)\left(c\lambda_iO_1A + J^T(t)O_3^T\right)w_i(t-h) \\
&\qquad + 2w_i^T(t)\left(c\lambda_iO_1B + J^T(t)O_4^T\right)\dot{w}_i(t-\tau) \\
&\qquad + 2\dot{w}_i^T(t)\left(c\lambda_iO_2A - O_3^T\right)w_i(t-h) \\
&\qquad + 2\dot{w}_i^T(t)\left(c\lambda_iO_2B - O_4^T\right)\dot{w}_i(t-\tau) \\
&\qquad + 2c\lambda_iw_i^T(t-h)O_3Aw_i(t-h) \\
&\qquad + 2c\lambda_iw_i^T(t-h)\left(O_3B + A^TO_4^T\right)\dot{w}_i(t-\tau) \\
&\qquad + 2c\lambda_i\dot{w}_i^T(t-\tau)O_4B\dot{w}_i(t-\tau)
\end{aligned}
\tag{13}
$$

Adding (8)–(13), by Lemma 3, we immediately obtain

$$\dot{V}(y(t)) \leq \eta_1^T(t)\Omega_1\eta_1(t),$$

where

$$
\begin{aligned}
\eta_1^T(t) = \Big[&w_i^T(t), w_i^T\left(t - \frac{h}{m}\right), w_i^T\left(t - \frac{2h}{m}\right), w_i^T\left(t - \frac{3h}{m}\right), \ldots, \\
&w_i^T\left(t - \frac{(m-1)h}{m}\right), w_i^T(t-h), w_i^T(t-\tau), \dot{w}_i^T(t), \dot{w}_i^T(t-\tau)\Big].
\end{aligned}
$$

If Ω_1 holds, then $\dot{V}_i(w_i(t)) \leq 0$, $V_i(w_i(t)) \leq V_i(w_i(0))$, and it implies the global asymptotic stability of the system (5). So by Lemma 4, the synchronized states (2) of network (1) are asymptotically stable. The proof is thus completed.

Theorem 2 The synchronous state $s(t)$ of complex networks with neutral-type coupling (1) is globally exponentially stable and has the exponential synchronization rake α if there exist matrices Q,N_i, $i = 1,2,3,\ldots,m$, O_i, $i = 1,2$, symmetrical positive definite matrices P, positive definite matrices Z, M_i, $i = 1,2,3,\ldots,m$, R_i, $i = 1,2,3$, where m is to be determined, and

$$\Omega_2 \in R^{(2m+3)\times(2m+3)}, \Omega_3 \in R^{(m+!)\times(m+!)}, X_i \in R^{m+1}, \ i = 1,2,\ldots,m+2;$$

such that

$$\Omega_2 = \begin{bmatrix} \Omega_3 & X_1 & X_2 & X_3 & X_4 & X_5 & X_6 & X_7 & \cdots & X_{m+2} \\ X_1^T & -e^{-2\beta\tau}R_2 & 0 & 0 & 0 & 0 & 0 & 0 & \cdots & 0 \\ X_2^T & 0 & -2O_2 & c\lambda_i O_2 B & 0 & 0 & 0 & 0 & \cdots & 0 \\ X_3^T & 0 & 0 & \begin{matrix} (c\lambda_i)^2 B^T R_3 B - e^{-2\beta\tau} R_3 \\ + (c\lambda_i)^2 B^T Z B \end{matrix} & 0 & 0 & 0 & 0 & \cdots & 0 \\ X_4^T & 0 & 0 & 0 & -Z & 0 & 0 & 0 & \cdots & 0 \\ X_5^T & 0 & 0 & 0 & 0 & -Z & 0 & 0 & \cdots & 0 \\ X_5^T & 0 & 0 & 0 & 0 & 0 & -Z & 0 & \cdots & 0 \\ X_7^T & 0 & 0 & 0 & 0 & 0 & 0 & -Z & \cdots & 0 \\ \vdots & \vdots & \vdots & \vdots & \vdots & \vdots & \vdots & \vdots & \ddots & \vdots \\ X_{m+2}^T & 0 & 0 & 0 & 0 & 0 & 0 & 0 & \cdots & -Z \end{bmatrix} < 0, \tag{14}$$

where

$$X_1 = \begin{bmatrix} c\lambda_i PB \\ 0 \\ 0 \\ \vdots \\ 0 \\ c\lambda_i O_2 A \end{bmatrix}, X_2 = \begin{bmatrix} J^T(t) O_2^T - O_1 \\ 0 \\ 0 \\ \vdots \\ 0 \\ 0 \end{bmatrix},$$

$$X_3 = \begin{bmatrix} c\lambda_i PB + c\lambda_i J^T(t) R_3 B + c\lambda_i J^T(t) Z B + c\lambda_i O_1 B \\ 0 \\ 0 \\ \vdots \\ 0 \\ (c\lambda_i)^2 A^T R_3 B + (c\lambda_i)^2 A^T Z B \end{bmatrix},$$

$$X_4 = \begin{bmatrix} \sqrt{h(e^{2\beta h} - 1)} Q \\ 0 \\ 0 \\ \vdots \\ 0 \\ 0 \end{bmatrix}, X_5 = \begin{bmatrix} \sqrt{h\left(e^{2\beta h} - e^{2\beta\frac{(m-1)h}{m}}\right)} N_1 \\ 0 \\ 0 \\ \vdots \\ 0 \\ 0 \end{bmatrix},$$

$$X_6 = \begin{bmatrix} \sqrt{h\left(e^{2\beta\frac{(m-1)h}{m}} - e^{2\beta\frac{(m-2)h}{m}}\right)} N_2 \\ 0 \\ 0 \\ \vdots \\ 0 \\ 0 \end{bmatrix},$$

$$X_7 = \begin{bmatrix} \sqrt{h\left(e^{2\beta\frac{(m-2)h}{m}} - e^{2\beta\frac{(m-3)h}{m}}\right)}N_3 \\ 0 \\ 0 \\ \vdots \\ 0 \\ 0 \end{bmatrix} X_{m+2} = \begin{bmatrix} \sqrt{h\left(e^{2\beta\frac{h}{m}} - 1\right)}N_m \\ 0 \\ 0 \\ \vdots \\ 0 \\ 0 \end{bmatrix},$$

and

$$\Omega_3 = \left[\begin{array}{cc} \begin{array}{c} 2\beta P + 2PJ(t) + R_1 + R_2 \\ +J^T(t)R_3J(t) + hM_m + hQ + 2\sqrt{2\beta} \\ h(N_m - Q) + 2h^2J^T(t)ZJ(t) + 2O_1J(t) \end{array} & \sqrt{2\beta}h(N_{m-1} - N_m) \\ 2\sqrt{2\beta}h(N_{m-1}^T - N_m^T) & he^{-2\beta\frac{h}{m}}(M_{m-1} - M_m) \\ 2\sqrt{2\beta}h(N_{m-2}^T - N_{m-1}^T) & 0 \\ \vdots & \vdots \\ 2\sqrt{2\beta}h(N_1^T - N_2^T) & 0 \\ c\lambda_iA^TR_3J(t) + \sqrt{2\beta}h(Q^T - N_1^T) & 0 \end{array}\right.$$

$$\left.\begin{array}{cccc} \sqrt{2\beta}h(N_{m-2} - N_{m-1}) & \cdots & \sqrt{2\beta}h(N_1 - N_2) & \begin{array}{r} c\lambda_iPA + \sqrt{2\beta}h(Q - N_1) \\ +c\lambda_iJ^T(t)R_3A \\ +c\lambda_iJ^T(t)ZA + c\lambda_iO_1A \end{array} \\ 0 & \cdots & 0 & 0 \\ he^{-2\beta\frac{2h}{m}}(M_{m-2} - M_{m-1}) & \cdots & 0 & 0 \\ \vdots & \ddots & \vdots & \vdots \\ 0 & \cdots & he^{-2\beta\frac{(m-1)h}{m}}(M_1 - M_2) & 0 \\ 0 & \cdots & 0 & \begin{array}{r} -R_1e^{-2\beta h} \\ +(c\lambda_i)^2A^TR_3A \\ -he^{-2\beta h}(M_1 + Q) \\ +(c\lambda_i)^2A^TZA \end{array} \end{array}\right]$$

Proof Select a Lyapunov–Krasovskii functional as

$$\begin{aligned} V_i(w_i(t)) = {} & V_{i1}(w_i(t)) + V_{i2}(w_i(t)) + V_{i3}(w_i(t)) \\ & + V_{i4}(w_i(t)) + V_{i5}(w_i(t)) + V_{i6}(w_i(t)), \end{aligned} \tag{15}$$

where

$$V_{i1}(w_i(t)) = e^{2\beta t} w_i^T(t) P w_i(t),$$

$$V_{i2}(w_i(t)) = \int_{-\tau}^{0} e^{2\beta(t+\xi)} w_i^T(t+\xi) R_1 w_i(t+\xi) \mathrm{d}\xi,$$

$$V_{i3}(w_i(t)) = \int_{-h}^{0} e^{2\beta(t+\xi)} w_i^T(t+\xi) R_2 w_i(t+\xi) \mathrm{d}\xi,$$

$$V_{i4}(w_i(t)) = \int_{-\tau}^{0} e^{2\beta(t+\xi)} \dot{w}_i^T(t+\xi) R_3 \dot{w}_i(t+\xi) \mathrm{d}\xi,$$

$$\begin{aligned} V_{i5}(w_i(t)) = {} & h \int_{-h}^{-\frac{(m-1)h}{m}} e^{2\beta(t+\xi)} w_i^T(t+\xi) M_1 w_i(t+\xi) \mathrm{d}\xi \\ & + h \int_{-\frac{(m-1)h}{m}}^{-\frac{(m-2)h}{m}} e^{2\beta(t+\xi)} w_i^T(t+\xi) M_2 w_i(t+\xi) \mathrm{d}\xi \\ & + h \int_{-\frac{(m-2)h}{m}}^{-\frac{(m-3)h}{m}} e^{2\beta(t+\xi)} w_i^T(t+\xi) M_3 w_i(t+\xi) \mathrm{d}\xi \\ & + \cdots + h \int_{-\frac{2h}{m}}^{-\frac{h}{m}} e^{2\beta(t+\xi)} w_i^T(t+\xi) M_{m-1} w_i(t+\xi) \mathrm{d}\xi \\ & + h \int_{-\frac{h}{m}}^{0} e^{2\beta(t+\xi)} w_i^T(t+\xi) M_m w_i(t+\xi) \mathrm{d}\xi \\ & + h \int_{-h}^{0} e^{2\beta(t+\xi)} w_i^T(t+\xi) Q w_i(t+\xi) \mathrm{d}\xi, \end{aligned}$$

$$\begin{aligned} V_{i6}(w_i(t)) = {} & h \int_{-h}^{-\frac{(m-1)h}{m}} \int_{t+\theta}^{t} e^{2\beta\xi} \dot{w}_i^T(\xi) Z \dot{w}_i(\xi) \mathrm{d}\xi \mathrm{d}\theta \\ & + h \int_{-\frac{(m-1)h}{m}}^{-\frac{(m-2)h}{m}} \int_{t+\theta}^{t} e^{2\beta\xi} \dot{w}_i^T(\xi) Z \dot{w}_i(\xi) \mathrm{d}\xi \mathrm{d}\theta \\ & + h \int_{-\frac{(m-2)h}{m}}^{-\frac{(m-3)h}{m}} \int_{t+\theta}^{t} e^{2\beta\xi} \dot{w}_i^T(\xi) Z \dot{w}_i(\xi) \mathrm{d}\xi \mathrm{d}\theta + \cdots \\ & + h \int_{-\frac{2h}{m}}^{-\frac{h}{m}} \int_{t+\theta}^{t} e^{2\beta\xi} \dot{w}_i^T(\xi) Z \dot{w}_i(\xi) \mathrm{d}\xi \mathrm{d}\theta \\ & + h \int_{-\frac{h}{m}}^{0} \int_{t+\theta}^{t} e^{2\beta\xi} \dot{w}_i^T(\xi) Z \dot{w}_i(\xi) \mathrm{d}\xi \mathrm{d}\theta, \end{aligned}$$

the time derivative of Lyapunov–Krasovskii functional along the trajectories of system (5)

$$\begin{aligned} \dot{V}_{i1}(w_i(t)) &= 2\beta e^{2\beta t} w_i(t)^T P w_i(t) + 2e^{2\beta t} w_i(t)^T P(J(t)w_i(t) \\ &\quad + c\lambda_i A w_i(t-h) + c\lambda_i B\dot{w}_i(t-\tau)) \\ &= e^{2\beta t} w_i(t)^T (2\beta P + 2PJ(t))w_i(t) + 2c\lambda_i e^{2\beta t} w_i(t)^T PAw_i(t-h) \\ &\quad + 2c\lambda_i e^{2\beta t} w_i(t)^T PB\dot{w}_i(t-\tau) \end{aligned} \tag{16}$$

$$\dot{V}_{i2}(w_i(t)) = e^{2\beta t} w_i^T(t) R_1 w_i(t) - e^{2\beta(t-h)} w_i^T(t-h) R_1 w_i(t-h) \tag{17}$$

$$\dot{V}_{i3}(w_i(t)) = e^{2\beta t} w_i^T(t) R_2 w_i(t) - e^{2\beta(t-\tau)} w_i^T(t-\tau) R_2 w_i(t-\tau) \tag{18}$$

$$\begin{aligned} \dot{V}_{i4}(w_i(t)) &= e^{2\beta t} \dot{w}_i^T(t) R_3 \dot{w}_i(t) - e^{2\beta(t-\tau)} \dot{w}_i^T(t-\tau) R_3 \dot{w}_i(t-\tau) \\ &= e^{2\beta t}[J(t)w_i(t) + c\lambda_i A w_i(t-h) + c\lambda_i B\dot{w}_i(t-\tau)]^T \\ &\qquad R_3[J(t)w_i(t) + c\lambda_i A w_i(t-h) + c\lambda_i B\dot{w}_i(t-\tau)] \\ &\quad - e^{2\beta(t-\tau)} \dot{w}_i^T(t-\tau) R_3 \dot{w}_i(t-\tau) \\ &= e^{2\beta t} w_i^T(t) J^T(t) R_3 J(t) w_i(t) \\ &\quad + e^{2\beta t}(c\lambda_i)^2 w_i^T(t-h) A^T R_3 A w_i(t-h) \\ &\quad + \dot{w}_i^T(t-\tau)\left((c\lambda_i)^2 e^{2\beta t} B^T R_3 B - e^{2\beta(t-\tau)} R_3\right) \\ &\qquad \dot{w}_i(t-\tau) + 2c\lambda_i e^{2\beta t} w_i^T(t) J^T(t) R_3 A w_i(t-h) \\ &\quad + 2c\lambda_i e^{2\beta t} w_i^T(t) J^T(t) R_3 B\dot{w}_i(t-\tau) \\ &\quad + 2(c\lambda_i)^2 e^{2\beta t} w_i^T(t-h) A^T R_3 B\dot{w}_i(t-\tau) \end{aligned} \tag{19}$$

$$\begin{aligned} \dot{V}_{i5}(w_i(t)) &= h\Bigg[e^{2\beta\left(t-\frac{(m-1)h}{m}\right)} w_i^T\left(t-\frac{(m-1)h}{m}\right)(M_1-M_2)w_i\left(t-\frac{(m-1)h}{m}\right) \\ &\quad + e^{2\beta\left(t-\frac{(m-2)h}{m}\right)} w_i^T\left(t-\frac{(m-2)h}{m}\right)(M_2-M_3)w_i\left(t-\frac{(m-2)h}{m}\right) \\ &\quad + \cdots + e^{2\beta\left(t-\frac{h}{m}\right)} w_i^T\left(t-\frac{h}{m}\right)(M_{m-1}-M_m)w_i\left(t-\frac{h}{m}\right) \\ &\quad + e^{2\beta t} w_i^T(t)(M_m+Q)w_i(t) \\ &\quad - e^{2\beta(t-h)} w_i^T(t-h)(M_1+Q)w_i(t-h)\Bigg], \end{aligned} \tag{20}$$

The popular way of introducing free-weighting matrices is to denote the relationship between the items in the Leibniz–Newton formula. Here, we introduced Q, N_i, $i = 1, 2, 3, \ldots, m$, to less comparatively conservativeness condition.

$$
\begin{aligned}
\dot{V}_{i6}(w_i(t)) = \dot{V}_{i6}(w_i(t)) \\
& + 2h\sqrt{2\beta}e^{2\beta t}w_i^T(t)N_1\left[w_i\left(t - \frac{(m-1)h}{m}\right)\right. \\
& \left. -w_i(t-h) - \int_{-h_M}^{-\frac{(m-1)h}{m}} \dot{w}_i^T(t+\xi)\mathrm{d}\xi\right] \\
& + 2h\sqrt{2\beta}e^{2\beta t}w_i^T(t)N_2\left[w_i\left(t - \frac{(m-2)h}{m}\right)\right. \\
& \left. -w_i\left(t - \frac{(m-1)h}{m}\right) - \int_{-\frac{(m-1)h}{m}}^{-\frac{(m-2)h}{m}} \dot{w}_i^T(t+\xi)\mathrm{d}\xi\right] \\
& + 2h\sqrt{2\beta}e^{2\beta t}w_i^T(t)N_3\left[w_i\left(t - \frac{(m-3)h}{m}\right)\right. \\
& \left. -w_i\left(t - \frac{(m-2)h}{m}\right) - \int_{-\frac{(m-2)h}{m}}^{-\frac{(m-3)h}{m}} \dot{w}_i^T(t+\xi)\mathrm{d}\xi\right] \\
& + \cdots + 2h\sqrt{2\beta}e^{2\beta t}w_i^T(t)N_{m-1}\left[w_i\left(t - \frac{h}{m}\right)\right. \\
& \left. -w_i\left(t - \frac{2h}{m}\right) - \int_{-\frac{2h}{m}}^{-\frac{h}{m}} \dot{w}_i^T(t+\xi)\mathrm{d}\xi\right] \\
& + 2h\sqrt{2\beta}e^{2\beta t}w_i^T(t)N_m[w_i(t) \\
& \left. -w_i\left(t - \frac{h}{m}\right) - \int_{-\frac{h}{m}}^{0} \dot{w}_i^T(t+\xi)\mathrm{d}\xi\right] \\
& - 2h\sqrt{2\beta}e^{2\beta t}w_i^T(t)Q[w_i(t) \\
& \left. -w_i(t-h) - \int_{-h}^{0} \dot{w}_i^T(t+\xi)\mathrm{d}\xi\right],
\end{aligned}
$$

then

$$
\begin{aligned}
\dot{V}_{i6}\big(w_i^T(t)\big) \leq\ & he^{2\beta t}w_i^T(t)\Big[\Big(e^{2\beta h} - e^{2\beta\frac{(m-1)h}{m}}\Big)N_1^T Z^{-1} N_1 \\
& + \Big(e^{2\beta\frac{(m-1)h}{m}} - e^{2\beta\frac{(m-2)h}{m}}\Big)N_2^T Z^{-1} N_2 \\
& + \Big(e^{2\beta\frac{(m-2)h}{m}} - e^{2\beta\frac{(m-3)h}{m}}\Big)N_3^T Z^{-1} N_3 + \cdots \\
& + \Big(e^{2\beta\frac{2h}{m}} - e^{2\beta\frac{h}{m}}\Big)N_{m-1}^T Z^{-1} N_{m-1} \\
& + \Big(e^{2\beta\frac{h}{m}} - 1\Big)N_m^T Z^{-1} N_m \\
& + (e^{2\beta h_M} - 1)N^T Z^{-1} N\Big] w_i(t) \\
& + 2h^2 e^{2\beta t}[J(t)w_i(t) + c\lambda_i A w_i(t-h) + c\lambda_i B\dot{w}_i(t-\tau)]^T \\
& \quad Z[J(t)w_i(t) + c\lambda_i A w_i(t-h) + c\lambda_i B\dot{w}_i(t-\tau)] \\
& + 2\sqrt{2\beta}he^{2\beta t}w_i^T(t)(N_1 - N_2)w_i\left(t - \frac{(m-1)h}{m}\right) \\
& + 2\sqrt{2\beta}he^{2\beta t}w_i^T(t)(N_2 - N_3)w_i\left(t - \frac{(m-2)h}{m}\right) \\
& + 2\sqrt{2\beta}he^{2\beta t}w_i^T(t)(N_3 - N_4)w_i\left(t - \frac{(m-3)h}{m}\right) \\
& + \cdots + 2\sqrt{2\beta}he^{2\beta t}w_i^T(t)(N_{m-1} - N_m)w_i\left(t - \frac{h}{m}\right) \\
& + 2\sqrt{2\beta}he^{2\beta t}w_i^T(t)(N_m - Q)w_i(t) \\
& + 2\sqrt{2\beta}he^{2\beta t}w_i^T(t)(Q - N_1)w_i(t-h) \\
& - e^{2\beta t}h\int_{-h}^{-\frac{(m-1)h}{m}}\Big[\sqrt{2\beta}e^{-\beta\xi}w_i^T(t)N_1 + e^{\beta\xi}\dot{w}_i^T(t+\xi)Z\Big] \\
& \quad Z^{-1}\Big[\sqrt{2\beta}e^{-\beta\xi}w_i^T(t)N_1 + e^{\beta\xi}\dot{w}_i^T(t+\xi)Z\Big]^T \mathrm{d}\xi \\
& - e^{2\beta t}h\int_{-\frac{(m-1)h}{m}}^{-\frac{(m-2)h}{m}}\Big[\sqrt{2\beta}e^{-\beta\xi}w_i^T(t)N_2 + e^{\beta\xi}\dot{w}_i^T(t+\xi)Z\Big] \\
& \quad Z^{-1}\Big[\sqrt{2\beta}e^{-\beta\xi}w_i^T(t)N_2 + e^{\beta\xi}\dot{w}_i^T(t+\xi)Z\Big]^T \mathrm{d}\xi \\
& - \cdots - e^{2\beta t}h\int_{-\frac{2h}{m}}^{-\frac{h}{m}}\Big[\sqrt{2\beta}e^{-\beta\xi}w_i^T(t)N_{m-1} + e^{\beta\xi}\dot{w}_i^T(t+\xi)Z\Big] \\
& \quad Z^{-1}\Big[\sqrt{2\beta}e^{-\beta\xi}w_i^T(t)N_{m-1} + e^{\beta\xi}\dot{w}_i^T(t+\xi)Z\Big]^T \mathrm{d}\xi \\
& - e^{2\beta t}h\int_{-\frac{h}{m}}^{0}\Big[\sqrt{2\beta}e^{-\beta\xi}w_i^T(t)N_m + e^{\beta\xi}\dot{w}_i^T(t+\xi)Z\Big] \\
& \quad Z^{-1}\Big[\sqrt{2\beta}e^{-\beta\xi}w_i^T(t)N_m + e^{\beta\xi}\dot{w}_i^T(t+\xi)Z\Big]^T \mathrm{d}\xi \\
& - e^{2\beta t}h\int_{-h}^{0}\Big[\sqrt{2\beta}e^{-\beta\xi}w_i^T(t)Q - e^{\beta\xi}\dot{w}_i^T(t+\xi)Z\Big] \\
& \quad Z^{-1}\Big[\sqrt{2\beta}e^{-\beta\xi}w_i^T(t)Q - e^{\beta\xi}\dot{w}_i^T(t+\xi)Z\Big]^T \mathrm{d}\xi,
\end{aligned} \tag{21}
$$

Since $Z > 0$, the last $m + 1$ parts are less than 0. We can omit them here for the LMI's simplicity, although it may bring more conservatism.

$$
\begin{aligned}
\dot{V}_{i6}(w_i(t)) \leq\ & he^{2\beta t}w_i^T(t)\Big[\Big(e^{2\beta h} - e^{2\beta\frac{(m-1)h}{m}}\Big)N_1^T Z^{-1}N_1 + \Big(e^{2\beta\frac{(m-1)h}{m}} - e^{2\beta\frac{(m-2)h}{m}}\Big)N_2^T Z^{-1}N_2 \\
& + \Big(e^{2\beta\frac{(m-2)h}{m}} - e^{2\beta\frac{(m-3)h}{m}}\Big)N_3^T Z^{-1}N_3 + \cdots + \Big(e^{2\beta\frac{2h}{m}} - e^{2\beta\frac{h}{m}}\Big)N_{m-1}^T Z^{-1}N_{m-1} \\
& + \Big(e^{2\beta\frac{h}{m}} - 1\Big)N_m^T Z^{-1}N_m + (e^{2\beta h} - 1)Q^T Z^{-1}Q\Big]w_i(t) \\
& + 2\sqrt{2\beta}he^{2\beta t}w_i^T(t)(N_1 - N_2)w_i\left(t - \frac{(m-1)h}{m}\right) \\
& + 2\sqrt{2\beta}he^{2\beta t}w_i^T(t)(N_2 - N_3)w_i\left(t - \frac{(m-2)h}{m}\right) \\
& + 2\sqrt{2\beta}he^{2\beta t}w_i^T(t)(N_3 - N_4)w_i\left(t - \frac{(m-3)h}{m}\right) \\
& + \cdots + 2\sqrt{2\beta}he^{2\beta t}w_i^T(t)(N_{m-1} - N_m)w_i\left(t - \frac{h}{m}\right) \\
& + 2\sqrt{2\beta}he^{2\beta t}w_i^T(t)(N_m - Q)w_i(t) + 2\sqrt{2\beta}he^{2\beta t}w_i^T(t)(Q - N_1)w_i(t-h) \\
& + 2h^2e^{2\beta t}\Big[w_i^T(t)J^T(t)ZJ(t)w_i(t) + (c\lambda_i)^2 w_i^T(t-h)A^T ZAw_i(t-h) \\
& + (c\lambda_i)^2\dot{w}_i^T(t-\tau)B^T ZB\dot{w}_i(t-\tau) + 2c\lambda_i w_i^T(t)J^T(t)ZAw_i(t-h) \\
& + 2c\lambda_i w_i^T(t)J^T(t)ZB\dot{w}_i(t-\tau) + 2(c\lambda_i)^2 w_i^T(t-h)A^T ZB\dot{w}_i(t-\tau)\Big]
\end{aligned}
\tag{22}
$$

The another popular way of introducing free-weighting matrices is to denote the relationship between the items in the dynamic systems.

Here, we introduced $O_i, i = 1, 2,$

$$
\begin{aligned}
& 2 \times e^{2\beta t}\begin{bmatrix} w_i^T(t) & \dot{w}_i^T(t) \end{bmatrix}\begin{bmatrix} O_1 \\ O_2 \end{bmatrix}(-\dot{w}_i(t) + J(t)w_i(t) + c\lambda_i Aw_i(t-h) + c\lambda_i B\dot{w}_i(t-\tau)) \\
&= 2 \times e^{2\beta t}(w_i^T(t)O_1 + \dot{w}_i^T(t)O_2)(-\dot{w}_i(t) + J(t)w_i(t) + c\lambda_i Aw_i(t-h) + c\lambda_i B\dot{w}_i(t-\tau)) \\
&= 2e^{2\beta t}w_i^T(t)O_1 J(t)w_i(t) + 2e^{2\beta t}w_i^T(t)\big(J^T(t)O_2^T - O_1\big)\dot{w}_i(t) - 2e^{2\beta t}\dot{w}_i^T(t)O_2\dot{w}_i(t) \\
&\quad + 2c\lambda_i e^{2\beta t}w_i^T(t)O_1 Aw_i(t-h) + 2c\lambda_i e^{2\beta t}w_i^T(t)O_1 B\dot{w}_i(t-\tau) \\
&\quad + 2c\lambda_i e^{2\beta t}\dot{w}_i^T(t)O_2 Aw_i(t-h) + 2c\lambda_i e^{2\beta t}\dot{w}_i^T(t)O_2 B\dot{w}_i(t-\tau).
\end{aligned}
\tag{23}
$$

Adding (16)–(20), (22)–(23), by Lemma 1, we give

$$
\dot{V}_i(w(t)) \leq e^{2\beta t}\eta_2^T(t)\Omega_3\eta_2(t),
$$

where

$$
\begin{aligned}
\eta_2^T(t) = \Big[& w_i^T(t), w_i^T\left(t - \frac{h}{m}\right), w_i^T\left(t - \frac{2h}{m}\right), w_i^T\left(t - \frac{3h}{m}\right), \ldots, w_i^T\left(t - \frac{(m-1)h}{m}\right), w_i^T(t-h), \\
& w_i^T(t-\tau), \dot{w}_i^T(t), \dot{w}_i^T(t-\tau), w_i^T(t) \times 1, w_i^T(t) \times 1^2, \ldots, w_i^T(t) \times 1^{m+1}\Big].
\end{aligned}
$$

However

$$
\begin{aligned}
V_i(w_i(0)) &= V_{i1}(w_i(0)) + V_{i2}(w_i(0)) + V_{i3}(w_i(0)) + V_{i4}(w_i(0)) + V_{i5}(w_i(0)) + V_{i6}(w_i(0)) \\
&= w_i^T(0)Pw_i(0) + \int_{-\tau}^{0} e^{2\beta\xi} w_i^T(\xi)R_1 w_i(\xi)d\xi + \int_{-h}^{0} e^{2\beta\xi} w_i^T(\xi)R_2 w_i(\xi)d\xi \\
&\quad + \int_{-\tau}^{0} e^{2\beta\xi} \dot{w}_i^T(\xi)R_3 \dot{w}_i(\xi)d\xi + h\int_{-h}^{-\frac{(m-1)h}{m}} e^{2\beta\xi} w_i^T(\xi)M_1 \dot{w}_i(\xi)d\xi \\
&\quad + h\int_{-\frac{(m-1)h}{m}}^{-\frac{(m-2)h}{m}} e^{2\beta\xi} w_i^T(\xi)M_2 w_i(\xi)d\xi \\
&\quad + h\int_{-\frac{(m-2)h}{m}}^{-\frac{(m-3)h}{m}} e^{2\beta\xi} w_i^T(\xi)M_3 w_i(\xi)d\xi + \cdots + h\int_{-\frac{2h}{m}}^{-\frac{h}{m}} e^{2\beta\xi} w_i^T(\xi)M_{m-1} w_i(\xi)d\xi \\
&\quad + h\int_{-\frac{h}{m}}^{0} e^{2\beta\xi} w_i^T(\xi)M_m w_i(\xi)d\xi + h\int_{-h}^{0} e^{2\beta\xi} w_i^T(\xi)Q w_i(\xi)d\xi \\
&\quad + h\int_{-h}^{-\frac{(m-1)h}{m}}\int_{\theta}^{0} e^{2\beta\xi} \dot{w}_i^T(\xi)Z\dot{w}_i(\xi)d\xi d\theta + h\int_{-\frac{(m-1)h}{m}}^{-\frac{(m-2)h}{m}}\int_{\theta}^{0} e^{2\beta\xi} \dot{w}_i^T(\xi)Z\dot{w}_i(\xi)d\xi d\theta \\
&\quad + h\int_{-\frac{(m-2)h}{m}}^{-\frac{(m-3)h}{m}}\int_{\theta}^{0} e^{2\beta\xi} \dot{w}_i^T(\xi)Z\dot{w}_i(\xi)d\xi d\theta + \cdots + h\int_{-\frac{2h}{m}}^{-\frac{h}{m}}\int_{\theta}^{0} e^{2\beta\xi} \dot{w}_i^T(\xi)Z\dot{w}_i(\xi)d\xi d\theta \\
&\quad + h\int_{-\frac{h}{m}}^{0}\int_{\theta}^{0} e^{2\beta\xi} \dot{w}_i^T(\xi)Z\dot{w}_i(\xi)d\xi d\theta \\
&\le \Bigg[\lambda_{\text{Max}}(P) + \lambda_{\text{Max}}(R_1)\int_{-\tau}^{0} e^{2\beta\xi}d\xi + \lambda_{\text{Max}}(R_2)\int_{-h}^{0} e^{2\beta\xi}d\xi + h\lambda_{\text{Max}}(M_{\text{Max}})\int_{-h}^{0} e^{2\beta\xi}d\xi \\
&\quad + h\lambda_{\text{Max}}(Q)\int_{-h}^{0} e^{2\beta\xi}d\xi\Bigg]||\phi||_\tau^2 + 2\Bigg[\lambda_{\text{Max}}(R_3)\int_{-\tau}^{0} e^{2\beta\xi}d\xi + \lambda_{\text{Max}}(Z)\int_{-\tau}^{0}\int_{\theta}^{0} e^{2\beta\xi}d\xi d\theta\Bigg]||\phi^*||_\tau^2 \\
&= \left[\lambda_{\text{Max}}(P) + \frac{1-e^{-2\beta\tau}}{2\beta}(\lambda_{\text{Max}}(R_1) + \lambda_{\text{Max}}(R_2) + h\lambda_{\text{Max}}(M_{\text{Max}}) + h\lambda_{\text{Max}}(Q))\right]||\phi||_\tau^2 \\
&\quad + \left[2\lambda_{\text{Max}}(Z)\frac{2\beta\tau - 1 + e^{-2\beta\tau}}{4\beta^2} + \frac{1-e^{-2\beta\tau}}{2\beta}(\lambda_{\text{Max}}(R_3)\right]||\phi^*||_\tau.
\end{aligned}
$$

Since

$$e^{2\beta t}\lambda_{\text{Min}}(P)||w_i(t)||^2 \le V_i(w_i(t)),$$

then, we immediately obtain that

$$w_i(t) \leq \frac{1}{\sqrt{\lambda_{\mathrm{Min}}(P)}} \left[\lambda_{\mathrm{Max}}(P) + \frac{1-e^{-2\beta\tau}}{2\beta} (\lambda_{\mathrm{Max}}(R_1) + \lambda_{\mathrm{Max}}(R_2) + h\lambda_{\mathrm{Max}}(M_{\mathrm{Max}}) + h\lambda_{\mathrm{Max}}(Q)) \right.$$
$$\left. + 2\lambda_{\mathrm{Max}}(Z) \frac{2\beta\tau - 1 + e^{-2\beta\tau}}{4\beta^2} \right]^{\frac{1}{2}} \max\{||\phi||_\tau, ||\phi^*||_\tau\} e^{-\beta t}.$$

And

$$\lim_{t\to\infty} ||x_i(t) - s(t)|| \leq \mu e^{-\alpha t} \max\{||\phi||_\tau, ||\phi^*||_\tau\}, \quad i = 1, 2, \ldots, n. \tag{24}$$

where

$$\mu = \frac{1}{\sqrt{\lambda_{Min}(P)}} \left[\lambda_{Max}(P) + \frac{1-e^{-2\beta\tau}}{2\beta} (\lambda_{Max}(R_1) + \lambda_{Max}(R_2) + h\lambda_{Max}(M_{Max}) \right.$$
$$\left. + h\lambda_{Max}(Q)) + 2\lambda_{Max}(Z) \frac{2\beta\tau - 1 + e^{-2\beta\tau}}{4\beta^2} \right]^{\frac{1}{2}},$$
$$\alpha = \beta$$

Finally, by Definition 3 and (24), it is obvious that the globally exponential stability of the system (5). So by Lemma 4, the synchronized states of network (1) are asymptotically stable. The proof is thus completed.

Remark 1 Both Theorem 1 and Theorem 2's assumptions are in forms of LMI. The conditions 7 are linear to O_i, $i = 1, 2, 3, 4$ P, Z, R_i, $i = 1, 2, 3$, and conditions 15 are linear to $Q, N_i, i = 1, 2, 3, \ldots, m, O_i, i = 1, 2,$ P, Z, M_i, $i = 1, 2, 3, \ldots, m$, R_i, $i = 1, 2, 3$.

When we use the matlab LMI toolbox, we always assume that the m and α is to be a specific value.

Remark 2 Both theorem 1 and theorem 2 use the delay fraction method. In Theorem 2, the number of the free-weighting matrices that denote the relationship between the items in the Leibniz–Newton formula is not any equal to fraction number m, but also is equal to the missed number of negative part

$$-\cdots - e^{2\beta t} h \int_{-\frac{(i+!)h}{m}}^{-\frac{ih}{m}} \left[\sqrt{2\beta} e^{-\beta\xi} w^T(t) N_{m-i} + e^{\beta\xi} \dot{w}^T(t+\xi) Z \right]$$
$$Z^{-1} \left[\sqrt{2\beta} e^{-\beta\xi} w^T(t) N_{m-i} + e^{\beta\xi} \dot{w}^T(t+\xi) Z \right]^T d\xi$$

We are not sure that the bigger value of m will lead less conservative results.
We want to depend on experiments to find the right m so that we can get the better results.

4 A Numerical Example

Consider a three-dimensional stable linear system described by Dai et al. [11]

$$\begin{bmatrix} \dot{x}_1 \\ \dot{x}_2 \\ \dot{x}_3 \end{bmatrix} = \begin{bmatrix} -x_1 \\ -2x_2 \\ -3x_3 \end{bmatrix}$$

which is asymptotically stable at $s(t) = 0$, and its Jacobian is

$$J = \begin{bmatrix} -1 & 0 & 0 \\ 0 & -2 & 0 \\ 0 & 0 & -3 \end{bmatrix}$$

Case 1 Assume that the inner-coupling matrices are

$$A = \begin{bmatrix} 1 & 0 & 0 \\ 0 & 1 & 0 \\ 0 & 0 & 1 \end{bmatrix}, \quad B = \begin{bmatrix} 0.2 & -0.1 & 0.5 \\ -0.3 & 0.09 & -0.15 \\ 0.3 & 0.1 & 0.2 \end{bmatrix}$$

The outer-coupling matrix is

$$G_1 = \begin{bmatrix} -2 & 1 & 0 & 0 & 1 \\ 1 & -3 & 1 & 1 & 0 \\ 0 & 1 & -2 & 1 & 0 \\ 0 & 1 & 1 & -3 & 1 \\ 1 & 0 & 0 & 1 & -2 \end{bmatrix}$$

Obviously, G_1 is an irreducible symmetrical matrix. The eigenvalues of G_1 are $\lambda_i = 0, -1.382, -2.382, -3.618, -4.618$.

Case 2 Assume that the inner-coupling matrices are

$$A = \begin{bmatrix} 1 & 0 & 0 \\ 0 & 1 & 0 \\ 0 & 0 & 1 \end{bmatrix}, \quad B = \begin{bmatrix} 0.2 & -0.1 & 0.5 \\ -0.3 & 0.09 & -0.15 \\ 0.3 & 0.1 & 0.2 \end{bmatrix}$$

The outer-coupling matrix is

$$G_2 = \begin{bmatrix} -4 & 1 & 1 & 0 & 0 & 0 & 0 & 0 & 1 & 1 \\ 1 & -4 & 1 & 1 & 0 & 0 & 0 & 0 & 0 & 1 \\ 1 & 1 & -4 & 1 & 1 & 0 & 0 & 0 & 0 & 0 \\ 0 & 1 & 1 & -4 & 1 & 1 & 0 & 0 & 0 & 0 \\ 0 & 0 & 1 & 1 & -4 & 1 & 1 & 0 & 0 & 0 \\ 0 & 0 & 0 & 1 & 1 & -4 & 1 & 1 & 0 & 0 \\ 0 & 0 & 0 & 0 & 1 & 1 & -4 & 1 & 1 & 0 \\ 0 & 0 & 0 & 0 & 0 & 1 & 1 & -4 & 1 & 1 \\ 1 & 0 & 0 & 0 & 0 & 0 & 1 & 1 & -4 & 1 \\ 1 & 1 & 0 & 0 & 0 & 0 & 0 & 1 & 1 & -4 \end{bmatrix}$$

Table 1 Simulation result for $c = 0.3$ with the outer-coupling matrix G_1

h	τ	[11]	Theorem 1
0–1	0.1	S	S
1.1	0.1	U	S
0.1	0–∞	S	S
0.15	0.15	S	S
0.15	0.19	S	S
0.15	0.20	U	S
0.16	0.19	U	S

Table 2 Simulation result for $c = 0.2$ with the outer-coupling matrix G_2

h	τ	[11]	Theorem 1
0.2	0–∞	S	S
0.22	0–∞	S	S
0.23	0.7	S	S
0.23	0.8	U	S
0.3	0.3	S	S
0.3	0.31	U	S
0.34	0.3	S	S
0.35	0.3	U	S

Table 3 Simulation result of theorem 1

	$m = 2$	$m = 3$	$m = 4$	$m = 5$	$m = 6$
Case 1 with $c = 0.3$	$h = \tau = 0.3173$	$h = \tau = 0.3173$	$h = \tau = 0.3173$	$h = \tau = 0.3235$	$h = \tau = 0.3235$
Case 2 with $c = 0.2$	$h = \tau = 0.6849$	$h = \tau = 0.6849$	$h = \tau = 0.6849$	$h = \tau = 0.6869$	$h = \tau = 0.6869$

The eigenvalues of G_1 are

$$\lambda_i = 0, -1.7639, -1.7639, -4, -5, -5, -5, -5, -6.2361, -6.2361.$$

The results of Theorems in this letter and those in [11] are listed in Table 1, where "S" means that the criterion is applicable to the corresponding case and "U" means that the criterion is not applicable to the corresponding case.

Obviously, both Tables 1 and 2 illustrated the correctness and efficiency of our results. Furthermore, if we assume that the $h = \tau$, the maximum bound of the delays obtained by Theorem 1 and Theorem 2 are listed as in the Tables 3 and 4, respectively.

Form these two tables, we can see that the $m = 5, 6$ in Theorem 1 and $m = 2$ in Theorem 2 are better choices. It may be concluded that in Theorem 1 which omitted nothing, the bigger value of fraction number leads better results, whereas it isn't work in Theorem 2.

Table 4 Simulation result of theorem 2 with $\alpha = \beta = 0.01$

	$m = 2$	$m = 3$	$m = 4$	$m = 5$	$m = 6$
Case 1 with $c = 0.3$	$h = \tau = 0.0658$	$h = \tau = 0.0656$	$h = \tau = 0.0656$	$h = \tau = 0.0656$	$h = \tau = 0.0656$
Case 2 with $c = 0.2$	$h = \tau = 0.2338$	$h = \tau = 0.2337$	$h = \tau = 0.2337$	$h = \tau = 0.2337$	$h = \tau = 0.2337$

5 Conclusion

In this chapter, we have investigated the globally asymptotically synchronization and the globally exponentially synchronization of complex networks with neutral-type coupling by combining several techniques such as delay fraction method, the free-weighting matrices approach, Lyapunov–Krasovskii functional, and LMI. Numerical examples are given to show their effectiveness and advantages over others.

Acknowledgments Thanks to the support by National Natural Science Foundation (61363032). And this work is supported by International Science & Technology Cooperation Program of China (2012DFA1127); Hainan International Cooperation Key Project(GJXM201105).

References

1. Watts, D.J., Strogatz, S.H.: Collective dynamics of 'small-world' networks. Nature **393**, 440–442 (1998)
2. Li, Z., Chen, G.: Global synchronization and asymptotic stability of complex dynamical networks. IEEE Trans. CAS-II **53**, 28–33 (2006)
3. Lu, W., Chen, T.: Synchronization analysis of linearly coupled networks of discrete time systems. Phys. D **198**, 148–168 (2004)
4. Wang, X.F., Chen, G.: Synchronization in scale-free dynamical networks: Robustness and fragility. IEEE Trans. CAS-II **49**, 54–62 (2002)
5. Li, C., Chen, G.: Synchronization in general complex dynamical networks with coupling delays. Phys. A **343**, 263–278 (2004)
6. Li, K., Guan, S., Gong, X., Lai, C.H.: Synchronization stability of general complex dynamical networks with time-varying delays. Phys. Lett. A **372**, 7133–7139 (2008)
7. Li, P., Yi, Z.: Synchronization analysis of delayed complex networks with time-varying couplings. Phys. A **387**, 3729–3737 (2008)
8. Tu, L., Lu, J.-A.: Delay-dependent synchronization in general complex delayed dynamical networks. Comput. Math. Appl. **57**, 28–36 (2009)
9. Wang, Y., Wang, Z., Liang, J.: A delay fractioning approach to global synchronization of delayed complex networks with stochastic disturbances. Phys. Lett. A **372**, 6066–6073 (2008)
10. Wen, S., Chen, S., Guo, W.: Adaptive global synchronization of a general complex dynamical network with non-delayed and delayed coupling. Phys. Lett. A **372**, 6340–6346 (2008)
11. Dai, Y., Cai, Y., Xu, X.: Synchronization criteria for complex dynamical networks with neutral-type coupling delay. Phys. A **387**, 4673–4682 (2008)

12. Solís-Perales, G., Ruiz-Velázquez, E., Valle-Rodríguez, D.: Synchronization in complex networks with distinct chaotic nodes. Commun. Nonlinear Sci. Numer. Simul **14**(6):2528–2535 (2009)
13. Gu, K.Q., Kharitonov, V.L., Chen, J.: Stability of Time-Delay Systems. Birkhauser, Boston (2003)
14. Boyd, S., ElGhaoui, L., Feron, E., Balakrishnan, A.V.: Linear Matrix Inequalities in System and Control Theory. SIAM, Philadelphia (1994)

Research of Image Watermarking Algorithm and Application in Eco-Tourism Digital Museums Copyrights Protection

Li-hua Wu, Wen-juan Jiang and Junkuo Cao

Abstract Digital watermarking technology being an important application of information hiding technology has become currently a hot issue in the field of digital information security. This chapter first discusses the digital copyright protection and safety requirements of the digital museum and then describes a common model of information hiding, digital watermarking technology of discrete cosine transform (referred as DCT), and its features in detail. At the same time, the chapter gives a DCT-based JPEG color image invisible watermarking algorithm, as well as swf animation file information hiding watermarking algorithm, in order to improve the information watermark security and attack resistance in the digital museum. The experimental tests show that the algorithm can better balance between imperceptibility, robustness, and security.

Keywords Digital media rights · Discrete cosine transform domain · Information hiding · Image watermarking

1 Introduction

Along with the constant development of network technology, digital museum is widely used in Internet and has its unique advantages. However, today's digital exhibits are likely to face a series of problems such as being copied, distributed, juggled, illegally obtained, and copyright infringement. Thus, the digital image copyright protection issues must be resolved under such background.

Usually, invisible image watermarks have the following two basic characteristics. The first is invisibility, and it is difficult to visually perceive the difference

L. Wu (✉) · W. Jiang · J. Cao
School of Computer Science and Technology, Hainan Normal University, HaiKou 571158, China
e-mail: lihuawu63@163.com

B.-Y. Cao et al. (eds.), *Ecosystem Assessment and Fuzzy Systems Management*, Advances in Intelligent Systems and Computing 254, DOI: 10.1007/978-3-319-03449-2_23,

between the watermarked image and the original image. The second is robustness, and also, the embedded watermark is robust against various signal processing. Since these two characteristics contradict each other. How to take a compromise, to ensure watermark invisibility under the premise, but also embedded the powerful watermark information possible. This is an important issue in the field of digital watermarking research.

2 The General Model of Information Hiding System

The digital watermark is interdisciplinary involving digital signal processing, image processing, cryptography, communication theory and algorithm design and other fields. Its basic idea is that the original copyright information in the works (source version, the original author, owner) instead as watermark information, and it is embedded into the image, text, video, audio and other digital media works (original carrier) through a certain algorithm, but does not affect the value in use or commercial value of the embedded watermark digital media works. Digital watermarking algorithm not only can identify the relevant information of owners which is embedded in the carrier object, but also can extract the information when needed [1].

In general, the generic model for information hiding core system consists of two phases: the watermark embedding and watermarking detection or extracting, as shown in Fig. 1.

In the above model, watermarking information can be copyright message or secret data or a serial number. The public data without watermark are called original carrier such as video and audio clips. Generally, the information hiding process is controlled by the secret key. The watermarking information is hidden in public information via embedding algorithm. And detection algorithm can extract the secret watermarking information from the watermarked carrier by means of a secret key.

For all this, the digital watermark embedded into digital exhibits of the digital museum must have the following basic characteristics [2, 3]:

1. **Imperceptibility or concealment.**

The digital watermark is invisible to anyone by seeing or hearing. The ideal watermark carriers (digital media with watermark) are exactly the same as the original carrier visually.

2. **Security.**

The procedure of the watermark embedding should be confidential. The methods of embedding and detecting digital watermarking are confidential to unauthorized third parties and cannot be easily cracked. Besides, unauthorized

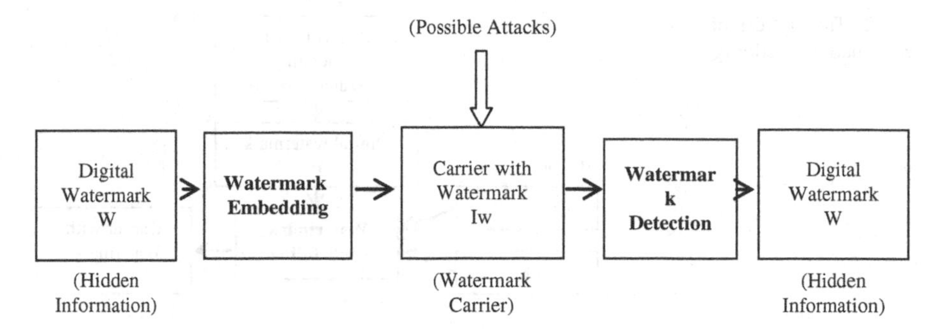

Fig. 1 The general model for information hiding core system

users cannot eliminate or destroy the watermark by changing the watermark carrier.

3. **Robustness or anti-aggressive capability.**

The digital watermark should be able to withstand a large number of physical and geometrical distortions, including intentional malicious attacks or image compression, filtering, scanning, copying, and size conversion After these operations, digital watermarking algorithm should still be able to extract watermark embedded from the watermark carrier or prove the presence of the watermark.

4. **Provability.**

The message carried by the digital watermark should be identified uniquely and certainly, so as to provide the ownership of digital exhibits with completely reliable evidence.

2.1 Digital Watermark Embedding

Assuming that original carrier is *I*, the digital watermarking is *W*, the carrier with watermark I_w, and the secret key *K*. Then, the process of watermark embedding is shown in Fig. 2.

As shown here, the common formula for the process of digital watermark embedding can be defined as follows:

$$I_W = E(I, W, K) \tag{1}$$

Among them, I_w means the data after the embedding watermark, *I* stands for the original carriers of data, and *W* refers to the watermark set. The algorithm of generating watermark *G* should ensure the properties of the watermark such as uniqueness, validity, and irreversibility. *K* stands for the secret key set (you can

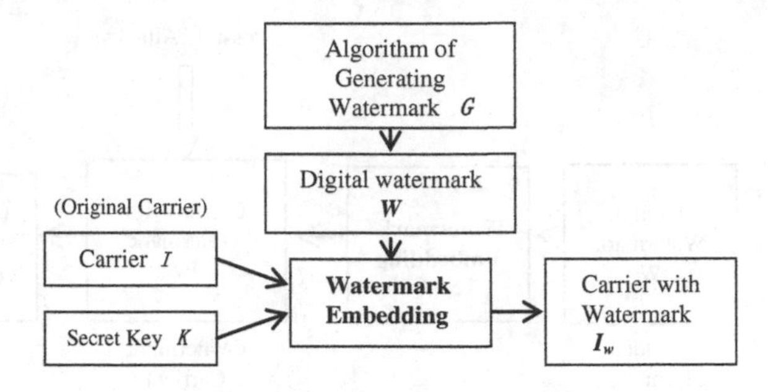

Fig. 2 The process of watermark embedding

choose the private key or public key usually). All utility systems must use a secret key and even the combination of several keys in some case. The watermarking information W is the data that can be existed in any form, e.g., random sequence or pseudo-random numbers, character or grid, binary image, gray image or color image, and 3D image. The watermarking information W can be generated by random number generator. When the digital watermark embedding stage, to find a better medium between and among intangibility, safe reliability and robustness for the digital watermark is the aim of embedding algorithm. Secret key K can be used to strengthen security, so as to avoid unauthorized restoration and rehabilitation of watermark [4].

2.2 Digital Watermarking Detection

The process of watermarking detection can be divided into two parts, as shown in Fig. 3.

Figure 3 is a sort of schematic diagram of the process of watermarking detection. Among them, remarks the extracted watermark, D indicates the watermark detection algorithm.

$\widehat{I}_W$ means the watermarking carriers of data after being attacked during transmission. The means of detecting a watermark can be divided into two kinds. The first one is extracting the embedded signal or testing the correlation verification of the embedded signal with the original information, and second is that the embedded information must be fully searching method or hypothesis testing without the original information. Generally, the result is presented in two forms: the one is exporting an extracted watermark and the other is exporting whether the monitored information contains the specified watermark [4].

As shown in Fig. 3, the common formula for the process of watermarking detection can be defined as follows:

- When there is the original carrier data,

$$\widehat{W} = D(\widehat{I}_W, I, K) \tag{2}$$

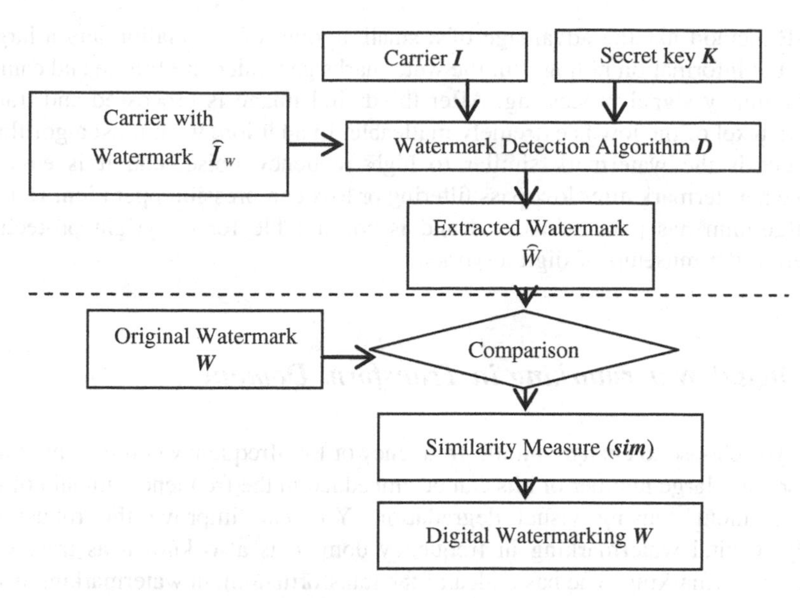

Fig. 3 The process of watermarking detection

- When there is no original information,

$$\widehat{W} = D(\widehat{I}_W, K) \tag{3}$$

3 Comparing Image Invisible Watermarking Algorithm

Invisible image watermarking can be classified as digital watermarking in space domain and digital watermarking in transform domain in accordance with its hiding location. Currently, the research of invisible image watermarking focuses on the technology of digital watermarking in transform domain.

3.1 Digital Watermarking in Space Domain

Essentially, the earlier digital watermarking algorithm is adopted in the space region. The digital watermarking is loaded directly in the data. The known algorithm in space is least significant bit (referred LSB) method, and its principle is that adjust the watermarking information conveyed by resolution unimportant to perception by modifying the color bit plane of digital images, to achieve the purpose of embedding the watermark.

LSB method has the advantage of a small amount of calculation and a larger amount of information hiding. But the watermarks generated are fragile and cannot resist ordinary signal processing. After the digital image is processed and transformed, pixel of the low is extremely malleable. In addition, what most algorithms introduce is the watermark similar to high-frequency noise, and it is easy to remove a watermark after low-pass filtering or loss compression operation. Hence, this algorithm has poor robustness and is not suitable for copyright protection applied to the museum of digital works.

3.2 Digital Watermarking in Transform Domain

When you choose to change middle-frequency or low-frequency component to add a watermark, large number of bits can be embedded in the frequency domain of the image without causing visual degradation. You can improve the robustness greatly. Digital watermarking in frequency domain is also known as transform domain watermarking. The basic idea of the transform domain watermarking is the first carrier signal of the data by a certain method to the frequency domain, and again after the watermark information is embedded in the frequency domain, the last of the processed signal for inversely converting the data signal back to the level of the carrier, thereby forming a final watermarked. More commonly used transform consisted of discrete Fourier transform (referred as DFT), discrete cosine transform (referred as DCT), and discrete wavelet transform (referred as DWT). Since media messages after transformation have obvious virtues of concentrated distribution of energy, good frequency division, etc., it can fit in easily with the awareness model of human visual system. Thus, the media messages after transformation can regulate the balance between the robustness and the imperceptivity easily [5].

4 Image Invisible Watermarking Algorithm of DCT

A watermarking algorithm of DCT transform domain is a research hot spot in the study of watermarking for the moment. Because DCT transform is second only to quadrature transform, and its algorithm is relatively easy to implement, DCT transform is commonly applied to image compression technology (e.g., JPEG). This also makes the watermarking based on the DCT method more resistant to JPEG compressing attack. Moreover, it makes the final watermark carrier to have good robustness.

Two-dimensional DCT transform is often used for digital picture processing. An image digital watermarking based on 2D-DCT is basically the idea that firstly divides an image into $N \times N$ parts. And then two-dimensional DCT transform is performed on each part. Let $F(x, y)$ denote the $N \times N$ image, and then, the formula of two-dimensional DCT transform is given as follows:

$$F(u,v) = \frac{1}{N}\sum_{x=0}^{N-1}\sum_{y=0}^{M-1} f(x,y) \quad (u,v=0) \tag{4}$$

$$F(u,v) = \frac{2}{N}\sum_{x=0}^{N-1}\sum_{y=0}^{M-1} f(x,y)\ \cos\left[\frac{\pi}{2N}(2y+1)v\right] \quad (u,v=1,2,\ldots,N-1). \tag{5}$$

Among them, $F(u,v)$ means the high-frequency portion of transform domain based on a DCT image digital watermarking, and $F(0, 0)$ means the low-frequency portion of transform domain based on a DCT image digital watermarking;

The original image can be restored by two-dimensional DCT inverse transform. The formula of two-dimensional DCT inverse transform is given as follows:

$$\begin{aligned} F(x,y) = {} & \frac{1}{N}F(0,0) + \frac{2}{\sqrt{N}}\sum_{x=1}^{N-1} F(u,0)\cos\left[\frac{\pi}{2N}(2x+1)u\right] \\ & + \frac{2}{\sqrt{N}}\sum_{y=1}^{N-1} F(0,v)\cos\left[\frac{\pi}{\sqrt{2N}}(2y+1)v\right] \\ & + \frac{2}{N}\sum_{x=1}^{N-1}\sum_{y=1}^{N-1} F(u,v)\cos\left[\frac{\pi}{2N}(2x+1)u\right]\cos\left[\frac{\pi}{2N}(2y+1)v\right] \\ & (u,v=1,2,\ldots,N-1) \end{aligned} \tag{6}$$

After two-dimensional DCT transform, most of the energy information about images focuses on middle-frequency and low-frequency coefficients. The watermark embedded in the domain of low-frequency coefficients has good robustness. While the watermark embedded in the domain of high-frequency coefficients has good imperceptibility but may has the data lost when processing the images. Taken together, the watermark embedded in the domain of high-frequency coefficients is usually not considered. Instead as a compromise, the domain of middle frequency and low frequency is regarded as coefficient of embedment.

DCT transform converts in blocks. The size of the blocks can be flexibly determined, and we usually choose the size of 8 × 8. The whole image can be treated as a block to perform DCT transform, and it too may be considered as different word blocks (divided from the whole image) to perform DCT transform independently. Positive DCT transform decomposes images into spatial frequency. The type of frequency-domain coefficients represents the proportion of the frequency components in the original image.

Compared with the algorithm in space domain, the algorithm in transform domain has the following advantages:

- The embedded watermark signal in transform domain can be distributed to all pixels of the spatial domain and is favourable to ensuring invisibility of the watermark.
- In transform domain, some characteristics of the human visual system can be more easily coupled to the watermark embedding process.
- The method of transform domain can make compatible with the image compression standard of international mainstream (e.g., JPEG), thus achieving watermark embedding on compressed domain.
- Generally, the method of transform domain has very good robustness. Large number of bits can be embedded in the frequency domain of the image without causing visual degradation.
- This algorithm has the anti-attack ability for image compression the popular image filtering and noise. So it is very suitable for copyright protection of digital works.

From this, digital watermarking in transform domain is the main technological measures applied to protect the copyright of museum numeral works.

5 Experiment and Results

The digital exhibits in the digital museum are digital project of various mediums' information such as picture, video, and vector animation. To protect the copyright and the integrity of the digital exhibits in the digital museum, a variety of digital watermarking technologies should be applied comprehensively.

In the digitized data of our project, the main format is plentiful JPEG image and Flash animation generated by the panorama. In safety program of this project, the protection of the digital exhibits in the digital museum is mainly the watermarking technology for the data in the two style, they are JPE image digital watermarking and Flash animation digital watermarking. And MATLAB is selected as the simulation platform to operate digital watermark operation on the digital exhibits.

5.1 JPEG Color Image Watermark Algorithm

In MATLAB, JPEG color image is taken as an index image or a RGB image. A single RGB image can be thought of as three gray images of three component images called red, green, and blue, respectively. According to the human visual system theorem, the human eye reacts differently to the three colors, the green component image is the clearest, and the preserving image detail is the best. Therefore, in this safety program, we choose to embed the green component image in watermark. Before adding watermark to color image, we need to extract three component images via the channel splitting technique [7, 8].

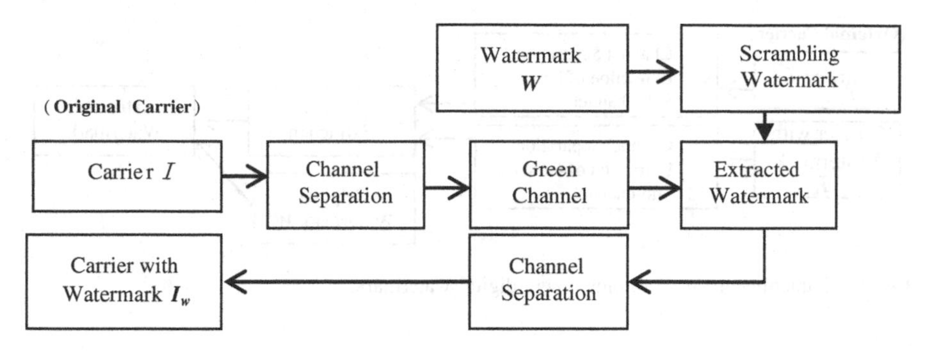

Fig. 4 Embedding process of color image digital watermark

6 Image Watermarking Embedding Algorithm

The process is illustrated in Fig. 4. During the embedding process of the watermark based on DCT transform domain algorithm. First, isolate the green component image from the original carrier image by using the channel splitting technique and then chunk this component image in the size of 8 × 8. Finally, perform the DCT transform. On account of focusing a great deal of energy in the low frequency by DCT transform, this DCT transform domain algorithm has a good robustness. But micro-chunking makes this advantage less obvious. At the same time considering the factors such as loss compression, we adopt the 8 × 8 block DCT transform.

7 Image Watermarking Extracting Algorithm

The process is illustrated in Fig. 5. When extracting the watermark, we need the participation of the original image. That is why we call it non-blind watermarking algorithm.

8 Algorithm Implementation and Results

The results of watermark embedding are shown in Figs. 6 and 7.

The results of watermark extracting are shown in Figs. 8 and 9.

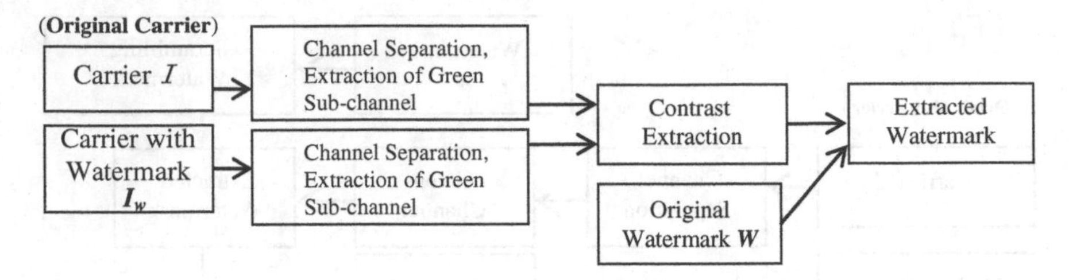

Fig. 5 Extracting process of color image digital watermark

Fig. 6 Original image

Fig. 7 Watermarked image

Fig. 8 Original watermark

Fig. 9 Extracted watermark

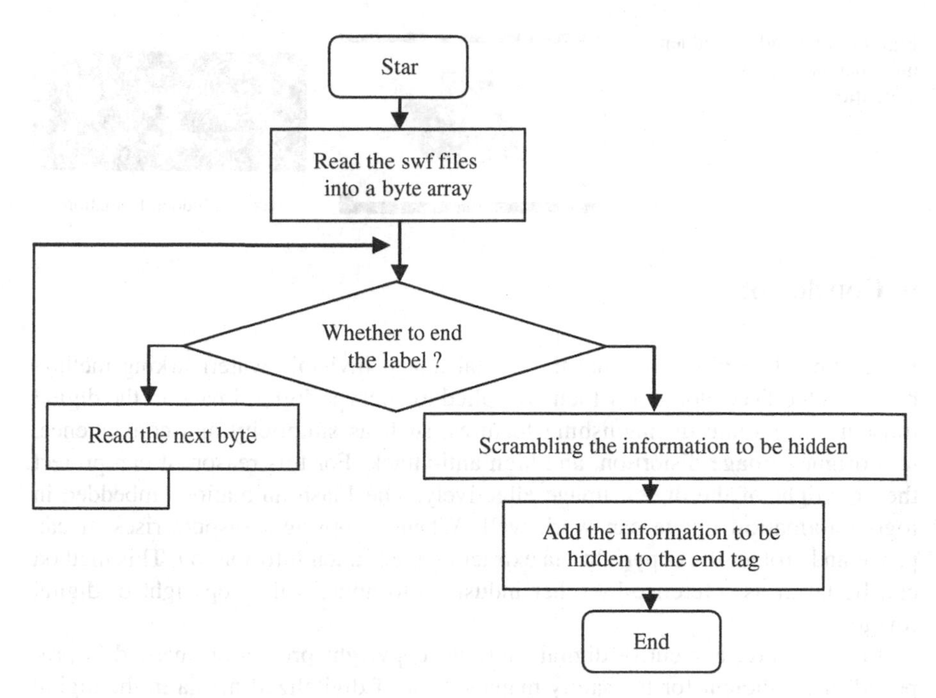

Fig. 10 The flow diagram of *swf* file information hiding

8.1 The SWF Animation Information Hiding Algorithm

Traditional methods of protecting Flash file include embedding hidden information, encrypting, and generating its exe file. These methods are unfavorahie for protecting the copyright of Flash file. While in the digital museum, the main purpose of protecting Flash file is to prevent copyright violations. Therefore, this chapter chooses the meaningful logo image information hiding technique as the topic and aims to protect its copyright and integrity.

The basic principle of the information hiding algorithm for swf animation files is that embed the hiding information in swf files through the analysis of swf file format. e.g., hiding information in the label, hiding information in the frame of the animation, hiding information by adding additional labels after the closing tag, replacing the object properties. In this chapter, we select the information hiding algorithm in which watermark image can be added after the end tag of the Flash animation to protect the Flash exhibits in digital exhibits. The player can ignore the content following the end tag; thus, the Flash animation can work well [9–13].

Watermark embedding and extraction block diagram is shown in Fig. 10, wherein the hidden information extraction process is the inverse of the information hiding.

The results of embedding hidden information of Flash animation are shown in Fig. 11.

(Extract hidden formation)

Fig. 11 Embedding hidden information of Flash animation

9 Conclusion

Experimental results show that the digital image invisible watermarking method based on the DTC domain, which is applied to encrypt digital image in the digital museum, has some distinguishing features, such as simplicity and convenience, less original image distortion, and high anti-attack. For this reason, it can protect the copyright of the digital image effectively. The Flash animation embedded in logo information image can work well. When a copyright dispute rises, it can prove and protect the copyright via extracting the hidden information. This method can be taken as reference by other industries to address the copyright of digital image.

In this study, a security digital museum copyright protection method is proposed. It is efficient for the safety management of digitalized media in the digital museum. But this method still has many could be further improved and perfected. In our digital museum, the exhibits are presented in various types, but in this security scheme, we only consider two the largest and the most important types. In further studies, various types of exhibits should be taken into consideration and require further study. Only then, it was possible to ensure the safety of the entire system.

Acknowledgments This study was supported by the Hainan Natural Science Foundation (No: 612120, 612126), the key Science and Technology Projects of Haikou City in 2010(No: 2012-017), and disciplines project of Hainan Normal University. The authors expressed their thanks together.

References

1. Zhao, X., Hao, L.: The digital watermarking summary. Comput. Eng. Des. 81–85 (2010)
2. Zong, Y.: Digital watermarking technology in digital Museum. Southeast Cult. 81–85 (2010)
3. Cao, LG., Chen, H., Cao, P.: Research on copyright protection of digital museum based on digital watermarking technology. **22**(4), 162–165 (2005)
4. Sun, SH., Lu, ZM., Niao, XM.: Digital Watermarking Technology and Application, pp. 205–207. The Science Press (2004)
5. Junling, Z.: DCT Digital Watermarking-Based Research, pp. 35–39 (2009)
6. Zhang, X., Peiliang, C.: Scrambling technology in digital watermarking. J. Circ. Syst. **6**(3), 32–36 (2001)

7. Kang, Y., Hui, X., Shi, J.: Multimedia data watermarking system and its attack. Comput. Sci. **26** (10), 44–48 (1999)
8. Fu, J.: Static image watermarking algorithm based on DCT domain. **4** (2007)
9. Zhang, X., Zhang, X.: Information hiding algorithm based flash animation. Comput. Eng. **36**(1), 181–183 (2010)
10. Dai, M.: Vector animation file data structure analysis. Jiamusi Educ. Inst. **10**, 413–414 (2012)
11. Shi, R., Qinmao, L., Liu, H.: Vector digital watermarking technology. J. Comput. Appl. **24**(8), 22–24 (2007)
12. Ding, L.: Vector digital watermarking technology research. **6** (2010)
13. Chen, Y.: MATLAB 6 the X graphical programming and the image processing. Xi'an University of Electronic Science and Technology Press (2002)

Research of Travel Route Programming on Hainan Island

De-jun Peng and Cheng-yi Zhang

Abstract Hainan Island is the one of the most famous tropic tourist attractions in our country. Its beautiful scene, pleasant climate, and ample natural resources attract more and more tourists domestic and foreign. This chapter studies how to design reasonable travel route for traveler and develops two models for this travel route programming problem.

Keywords Hainan Island · Travel route programming · Model

1 Introduction

Hainan Island locates in the south end of China, north to Qiongzhou channel she looks each other with Guangdong, south to the widest South China Sea, she is the unique tropic Island province of China, involves the Hainan Island, the Xisha Islands, the Zhongsha Islands, the Nansha Islands and their reef and the sea field. Hainan Island is located between $18^{\circ}10'$–$20^{\circ}10'$N and $108^{\circ}37'$–$111^{\circ}05'$E, and its latitude is similar to that of Hawaii of USA. Hainan Island possesses beautiful scene and pleasant climate. In recent years, along with determination of the national strategic planning of construction international tourism Island, Hainan's tourism industry is booming. Hainan Island attracts more and more domestic and foreign tourists coming to visit scenic spot here, such as the Tianyahaijiao, the Nanshan, and the Yalong Bay. But there are hundreds of scenic spots on Hainan Island, and therefore, how to set reasonable tourism scheme for tourists is a very worth problem to study. In this chapter, the travel route planning problem is studied, and tourism route planning model is established.

D. Peng (✉) · C. Zhang
Hainan Normal University, Haikou, Hainan, People's Republic of China
e-mail: baihepdj@163.com

B.-Y. Cao et al. (eds.), *Ecosystem Assessment and Fuzzy Systems Management*,
Advances in Intelligent Systems and Computing 254, DOI: 10.1007/978-3-319-03449-2_24,

2 Programming Problem of Tour Route

Programming of tour route, which is the problem for several scenic spots, according to the target given beforehand, upbuilds the sequence to travel the scenic spots. The given targets may be the shortest path and time, or may be the lowest expenses. According to the statistics, the most part of tourists coming to Hainan Island will choose two kinds of touring model: one is to circle the Island, thus choosing several scenic spots which are distributed throughout the Island: setting out from *A* city, and after visiting all the scenic spots, coming back to *A* city; the other model is setting out from *A* city, and after visiting all the scenic spots, leaving the Island from *B* city. The first model uses the Hamilton loop problem in graph theory, and the second model uses modified Hamilton loop problem. Next, we will research those two models.

Model 1: Around the Island tour mode, in which the tourist starts from one of the chosen scenic spots *A* and after visiting all the scenic spots, the tourist comes back to *A* city; this mode can be summed up in Hamilton loop problem in graph theory. It is well known that Hamilton loop problem is an open problem, its optimal solution has not been solved yet, and researches of experts and scholars at home and abroad were all concentrated on seeking approximate optimal solution for this problem. Annotation [1] proposed the principle of one-by-one reversion of two sides to obtain the approximate optimum solution, but the traditional solving process of this principle was very complicated, especially in the case of many vertexes, almost unsolvable. Accordingly, annotation [2] puts forward the matrix turning method to realize the solution-seeking process of the principle of one-by-one reversion of two sides and the improved method is relatively easy in programming, concise and independent in procedures, and fast in computing; furthermore, it is suitable for the case of many vertexes. The following simple introduced this method.

3 Basic Conception

Assume that $G = (V, E)$ is a weighted undirected graph, wherein $V = \{v_1, v_2, \ldots, v_n\}$ is the set of vertexes and E is the set of edges. In graph G, each edge corresponds to a real number $w(e)$, which is called the weight of this edge. If two arbitrarily vertexes in graph G were linked by an edge, the graph G is a complete graph.

Distance matrix: For undirected graph G, $A = (a_{ij})_{n \times n}$ is its distance matrix,wherein

$$a_{ij} = a_{ji} = \begin{cases} w_{ij}, & (v_i, v_j) \in E \\ 0, & i = j \end{cases}$$

where w_{ij} be the weight between i and j.

Hamilton loop: Let $G = (V, E)$ be a connected and undirected graph, and the circle which passes every point of G once and only once is called a Hamilton circle of G, called H circle for short.

Optimum H loop: In weighted graph $G = (V, E)$, the Hamilton circle with minimum weight is called the optimum H loop.

4 The Principle of One-by-One Reversion of Two Sides

The principles of one-by-one reversion of two sides [1] are as follows:

1. Arbitrarily take the initial H loop: $C_0 = v_1, v_2, \ldots, v_i, \ldots, v_p, \ldots, v_n, v_1$
2. For all i,j, $1 < i+1 < j < n$, if $w(v_i, v_j) + w(v_{i+1}, v_{j+1}) < w(v_i, v_{i+1}) + w(v_j, v_{j+1})$, then delete edges $w(v_i, v_j)$ and (v_{i+1}, v_{j+1}) in C_0 and add new edges $w(v_i, v_{i+1})$ and $w(v_j, v_{j+1})$, to form a new H loop C, namely $C = v_1, v_2, \ldots, v_{i-1}, v_{j-1}, \ldots, v_{i+1}, v_{j+1}, \ldots, v_n, v_1$
3. Execute Step 2 repeatedly until condition cannot be satisfied. Then, the C obtained is the approximate optimum route.

The principle of one-by-one reversion of two sides is concise and easy to understand, and it implements more convenient, but in the case of more vertexes, it implements more complicated, even unsolvable. Annotation [2] puts forward the matrix turning method to realize the solution-seeking process of the principle of one-by-one reversion of two sides, and the improved method is relatively easy in programming, concise and independent in procedures, and fast in computing; the following simply introduced the step for seeking optimum H loop by the matrix turning method.

Step 1: Construct an initial fictitious H circle in a complete weighted graph: $C_0 = v_1, v_2, \ldots, v_i, \ldots, v_p, \ldots, v_n, v_1$, wherein v_1 is the starting point, v_n is the leaving point, and the sequence of points in between is arbitrary. According to the sequence of those points, we get a distance matrix $A = (a_{ij})_{n+1,n+1}$, wherein

$$a_{ij} = a_{ji} = \begin{cases} w_{ij}, & (v_i, v_j) \in E \\ 0, & i = j \end{cases}$$

where w_{ij} be the weight between i and j.

Step 2: Create a permutation sequence frame for points above the first row of A and under the last row of A; meanwhile, add a zero rank before the first rank and after the last rank each. Then, we get matrix A, and the total weight of C_0 is $\sum_{i=2}^{n-2} A(i, i+1)$.

Step 3: In matrix A, for all i, j, $2 < i+1 < j < n-2$, if $A(i,j) + A(i+1, j+1) < A(i, i+1) + A(j, j+1)$, we turn rows from $(i+1)$ to j and those from $(i+1)$ to j, thus forming a new distance matrix. The sequence of points in A changes to $C = v_1, v_2, \ldots, v_{i-1}, v_{j-1}, \ldots, v_{i+1}, v_{j+1}, \ldots, v_n$.

Step 4: Execute Step 3 repeatedly until condition $A(i,j)+A(i+1,j+1) < A(i,i+1)+A(j,j+1)$ cannot be satisfied. Then, the C obtained is the approximate optimum route.

This process can be realized by computer, and the MATLAB program is as follows:

Program: seek for optimum H circle function $[a, b] = h(e)$ % e is distance matrix with points sequence frame of initial H circle.

```
n = size(e); % compute the dimension of distance matrix.
    for i = 2: n − 2;
    for j = i + 1: n − 2; % circulation for 2 to n − 2.
    if e(i, j) + e(i + 1, j + 1) < e(i, i + 1) + e(j, j + 1);
    a = horzcat(e(:, 1: i), e(:, j:−1: i + 1), e(:, j + 1: n)); % turn from the (i + 1)th
    rank to the jth rank of e.
    b = vertcat(a (1: i, :), a(j:−1: i + 1, :), a(j + 1: n, :)); % turn from the
    (i + 1)th row to the jth row of a.
    e = b;
                    end
            end
    end
    s = 0;
    for i = 2: n − 2;
    s = s + e(i, i + 1); % calculates total weight of H circle after optimization.
    end
    e
    s
```

Model 2: Setting out from A city and, after visit all of the scenic spots, leave from B city. This mode is not the problem of circle route, to which the Hamilton loop problem is not applicable. At present, the second model is rarely researched on by scholars, and, through analogy, we find that it can be solved easily by slightly improving the principle of one-by-one reversion of two sides. The improved method is illustrated as follows:

Step 1: Construct an initial fictitious H circle in a complete weighted graph $C_0 = v_1, v_2, \ldots, v_i, \ldots, v_p, \ldots, v_n, v_1$, wherein v_1 is the starting point, v_n is the leaving point, the sequence of points in between is arbitrary, and edge v_n, v_1 is fictitious. Note that in the whole computing process, this fictitious side is kept unchanged, and the real route is $C_0' = v_1, v_2, \ldots, v_i, \ldots, v_p, \ldots, v_n$. According to the sequence of those points, we get a distance matrix $A = (a_{ij})_{n+1,n+1}$, wherein

$$a_{ij} = a_{ji} = \begin{cases} w_{ij}, & (v_i, v_j) \in E \\ 0, & i = j \end{cases}$$

where w_{ij} be the weight between i and j.

Step 2: Create a permutation sequence frame for points above the first row of A and under the last row of A; meanwhile, add a zero rank before the first rank and after the last rank each. Then, we get matrix A', and the total weight of $C_0{}'$ is $\sum_{i=2}^{n-3} A'(i, i+1)$.

Step 3: In matrix A', for all i, j, $2<i+1<j<n-3$, if $A'(i, j)+A'(i+1, j+1)<A'(i, i+1)+A'(j, j+1)$, we turn rows from $(i+1)$ to j and those from $(i+1)$ to jth, thus forming a new distance matrix. The sequence of points in A' changes to $C' = v_1, v_2, \ldots, v_{i-1}, v_{j-1}, \ldots, v_{i+1}, v_{j+1}, \ldots, v_n$.

Step 4: Execute Step 3 repeatedly until condition $A'(i, j)+A'(i+1, j+1) < A'(i, i+1)+A'(j, j+1)$ cannot be satisfied. Then, the C' obtained is the approximate optimum route. This process also can be realized by computer, and it only has to do little change to a parameter of the loop mode program, which we omit here.

5 Practical Application

Case: Among the two travelers visiting Hainan Island, one plans to circle travel the Island, starting from Haikou city and then coming back to Haikou and finally leaving the Island from Haikou after a circle traveling; the another traveler plans to start from Haikou city and depart the Island from Sanya city after traveling all over the scenic spots chosen. Both travelers choose the following seven scenic spots, which are the typical of Hainan Island: the Volcanic Vent Geology Park (v_1), Seven Fairy Mountain (v_2), Boao Thousand Boats Bay (v_3), Stone Flower Water Tunnel (v_4), Xinglong Tropic Botanic Garden (v_5), Five Fingers Mountain (v_6), and Ends of the Earth (v_7).v_i, $i = 1, 2, \ldots, 7$ stands for those seven scenic spots, respectively. Suppose that there is a path connecting every two scenic spots, Table 1 shows the distance between every two spots (unit: km).

Please design a reasonable travel route for those two travelers.

Since the seven scenic spots scattered in different cities or counties of Hainan Island, Table 2 lists the cities or counties where the scenic spots belong to.

For the first traveler, his choice is circle travel, namely the model one mentioned above. As the Volcanic Vent Geology Park belongs to Haikou city, we can set up the initial circle $C_0 = v_1, v_2, \ldots, v_7, v_1$. The corresponding distance matrix after edging A is as follows:

Table 1 Distance between every two spots (unit: km)

	v_1	v_2	v_3	v_4	v_5	v_6	v_7
v_1	0	258.8	124.6	153	182.2	319	287.2
v_2	258.8	0	162.5	178.1	99.4	69.8	97.2
v_3	124.6	162.5	0	227	87.4	222.6	191.8
v_4	153	178.1	227	0	334	134	230.4
v_5	182.2	99.4	87.4	334	0	159.5	128.1
v_6	319	69.8	222.6	134	159.5	0	133.2
v_7	287.2	97.2	191.8	230.4	128.1	133.2	0

Table 2 Cities or counties where the scenic spots belong to

Scenic spot	v_1	v_2	v_3	v_4	v_5	v_6	v_7
City/county	Haikou	Baoting	Qionghai	Danzhou	Wanning	Wuzhishan	Sanya

$$A = \begin{bmatrix} 0 & 1 & 2 & 3 & 4 & 5 & 6 & 7 & 1 & 0 \\ 0 & 0 & 258.8 & 124.6 & 153 & 182.2 & 319 & 287.2 & 0 & 0 \\ 0 & 258.8 & 0 & 162.5 & 178.1 & 99.4 & 69.8 & 97.2 & 258.8 & 0 \\ 0 & 124.6 & 162.5 & 0 & 227 & 87.4 & 222.6 & 191.8 & 124.6 & 0 \\ 0 & 153 & 178.1 & 227 & 0 & 334 & 134 & 230.4 & 153 & 0 \\ 0 & 182.2 & 99.4 & 87.4 & 334 & 0 & 159.5 & 128.1 & 182.2 & 0 \\ 0 & 319 & 69.8 & 222.6 & 134 & 159.5 & 0 & 133.2 & 319 & 0 \\ 0 & 287.2 & 97.2 & 191.8 & 230.4 & 128.1 & 133.2 & 0 & 287.2 & 0 \\ 0 & 0 & 258.8 & 123.6 & 153 & 182.2 & 319 & 287.2 & 0 & 0 \\ 0 & 1 & 2 & 3 & 4 & 5 & 6 & 7 & 1 & 0 \end{bmatrix}$$

Deploying the MATLAB program, we can get the approximate optimum H circle $C = v_1, v_4, v_6, v_2, v_7, v_5, v_3, v_1$ and the travel route is as follows: Volcanic Vent Geology Park (v_1) → Stone Flower Water Tunnel (v_4) → Five Fingers Mountain (v_6)→ Seven Fairy Mountain (v_2) → the Ends of the Earth (v_7) → Xinglong Tropic Botanic Garden (v_5) → Boao Thousand Boats Bay (v_3) → Volcanic Vent Geology Park (v_1). The total mileage is 794.1 km. Through analyzing the computing result and the traffic net of Hainan Island, we know that this travel route is an optimized scheme for circle travel indeed.

For the other traveler, he chooses model 2, starting from Haikou city and then depart from Sanya city after traveling over the scenic spots. As the Volcanic Vent Geology Park is located in Haikou, we can start from Volcanic Vent Geology Park and finally depart from the Ends of the Earth in Sanya. Set up the initial fictitious circle $C_0{}' = v_1, v_2, \ldots, v_7, v_1$, wherein the fictitious edge v_7, v_1 stands for the fractious edge between Sanya and Haikou, which maintains unchanged in the whole computing process. The order of other edges can be adjusted with the principle of one-by-one reversion of two sides before finally determining the approximate optimum route. We can also employ the MATLAB program to obtain the approximate optimum fictitious H circle $C' = v_1, v_4, v_6, v_2, v_3, v_5, v_7, v_1$. The real route is

$C = v_1, v_4, v_6, v_2, v_3, v_5, v_7$, Volcanic Vent Geology Park (v_1) → Stone Flower Water Tunnel (v_4) → Five Fingers Mountain (v_6) → Seven Fairy Mountain (v_2) → Boao Thousand Boats Bay (v_3) → Xinglong Tropic Botanic Garden (v_5) → the Ends of the Earth (v_7). The total mileage is 734.8 km. It is a reasonable route.

6 Conclusion

This intelligent travel route programming model boasts widely application, which is verified with the example quoted in this chapter. Besides, if the weight of the weighted graph of scenic spots changes to time, then this model can be used to solve the problem of optimum traveling timetable. If the weight changes to charge, then this model can be applied to solve the problem of optimum traveling cost as well.

References

1. Zhao, J., Dan, Q.: Mathematical Modeling and Mathematical Experiments. Higher Education Press, Beijing (2004)
2. Yang, X., Chen, Z.: Seeking the best Hamilton circle through matrix turning. J. Logistical Eng. Univ. **24**(1), 102–106 (2008)
3. Liu, Z.: Research for present development situation and countermeasure of tourism in Hainan. Sci. Technolo. Inf. **6**, 54 (2013)

Path Optimization Method of Logistics Distribution Based on Mixed Multi-Intelligence Algorithms

Rui-qiong Zhou and Jun-kuo Cao

Abstract In order to improve the efficiency and benefit for logistics distribution system, a path optimization method based on mixed multi-intelligence algorithms for vehicle routing problem was developed. In the process of proposed method, it firstly calculates the shortest paths between nodes of road network by Floyd algorithm and then obtains the merge distribution paths by means of saving method. Finally, genetic algorithm is employed to optimize the merge distribution paths. The experimental results show that the proposed method of logistics distribution can well solve the complex network conditions and lower the cost of logistics distribution, illustrating that the proposed model is practical and effective.

Keywords Logistics distribution · Floyd algorithm · Shortest path · Genetic algorithm

1 Introduction

In modern society, logistics, business flow, and information flow are called the three pillars of the economy, logistics management system, reasonably can create considerable economic benefits for the enterprise. At the same time, the development of economic globalization makes the enterprise procurement, warehousing, sales and distribution becomes more and more complex, the competition between enterprises is not only the quality and performance of competing

R. Zhou
Center of Network, Hainan Normal University, Haikou 571158, China

J. Cao (✉)
College of Information Science and Technology, Hainan Normal University,
Haikou 571158, China
e-mail: junkuocao@gmail.com

B.-Y. Cao et al. (eds.), *Ecosystem Assessment and Fuzzy Systems Management*,
Advances in Intelligent Systems and Computing 254, DOI: 10.1007/978-3-319-03449-2_25,

products, but also including logistics capability of competition. As an important part of enterprise management, logistics is based on the user's orders, in the distribution center to distribute goods to the consignee activity [1].

In recent years, many scholars have studied the logistics distribution system, as the distribution route optimization in response to a series of customer demand design appropriate, thereby reducing distribution costs, shortening the total mileage of vehicles, and improving vehicle utilization [2]. At present, to solve the problem of logistics distribution path optimization algorithm has two categories: exact algorithms and heuristic algorithm for [3]. Precision of the algorithm is the algorithm that can find the optimal solutions, including the branch and bound method, the network flow algorithm, and dynamic programming. The exact algorithm is only suitable for small problem or for solving the local optimal problem. Heuristic algorithm is a kind of intuition or experience structure algorithm based, to obtain a satisfactory solution, rather than the optimal solution. The heuristic algorithm can quickly obtain a satisfactory solution and can provide more effective solution of logistics distribution. However, a heuristic algorithm, a series of example, conservation law, cannot solve the complex node problem; genetic algorithm is an effective adaptive iterative heuristic probabilistic global search method, but it was "premature," and the search efficient of [4, 5].

In this case, the paper puts forward an optimization method of logistics distribution route based on mixed multi-intelligence algorithms, which wants to logistics enterprise to provide practical and efficient logistics distribution route optimization scheme.

2 Logistics Distribution Vehicle Scheduling Description

Logistics distribution vehicle scheduling problem that how to reasonably, higher efficiently, and lower costly solves freight. Distribution scheme should include two related links: [4, 5]

1. Which customers need to be assigned to the same path, namely what cargo boxes need to be arranged in the same car?
2. The connection order of the customer in each delivery routes. The optimal solution of distribution vehicle routing problem is actually a most efficient transport scheme. It should be made clear that sent vehicle model, the number of vehicles, and the route of each car. Implementing this distribution scheme cannot only meet customer demand, but also make the lowest total transport cost.

Combined with reality in general, small- and medium-sized logistics enterprise actual situation, given objective logistics center has a distribution center and multiple distribution centers, and each distribution center for most of the business is focused on distribution center unified delivery, only to meet the following conditions, is sent directly by the distribution center:

1. The distribution center close a hub, and the freight rate $r >= Z_0$.
2. A distribution line of freight rate $r >= Z_0$, and the distribution costs of the distribution center less than the cost which the distribution center delivers to hub.

Assume the target logistics center has K vehicle models, k represents a vehicle model ($k = 1, \ldots, K$), the distribution center number is 0, the rest of the distribution center number with $i(i = 1, \ldots, m)$, each task point number with $j(j = 1, \ldots, n)$, the mission point j demand for q_j. If the minimal total cost is final target, then the mathematical model of logistics distribution problem is:

$$\min Z = \sum_{I=1}^{m} \sum_{J=1}^{n} \sum_{K=1}^{K} c_{ij} x_{ij}^{(k)}$$

$$\begin{cases} \sum_{j=1}^{n} x_{ij}^{(k)} < a_j^{(k)}, & (i = 1, 2, \ldots, m), \\ \sum_{j=1}^{n} q_j < \sum_{i=1}^{n} \sum_{k=1}^{K} w^{(k)}, & (j = 1, 2, \ldots, n;\ i = 1, 2, \ldots, m;\ k = 1, 2, \ldots, K), \\ x_{ij}^{(k)} \geq 0 \text{ and is an integer}, & (i = 1, 2, \ldots, m;\ j = 1, 2, \ldots, n;\ k = 1, 2, \ldots, K). \end{cases} \quad (1)$$

In formula (1), $x_{ij}^{(k)}$ represents k model vehicle which is sent from point i to point j, $a_i^{(k)}$ represents the quantity of k model vehicle in distribution center i, q_j represents the demand of j, $w^{(k)}$ represents load capacity of k model vehicle, c_{ij} represents the cost from point i to point j, which includes distribution vehicle fixed cost and vehicle operating cost: $c_{ij} = c_0 + c_1 t_{ij}$, C_0 represents distribution vehicle fixed cost, c_1 represents the coefficient which is relative to the running mileage charge. If the $c_1 = 0$ and $c_0 > 0$, then the model objective is to minimize the number of used vehicles.

3 Proposed Logistics Distribution Path Optimization Method

This method adopts precision algorithm and heuristic algorithm to solve logistics distribution vehicle scheduling problem. To overcome the lack of saving method, this method firstly uses Floyd algorithm to get the shortest path between each point, then employs the saving method expand loop, subsequently adopts genetic algorithm to seek optimal solution, which can overcome the disadvantages of premature and low searching efficiency of genetic algorithm. In this case, the method includes three stages to find the optimal distribution scheme.

First stage: Using Floyd algorithm to calculate the entire shortest paths and their path toward between two nodes to obtain an effective distribution vehicle route scheme.

Second stage: Employing saving method to expand the loop in permitted range of distribution vehicle load to optimize line scheme.

Third stage: Adopting genetic algorithm in moderate feasible solution to look for the optimal goal in the solution of adjacent domain to overcome the premature convergence and the lower searching efficiency of genetic algorithm.

3.1 Solving the Shortest Path Based on Floyd Algorithm

Floyd algorithm is one of the most effective algorithms which solves the shortest path between any two nodes in a general network, if find the shortest path from vertex i to j. If the edge from i to j, from Vi to Vj there is a path of length edges (i) (j), the path is not necessarily the most short path, the n test is required. First, consider the path (i, 0, j), if it exists, is (i, j) and (i, 0, j) path length and short length is from i to intermediate vertex j numbers less than 0 shortest path. If the path to add a vertex 1, and that is to say, if (i, …, 1) and (1, …, j) are intermediate vertex currently, find number of not more than 0 shortest path, then (i, …, 1, …, j), there may be from I to intermediate vertex j numbers less than 1 shortest path. If it is got from the i to the j intermediate vertex, the number is not greater than the 0, then the shortest path is compared and selected intermediate vertex from the serial number is not greater than the shortest path 1; add a vertex 2 and continue to test by analogy. So, after n times, finally obtained will be the shortest path from i to j. According to this method, can obtain the shortest path between points.

After getting the shortest path between any two points, the proposed method, respectively, judges $x(x = 1, 2, \ldots, n)$ whether is leaf node of spanning tree which is composed by distribution sites and client nodes, and then according to the principle, which the vehicle runs from distribution center to leaf node and back-track, distributes customers, respectively.

3.2 Merging Paths Based on Saving Method

Saving method which was presented by Hossain et al. [6] aims to solve the problem of logistics distribution, and its basic principle is as follows:

Given the distances between distribution center and the two customers are d_{oi} and the d_{oj}, respectively, and the distance of between i and j is d_{ij}. If a distribution center, respectively, delivers to two customer i and j, then it needs two vehicles: round-trip route are (0, i, 0) and (0, j, 0) and a total distance is $y_1 = 2(d_{oi} + d_{oj})$.

But, if a vehicles is used circuitly as delivery to two customer i and j, then it needs only a vehicle and the route is (0, i, j, 0), total distance is $y_1 = d_{oi} + d_{oj} + d_{ij}$. We assume that if it was not a straight line at the distribution center, i and j, then $(d_{oi} + d_{oj}) > d_{ij}$. So, the second distribution scheme is better than the first, and the total savings is $y = (d_{oi} + d_{oj} - d_{ij}) > 0$.

By means of Floyd algorithm of the first stage, we obtain all shortest path value d_{ij} between two nodes value and related path toward. If the shortest path between any two nodes is considered as the edge of the two nodes, then we can get a new map, and then use the saving method merge paths. We firstly calculate all the savings $d_{oi} + d_{oj} - d_{ij}(i, j = 0, 1, \ldots, n)$ according to the principle of the save method, and sorted from small to large, and then expand its circuit. In process of loop expansion, we firstly select the node i and the node j in which the savings is largest and form a loop (0, i, j, 0) and then choose the second large savings. If the load conditions permit, we constantly merge loop, until the entire loop. Obviously, each with the total path the greatest savings, path length of each increase is the minimum current, and the distribution route of the total distance has been optimized.

3.3 Path Optimization Based on Genetic Algorithm

Genetic algorithm, which is similar to the natural evolution, acts on the gene on the chromosome to find good chromosomes to solve the problem. In genetic algorithm, it uses random way to produce several codes of the problem to be solved, namely chromosome, and forms the initial group. Through fitness function which gives each individual a numerical evaluation, it obsoletes lower-fitness individuals and chooses higher-fitness individuals to participate in the genetic operation. After genetic operation, individual species forms the next generation new species [7–9].

After the first two stages of solving method, we get merged distribution path set, and record the number of all kinds of box according to the merged record route, and count the number of box in vehicle. This stage uses genetic algorithm optimize the distribution path, the process is as follows:

1. Randomly generating N paths of initial gene sequence, the N paths which calculated the total costs (estimated costs) are set as the initial population s.
2. Sorting population according to the estimated cost of the path.
3. Randomly selecting two paths a and b in the group s.
4. According to a predefined probability, a and b are act on random mutation operation, a new distribution path ($s(i)$, $s(i + 1)$) is generated.
5. Calculating the cost of path ($s(i)$, s($i + 1$)) and comparing with paths of population, obsolescing the two paths which distribution costs are the largest.
6. Determining whether a population reaches the preset evolution times, if the termination condition is met, then algorithm is stopped, otherwise, go to (3).

7. Selecting the optimal path $s[0]$ of the offspring population as the returned results.
8. According to the optimal path, assigning all the cargo boxes to vehicles.

3.4 Optimized Logistics Distribution Path Process

Assuming that logistics center consists of n distribution centers and m destinations, first, we employ Floyd algorithm iteratively to calculate the shortest distances between any two nodes (including from distribution center to the destination and from one destination to other destination). Second, according to the preset delivery rules, we assign different cargo boxes to n distribution centers. For the distribution center k (the initial value of k is 1), we initialize the delivery routes and costs which the distribution center k assigns every cargo box to m destinations and calculate merged paths and distribution costs $c(i, j)$ which the distribution center k assigns to every cargo box destination i and j based on saving method. Subsequently, we choose the next delivery destination $i + 1$ (the initial value of i is 1) of the distribution center k do the same above operations, until all the delivery destination are merged into path set. And then, we adopt genetic algorithm to optimize the merged path set and obtain the best distribution paths from the distribution center k to m destinations. Finally, we select the next distribution center $k + 1$ to do the same above operations, until the best distribution paths are gotten from n distribution centers to m destinations. Figure 1 shows the optimized logistics distribution path process.

4 Experimental Verification

In order to test the effectiveness of the path optimized method of logistics distribution, the experiment simulates logistics center to validate the method, the relevant data is as follows:

1. Logistics center has two distribution centers A and B, which A is hub center and B is the common distribution center.
2. Each distribution center has large, medium, and small three kinds of vehicles; length, width, and height are, respectively, $12 \times 2 \times 1.8$ m, $8 \times 2 \times 1.8$ m, and $4 \times 2 \times 1.8$ m;
3. Each distribution center has large, medium, and small three kinds of cargo boxes; length, width, and height are, respectively, $1 \times 2 \times 0.9$ m, $1 \times 1 \times 0.9$ m, $1 \times 0.5 \times 0.5$ m;
4. Cost per kilometer of large, medium, and small three kinds of vehicles are, respectively, 4, 3, and 2 RMB.

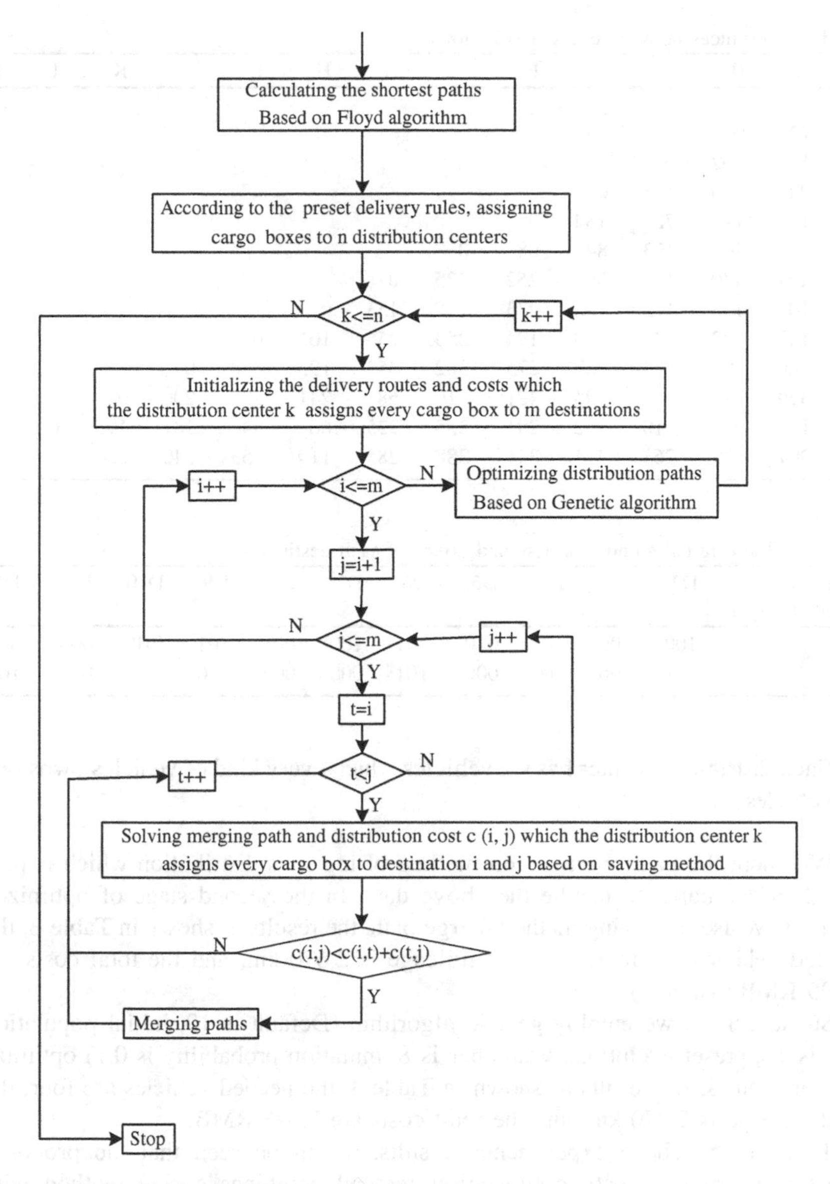

Fig. 1 Optimized logistics distribution path flowchart

5. Fixed send cost of large, medium, and small three kinds of vehicles are, respectively, 400, 350, and 300 RMB.
6. Logistics center is responsible for around 11 (*C*, *D*, *E*, …, *H*) destinations.
7. Distances between every destination are shown in Table 1.
8. The number of assigned boxes of each destination is shown in Table 2, and the number of each box which is sent to the same destination is less than 9.

Table 1 Distances between every destinations

	A	B	C	D	E	F	G	H	I	J	K	L	M
A	0												
B	226	0											
C	145	302	0										
D	92	286	195	0									
E	152	140	72	184	0								
F	91	307	123	84	68	0							
G	249	430	144	306	252	235	0						
H	102	138	172	153	233	178	153	0					
I	143	172	139	228	194	253	256	106	0				
J	269	192	225	337	275	362	352	193	148	0			
K	329	124	385	495	421	410	68	241	236	200	0		
L	155	96	210	352	291	235	225	66	94	181	195	0	
M	207	129	263	411	341	288	285	119	153	240	230	72	0

Table 2 The original number of assigned boxes of each destination

Destination Point of origin	D2	D3	D4	D5	D6	D7	D8	D9	D10	D11	D12
D0	100	200	101	010	201	100	001	101	010	000	000
D1	01	000	000	000	101	000	000	010	100	010	100

9. Each distribution center has six vehicles which every kind of vehicles owns two vehicles.

We adopt the path optimization method of logistics distribution which is proposed in this paper to handle the above data. In the second stage of optimized method, we use the saving method merge path, the results as shown in Table 3, the needed vehicles are four, the total mileage is 2,670 km, and the total costs are 8,795 RMB (Table 4).

Subsequently, we employ genetic algorithm (Default is 10, initial population size is 10, preset evolutionary number is 8, mutation probability is 0.1) optimize delivery routes, the results as shown in Table 3, the needed vehicles are four, the total mileage is 2,370 km, and the total costs are 7,648 RMB.

Through the above experimental results, it can be seen that the proposed logistics distribution path optimization method combines saving method with genetic algorithm to optimize distribution paths and saves logistics cost, it shows this method the advantages in solving logistics distribution path selection problem. The distribution route scheme which is obtained by the proposed method can provide more intuitive and specific decision support for logistics companies and has certain practical value.

Table 3 The number of assigned boxes of each destination after merging paths based on saving method

Path number	Vehicle type	Delivery routes	Total mileage	Total volume	Cost
1	Large	0->2->6->4->5->3->0	785	33.1	3,540
2	Medium	0->7->8->9->0	625	20.7	2,225
3	Small	0->10->1->0	679	7.2	1,568
4	Small	1->10->12->11->0	581	13.72	1,462

Table 4 The number of assigned boxes of each destination after optimizing paths based on genetic algorithm

Path number	Vehicle type	Delivery routes	Total mileage	Total volume	Cost
1	Medium	0->2->6->4->5->0	700	25.65	2,450
2	Small	0->3->0	184	7.42	668
3	Medium	0->7->8->9->10->1->0	906	27.9	3,068
4	Small	1->10->12->11->0	581	13.72	1,462

5 Conclusion

This chapter studies logistics distribution path selection problem and establishes a new logistics distribution path optimization method. The method firstly uses Floyd algorithm to solve the shortest path between distribution sites and then employs saving method expand its circuits, subsequently, adopts genetic algorithms to seek optimal distribution route, which not only makes up for the deficiency of saving method, but also overcomes the problem of lower search efficiency and premature convergence of genetic algorithm. The experimental results show that the proposed method can realize complex logistics distribution route optimization, to a certain extent, reduce logistics cost, so that it can achieve good effect in practical application.

An effective adaptive iterative heuristic probabilistic global search method, but it was “premature”, and the search efficiency of [4].

Acknowledgments This work is supported by International Science & Technology Cooperation Program of China (2012DFA11270); Hainan International Cooperation Key Project (GJXM201105); Hainan Provincial Natural Science Fundunder (612120, 612126); and National Natural Science Fund (No. 61362016).

References:

1. Deng, X., Wang, X., Ada, S.F.N., Lin, Y.: Analysis of the total factor productivity of China logistics enterprises. Syst. Eng. **26**, 1–9 (2008)
2. Li, Y.: Research on Logistics Distribution Vehicle Scheduling Decision Method Based on Multi Agent. Central South University, Traffic and Transportation Engineering (2012)

3. Qi, C.: Application of improved discrete particle swarm optimization in logistics distribution routing problem. Procedia Eng. **15**, 3673–3677 (2011)
4. Lang, M., Hu, S.: Study on optimization of logistics distribution routing problem with hybrid genetic algorithm. Chin. J. Manage. Sci. 51–56 (2012)
5. Asgharzadeh, A., Valiollahi, R., Raqab, M.Z.: Estimation of the stress–strength reliability for the generalized logistic distribution. Stat. Methodol. **15**(1), 73–94 (2013)
6. Hossain, M., Wright, S., Petersen, L.A.: Comparing performance of multinomial logistic regression and discriminant analysis for monitoring access to care for acute myocardial infarction. J. Clin. Epidemiol. **55**(4), 400–406 (2012)
7. Perrier, N., Agard, B., Baptiste, P.: A survey of models and algorithms for emergency response logistics in electric distribution systems, Part I: Reliability planning with fault considerations. Comput. Oper. Res. **40**(7), 1895–1906 (2013)
8. Balakrishnan, N., Saleh, H.M.: Relations for moments of progressively type-II censored order statistics from half-logistics distribution with applications to inference. Comput. Stat. Data Anal. **55**(10), 2775–2792 (2011)
9. Nassar, M.M., Elmasry, A.: A study of generalized logistics distributions. J. Egypt. Math. Soc. **20**(2), 126–133 (2012)

Association Rules Mining Based on Minimal Generator of Frequent Closed Itemset

Xiao-mei Chen, Chang-ying Wang and Han Cao

Abstract As there is rules redundancy in mining association rules with traditional methods, an improvement is presented by increasing combinations of itemsets based on minimal generator method; thus, the loss of minimal generator items of frequent closed itemsets is avoided, and the concept and process of rule generation algorithm are further simplified, which makes the algorithm more readable and easier to implement. Experimental results show that the algorithm provide better integrity of minimal generator items and more effective association rules without increasing timing cost.

Keywords Frequent closed itemset · Adjacent map · Minimal generator

1 Introduction

When mining association rules with traditional frequent itemset, plenty of redundancy association rules may be generated, leading to low mining efficiency [1]. To solve the problem, Pasquier and others propose mining association rules based on frequent closed itemset [2], which decreases the redundancy of rules and thus achieve improvement of mining efficiency. However, people still puzzles when choosing the required association rules. Thus, the Ref. [3] proposes the notion of simplest association rules and corresponding mining algorithm. On the other hand, some people propose to get what they called the degree of interest by

X. Chen · C. Wang (✉)
College of Computer and Information Science, Fujian Agriculture and Forestry University, Fuzhou 350002, China
e-mail: wangchangying@fjau.edu.cn

H. Cao
Department of Computer Science, Shaanxi Normal University, Xian 710062, China

B.-Y. Cao et al. (eds.), *Ecosystem Assessment and Fuzzy Systems Management*, Advances in Intelligent Systems and Computing 254, DOI: 10.1007/978-3-319-03449-2_26,

means of statistic method, which are than used for weights of restriction to filter those uninterested association rules [4]. This is mostly a subjective scheme. Therefore, there is an idea that more effective rules can probably be obtained by combining objective approaches with subjective ones. Reference [1] proposes an association rules mining algorithm based on adjacent graph of frequent closed itemset, which greatly eliminates the redundancy of mined association rules. However, some useful association rules are missed with the algorithm. Thus, we propose a simple yet effective association rule mining algorithm by means of adjacent table and minimal generating subset of frequent closed itemset.

2 Generating Minimal Generator of Frequent Closed Itemset

2.1 Basic Concept

Definition 1 Frequent Closed Itemset: For an itemset X, if there exists no proper superset $Y \supset X$ that enable $\sup(Y) = \sup(X)$, X may be referred to as a frequent closed itemset.

Definition 2 Association Rule: Assume that $I = \{i_1, i_2, \ldots, i_n\}$ is a set of items, $D = \{d_1, d_2, \ldots, d_m\}$ is a transaction set, $X, Y \subset I$ and $X \cap Y = \phi$, $X \Rightarrow Y$ may be referred to as an association rule while X, Y may be referred to as the premise and conclusion of $X \Rightarrow Y$, respectively. The support degree of itemset $X \cup Y$ may be referred to as the support degree of the association rule and recorded as $\sup(X \Rightarrow Y)$. The confidence coefficient of association rule $X \Rightarrow Y$ may be recorded as [5]

$$\begin{aligned}
&\mathrm{conf}(X \Rightarrow Y) \text{ where,}\\
&\sup(X \Rightarrow Y) = \sup(X \cup Y)\\
&\mathrm{conf}(X \Rightarrow Y) = \frac{\sup(X \cup Y)}{\sup(X)} \times 100\,\%
\end{aligned}$$

Definition 3 Adjacency List of Frequent Closed Itemset (FAL): Data structure for storing the adjacency graph of frequent closed itemset.

Definition 4 Adjacency Graph of Frequent Closed Itemset Digraph [1] with the following special structures:

1. Combination of one root node and nodes constituted by all frequent closed itemset;
2. Each node includes three domains: frequent closed itemset, support degree for this itemset and location of forward node;

3. Each node is the direct superset of its direct predecessor (direct superset means that if $X \subset Y$, the itemset Z does not exist, which results in $X \subset Z \subset Y$, and Y is referred as the direct superset of X).

Definition 5 Minimal Generator (MG): Minimal Generator: Itemset g is the minimal generator of frequent closed itemset L, if and only if g can generate the maximal closed itemset L, i.e., support degree of itemset g and that of L are equal, and L does not act as the maximal closed itemset for itemset $g' \subset g$ [6].

Definition 6 Minimal Generator Set: Set for all minimal generator of frequent closed itemset L [6].

2.2 Algorithm for Mining Minimal Generator of Frequent Closed Itemset

Exhaustion method can be adopted for the generation of minimal generator itemset, but it is low in efficiency. To overcome this weakness, this chapter introduces three properties of the minimal generator and conducts mining with the help of adjacency graph of frequent closed itemset. In this way, adjacency graph should be constituted at first which provides the base for the generation of minimal generator.

Property 1 *Assume that f is one node in the adjacency graph, and the direct predecessor of frequent closed itemset, if the direct predecessor of f is a root node, each item of f is its minimal generator.*

Property 2 *Assume that f is one node in the adjacency graph, U is the union set of all direct predecessors of frequent closed itemset, U is the union set of all direct predecessors of f, and if* $f/U \neq \phi$*, each item of f/U is minimal generator of f.*

Property 3 *Assume that f is one node in the adjacency graph and frequent closed itemset and* $X1, X2, \ldots$ *are its all direct predecessors; then, any itemset reorganized by any two or more non-void subsets from* $S1 \subseteq f/X1,\ S2 \subseteq f/X2, \ldots$ *and excluded in any direct predecessor of f is the minimal generator of f.*

2.2.1 Organiztion of the Adjacency Graph for Frequent Closed Itemset Adjacency Graph of Frequent Closed Itemset

As indicated in Ref. [1], all association rules between non-adjacent frequent closed itemset can be reduced according to the transitivity of rules, as long as rules between adjacent frequent closed itemset have been mined. So we propose an approach of making adjacency graph and restore it by adjacency list according to definition of adjacency graph. The adjacency lists what we adopt contain structure as follows:

1. Each row represents each node in the adjacency graph;
2. Each adjacency list includes five fields, namely itemset element number, frequent closed itemset, corresponding support of itemset, location of direct predecessor of itemset (which can be deemed as the location of father node);
3. Root node in the adjacency graph is displayed with void set.

2.2.2 Algorithm Framework for Generation of Minimal Generator

```
Input: Adjacency List of Frequent Closed Itemset FAL FAG-adjacency graph of
frequent closed itemset
Output: Minimal generator Set MG for every node MG-minimal generator itemset for
each node
for every f ∈ FAL
    if direct predecessor of f is root node
          Add each item of f to MG as the minimal generator of f, and mark
        corresponding support and node location.
    else
          U is the union set of all direct predecessors;
          diff_U is the union set of f/ every direct predecessors;
          n_u is the number of item of diff_U;
          if f/U ≠∅
                Add each item of f / U to MG, and mark corresponding support
          and node location;
          end
          if direct predecessor number of f, i.e. f_pre_n is larger than 1
                    Initialize the temporary Minimal generator Set TMG
            for every cir_i from 2 to n_u
                         Combine cir_i items selected from diff_U;
                       Select item sets from the combined item sets which
            are not included in any direct predecessor of f and are not the super
            sets of item sets in the temporary Minimal generator Set TMG, add
            into TMG;
                    end
                Add TMG to MG;
          end
      end
end
```

2.3 *Mining Simple Association Rule*

Because all association rules between non-adjacent frequent closed itemsets can be deduced by mining the association rules of adjacent frequent closed itemsets [5] as long as rules between adjacent frequent closed itemsets have been mined, all association rules between the non-adjacent can be reduced, and all association rules in any database can be generated by its simple association rules. This chapter conducts the mining based on the following association rules and the data from adjacency graph of frequent closed itemset and minimal generator.

Property 4 *Assume that f is one node on the adjacency graph, then rules generated between each item in $f/\cup g$ (g is the direct predecessor of f) is two-way association rule with 100 % confidence coefficient.*

Property 5 *In adjacency graph, confidence coefficient for association rule generated form successor to predecessor is 100 %, while confidence coefficient for association rule generated form predecessor to successor is less than 100 %.*

Property 6 *Association rule generated between minimal generator is simple association rule.*

2.3.1 Description of Mining Algorithm Framework

```
Input: Adjacency List FAL of frequent closed itemset and minimal generator itemset
MG for each frequent closed itemset
Output: R
for node f of each frequent closed itemset
        if | f / ∪g |>1
                R= R∪{ i↔ j,sup(i ∪ j) = sup f,sup  j=sup f,1};
        end
        if predecessor of f is not void
                R = R∪  {f_pre_g, f/f_pre_g, sup f, support degree of predecessor,
         sup f /support degree of predecessor }
                Make z the intersection set of minimal generator and predecessor
            if f_pre/z is not void
                    if z is not included in direct predecessor of f
                            R = R ∪  {minimal generator of f, f_pre/z, sup f, sup f,
        1};
                    end
            end
    end
end
```

3 Exemplification

We adopt the example in Ref. [3] and use the algorithms suggested in Ref. [1] and the paper to generate minimal generator of frequent closed itemset and mining association rule. The transaction record is indicated in Table 1 and the minimal generator itemset and association rule of the two algorithms obtained with minimal

Table 1 Transaction record

Transaction number	Product item
#1	AD
#2	BE
#3	ABDE
#4	BDE
#5	BCDE
#6	ABE
#7	ABCDE

Table 2 Minimal generator itemset generated by two algorithms

Frequent closed itemset	This chapter minimal generator	Reference [1] minimal generator
D	D	D
A	A	A
BE	B	B
	E	E
AD	AD	AD
BDE	BD	BD
	DE	DE
ABE	AB	AB
	AE	AE
ABDE	ABD	
	ADE	
BCDE	C	C

support degree of 2/7 and the minimal confidence coefficient of 65 % are shown in Tables 2 and 3.

From Table 2, we know that there is omission in generator obtained in Ref. [1] in that the combination of any two direct predecessors is subject to limitation in the algorithm while the available combination may be incomplete, resulting in the aforesaid omission. From Table 3, we found that the number of association rules mined through the algorithm provided by Ref. [1] is less than the ones mined through the algorithm proposed by this chapter. In addition, there is redundancy in association rules mined through the algorithm in Ref. [1]. Take association rules $C \Rightarrow BD$ and $C \Rightarrow DE$ as an example, we have only the association rule $C \Rightarrow BDE$ to express the algorithm in this chapter and support degree and confidence coefficient of the three association rules are all 2/7 and 65 %, respectively. Therefore, compared with the algorithm suggested in Ref. [1], relative simpler association rule may be mined with the algorithm put forward in this chapter and the minimal generator itemset generated is more complete.

Table 3 Association rules mined through two algorithms

Mining association rules of this chapter				Mining association rules of Ref. [1]		
No.	Association rule	Support degree	Confidence coefficient (%)	Association rule	Support degree	Confidence coefficient (%)
1	$B \Rightarrow E$	6/7	100	$B \Rightarrow E$	6/7	100
2	$E \Rightarrow B$	6/7	100	$E \Rightarrow B$	6/7	100
3	$A \Rightarrow D$	3/7	73	$BD \Rightarrow E$	4/7	100
4	$D \Rightarrow BE$	4/7	80	$E \Rightarrow BD$	4/7	66.7
5	$B \Rightarrow DE$	4/7	66.7	$DE \Rightarrow B$	4/7	100
6	$E \Rightarrow BD$	4/7	66.7	$B \Rightarrow DE$	4/7	66.7
7	$A \Rightarrow BE$	3/7	75	$AB \Rightarrow E$	3/7	100
8	$AD \Rightarrow BE$	2/7	66.7	$AE \Rightarrow B$	3/7	100
9	$ABD \Rightarrow E$	2/7	100	$C \Rightarrow BD$	2/7	100
10	$ADE \Rightarrow B$	2/7	100	$C \Rightarrow DE$	2/7	100
11	$AB \Rightarrow DE$	2/7	66.7			
12	$AE \Rightarrow BD$	2/7	66.7			
13	$ABD \Rightarrow E$	2/7	100			
14	$ADE \Rightarrow B$	2/7	100			
15	$C \Rightarrow BDE$	2/7	100			

4 Conclusion

As an improvement on the algorithm of Ref. [1], the algorithm proposed obtains better integrity and effectiveness of rules with the same dataset, minimal support degree and minimal confidence coefficient and without paying extra timing cost. Both algorithms of Ref. [1] proposed in this chapter are superior than traditional algorithms in that only adjacent nodes of the adjacency graph in the frequent closed itemset are searched and association rules are generated with minimal generator, which greatly reduce the searching space and time.

When generating minimal generator.

Acknowledgments Thanks to the support by The Introduction of International Advanced Forestry Science and Technology project of the State Forestry Administration of China (No. 2013-4-65), the National Natural Science Foundation of China (No. 41271387), the tourism constructing study soft science project of Xian city (No. SF1228-3), and the academician innovation project of Shaanxi Normal University (No. 999521).

References

1. Luo, G., Liu, J.: Algorithm based on association rule for adjacency graph of frequent closed itemset. Comput. Eng. **36**(12), 36–42 (2010)
2. Pasquier, N., Bastide, Y.: Discovering frequent closed itemsets for association rules. In: Proceedings of the 7th International Conference on Database Theory Jerusalem, pp. 398–416. Springer (1999)

3. Li, J., Xu, Y., Wang, Y., Wang, Y.: The simplest association rule and mining algorithm. Comput. Eng. **33**(13), 46–48 (2007)
4. Geng, L., Hamiton, J.: Interestingness measure for data mining: a survey. ACM Comput. Surv. **38**(3), 1145–1177 (2006)
5. Shao, F., et al.: Data Mining: Principles and Algorithms, 2nd edn, vol. 9, pp. 90–91. Science Press (2009)
6. Bastide, Y., Pasquier, N., Taouil, R., et al.: Mining minimal non-redundant association rules using frequent closed itemsets. In: Proceedings of the 1st International Conference on Computational Logic, pp. 927–986. Springer, London (2000)

Part III
Spatial Data Analysis and Intelligent Information Processing

Evaluation of Suitability Areas for Maize in China Based on GIS and Its Variation Trend on the Future Climate Condition

Chao-jie Jia, Le-le Wang, Xiao-li Luo, Wei-hong Zhou, Ya-xiong Chen and Guo-jun Sun

Abstract Climate change is one of the most significant factors for the migration of the suitable planted areas of maize. The hyper-resolution data of daily mean precipitation, temperature, soil, and topography that supported by Institute of Soil Science and National Climate Center were collected to assess the effect of climate change on migration of the suitable planted maize areas, and it provides a foundation for macro-management decisions of agriculture. Climate data mentioned above was simulated by regional climate model, which showed the experiments of the twentieth century and the forecast tests of twenty-first century. Based on the factors above, relevant criteria, suitability levels, and their weights for each factor were defined. And the database of climate and soil was established. In this article, we use the multi-criteria evaluation (MCE) approach based on GIS to identify the suitable planting areas for maize in China. The results demonstrated that the most suitable degrees of active accumulated temperature, maximum and minimum temperature in growth period moving north dramatically, and the precipitation moving slightly in the future, which will lead to the planting areas of maize moving northward, and the areas were increased by 4.61 × 105 km^2.

Keywords Maize · Geographic information system (GIS) · Climate change · Multi-criteria evaluation · Kriging interpolation · Analytical hierarchy process

1 Introduction

Any change in climate will have implications for climate-sensitive systems such as agriculture, forestry, and some other natural resources. With respect to agriculture, change in temperature and precipitation will produce change in scheduling of field

---

C. Jia · L. Wang · X. Luo · W. Zhou · Y. Chen · G. Sun (✉)
Institute of Arid Agroecology, School of Life Sciences, Lanzhou University, Lanzhou 730000, China
e-mail: sungj@lzu.edu.cn

B.-Y. Cao et al. (eds.), *Ecosystem Assessment and Fuzzy Systems Management*, Advances in Intelligent Systems and Computing 254, DOI: 10.1007/978-3-319-03449-2_27,

operations, crop yield, cropping systems, and so on [1]. To research and develop agriculture on condition of climate change, the accurate identification and the characterization of current production areas and potential areas in future are necessary [2]. Multi-criteria evaluation (MCE) approach was used to determine relevant criteria (factors) as the biophysical restraints and to define the suitability levels for each factor. MCE was defined as "an umbrella term to describe a collection of formal approaches which seek to take explicit account of multiple criteria in helping individuals or groups explore decisions that matter" [3] could be understood as a world of concepts, approaches, models, and methods that aid an evaluation (expressed by weights, values, or intensities of preference) according to several criteria [4]. MCE has been one of the most widely applied models in management and planning because (1) it was the formal approach, (2) the presence of multiple criteria, and (3) the evaluation are made either by individuals or by groups [5]. Geographic information system (GIS) has a powerful function in spatial analyses such as predicting the distribution of the wild relatives of bean by analyzing climate conditions that favor bean's growth [6] and for planning potential conservation areas by using relationships between environmental factors and the distribution of birds [7]. However, the utility or GIS functionality in the management of the above areas has been limited by the restrictions inherent in overlaying of digital information maps. Some of these restrictions are as follows: (1) overlays are difficult to use when there are many underlying variables (more than 4), (2) the overlay procedure does not enable one to take into account that the underlying variables are not of equal importance [8]. From the 1990s, integration of the MCE approach with GIS for solving spatial planning problems has received considerable attentions among urban planners. The ability of GIS to integrate with the MCE approach has been shown in studies related to site determination for a nuclear waste facility [9] and for a noxious waste facility [10]. And the GIS-based MCE has also extended to solving planning problems that involve conflicting multi-objectives such as land use allocation problems [8, 11, 12]. MCE seems to be applicable to GIS-based land suitability analysis [13] and help us to carry out the delineation of suitable areas for crops [10, 14, 15]. The maize is one of the most important grain crops in China, playing a crucial role in protect food security [16]. The main goal of this research is to study the change of the suitable areas for the maize with the GIS-based MCE in current climate condition and the future climate condition, using the climate data from National Climate Center and the soil data from Institute of soil Science in Nanjing.

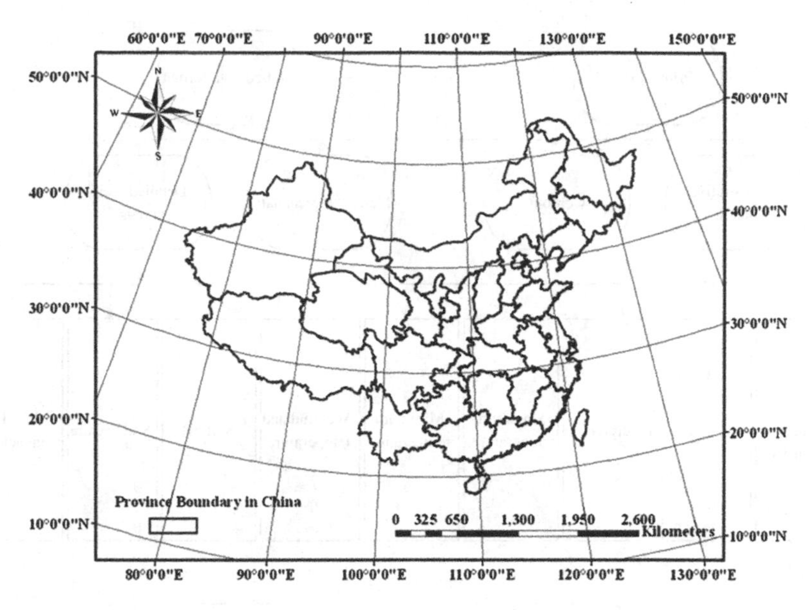

Fig. 1 Study area

2 Methods

2.1 Study Area

The study area throughout the land of China, include mainland of China, Hong Kong Special Administrative Region, the Macao Special Administrative Region, and Taiwan Province. It is located between latitude 3°52′N and 53°33′N and longitude 73°40′E and 135°2′30″E. The total area is approximately 9,600,000 km^2 [17] (Fig. 1).

2.1.1 Procedures

The distribution and growth of crops are decided by water, temperature, light, and soil, according to the leading dominant (the main factor affecting the growth of maize), incompatibility (the factor affecting the maize independence), diversity (selected factors were significantly different and can express the threshold), and marketability (factors which be chosen have a corresponding data) [18]. In light of expert opinions, as well as literatures, eight factors, which are closely connected with the growth of maize, were selected to establish the relevant criteria of MCE. They were as follows: accumulated temperature, average temperature in April which was regarded as the minimum temperature, the average temperature in July

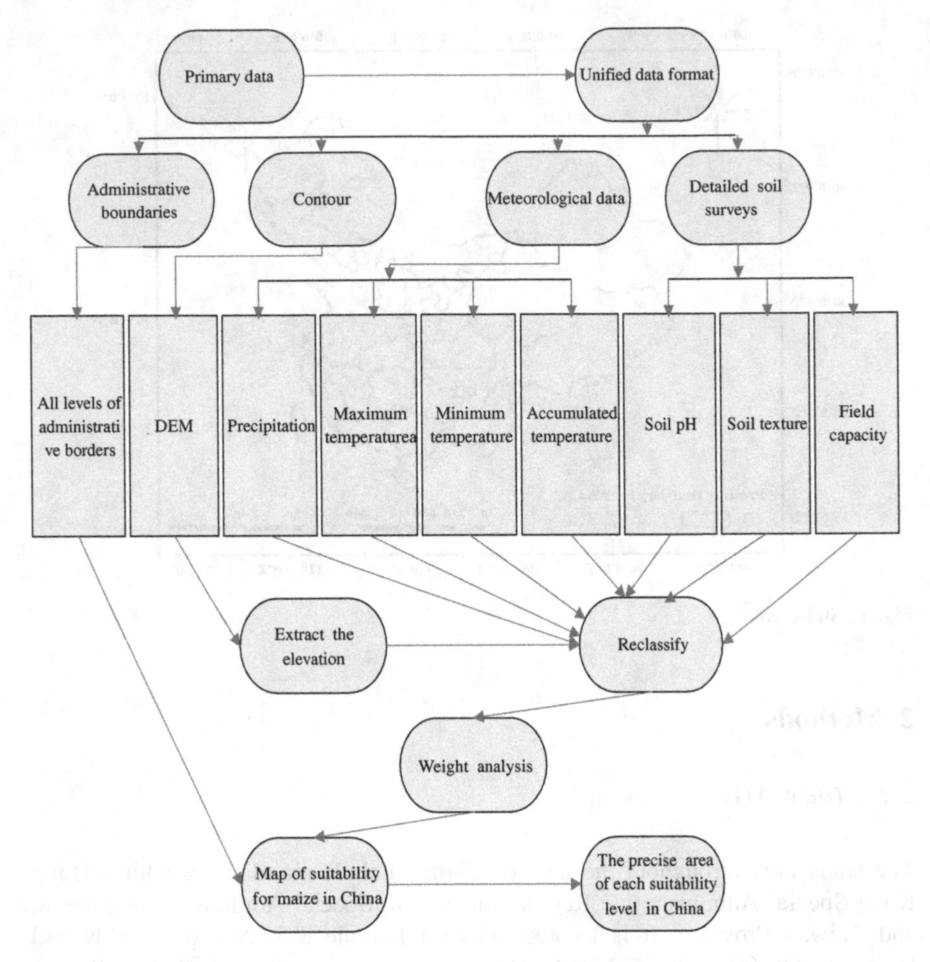

Fig. 2 Flow chart shows the main procedures applied in this study

was the maximum temperature, the annual precipitation, soil texture, soil pH, elevation, and field water-holding capacity. Then, the database was established (Fig. 2).

2.1.2 Establish of Spatial Databases

Climate Database

Climate data from the simulation with regional climate model by National Climate Center, including the control experiments data of the twentieth century (1961–1990) and the forecast tests data of twenty-first century (2071–2100). The format of the original data was text, calculated with the R software; the daily

Table 1 Mean relative errors (MRE) results of four interpolation methods

	IDW	OK	SP	TR
Accumulated temperature	0.142731	0.101632	0.175351	0.147622
Minimum temperature	0.162719	0.151387	0.189255	0.157697
Maximum temperature	0.085669	0.069609	0.098144	0.102951
Precipitation	0.077262	0.449331	0.477557	0.3924253

average data of 30 years were obtained. Based on these data, annual accumulated temperature (≥10 °C), the minimum temperature, the maximum temperature, and the annual precipitation were extracted and calculated separately. Convert the data to grid in ArcGIS9.2, extract by point, and then interpolate the temperature data and the precipitation data with the kriging and the IDW method. Before interpolation for the spatial data, compare the accuracy of various interpolation is necessary. 90 % percent of the points were used as the training points while the others were test points, and use mean relative error (MRE) between the measured (Zoi) and the value after interpolation to test the results of the inverse distance weighting (IDW), ordinary kriging (OK), spline (SP), and trend surface (TR). The formula is as follows (Table 1):

$$\mathrm{MRE} = \frac{1}{n}\sum_{i=1}^{n}\left|\frac{Z_{oi} - Z_{ei}}{Z_{oi}}\right| \tag{1}$$

The results showed that OK was the best interpolation for the temperature and the IDW for the precipitation in this study.

Soil Database

Soil characteristics data were taken from digital Soil Type Maps (from ISSCAS) using a scale of 1:1,000,000. Sampling points was created, and the total number of points was 99,034. The information of soil texture and soil pH was obtained from soil type. Then, the soil texture point data and the soil pH point data were interpolated into grid maps within ArcGIS.

Relief Database

The altitude data were obtained from the digital elevation model (DEM). DEM has characteristics of space location and attribute of terrain, and it is an indispensability part when establishing resource and environment information in different levels [19]. National digital contour map at the scale of 1:250,000 was obtained from the State Bureau of Surveying and Mapping. These contour data were used to create DEM within ArcGIS, and the process was contours → TIN → lattice → DEM.

Unification of the Data Format

All the data were converted to the same format, the spatial resolution was 5,000 m per pixel, the Krasovsky_1940_Albers coordinate system was used as the projected coordinate system, and GCS_Krasovsky_1940 was the geographic coordinate system. The spheroid of the maps was the Krasovsky_1940, and the datum was D_Krasovsky_1940.

2.1.3 MCE Process for Suitable Areas of Maize in Current and Future Climatic Conditions

Define Specific Suitability Level of the Factors

By means of expert opinion and literatures [20], a specific suitability level per factor for maize was defined (Table 2). These levels were used as a base to construct the criteria maps. The levels were as follows: most suitable, moderate suitable, medium, moderate unsuitable, and unsuitable. Then, factor maps were constructed in the ArcGIS environment.

Weighting for the Factors

A variety of techniques exist for the development of weights, but the promising technique is the pairwise comparison developed by Saaty (1977) in the context of a decision-making process known as the analytical hierarchy process [11]. It has characterized of systematization, flexibility, and practicality. Relative importance of each factor was compared, according to regression between them and final yield of maize, the score of each factor were given, and amend the score with the relationship between them. The comparison of first pairwise matrixes was made. Ratings were provided on a nine-point continuous scale, which ranges from 9 to 1/9. A 9 indicates that relative to the column variable, the row variable is significantly more important. A 1/9 indicates that relative to the column variable, the row variable is significantly less important. Statistics the number of each factor appeared in the literature which described the yield and the distribution of maize, establish another pairwise comparison matrixes. Test two matrix use the CRJ (Matrix Consistency ratio), if the CRJ < 0.10 matrix was available; if the CRJ > 0.10, we must amend the matrix on the opinion of experts who research the maize. Weight was calculated in the MATLAB software using the pairwise comparison matrixes above (Table 3).

3 Results

Once the factors and constraints maps have been obtained, and the weights of each factor were calculated, the next step was to overlay each factor map by its weight to construct the suitability map for maize in China. These works were conducted with the weighted overlay module in the ArcGIS environment.

Table 2 Suitability level per factor for the maize

	Accumulated temperature (°C)	Precipitation (mm)	Max temperature (°C)	Min temperature (°C)	Soil texture	Soil pH	Elevation (m)	Field capacity
Most suitable	2,500–3,300	500–800	20–25	10–17	Loam	6.5–7.0	<1,200	030–0.46
Moderate suitable	3,300–5,000	300–500 or 800–1,500	15–20 or 25–30	17–20	Sand	5.0–6.5	1,200–1,500	0.25–0.30
Medium	5,000–7,000	1,500–2,000	5–15	0–10	Sand clay or loam	7.0–8.0	1,500–3,000	0.23–0.25 or 0.46–0.50
Moderate unsuitable	1,000–2,500	2,000–2,500	>25	>20	Other	<5.0	3,000–3,600	0.18–0.23
Unsuitable	<1,000 or 7,000	0–300	<0	<–5	Clay	>8.0	>3,600	<–0.18 or >0.53

Table 3 Weight of various factors

Factor	Accumulated temperature	Precipitation	Minimum temperature	Maximum temperature	Field capacity	Soil texture	Soil pH	Elevation	Total weight
Weight	0.20948	0.19134	0.141142	0.163879	0.11368	0.075808	0.059333	0.04547	1

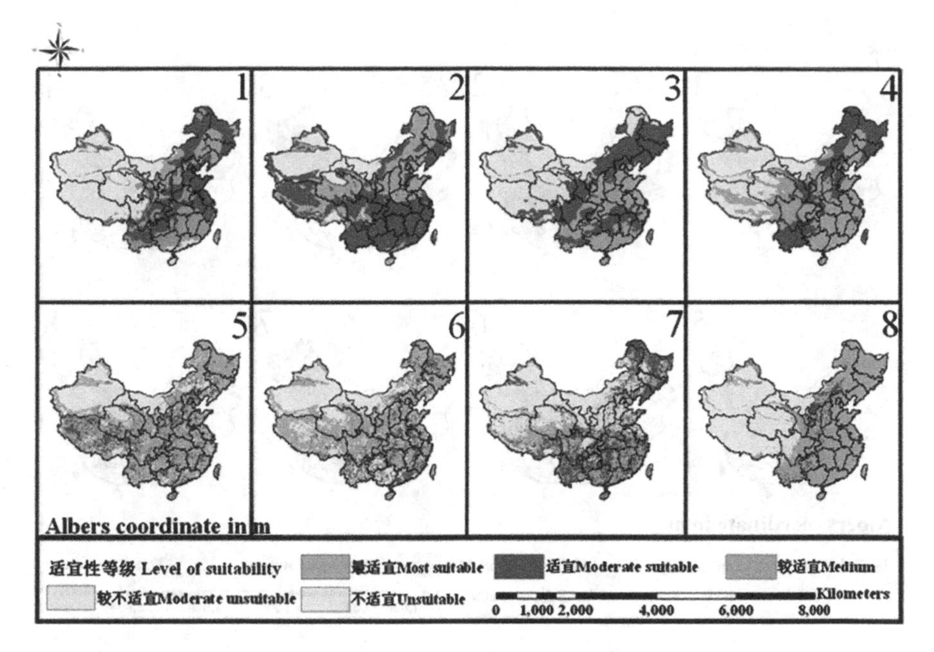

Fig. 3 Factor maps for the maize in current climate condition in China

The maps include suitability levels for each factor. Keys are as follows: (1) accumulated temperature, (2) precipitation, (3) minimum temperature, (4) maximum temperature, (5) field water-holding capacity, (6) soil texture, (7) soil pH, and (8) elevation.

We assume that the terrain and soil conditions remain unchanged. The result (Figs. 3, 4) indicates that under the same classification, (1) the most suitable levels of the accumulated temperature, minimum temperature, and the maximum temperature shift to northward significantly, and the precipitation has inconspicuous deviation to northward, and (2) compared with the accumulated temperature of the current climate condition and the accumulated temperature of the future, all of the China have the trend of becoming warmer in future, but more obvious change appeared in the north.

The results (Figs. 5, 6) identified the most suitability level for maize in current climate condition in China located in the Northeast China Plain, North China Plain, some areas of Ningxia, Gansu, Shanxi, and so on. Statistics show that the areas were approximately 7.82 × 105 km^2. The moderate suitability areas main appeared in the south of ecotone, and it is about 4.53×10^6 km^2. Results of the "Map of suitability for maize in future climate condition in China" indicated that the most suitability level mainly distributed around the Da Hinggan Mountains and the Xiao Hinggan Mountains, and the total area was 8.03×10^5 km^2, showed increasing trend.

The ecotone was the boundary to the influence of climate change (Fig. 7), and decreased suitability levels appeared in the south of ecotone, including Northeast

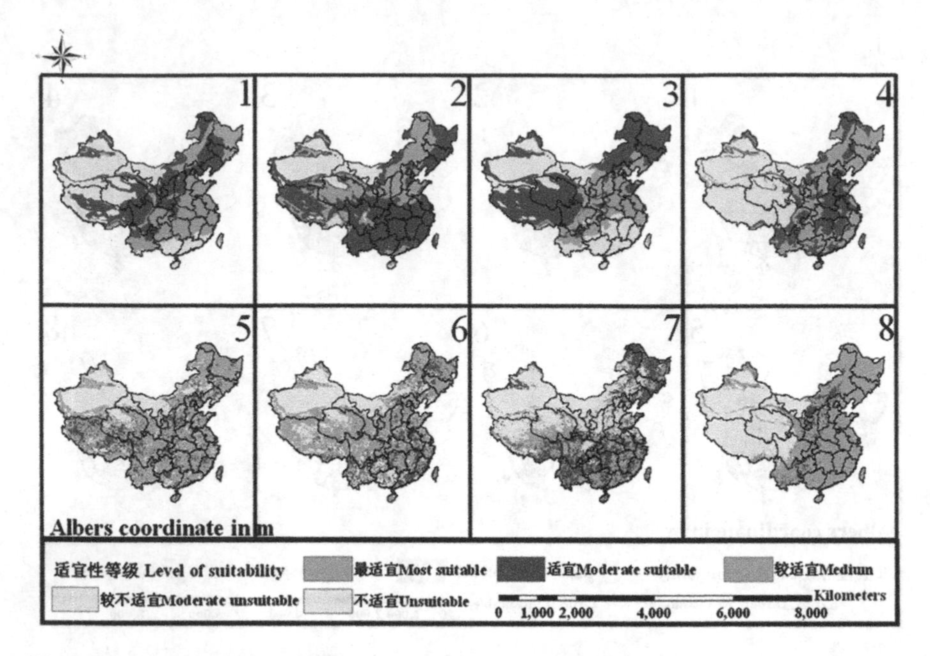

Fig. 4 Factor maps for the maize in future climate condition in China

China Plain, North China Plain, and so on. The increased levels were located in the Qinghai–Tibet Plateau, the Da Hinggan Mountains, the Xiao Hinggan Mountains, the piedmont of Tianshan, and so on. These places will be more appropriate for growth of maize in future. The most suitability level and the moderate suitability level were regarded as the areas of the potential high yield. Table 4 shows change in the areas of the potential high yield in every province from current to future. "+" refers to the increase in areas, "−" the decrease, and "0" means that there was no change in that province. We can find that the areas of the potential high yield were increased for 4.61×10^5 km^2.

4 Discussion

Production and distribution of maize were the result of multifactor effect [21], and the eight factors that have been chosen in this research were the restrictive factor in various stage of maize growth in the natural conditions. Based on the report of Ceballos-Silva et al., the spatial data input, extraction, analysis, and visualization functions of GIS were used to establish the national spatial database of climate, topography, and soil. MCE procedure in this research was useful to evaluate the suitable areas for certain crops. As the first phase of MCE, the factors were selected based on agronomic knowledge of local experts and reviews of existing literature. And then the comparison pairwise matrixes in the context of analytical

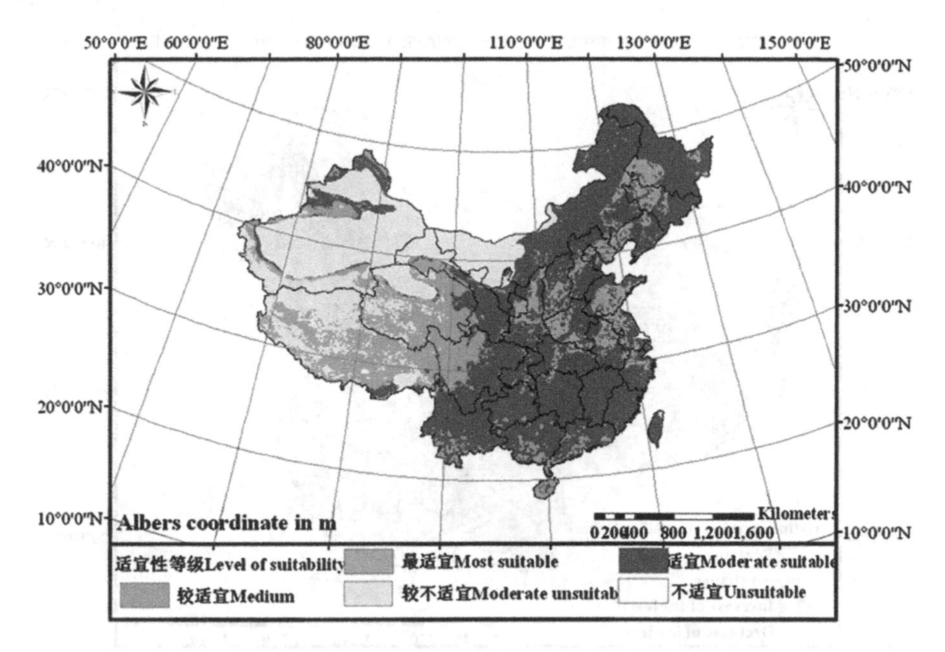

Fig. 5 Map of suitability for maize in current climate condition in China

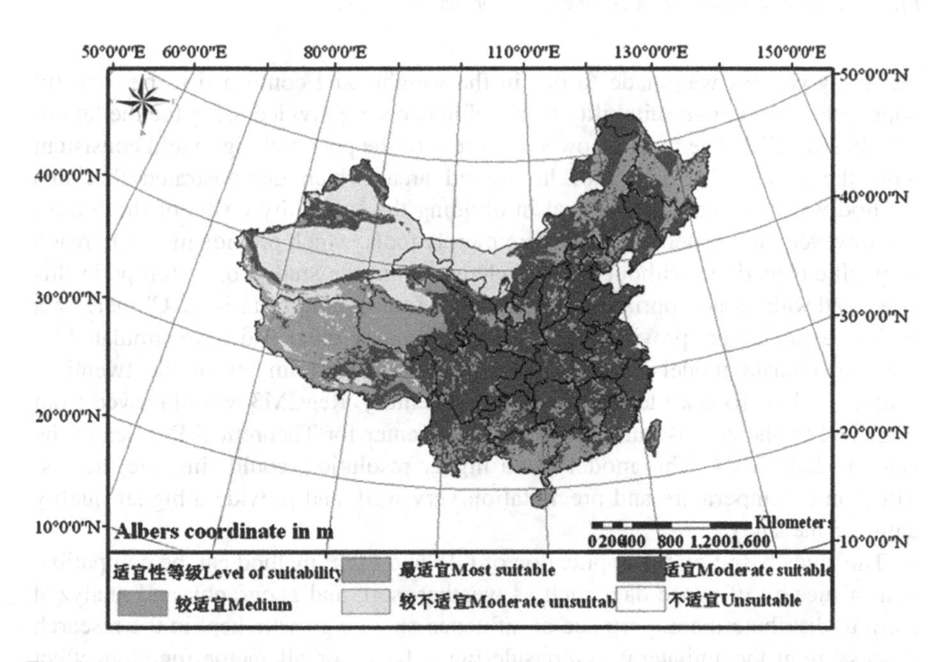

Fig. 6 Map of suitability for maize in future climate condition in China

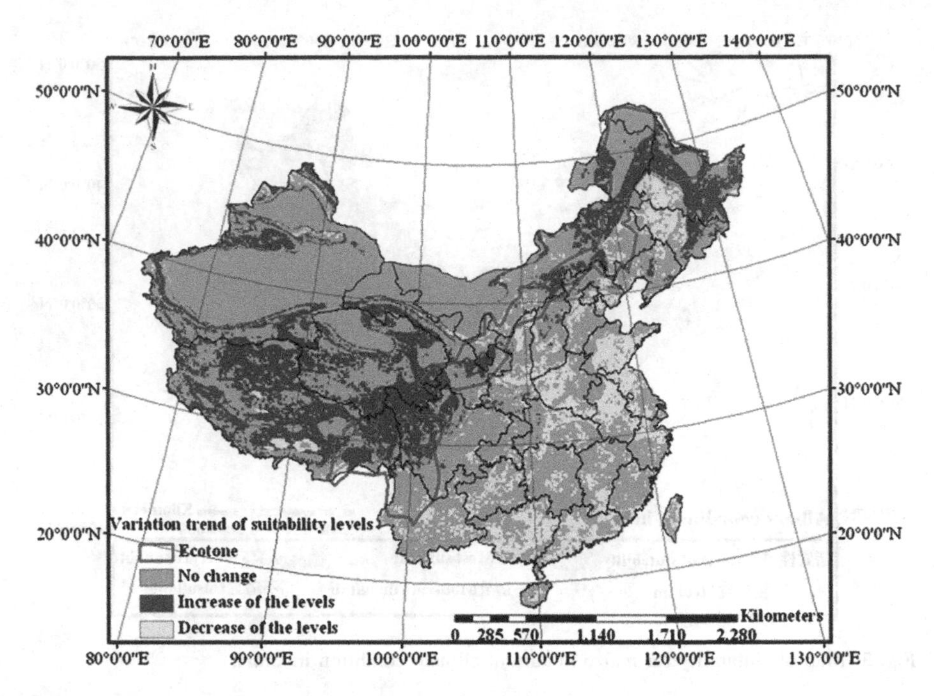

Fig. 7 Map of suitability area change trend for maize in China

hierarchy process was made to obtain the weights and confirmed to be a useful approach. Finally, five suitability levels of maize were divided using the method of GIS-based MCE. The result showed the areas of the potential high yield consistent with the actual distribution of high-yield area, which demonstrated that this method was credible and practical in dividing the suitability levels of the future.

However, the former techniques are merely tools, which provide means to reach a specific objective. Without the knowledge about the study object (crops in this case) and without appropriate databases, such tools would be useless. Climate data in this research were provided by National Climate Center and were simulated by regional climate model (RegCM3) to the control experiments of the twentieth century and the forecast tests of twenty-first century. RegCM3 was improved from RegCM2 by the Abdus Salam International Center for Theoretical Physics in the year of 2003–2004. This model has a higher resolution, could simulate thc distribution of temperature and precipitation very well, and provide a higher quality data for the study.

The study based on the application of GIS and MCE method can be compatible with a nice bit of space data such as weather, soil, and geography and analyzed them to distribute the importance of influence on corn growth, kept in the research process from the unilateral of considering a factor or all factor function effect equal in value. In China, the suitability for maize moves toward the north, so the distributing of forest is in the current climate condition replace by grow maize in

Table 4 Analysis the areas of the potential high yield in 32 provinces

Province	The areas of the potential high yield in current climate ($\times 10^5$ km^2)	The areas of the potential high yield in future climate ($\times 10^5$ km^2)	The areas of the potential high yield changes ($\times 10^5$ km^2)
Xizang	0.44	3.17	+2.73
Qinghai	0.49	2.44	+1.95
Xinjiang	2.14	3.80	+1.67
Sichuan	2.81	4.37	+1.57
Gansu	2.00	2.67	+0.66
Neimenggu	8.20	8.36	+0.16
Heilongjiang	4.48	4.48	0.00
Shanghai	0.05	0.05	0.00
Liaoning	1.44	1.44	0.00
Jilin	1.90	1.90	0.00
Beijing	0.16	0.16	0.00
Shanxi	1.57	1.57	0.00
Tianjin	0.12	0.11	−0.01
Shananxi	2.05	2.04	−0.01
Chongqing	0.82	0.81	−0.01
Zhejiang	0.98	0.97	−0.02
Hebei	1.87	1.85	−0.02
Ningxia	0.43	0.40	−0.03
Shandong	1.52	1.48	−0.04
Jiangsu	0.93	0.89	−0.05
Hainan	0.07	0.02	−0.05
Henan	1.65	1.54	−0.11
Taiwan	0.32	0.19	−0.13
Anhui	1.40	1.25	−0.15
Hubei	1.86	1.69	−0.17
Fujian	1.17	0.98	−0.19
Hunan	2.12	1.93	−0.19
Jiangxi	1.67	1.33	−0.34
Guizhou	1.72	1.35	−0.37
Yunnan	3.38	2.90	−0.48
Guangdong	1.36	0.64	−0.72
Guangxi	2.01	0.95	−1.06
Total	53.12	57.73	4.61

the future climate condition, the area increased 4.61×10^5 km^2. Reflected that climate variety lead forest to degeneration or distribute area deviation.

The agriculture manages measure, socioeconomic income, and soil tiny environment, etc. to have the function that cannot underrate to the maize's planting, and collected the data of this date in the further research and carried on an evaluation to the planting of small dimensions scope area first. Later expand a whole country further again

This study was the earlier stage of ecosystem-agriculture valuation management (EAM) information platform technique cooperation work, also be placed in the processing analysis to the static state data up, when environment data of database occurrence variety the whole system need to re-tidy up, its inconvenience links at the Internet, data and handing over of the user's with each other, so at after expect and work in, would emphasized looking for the dynamic state model, finally build up the dynamic valuation management system.

Acknowledgments This work is financial supported by the ISTCP 2010DFA92800 and 2012DFG31450.

References

1. Southworth, J., Randolph, J.C., Habeck, M., Doering, O.C., Pfeifer, R.A., Rao, D.G., Johnston, J.J.: Consequences of future climate change and changing climate variability on maize yields in the midwestern United States. Agric. Ecosyst. Environ. **82**(1–3), 139–158 (2000)
2. Corbett, J.H.: GIS and environmental modeling: progress research issues. In: Goodchild, M.F., Steyaert, T.L., Parks, O.B. (eds.) pp. 117–122. USA (1996)
3. Belton, V., Stewart, T.J.: Multiple Criteria Decision Analysis: An Integrated Approach. Kluwer Academic Publishers, Boston (2002)
4. Barredo, C.J.I.: Systems Information Geographical Evaluation Multi-Criteria laordenacio n deltetritorio. Editorial RA-MA Editorial, Madrid, EsPana (1996)
5. Mendoza, G.A., Martins, H.: Multi-criteria decision analysis in natural resource management: a critical review of methods and new modelling paradigms. For. Ecol. Manage. **230**(1–3), 1–22 (2006)
6. Jones, P.G., Beebe, S.E., Tohme, J., Galwey, N.W.: The use of geographic information systems in biodiversity exploration and conservation. Biodivers. Conserv. **6**, 947–958 (1997)
7. Mariuki, J.N., De Klerk, H.M., Williams, P.H., Bennun, L.A., Crowe, T.M., Berge, E.V.: Using patterns of distribution and diversity of Kenyan birds to select and prioritize areas for conservation. Biodivers. Conserv. **6**, 191–210 (1997)
8. Janssen, R., Rietved, P.: Geographical Information Systems for Urban and Regional Planning, pp. 129–138. Kluwer press, Dordrecht (1990)
9. Carver, S.J.: Integrating multi-criteria evaluation with geographical information systems. Int. J. Geogr. Inf. Syst. **5**(3), 321–339 (1991)
10. Malczewski, J.A.: GIS-based approach to multiple criteria group decision-making. Int. J. Geog. Inf. Sci. **10**(8), 321–339 (1996)
11. Eastman, J.R., Jin, w., Kyem, A.K., Toledano, J.: Raster procedures for multi-criteria/multi-objective decisions. Photogram. Eng. Sens. **61**(5), 539–547 (1995)
12. Yeh, W.H., Li, L., Mansuripur, M.: Vector diffraction and polarization effects in an optical disk system. Appl. Opt. **37**(29), 6983–6988 (1998)
13. Pereira, J.M.C., Duckstein, L.: A multiple criteria decision-making approach to GIS-based land suitability evaluation. Int. J. Geog. Inf. Sci. **7**(5), 407–424 (1993)
14. Aguilera, H.N.: Panorama Agriculture Mexicana, pp. 58–69. De geografia, Mexico (1986)
15. Heywood, I., Oliver, J., Tomlinson, S.: Innovations of GIS, pp. 127–136. Taylor and Francis, UK (1995)
16. Huang, H.Q.: The Influence of global food shortage on Chinese economy. Contemp. Econ. **7**, 6–7 (2008)

17. Ceballos-silva, A., Lo'pez-blanco, J.: Delineation of suitable areas for crops using a multicriteria evaluation approach and land use/cover mapping: a case study in central Mexico. Agric. Syst. **77**(2), 117–136 (2003)
18. Shi, B.S., Wen, Z.P.: The weighted method for the score that experts give. China Acad. J. Electron. Publishing House (5), 57–58 (1996)
19. Guo, R., Chen, Y., Wan, F., Li, F., Liu, J., Sun, G.: Delineation of suitable areas for potato in China using a multi-criteria evaluation approach and geographic information system. Sens. Lett. **8**(1), 167–172 (2010)
20. Ceballos-Silva, A., Lopez-Blanco, J.: Delineation of suitable areas for crops using a multi-criteria evaluation approach and land use/cover mapping: a case study in Central Mexico. Agric. Syst. **95**, 117–136 (2003)
21. Li, Y.Z., Dong, X.W., Liu, G.L., Tao, F.: Effects of light and temperature factors on yield and its components in maize. Chin. J. Eco-Agri. **10**(2), 86–89 (2002)

City in Transition: The Structure of Social Space in Tianjin in 2000

Zi-wei Liu, Rui-bo Han and Li-qin Zhang

Abstract Adapting neighborhood-level census data from 2000, this chapter uses factorial ecological analysis method (principal component analysis, PCA) to examine Tianjin's social spatial structure and its formulation mechanisms in 2000. The result shows that principal factors under consideration for the components of social structure in Tianjin include the influx of the salaried class, economic status, retirees and minorities, marital status, and family structure. Every factor has its spatial distribution pattern in the city of Tianjin. Generally, the advanced geographical location, distinctive historical development process, rapid socioeconomic development, and ever-changing government policies and urban planning have worked together to compose the formulation mechanisms of Tianjin's social space in 2000.

Keywords Social space · Sociospatial dynamics · Factorial ecology · PCA · Tianjin metropolitan area

1 Introduction

With the rapid growth of urban populations and supporting infrastructure in the last century, cities have played a vital role in the development of modern society. In recent years, large number of people migrating from rural areas to cities seeking economic opportunities and better lives has caused remarkable urbanization rates. Such rapid urbanization process leads to increased complexity of cities'

Z. Liu (✉) · R. Han · L. Zhang
Department of Geography, University of Ottawa, Ottawa K1N6N5, Canada
e-mail: Zliu052@uottawwa.ca

L. Zhang
Faculty of Public Administration, China University of GeoSciences, Wuhan 430070, China

B.-Y. Cao et al. (eds.), *Ecosystem Assessment and Fuzzy Systems Management*, Advances in Intelligent Systems and Computing 254, DOI: 10.1007/978-3-319-03449-2_28,

sociospatial dynamics, which is affected not only by the physical environment, but also by individual behavior culture, politics, economics, as well as social organization (social environment). The emergence of human ecology (known as the "Chicago School") in the United States in the 1920s encouraged a developing awareness of the social, economic, and political significance of cities [1]. As early as the first half of the twentieth century, three classic models were proposed by western countries as the foundation of urban sociospatial structure studies. In the 1960s, following on the Chicago School's ecological approach, factorial ecology provided a new method to study urban sociospatial structure. The factorial ecology approach is widely applied to studies of urban sociospatial structure for it represents the complexity of urban social by space. The main theoretical contribution to the origin of this approach comes from Shevky and Bell [2], who developed a three-dimensional (socioeconomic status, family status, and ethnic status) model to describe how urban populations differ in industrial societies. Subsequent studies using a wide variety of measures confirmed the significance of the three statuses. Although it has criticisms, the factorial ecological approach of social-area analysis increases understanding of residential differentiation in cities and is an important instrument for studying intra-urban social–spatial structure. However, all these current studies are not universal: They are simply suitable to the given case cities, especially the cities in Western countries, which means most of the work took place in capitalist cities. Very few studies about urban transformation from an ecological perspective have focused on socialist countries, especially China, which is identified as the one on the transition.

China's economic reforms in 1978 ushered in a phase of accelerated urbanization, which has involved increases in social infrastructure and housing [3]. This is especially evident after the country's transition from a centrally planned economy to a market-oriented economy in 1992, which led to significant urban socioeconomic changes due to four particular policy adjustments. The adjustment of these four policies includes Hukou Reform, Housing Allocation System change, Land Use Policy adjustment, and Urban Planning adjustment. Furthermore, the significant urban sprawl and reorganized process also have impact on urban sociospatial structure transformation. Therefore, more and more scholars are dedicating themselves to the study on such process as well as its transformation mechanism. The history of when the Chinese scholars began to measure Chinese cities' sociospatial structure can be dated back to late 1980s. The first relevant studies related to China are Lo's work about Hong Kong [4, 5] and Hsu's research on Taipei [6]. Yu's study about the sociospatial characteristics of Shanghai urban area and its formation as well as its relevance to urban planning is the first one to quantitatively examine mainland Chinese cities' sociospatial structure [7]. Gan [8] studied Beijing's urban structure and its historic and cultural background in his doctoral dissertation. In 1989, Xu et al. [9] explained Guangzhou's urban sociospatial structure model and its mechanisms. Yang [10] applied census data from 1985 to 1990 to study Beijing's sociospatial structure. After that, many scholars began to apply and compare census data to research regarding the urban sociospatial structures of Beijing, Shanghai, and Guangzhou [11–13]. Due to inadequate or incomplete data, all these studies were limited. With the release of the fifth population census in 2000, more Chinese cities were examined by scholars, and the methodology for applying factor analysis to Chinese cities became more developed.

Tianjin, the second major metropolis in northern China, is one of the four provincial-level municipalities in China. As part of the economic center of northern China (together with Beijing) as well as a concession city, Tianjin has a distinctive cultural, economic, and historical importance. More particularly, after the central government decided to support the development of the port area (Binhai New Area), the overall social space of Tianjin has changed significantly. Increased levels of foreign investment have contributed to the construction of a World Financial Center, the hosting of the summer Davos Forum, and the idea of Tianjin as a global city. Due to its unique historical development, Tianjin certainly has its characteristic urban development process.

Therefore, adapting spatial analysis method, this research has as its objective to understand the urban sociospatial pattern and its mechanisms, in the case of Tianjin, using neighborhood-level census data from 2000.

2 Study Area, Data Sources, and Methodology

2.1 Study Area

Tianjin (the "Heavenly Ford"), a traditional industrial city near Beijing, is the second major metropolis in northern China. It is the third largest city in China after Shanghai and Beijing as well as one of four provincial-level municipalities in China. Tianjin is located 130 km southeast of Beijing in the North China Plain, bounded to the east by the Bohai Gulf portion of the Yellow Sea. As part of the economic center of northern China (together with Beijing) as well as a concession city, Tianjin has a distinctive cultural, economic, and historical importance.

Tianjin comprises 18 administrative subdivisions, which are county-level units governed directly by the municipality (second-level divisions). Of these, 15 are districts and three are counties. The study area of this research covers Tianjin's 15 districts (Heping District, Hedong District, Hexi District, Nankai District, Hebei District and Hongqiao District, Tanggu District, Hangu District, Dagang District, Dongli District, Xiqing District, Jinnan District, Beichen District, Wuqing District, and Baodi District), excluding the three counties (Jinghai, Ninghe, and Jixian). The study area had a total population of 7,746,700; the overall study area is 7,378 km^2. According to Tianjin's administrative division, Heping, Hedong, Hexi, Nankai, Hebei, and Hongqiao comprise the urban core of Tianjin; the region composed of Tanggu, Hangu, and Dagang is referred to as the "Binhai New Area"; Dongli, Xiqing, Jinnan, Beichen, Wuqing, and Baodi constitute the suburb of Tianjin (Fig. 1).

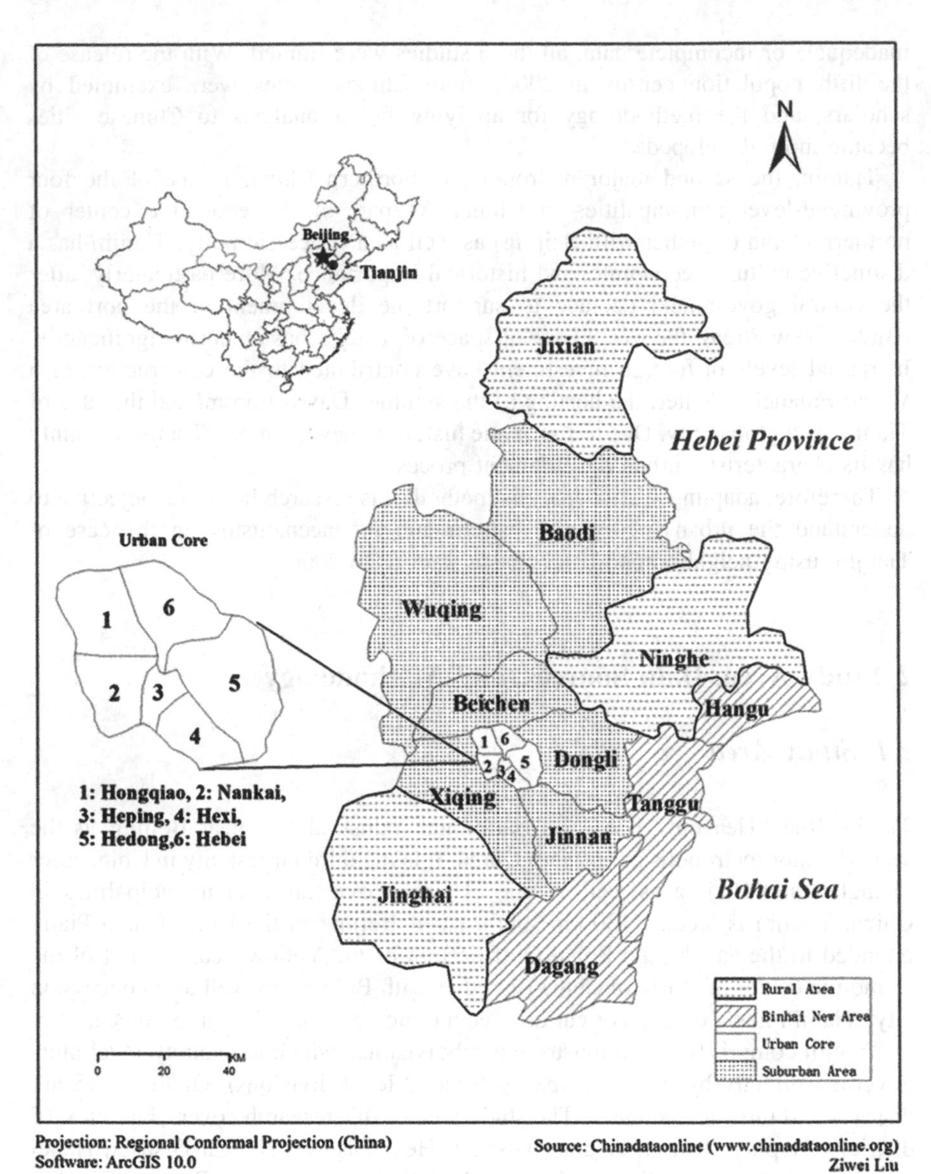

Fig. 1 Location and administrative divisions of Tianjin

Subdistricts or neighborhoods have been the basic administrative unit reported in publically accessible government reports in Chinese cities for decades [14]. As such, it was unit of analysis for this research. The GIS coverage of subdistricts was downloaded from Chinadataonline (www.chinadataonline.org).

2.2 Data Sources and Methodology

In this study, the "factorial ecology" approach is used to demonstrate the urban spatial structure of Tianjin in 2000. The factorial ecology as a mode of analysis of the sociospatial structure of cities has been used for more than 40 years [15]. Factor analysis, which is employed in data reduction process to identify a small number of factors to explain variances in the original data file, which is usually composed of a much larger number of variables [16], is the method adopted by factorial ecology. It provides a systematic way to identify the main dimensions that shape the social geography of cities. The principal components extraction (PCA) method is adopted in factor analysis to reduce the dimension of the data file. PCA is used to identify the factors (principal components) and give them scores relative to their importance for Tianjin's internal social transformation in 2000 at the neighborhood level [17]. In principal component analysis (PCA), the components are uncorrelated with each other, but they capture the same variance contained in the original data set. By ranking the components by the proportions of the total variation, one may see that few components account for the majority of the total variation.

Neighborhood-level data of 190 neighborhoods were collected from the 2000 census of Tianjin, with 63 variables initially. Based on previous literature, and in order to obtain a more significant result, some variables were regrouped and highly correlated variables are eliminated. Ultimately, 22 variables were selected, and a 190 × 22 matrix was made. Generally speaking, these 22 variables can be classified into six categories [14].

1. Population and demographic structure: In this category, four variables are selected: age group 15–64, age group 65+, non-agricultural population, and sex ratio.
2. Family structure: five variables are collected, including family of two generations, family of three generations, one-person family, two-person family, and three-person family.
3. Marital status: two variables are used: married rate and single rate.
4. Neighborhood stability: temporary population rate.
5. Education level: three integrated variables are included: basic education level (primary school and junior school), higher education level (higher than senior school), and illiteracy rate.
6. Occupation: three variables: unemployment rate, secondary industry workers, and tertiary industry workers.
7. Economic status: two variables are selected: higher expenditure on houses and higher monthly rent.
8. Ethnic status: three minorities are selected: Manchu, Uygur, and Hui.

Data on housing were collected for the first time in 2000. SPSS Statistics 19.0 was used to analyze the original matrix of factor analysis. Meanwhile, because there is correlation among some variables and due to inefficiency in mapping each variable individually, a technique is required to either remove highly correlated variables (correlation coefficient greater than 0.6) or represent the data file using a smaller number of uncorrelated factors [18].

3 Analysis Result

Factor analysis was used to examine underlying dimensions of residential differentiation from different variables. In this research, in order to interpret and label different components, the popular varimax rotation technique was used to maximize the loading of a variable on one factor and minimize the loadings on all others. According to a PCA requirement that only eigenvalues greater than one are significant as well as the scree plot (Fig. 2), we retained five factors. Table 1 reports the eigenvalues from the PCA.

The cumulative variance of the five factors has reached 72.350 %. Table 2 presents the rotated factor structure.

Factor 1: Flowing Salaried Class

This factor was by far the most important factor, explaining 22.202 % of the total variance as well as Tianjin's social space in 2000. It includes six variables, which are population living in the family with three generations and over, temporary population, population aged 15–64, population living in two-person family, secondary industry workers, and population living in three-person family. Since the scores of these variables are all positive except the first one, this factor represents the temporary labor force population working as the secondary industry workers and living in two- to three-person family, which is known as the "salaried class" (*gongxin jiecheng*) in China. Figure 3 A shows the spatial pattern of it, displaying a concentric zonal pattern which is, contrary to the common one, with the lowest score in the center. It demonstrates that the longer the distance from urban core, the more people illustrated by this factor distribute. Most of them live in the suburban areas and Binhai New Area. As explained in the classic urban land use theory, central areas possessed better locations with easy access to various services and the urban transportation network and thus corresponded to high housing prices. Therefore, considering temporary populations are more inclined to live in areas where housing prices are comparative low, they prefer to live in suburban areas. Binhai New Area is a high-tech industrial zone that attracts large amount of labor force to migrate to there.

Factor 2: Socioeconomic Status

This factor accounted for 17.138 % of the total variance and included six variables: population in higher education level, population spending more money on housing, illiteracy rate, tertiary industry workers, non-agricultural population, and population paying higher monthly rent. Among the scores of these variables,

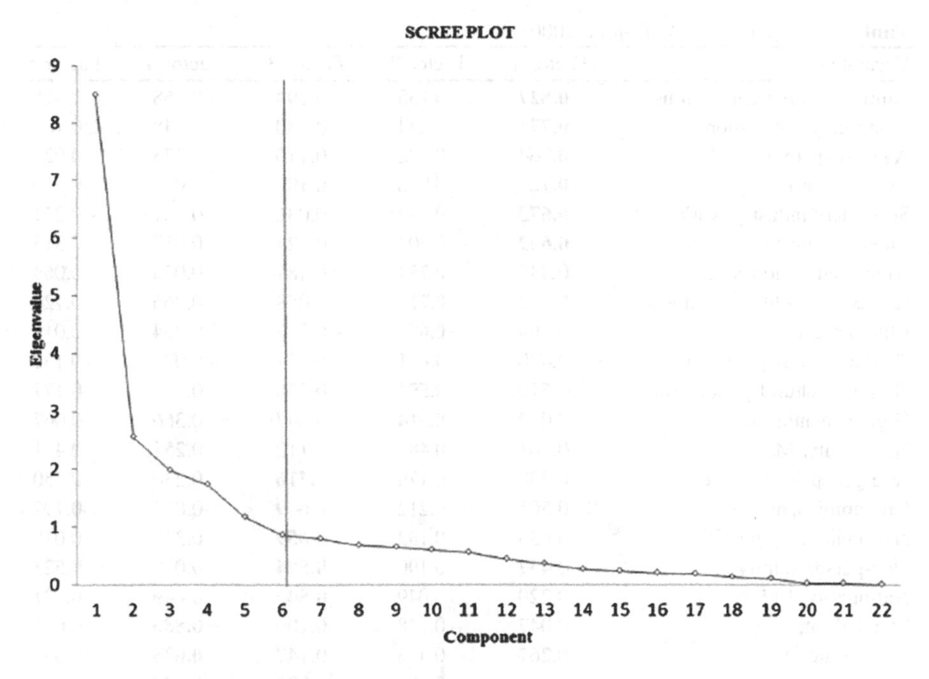

Fig. 2 Scree plot from principal component analysis

Table 1 Eigenvalues from principal component analysis

Component	Eigenvalue	Proportion (%)	Cumulative (%)
1	4.884	22.202	22.202
2	3.770	17.138	39.339
3	2.998	13.627	52.966
4	2.278	10.353	63.319
5	1.987	9.030	72.350

all are positive except the illiteracy rate. Therefore, this factor represents the well-educated population working in tertiary industry with a higher economic status. Figure 3b shows the spatial distribution of this factor, which is more concentrated at the urban core, especially the southwestern area and neighboring suburban area.

Factor 3: Retirees and Minorities

This factor accounted for 13.627 % of the total variance and included five variables: population aged 65 and over, unemployment rate, Uygur population, one-person family, and Hui population. The scores of all the variables are positive demonstrates this factor represents elderly people aged over 65 who live alone.

Table 2 Factor loading in Tianjin, 2000

Variable (%)	Factor 1	Factor 2	Factor 3	Factor 4	Factor 5
Family of three generations	**−0.827**	−0.135	−0.194	−0.068	−0.015
Temporary population	**0.771**	0.234	0.130	0.248	0.184
Age group 15–64	**0.764**	0.422	0.110	0.275	0.025
Two-person family	**0.726**	0.090	0.190	−0.070	0.510
Secondary industry workers	**0.673**	0.173	−0.036	0.101	−0.221
Three-person family	**0.642**	0.502	0.324	−0.052	−0.315
Higher education level	0.348	**0.754**	0.481	0.036	0.064
Higher expenditure on houses	0.234	**0.712**	0.058	−0.066	−0.128
Illiteracy rate	−0.069	**−0.697**	−0.223	−0.234	0.011
Tertiary industry workers	0.440	**0.661**	0.337	−0.027	−0.132
Non-agricultural population	0.519	**0.551**	0.538	−0.100	0.177
Higher monthly rent	−0.009	**0.514**	−0.047	0.366	0.067
Nationality, Manchu	0.301	0.483	−0.112	0.251	0.441
Age group 65 and over	−0.338	0.170	**0.716**	−0.230	0.350
Unemployment rate	0.505	0.212	**0.669**	−0.063	0.192
Nationality, Uygur	0.033	0.132	**0.659**	0.342	−0.018
One-person family	0.437	0.190	**0.594**	0.079	0.523
Nationality, Hui	0.229	0.049	**0.543**	0.049	−0.221
Married rate	0.047	−0.148	−0.109	**−0.880**	0.092
Single rate	0.267	0.463	0.147	**0.676**	0.058
Sex ratio	0.450	−0.191	−0.094	**0.653**	−0.034
Family of two generations	0.078	0.168	−0.040	0.093	**−0.884**

These people were mostly unemployed. Some of them are from Uygur minority and Hui minority. Figure 3c shows the spatial pattern of this factor. This factor mostly distributes mainly in the northwestern urban core and suburban area. The northwestern urban area is where the old city is located.

Factor 4: Marital Status

This factor accounted for 10.353 % of the total variance and included three variables: married population, single population, and sex ratio. It mainly represents the single population. Figure 3d shows the spatial pattern of this factor, displaying a sectoral pattern. This demographic is mostly located in the suburban area.

Factor 5: Family Structure

This factor accounted for 9.030 % of the total variance and included one variable: population living in family of two generations. Figure 3e shows the spatial pattern of this factor, displaying a concentric zonal pattern. The concentration of this factor in Baodi district and northern Wuqing district is comparatively high.

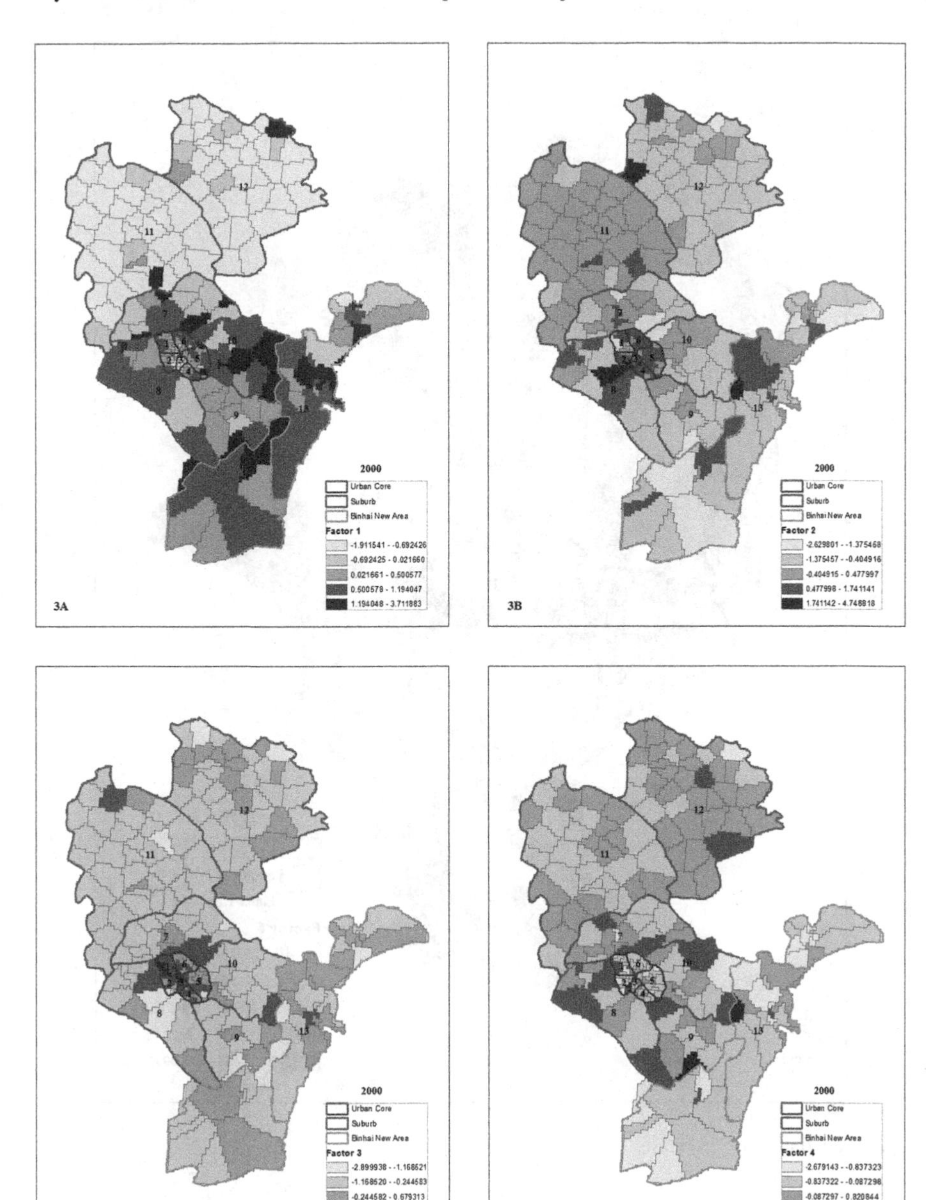

Fig. 3 Social space of Tianjin, 2000

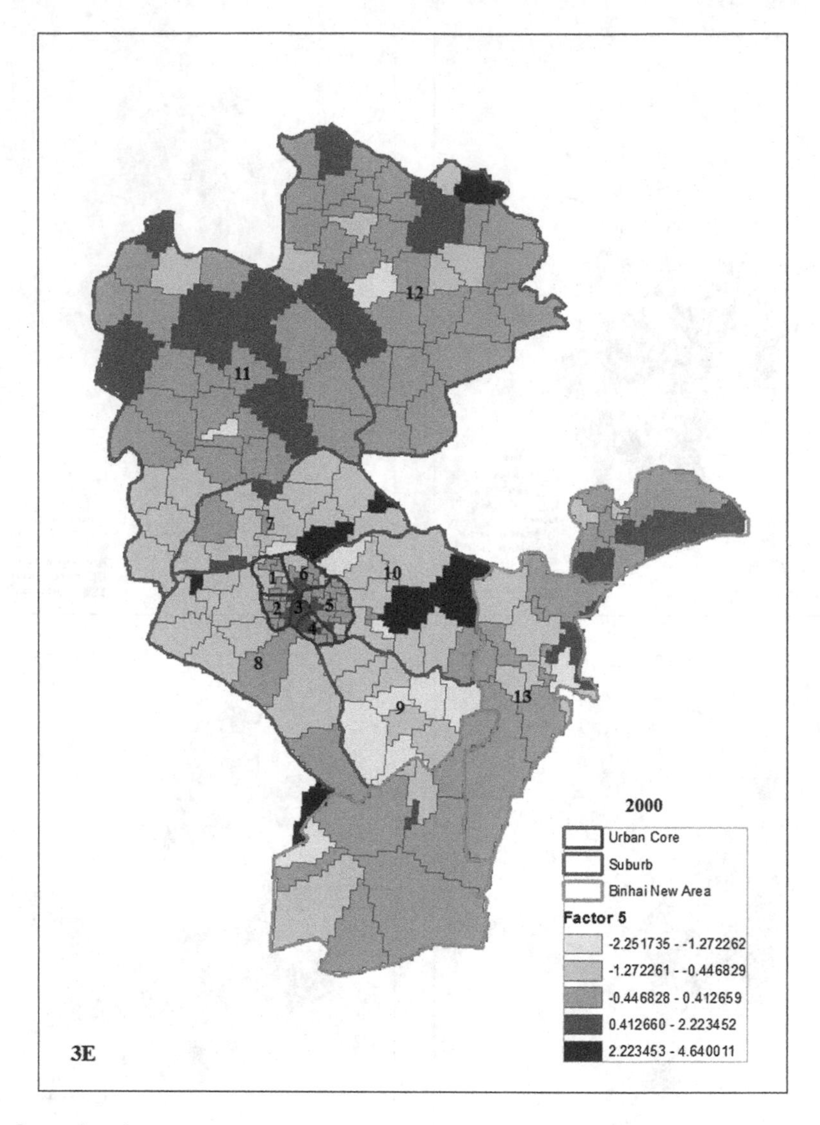

Fig. 3 continued

4 Conclusions and Discussion

As above discussion, the main factors that shape Tianjin's social space are the influx of the salaried class, economic status, retirees and minorities, marital status, and family structure. The overall social space of Tianjin can also be classified into five categories based on these factors. The salaried class are more heavily located in the suburban areas and Binhai New Area. On the contrary, the well-educated population working in tertiary industry and having good economic status are more concentrated at the urban core and suburban area that are close to urban core as well as the retirees and minorities.

4.1 Mechanisms of Tianjin's Social Space in 2000

Different from inland cities in China, Tianjin is built along the Hai River. The opening of the Grand Canal during the Sui Dynasty prompted the development of Tianjin. During the same period, a town along the Hai River and a rectangular "old city" nearby it with a crossed street inside began to shape their form. The main function of the town is commercial activities, while that of the old city is political activities. Until the creation of the PRC, the old town and old city areas had extremely high population densities. The population in these two areas is mainly composed of permanent working class whose income is relatively low, and the living conditions there are comparatively poor.

The invasion of the Western countries contributed to Tianjin's transformation. In 1860, with the signing of the Peking Treaty, Tianjin was forced to open up to colonists and foreign investors as a treaty port. This change influenced two aspects of Tianjin. First, it dramatically altered the industrial structure due to the increasingly flourishing foreign investment activities. Modern industries such as machinery, chemicals, metallurgy, textiles, and food processing industries were built in Tianjin. In the early twentieth century, Tianjin was the second largest city in China in terms of population, industrial production, banking, and port shipment (just behind Shanghai). During this period, Tianjin's industrial factories were majorly located in urban areas. Secondly, as a result of colonization, Tianjin's urban space expanded rapidly and was segregated by the establishment of nine foreign concessions (British, American, German, Japanese, French, Russian, Belgian, Italian, and Austrian), with a land area eight times greater than the old built-up area. The concessions not only stimulated the development of the real estate market and a reorganization of land resources, but also introduced the modern technologies of urban planning to the city. The construction of the "Hebei New Area" moved the urban center northward; meanwhile, the area between the old city and the concessions was not subject to any administration. At the same time, due to the construction of the concessions, people residing in the area were forcefully evicted.

Generally speaking, the populations living in concessions are from the upper class. In old cities and other sections outside the concessions, the middle class lived there. In the urban fringe, a squatter settlement was formed by the poorest people in the city. Therefore, during this period, the overall pattern of Tianjin is scattered and chaotic.

Chronologically speaking, three urban expansions occurred in Tianjin. The first one is during the colonial period. During this period, the commercial center of urban was located at the southeast part of the city; then, with the development of modern industry, the manufacturing district has been formed at the periphery of the urban area which was concentrated in the southeast and north. The second one is the "Great Leap Forward" (dayuejin), in which period, Tianjin was identified as an important industrial base of northern China [3]. In 1984, the third expansion period began with the opening up of 14 coastal cities to the outside. Drawing support from its geographic location and natural resources, being one of the 14 open coastal cities, the Tianjin Economic–Technological Development Area (known as TEDA) was established. In 1985, the Tianjin Master Plan announced that further industrial development be concentrated in TEDA. Then, in 1991, China announced the establishment of several duty-free districts in coastal cities. Tianjin's Duty-free District was established during this period. Emphasizing the importance of opening up of Tianjin in 1993, the city government announced the development of the Binhai District. In early 1994, the People's Congress of Tianjin approved the decision to develop the Binhai District as the key area of Tianjin's development [19]. According to Lichang Zhang, who is the Party Secretary of Tianjin at the time, "the key goal is to build the Binhai District into a modern industrial base…high level of agglomeration effect, with concentrated clusters and integrated production infrastructure." Obviously, with the process of the replacement of old industrial areas in urban spaces with residential and commercial areas, almost all the industries (both secondary industry and tertiary industry) were located in Binhai District. Furthermore, the district has established basic living conditions, including housing and schools, making the district potentially attractive to immigrations as well as young professionals and foreign investors. Therefore, more and more secondary industrial workers are now living in suburban areas, especially in Binhai New Area. Meanwhile, due to the urban reforms in 1984, an urban land market and a housing market have been introduced to Chinese cities accompanied by adjustment policies which laid out foundation for urban residential differentiation in China [14]. The adjustment of Hukou phenomenon is one of the policies' adjustments. Houkou is a strict household registration system used by Chinese government to control migration and resource allocation. Reforms to this household registration policy made it more flexible and made it easier for people to migrate. The migration of people from rural to urban areas strongly influenced the growth of the urban population and the increased complexity of Tianjin's urban sociospatial structure. After the Hukou reform, the amount of transient population in Tianjin increased rapidly. In the suburban area, the agglomeration of temporary labors has been formed. Besides, as discussed above, the building and development of Binhai New Area made the industrial zones move from the old city to the coastal area, which led to the transformation of industrial workers' residential locations.

At the same time, the implementation of the rehabilitation and the progress of gentrification in old city created the significant functional aggregation and segregation in the overall space. Much of the housing in the old city area of Tianjin is 140 years old. The concentration of population in these density areas is so high: Often, several families live sharing one internal courtyard. Many houses were found to be substandard and in danger of collapse. The living conditions are unimaginably poor, and therefore, land prices in this area are relatively low. Due to the "Rehabilitation Project," more and more administrative, cultural, and commercial buildings concentrated in the old city area, accompanying population mobility. After the project, land and housing prices in the old city doubled. The economic status of the population living in the urban core has improved. Meanwhile, according to Tianjin's Master Plan, an upper-scale residential area was built in Hexi District, which belongs to urban area neighboring Xiqing District that belongs to suburb. Therefore, in Tianjin's social space, people with a higher socioeconomic status are more concentrated at the urban core and neighboring suburban area.

In conclusion, taking its geographic and historical conditions as the foundation, Tianjin's social space pattern has formed by the interactions between the socioeconomic development and policy adjustments such as Hukou reform, house reform, urban land use policy adjustment, and urban planning change.

References

1. Grove, J., Burch, W.: A social ecology approach and applications of urban ecosystem and landscape analyses: a case study of Baltimore, Maryland. Urban Ecosyst. **1**(4), 259–275 (1997)
2. Shevky, E., Bell, W.: Social Area Analysis: Theory, Illustrative Application, and Computational Procedures. Stanford University Press, Palo Alto, California (1955)
3. Edgington, D.: Tianjin. Cities **3**(2), 117–124 (1986)
4. Lo, C.: Changes in the ecological structure of Hong Kong 1961–1971: a comparative analysis. Environ. Plann. A **7**, 941–963 (1975)
5. Lo, C.: The evolution of the ecological structure of Hong Kong: implications for planning and future development. Urban Geogr. **7**(4), 3311–3351 (1986)
6. Hsu, Y., Pannell, C.: Urbanization and residential spatial structure in Taiwan. Pac. Viewpoint **23**, 22–52 (1982)
7. Yu, W.: Studies on urban social space. Urban Plann. **10**(6), 25–28 (1986)
8. Gan, G.: Study on Urban Area Structure of Beijing City. Doctoral Dissertation of CAS, Beijing (1986)
9. Xu, X., Hu, H., Yeh, A.: A factorial ecological study of social spatial structure in Guangzhou. Acta Geogr. Sinica **44**(4), 385–399 (1989)
10. Yang, X.: Applying Factorial Ecology Analysis to Study Beijing Social Structure. Master Thesis of Peking University, Beijing (1992)
11. Gu, C.: The impact factors of the Beijing Social Spatial Structure. Urban Plann. **21**(4), 12–15 (1997)
12. Wu, J., Gu, C., Huang, Y.: Analysis of the urban social areas in Nanchang: analysis of the data based on the Fifth National Population Census. Geogr. Res. **24**(4), 611–619 (2005)
13. Zheng, J., Xu, X.: An ecological re-analysis on the factors related to social spatial structure in Guangzhou city. Geogr. Res. **14**(2), 15–26 (1995)
14. Gu, C., Wang, F., Liu, G.: The structure of social space in Beijing in 1998: a socialist city in transition. Urban Geogr. **26**(2), 167–192 (2005)
15. Roy, V., Cao, H.: Transformation ethnolinguistique de l'espace social du Grand Moncton, Nouveau-Brunswick (Canada), 1981–2006. Minorités linguistiques et société/Linguistic Minorities and Society **2**, 85–106 (2013)

16. Hair, J., Black, B., Babin, B., Anderson, R., Tatham, R.: Multivariate Data Analysis, 6th edn. Prentice Hall, Upper Saddle River (2005)
17. Cao, H., Villeneuve, P.: La localisation des garderies dans l'espace social de l'agglomération de Québec (Social Space and Daycare Centers in the Quebec Metropolitan Area). Cahiers de Géographie du Québec **42**(115), 35–65 (1998)
18. Ferguson, E., Cox, T.: Exploratory factor analysis: a users' guide. Int. J. Sel. Assess. **1**(2), 84–94 (1993)
19. Wei, Y., Jia, Y.: The geographical foundations of local state initiatives: globalizing Tianjin, China. Cities **20**(2), 101–114 (2003)

Hainan Virtual Tourism GIS Based on Speech Interface

Jun-kuo Cao, Rui-qiong Zhou and Li-hua Wu

Abstract Hainan virtual tourism system is a public-facing online travel service network platform based on Web GIS distributed architecture. Visitors' request is sent to the Web application server through the voice interface. Then, the Web server deals with the users' request by language processing and forwards the information to the GIS server for spatial analysis. Finally, the information is presented to the client. The user realizes the virtual tourism according to the operation tips. The GIS-based virtual tourism can achieve scenic tourist traffic and navigation using geographic information system and can achieve a three-dimensional simulation of the scene through computer multimedia technology.

Keywords Speech interface · Web GIS · Virtual tourism system · Spatial information system

1 Introduction

The twenty first century is the era of the information economy. It is the information knowledge not the capital that plays a decisive role in the information society. With the development and popularization of the Internet, network tourism has gradually become a kind of fashion and trend; people pay more attention to the development of network travel. On the one hand, the number of the domestic and foreign tourism Web site is increasing rapidly. Tourism information on the Internet

J. Cao · L. Wu
College of Information Science and Technology, Hainan Normal University, Haikou 571158, China

R. Zhou (✉)
Center of Network, Hainan Normal University, Haikou 571158, China
e-mail: 296089873@qq.com

B.-Y. Cao et al. (eds.), *Ecosystem Assessment and Fuzzy Systems Management*, Advances in Intelligent Systems and Computing 254, DOI: 10.1007/978-3-319-03449-2_29,

becomes more abundant. On the other hand, the existing tourist information is mainly reflected in the support of business information. And the introduction of the attractions is little with only a few still photos, which makes it difficult for visitors to have real feelings and interest of the attractions. Many users are no longer satisfied with simple text introduction and pictures and hope to see more interactive and authenticity data. Therefore, virtual reality and GIS technology are introduced into the network travel [1]. That is to say, the GIS-based virtual systems become a new tourist hot spot.

GIS-based virtual tourism can achieve scenic tourist traffic and navigation using geographic information system and achieve a three-dimensional simulation of the scene through the computer multimedia technology to build a virtual tour of the environment. This new way of travel experience has become the new choice of many travel enthusiasts. Currently, foreign virtual tour Web site has been very mature, and there are many success examples. Such as, the German Fraunhofer Image Processing Research Institute developed a tourism system of a single-computer mode, so that visitors can see the destroyed monuments of the original [2]; Italy Altair 4 creative team produced a set of ancient Egypt virtual tour system (http://www.altair4.com/) by 360° panoramic model and 3D model which vividly demonstrates the ancient Egypt; "the second life" in the United States has 4 or 5 million users. Visitors can tour the world famous thousands of attractions.

However, domestic virtual tour is still in its infancy. Most of the tourist's Web sites just provide a simple 360° panoramic model, photos, videos, and text for illustration. The expressive and interesting experience is so poor that it can just attract few people [3]. It is little far from abroad virtual tourism development. Although the virtual tourism cannot completely take the place of a field trip, it will be more and more close to the field of tourism as the technology is improved, especially the study of GIS technology, 3D technology, and speech technology. In this chapter, we introduce the concept of GIS on the basis of the traditional 3D virtual tourism, the knowledge of geographic information and tourism resources integration, so as to realize the simulation or restore the reality of tourism scenic spots, to construct a virtual three-dimensional 3D tourism environment. In addition, with the help of the language-processing technology, visitors can interact with the environment and talk to tourists in the virtual landscape, which can be personally on the scene tourism perception.

2 The System Structure

Hainan attractions of virtual tourism system are a public online travel service network platform based on Web GIS structure (shown as Fig. 1). Different from the traditional centralized mode of GIS, it is the browser/server (B/S) mode, using distributed architecture. Web GIS client is through a standard Web browser to interact with the user, send a request to the Web application server to explain the data returned by the Web server and displays it; the intermediate layer is a Web

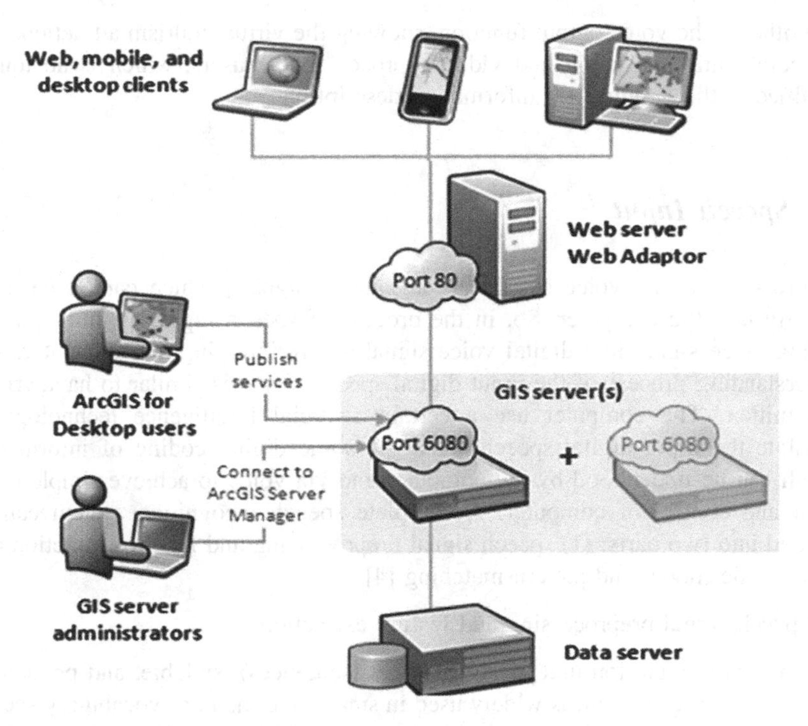

Fig. 1 Three layers of system architecture

server, which accepts requests coming from the browser. At the same time, it will process the request and transform the information into a form acceptable to the database (SQL). Then, they are sent to the database server. After the database server receives the query request, the appropriate action will be done and return the result to the Web application server. Web application server will deal with the results by GIS spatial analysis and processing. Meanwhile, it can be converted into a form acceptable to the browser (HTML). Then, it will send the HTML to the Web server. Finally, Web server sends an HTML document, including information back to the Web browser. Development and application of three-layer structure system has lots of advantages. The system is divided into different logic blocks, and every level is very clear, which is helpful to improve the development efficiency.

3 Design of Voice Interface

The speech interface Hainan attractions virtual tour system is divided into two parts. A voice input module, providing speech input and speech recognition function for the terminal user whose outdoor manual input is inconvenient.

The other is the voice output function, showing the virtual tourism attractions and the combination of audio and video approach to the users, which is no longer confined to the lengthy text information description.

3.1 Speech Input

Nature's voice and voice of speech are analog signals, which cannot be input directly into the computer. So, in the process of speech input, we must put the analog voice signal into digital voice signal before inputting into computer. The understanding process of the input digital speech signal is similar to handwriting recognition. The computer use a certain artificial intelligence technology to translate the input digital speech signal by some digital coding of information which can be understood by the computer and via voice to achieve simple operation and control of computer. A complete speech recognition system can be divided into two parts: (1) speech signal preprocessing and feature extraction and (2) acoustic model and pattern matching [4].

1. Speech signal preprocessing and feature extraction

The speech recognition unit includes words (sentences), syllable, and phonemes. The word (sentence) unit is widely used in small- and medium-vocabulary speech recognition system. It is not suitable for large-vocabulary system, because the model is too large and the model matching algorithm is complex and difficult to meet the real-time requirement, while the phoneme units are widely used in English speech recognition. Although large-vocabulary Mandarin speech recognition is widely used, it increases the number and complexity of the system model because the acoustic properties of Chinese vowel sound vary greatly. Therefore, the system adopts the Chinese speech recognition technology based on syllable unit.

Chinese is a single-syllable structure of language; although there are only about 1,300 syllables, if we do not consider the tone, there are only about 408 atonal syllables. Relatively speaking, the number of the syllable is not that much. Therefore, for medium- and large-vocabulary Chinese speech recognition system, the syllable is regarded as the basic recognition unit, is feasible. Speech signal contains abundant information. For nonspecific speech recognition, we hope the characteristic parameters can reflect the semantic information as much as possible trying to reduce the speaker's personal information. While for the specific voice recognition system, we will hope the characteristic parameters can reflect the semantic information as well as the speaker's personal information.

Feature extraction is a process of information compression. Linear prediction (LP) analysis technology is the current widely used characteristic parameters extraction technology. Many successful applications are based on the cepstrum parameters extracted by LP technology. But, LP model is purely mathematical model, which does not take into account the human auditory system for voice-processing characteristics. Based on the technology in the LP technology, the system introduces the Mel parameter and the perceptual linear prediction (PLP) analysis method. Thus, we can extract the perceptual linear predictive cepstrum, to some extent, the system simulate the human ear characteristics of voice processing. Compared with the traditional LP technology, the application of some research achievements in the field of human auditory perception improves the recognition rate greatly.

2. Acoustic model and pattern matching

Acoustic model is at the bottom of the speech recognition system model and is the most key part of the speech recognition system. Acoustic model is designed to provide an effective method to calculate phonetic pronunciation characteristic vector sequence, and the distance between templates each pronunciation. The size of the acoustic model unit can greatly affect the voice-training data size, system identification, and flexibility. Therefore, in order to improve the efficiency of recognition of a large-vocabulary continuously, we adopted the hidden Markov model (HMM) [5]. HMM is a statistical model of time sequence structure of speech signal, and it is regarded as a mathematical double random process: One is an implicit stochastic process with the number of finite state Markov chain to analog voice signal statistical characteristics changes, and the other is a random process associated with each state of the Markov chain of the observation sequence. Using HMM for speech recognition is essentially a kind of probability calculation [6]. According to the training set data model parameters, in order to get the test set data, we need to calculate the conditional probability of each model (Viterbi algorithm), and the maximal probability is the recognition result [7].

3.2 Speech Output

Speech synthesis (text-to-speech, TTS) includes acoustics, linguistics, digital signal processing, multimedia multiple disciplines, etc. It is a leading technology in the field of Chinese information processing. Speech synthesis system mainly includes three parts: text analysis, prosody generation module, and voice synthesis module. Among them, speech synthesis module is the most basic and the most important module. It uses the synthesis algorithms to synthesize the syllable waveform data which are required. And then, it splices idiom data into audio output module. At present, the commonly used speech synthesis technology is mainly formant synthesis, speech organs parameter combination, LPC synthesis, PSOLA techniques, and channel modeling techniques LMA. Among which, the formant synthesis techniques and algorithm-based PSOLA waveform concatenation synthesis technology are the mainstream technology.

As the advanced research in the field of information technology, the development of speech synthesis technology, especially the Chinese speech synthesis technology has gained high attention in the information industry. There are many domestic and foreign institutions engaged in research and development in this direction. Anhui Ustc Xunfei speech synthesis technology became a leader in this field after 10-year hard work with the support of the National major projects. Therefore, the system uses the local interface of IFLYTEK voice application service platform (VAP Express) as a module to develop our tourism GIS system. The advantage of this interface is that it can provide speech synthesis service directly using the TTS kit [8].

4 Construction of Hainan Virtual Tourism GIS

The emerging GIS technology can not only provide accurate and timely information but also take management capability, so that in the tourism industry, it can play an important role in the tourism geography information system (TGIS). Everything associated with tourism geographic information and data, such as scenic spots, transportation, accommodation, entertainment, shopping, and so on are TGIS research object, and will provide a full range of geographic information support for virtual tourism.

4.1 The Construction of Tourism GIS

4.1.1 Tourism GIS Structure

Essentially, GIS is an instrument of tourism, and it is quite helpful for the development of tourism. It can help the tourism adapt to the information society and contribute to the progress, which has become the inevitable choice with its powerful features. With the powerful geological analysis and spatial data management function of GIS technology, TGIS is widely used in all aspects. The function of this system is to provide support of electronic map of the famous tourist attractions in Hainan by means of geographic data input, database construction, implementation of spatial data query, roaming, management, and analysis (such as select the best path, etc.) [9].

System uses the classical multilayer software system construction (Fig. 2). It can not only logically divided the function of each module and mutual relationship but physically achieve truly independent components. Like the client applications, Web server extensions, GIS application servers, data servers, etc., each component can maintain and upgrade individually. The data stream is mainly composed of 2D GIS data and the model data. Among them, 2D GIS data are converted to shapefile layer, three-dimensional model data converted to GIS support multipatch format, eventually these two kinds of data and attribute data together into the spatial database access. Then, the users can scan the Web sites. Considering the stability and security of the system, the system uses the B/S structure development model, using ESRI ARCGIS10.1 as the company's Web2GIS development platform and adopting the Internet-oriented distributed computing technology.

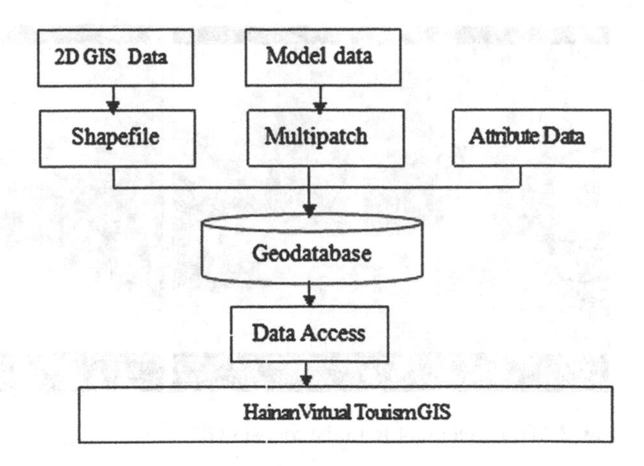

Fig. 2 The structure of tourism GIS

4.1.2 Data Organization

Geography Thematic Database

Geographic data are the data associated with the map layer, which are used to describe the distribution of geographic features. Digital layer includes the main traffic routes, Hainan province tourist attractions, hotel distribution, etc. The layer above, according to the data of the corresponding type, will be further divided into more layers, such as transport layer, subdivided into urban trunk road, national highway, provincial highway, highway, railway, etc. According to the classification system and granularity, the attractions' layer-size spots can be divided into class 5A, class 4A, class 3A, or divided into human landscape, natural landscape, and so on. For hotels, according to its type, they can be divided into the 5 star, 4 star hotels, etc.

Thematic Attribute Database

The classification of different tourism thematic database is based on the different people using the tourist information. Tourists mainly check some tourism information relevant to their trip, including tourism destination, tourism activities, tourism cost, tourist routes, traffic, accommodation, entertainment, hospitality of local customs and practices, local residents, and dietary habits, so that visitors can make reasonable tourism decision. The Planning Management Development Department is interested in how to exploit tourism resources, provide better service for tourists, and how to coordinate with each department to provide tourists with food, housing, transportation, travel, shopping, and entertainment. They have to balance the proportion of the various departments, so that tourism can be sustainable and harmoniously developed.

Fig. 3 Two screenshot from the tourism GIS

4.2 Design of 3D Scenes Virtual Tour

To construct tourism environment in cyberspace, the existing virtual reality system is far from meet the requirements. First of all is the network transmission speed cannot satisfy the large amount of data exchange; Followed by virtual reality technology used by the device is in the stage of development, not to the point of popularity. Therefore, in this chapter, the virtual reality which is based on real-image technology, namely directly using the live image obtained from the camera or camera (Real w Id Images) to construct the virtual landscape of the viewpoint space (View Point Space) [10]. The so-called viewpoint space can be observed in a point of observation, which is composed of a panoramic image of according to connections between the different focal length, reflecting different levels of detail of the scene space. Observer can view the 360° look around space, top, bottom, and focal length conversion and other means of observation. All the landscape that can be observed is defined as a complete panorama.

The virtual reality technical basic principle based on real image is: assuming that people observed in an interior space, and there are six surfaces in the interior space. If we can access many of these different distances and scenes in different directions, we can combine them according to their mutual relations in accordance with the organic linking. Then, we can form the overall understanding of the room from the perspective of visual space. When observing, we can arbitrarily turning to watch as well as change the point of view, or come to watch carefully. Because these pictures are interconnected, as long as there is sufficient accuracy in the photos, you can get the feeling of space. Accompanied by sound interface in our system, visitors can get a better casual observation and interactive access.

The implementation of a virtual tour system involves three aspects of technology: First, use Web GIS electronic map support functions to achieve the generation, management, display, and network sharing; Second is to generate a panoramic image using virtual reality technology; Third is the to use Java applet combined with Web GIS to complete the network-roaming panoramic images. Figures 3–4 are system design of Hainan mangrove panoramic virtual tour screenshots; visitors can click the forward direction and then walk into the corresponding scenic spots, so as to realize the full 360° of audio-visual tour.

5 Conclusion

With the development and popularization of Internet, network tourism has gradually become a kind of fashion and trend; people pay more and more attention to develop tourism network. But, the existing tourist information is mainly reflected in the support of business information. And the attractions are little with only a few still photos, which it makes difficult for visitors to have real feelings and interest of the attractions. As a result, many users hope scan more interactive and space information into the existing tourism information system.

Therefore, the virtual tourism system based on GIS becomes research hot spots, in which it can provide scenic tourist traffic and navigation using geographic information system and achieve a three-dimensional simulation of the scene through the computer multimedia technology. This chapter is dedicated to design Hainan virtual tour system providing a public-facing online travel service network platform based on the Web GIS architecture. Visitors request is sent to the Web application server through the voice interface. Then, the Web server deals with the users' request by language processing and forwards the information to the GIS server for spatial analysis. Finally, the valuable virtual tourism information will be presented to the client according to the operation tips.

Acknowledgments Thanks to the support by National Natural Science Foundation (61363032), National Natural Science Foundation of Hainan (No. 612120), Major Scientific Projects of Haikou (No. 2012-050). And this work is supported by International Science and Technology Cooperation Program of China (2012DFA1127) and Hainan International Cooperation Key Project (GJXM201105).

References

1. Huang, B., Jiang, B., Li, H.: An integration of GIS, virtual reality and the internet for visualization, analysis and exploration of spatial data. Int. J. Geogr. Inf. Sci. **15**(5), 439–456 (2001)
2. Schilling, A., Coors, V., Giersich, M., Aasgaard, R.: Introducing 3D GIS for the mobile community. Technical aspects in the case of TellMaris, IMC workshop, pp. 17–18 (2003)
3. Balogun, V.F., Thompson, A.F., Sarumi, O.A.: A 3D geo-spatial virtual reality system for virtual tourism. Pac. J. Sci. Technol. **11**(2), 601–609 (2010)
4. Meinedo, H., Neto, J.P.: Automatic speech annotation and transcription in a broadcast news task. In: ISCA Workshop on Multilingual Spoken Document Retrieval, Macau, China (2003)

5. Gales, M., Young, S.: The application of hidden Markov models in speech recognition. Found. Trends Signal Process. **1**(3), 195–304 (2007)
6. Tokuda, K., Yoshimura, T., Masuko, T., Kobayashi, T., Kitamura, T.: Speech parameter generation algorithms for HMM-based speech synthesis. In: Proceedings of ICASSP2000, vol. 3, pp. 1315–1318 (2000)
7. Morwal, S., Jahan, N., Chopra, D.: Named entity recognition using hidden Markov model (HMM). Int. J. Nat. Lang. Comput. (IJNLC) **1**(4), 15–23 (2012)
8. Avaya Solution & Interoperability Test Lab.: Application Notes for Anhui USTC iFLYTEK InterPhonic and iFLYTEK InterReco with Avaya Interactive Response using iFLYTEK MRCP Server—Issue 1.0, pp. 1–27 (2009)
9. Aubert, A., Csapo, J., Pirkhoffer, E., Puczko, L., Szabo, G.: A method for complex spatial delimitation of tourism destinations in South Transdanubia. Hung. Geogr. Bull. **59**(3), 271–287 (2010)
10. Lifeng, S., Li, Z.: Real-time virtual reality space roaming. Chin. J. Image Graph. **4**(6), 507–513 (1999)

The Trend of GIS-Based Suitable Planting Areas for Chinese Soybean Under the Future Climate Scenario

Wen-ying He, Le-le Wang, Xiao-li Luo and Guo-jun Sun

Abstract In this study, it gets the distribution of suitable planting areas for soybean by taking the soybean, one of the Chinese major food crops, as the object of study, adopting the multi-criteria evaluation (MCE) method, combining the climate, soil and crop growing conditions, and making use of the geographical information system (GIS) technology. The time zone for future climate chosen in this study is between 2071 and 2100. This chapter gets the distribution of suitable planting areas for soybean in the future presuming that the sunlight, soil, topography, and other factors are the same as those between 1971 and 2000 while the precipitation (PRE) and temperature change. In addition, it has compared it with the suitable planting areas at present. The results show that under the future climate conditions, the most suitable planting areas for spring soybean mainly concentrate in Inner Mongolia near to Jilin as well as the southeast of Gansu and southern Ningxia. Compared with the former suitable planting areas, the overall most suitable planting areas for spring soybean is moving to northwest. Under the future climate conditions, the total suitable planting area for Chinese spring soybean is 5,957,960 km^2, taking 79.35 % of the total area. Under the future climate conditions, the most suitable planting area for summer soybean mainly concentrate in western Jilin, western Liaoning, part of Hebei, Beijing, eastern Shandong, southern Jiangsu, Zhejiang, central of Fujian, part of Anhui, southern Hunan, northern Guangxi, southeast of Guizhou, Chongqing, Hubei, and part of Shanxi. The most suitable planting areas for summer soybean are scattered without obvious regularity and extend to Liaoning, Jilin, and northern Hebei to the north and to Shandong peninsula to the east and to the Yangtze River region to the south. The total suitable planting area of summer soybean is 6,118,013 km^2 taking 81.47 % of the total.

W. He · L. Wang · X. Luo · G. Sun (✉)
Institute of Arid Agroecology, School of Life Sciences, Lanzhou University,
Lanzhou 730000, China
e-mail: sungj@lzu.edu.cn

B.-Y. Cao et al. (eds.), *Ecosystem Assessment and Fuzzy Systems Management*,
Advances in Intelligent Systems and Computing 254, DOI: 10.1007/978-3-319-03449-2_30,

Keywords GIS · Multi-criteria evaluation · Soybean · Potential suitable planting area · Trend of change in the future

1 Introduction

The future climate change taking the global warming as the main characteristic has caused serious impact to the global ecological environment as well as economy and society. In particular with the global warming, the national agricultural planting area has changed a lot and the agricultural products production has decreased greatly in Brazil and other countries, which has caused the high concern of international society. China as a country with large population although has the third largest land area, it is in lack of per capita arable land area because of the large population. It is of great importance to the normal development of national economy to effectively evaluate the suitable planting area for a variety of good crops. With complicated climate conditions and fragile ecological environment in China, climate change will mainly cause the occurrence rates of water deficiency of major food crops, frequent natural disasters of agriculture, and fluctuation of food production increased a lot. Under the context of global warming, the occurrence degree of meteorological disasters of China's agriculture, water shortage and agricultural pests and diseases will be intensified while it is of great importance to make research on the effect of climate change to the crop planting in ensuring the national food security and promoting rural economic development [1] as well as the ecological environment protection and sustainable development.

This study makes evaluation to the suitable planting areas of Chinese soybean based on the multi-criteria evaluation (MCE) of GIS. MCE-GIS method, that is on the basis of geographical information, assists the decision-maker to find the most suitable one in the large amount of alternatives by using the eligible criteria and their weights and uses an evaluation matrix to solve a practical problem or target [2] and make the result visualized. This kind of method combining MCE and GIS has got extensive application in the species suitability evaluation and decision analysis.

The current researches on future climate mainly focus on the PRE and temperature change two aspects. The future climate time zone chosen in this research is between 2071 and 2100, presuming that the sunlight, soil, topography, and other factors are the same as those between 1971 and 2000 while the PRE and temperature change. The distribution of suitable planting area for soybean in the future with the current suitable planting areas was compared.

The future PRE and temperature data used in this research are the result of future regional climate change simulation and estimation performed by regional climate mode RegCM3 of International Center of Physical (ICTP, the Abdus Salam International Centre for Theoretical Physics, Italy) used by the researchers of National Climate Center and it is the ground temperature and PRE data within China in the context of SRES A2 greenhouse gas emission between 2071 and 2100 of later twenty-first century. Special report on emissions scenarios (SRES) is to get a future emission situation of greenhouse gases and sulfate aerosols through a series of factor hypothesizes (such as population growth, economic growth, technology development, environmental conditions, and global and fairness principle) and then estimate the climate changes in the global and regional in the future. Because of the differences in social and economic development that will appear in the future, it usually needs to make the corresponding different emission scenarios and SRES is one of them.

2 Research Method

This study establishes the spatial database of regional weather, topography, soil, and other environment conditions by taking the GIS as the main technological methods, choosing the soybean eligible factor, collecting the environmental condition data and making use of the spatial analysis module, spatial interpolation module, re-classification module, weight overlay, and other modules, and then, it uses MCE method to study the spatial distribution area and suitable levels of potential planting area for soybean.

The main steps are as follows:

1. Choose eight factors that affect the growth and development of soybean based on the expert advice and related literature records.
2. Establish the database of impact factors, mainly separate into database of meteorological factors and database of soil types.
3. In ArcGIS 9.2 software, use the spatial analysis module and spatial interpolation module to change the impact factors into raster layers.
4. Based on the expert advice and related literature records, use the re-classification module to make re-classification for the variable raster layer and get the suitable level figure of each factor that affects the soybean growth.
5. The agriculture experts make pairwise comparison for each factor by making use of analytic hierarchy process (AHP) and use MATLAB to calculate the weight of each impact factor.
6. Use the weight overlay module in the ArcGIS 9.2 software to make weight overlay for appropriate level figures of each impact factor and get the suitable planting area figure of Chinese soybean.

2.1 Data Collection and Collation

2.1.1 Eligible Factor Selection of Soybean

Among many impact factors that affect the soybean growth and development, based on the principles of domination, difference, and operability, etc. and referring to the expert [3–14] advice as well as the related literature records, it selects eight factors of representative and influential: accumulated temperature at the whole growth period (≥10 °C), PRE at the whole growth period, sunshine duration at the whole growth period, the maximum temperature at the whole growth period (August), the minimum temperature at the whole growth period (May), PH value of soil, field moisture capacity, and type of soil.

2.1.2 Data Collection of Environmental Conditions for Soybean Growth

The future PRE and temperature data used in this research are the result of future regional climate change simulation and estimation performed by regional climate mode RegCM3 of International Center of Physical (ICTP, the Abdus Salam International Centre for Theoretical Physics, Italy) used by the researchers of National Climate Center, and it is the ground temperature and PRE data within China in the context of SRES A2 greenhouse gas emission between 2071 and 2100 of later twenty-first century. The resolution is 20 km and the range covers the whole China as well as the surrounding East Asia. The temperature is surface temperature (TAS): It is the temperature near the ground (usually at the height of 2 m), unit: K; PRE: including all kinds (rain, snow, large-scale precipitation and convective precipitation, etc.), unit: mm/day; region: longitude 60°E–149°E and latitude 0.5°N–69.5°N.

The meteorological data under the current climate conditions needed in this study is downloaded from the Web site of China Meteorological Data Sharing Service System (http://cdc.cma.gov.cn/). The data can be mainly divided into three types:

1. Data of Month: including monthly average maximum temperature, monthly average minimum temperature, monthly average temperature, and monthly PRE etc. of the nationwide 794 meteorological stations from 1971 to 2000.
2. Data of Day: the time range of this data is also between 1971 and 2000 and also covers 794 meteorological stations of the whole mainland. The main contents of the data are daily temperature data within the soybean growth period, etc.
3. Data of Sunshine: the data of this part not only include the part of mainland, but also cover the meteorological stations of our surrounding areas, totally more than 2,300. The time range is also from 1971 to 2000 and the main content of data is average sunshine hours.

The topographical data used in this study are from National Administration of Surveying, Mapping, and Geoinformation. The resolution of this data is 1:250,000 and the main contents include the following: contour lines, data of national rivers and lakes, administrative boundary data of national borders and provincial boundaries.

The soil data of this study is from Institute of Soil Science, Chinese Academy of Science, including 0–20 cm soil properties of raster data and the national 1: 1,000,000 soil types data.

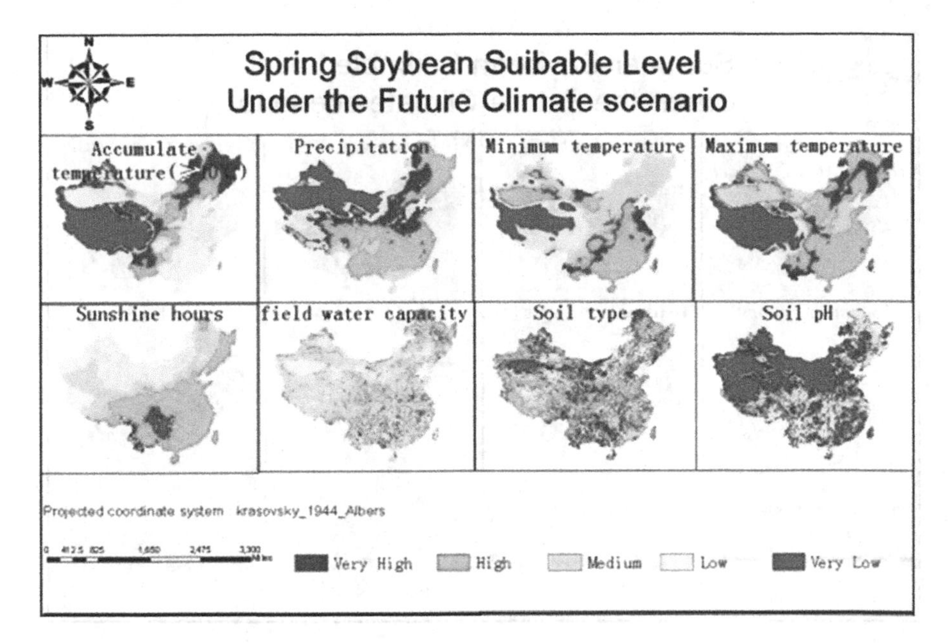

Fig. 1 Factor map for spring soybean under the future climate scenario, including suitability levels for each factors

The land-use data in this study are from Environmental and Ecological Science Data Center for West China and include the land-use map with national 1-km resolution.

2.2 Spatial Database Establishment of National Climate, Topography, and Soil

2.2.1 Establishment of Climate Database

According to the oat crop mapping from Fan Wan [15] and others as well as the interpolation method comparison of each meteorological factor in the potential suitable planting area of wheat from Yaxiong Chen and others [16], this study follows the results of their researches: the maximum temperature and minimum temperature are got from the average monthly data of 1971–2000 and 2071–2100, adopting the ordinary Kriging method to make the interpolation. ≥10 °C accumulate temperature is got from the average daily material of 1971–2000 and 2071–2100, adopting the IDW method to make the interpolation. PRE in the growing period is got from the average daily data of 1971–2000 and 2071–2100, adopting the UK method to make the interpolation and get the figures of appropriate level of each factor (Figs. 1, 2).

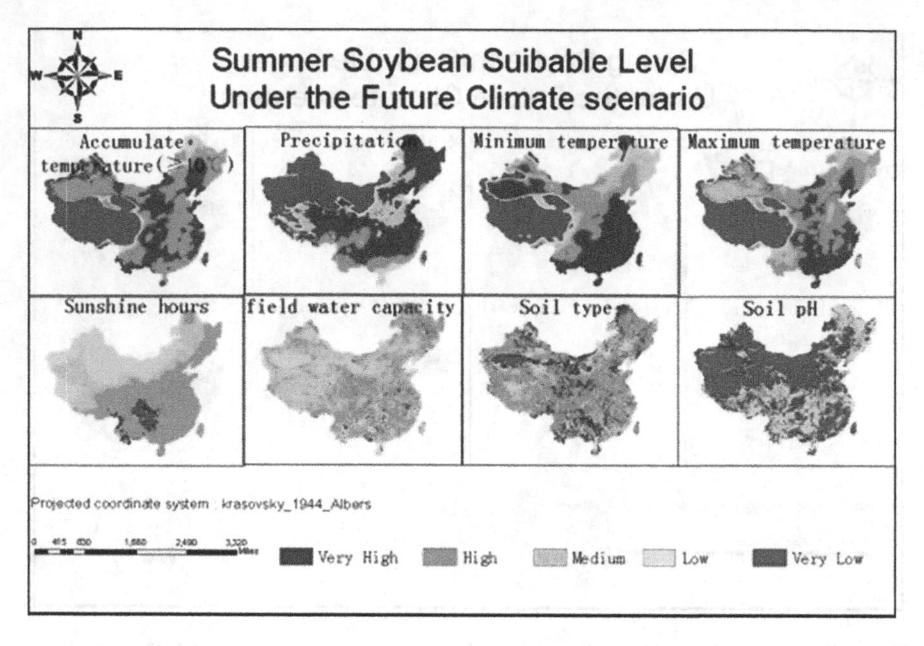

Fig. 2 Factor map for summer soybean under the future climate scenario, including suitability levels for each factors

2.2.2 Establishment of Topography Database

The national 1:250,000 contour lines data are from National Administration of Surveying, Mapping, and Geoinformation. Because the contour lines are only expressing the partly elevation values within the range, therefore in this application, we have made the contour lines into digital elevation model (DEM), which is finished in the ArcGIS 9.2 software. First, in the 3D analysis module of Arcinfo, change the original contour lines data into triangulated irregular network (TIN) and then change it into DEM. Merging the already-got scattered DEM in the spatial analysis module and get the national DEM, from which extract the elevation information. Usually, the DEM is expressed in elevation matrix composed of ground regular network unit, which has been widely applied to the description of topography and has replaced the contour representation in traditional topographical maps [17]. It adopts Albers projection and the size of raster is 1,000 × 1,000 m.

2.2.3 Establishment of Soil Database

The national 1:1 billion soil type data and 0–20 cm soil property raster data are from Institute of Soil Science, Chinese Academy of Science. Firstly, sample the data and get 99.034 point data. Make Kriging spatial interpolation for these point

Table 1 Specific level per factor for the spring soybean

Factor	Level of suitability				
	Very high	High	Medium	Low	Very low
Accumulate temperature (≥10)	3,300–2,400	2,400–2,200 or 3,300–3,800	2,200–1,900 or 3,800–4,000	>4,000	<1,900
Precipitation (mm)	540–370	370–320 or 540–1,000	320–250 or >1,000	320–180	<180
Minimum temperature (°C)	22–20	20–18	18–10	10–4	<4
Maximum temperature (°C)	25–22	22–18	18–16	16–4	<14 or >25
Sunshine hours (h)	750–700	750–1,200	1,200–1,350	1,350-1,700	<700
Field water capacity	0.80–0.80	0.7–0.5	0.5–04	0.4–0.2	<0.2
Soil type	Loam	Sandy/Clay/Silt loam	Sandy/Silt clay	Other class	Sandy
Soil pH	6.0–6.5	6.5–7.0	7.0–7.5	7.5–7.8	<6.0 or >7.8

data in ArcGIS 9.2 to get the soil raster layer and extract the soil texture and soil pH value and other information about each soil type by combining "Chinese Soil Genus Records" and then generate the raster layer of each type of soil in ArcGIS 9.2 adopting Albers projection and the size of raster is 1,000 × 1,000 m. National 0–20-cm soil layer property raster data, including organic matter content, total nitrogen, total phosphorus, total potassium content data, are obtained through projection conversion and re-sampling project kasovsky-1940-Albers equal area projection and the size of raster is 1,000 × 1,000 m.

2.3 *Classification of Suitable Area and Suitable Level of Soybean*

Based on the established database and refer to the related literature, the weather and soil data were classified into five levels according to suitability, respectively very high, high, medium, low and very low (Tables 1, 2), and through the reclassification in ArcGIS, the proper-level figure of each factor that affects the soybean growth (Figs. 1, 2) was got.

The weight of eligible factor was determined, and the agriculture experts make pairwise comparison for each factor by making use of AHP and use MATLAB to calculate the weight of each impact factor (Table 3).

Table 2 Specific level per factor for the summer soybean

Factor	Level of suitability				
	Very high	High	Medium	Low	Very low
Accumulate temperature (≥10 °C)	3,400–2,800	2,800–2,400 or >3,400	2,400–2,000	2,000–1,000	<1,900
Precipitation (mm)	650–360	650–1,000	360–250	>1,000	<250
Minimum temperature (°C)	30–20	20–15	15–10	10–8	<8
Maximum temperature (°C)	28–25	25–22	22–18	18–16	<16
Sunshine hours (h)	750–700	750–1,200	1,200–1,350	1,350–1,700	<700 or >1,700
Field water capacity	0.80–0.80	0.7–0.5	0.5–04	0.4–0.2	<0.2
Soil type	Loam	Sandy/Clay/ Silt loam	Sandy/Silt clay	Other class	Sandy
Soil pH	6.0–6.5	6.5–7.0	7.0–7.5	7.5–7.8	<6.0 or >7.8

Table 3 Analysis results of the comparison

	AT	PR	MI	MA	SH	WC	ST	PH
AT	1	9/8	3/2	6/5	6/1	7/1	8/1	8/1
PR	8/9	1	4/3	75	7/1	8/1	9/1	9/1
MI	2/3	3/4	1	23	2/1	3/1	4/1	4/1
MA	5/6	5/7	3/2	1	3/1	4/1	5/1	5/1
SH	1/6	1/7	1/2	1/3	1	7/6	9/6	9/6
WC	1/7	1/8	1/3	1/4	67	1	9/7	9/7
ST	1/8	1/9	1/4	1/5	6/7	7/9	1	1
PH	1/8	1/9	1/4	1/5	6/7	7/9	1	1

The corresponding weight coefficient of each factor after calculation includes the following: accumulated temperature 0.2657, PRE 0.2744, minimum temperature 0.1633, maximum temperature 0.1633, sunshine hours 0.0505, field water capacity 0.0410, soil pH 0.0328, and soil type 0.0328. Making use of the got weight coefficiencies as well as the weight overlay module in the ArcGIS 9.2 software, the weight overlay for the proper-level map of each influence factor was made and then the already-got level diagram and land-use map extracting water, sand, desert and alpine wilderness were overlaid, and finally, the potential suitable planting area diagram for Chinese soybean at present and in the future was got (Figs. 3, 4).

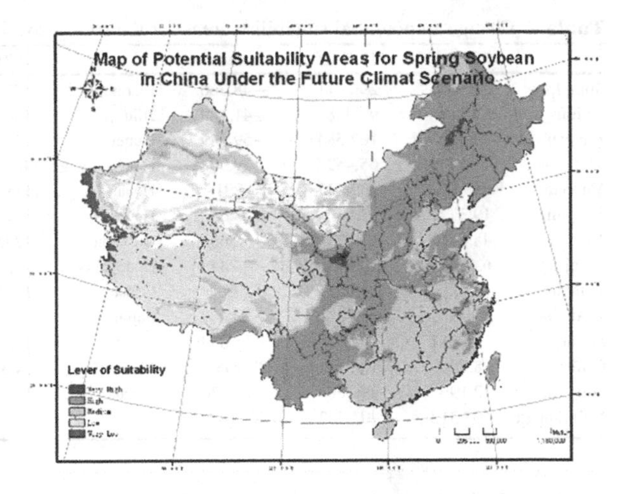

Fig. 3 Map of potential suitability areas for spring soybean in China under the future climate scenario

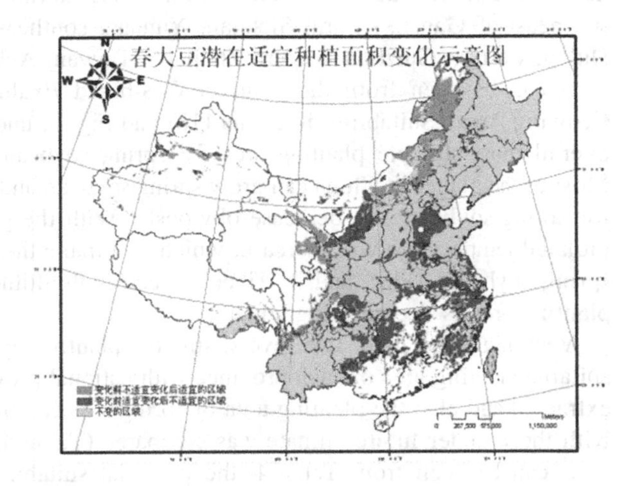

Fig. 4 The sketch map of change in potential suitability areas for the spring soybean

3 Results and Analysis

3.1 The Change in Potential Suitable Planting Area for Chinese Spring Soybean Under Future Climate

If very high, high, and medium suitable planting areas were chosen, the suitable planting area for Chinese spring soybean under the future climate is 5,960,271 km^2, which takes 79.35 % of the total area (Fig. 3).

It can be seen from Fig. 8, the most suitable planting areas for spring soybean under future climate change concentrate in the area of Inner Mongolia near to Jilin as well as southeast of Gansu and southern Ningxia. Relatively suitable planting areas mainly concentrate in Heilongjiang, Liaoning, Jilin, northeast of Mongolia,

Table 4 Change in potential suitability areas for the spring soybean in each province units: km^2

	2071–2100	1971–2000	Difference		2071–2100	1971–2000	Difference
Shandong	247,645	296,011	−48,366	Liaoning	142,448	142,422	26
Anhui	54,319	95,448	−41,129	Jilin	180,240	172,723	7,517
Jiangsu	48,035	107,584	−59,549	Hubei	107,050	217,002	−109,952
Zhejiang	45,544	65,452	−19,908	Heilongjiang	406,537	400,373	6,164
Yunnan	358,723	337,432	21,291	Henan	118,492	159,466	−40,974
Xinjiang	19,377	32,930	−13,553	Hebei	127,201	172,325	−45,124
Xizang	48,011	53,577	−5,566	Guizhou	129,977	162,693	−32,716
Tianjing	6,624	10,868	−4,244	Guangzhou	16,976	55,202	−38,226
Taiwan	25,630	18,732	6,898	Gansu	151,729	120,491	31,238
Shanghai	4,142	4,848	−706	Fujian	26,228	59,974	−33,746
Shannxi	181,211	197,020	−15,809	Beijing	14,368	16,077	−1,709
Qinhai	28,986	1,881	27,105	Chongqing	54,885	61,525	−6,640
Ningxia	12,196	29,166	−16,970	Sichuan	180,315	184,621	−4,306
Neimenggu	607,313	402,242	205,071				

northern Heibei, southwest of Henan, Shandong, northern Anhui, Shanxi, Shanxi, southeast of Gansu, central Sichuan, Yunnan, southeast of Chongqing, western Hubei, western Guizhou, eastern Zhejiang, Taiwan, Anhui, and part of Fujian.

It can be seen from the contract GIS-based Evaluation of spring Soybean Growing Areas Suitability in China [18] and Fig. 8, under the future climate, the overall most suitable planting area for spring soybean will move to northwest. Most of the places of China can grow spring soybean and the most unsuitable areas for spring soybean will decrease obviously. With the global warming, the accumulated temperature will increase, which will make the suitable planting area for spring soybean wider, and in Tibetan, such high-altitude region, the unsuitable planting area is decreasing gradually.

Very high and high, two levels, suitable planting areas were chosen, and the suitable planting area of each province with national provincial boundary map was extracted and also the planting area of spring soybean in each province at present with those under future climate was compared (Table 4, Fig. 4).

It can be seen from Table 4, the potential suitable planting areas of spring soybean under future climate change such as Neimenggu, Qinghai, and Gansu get the highest increase with Inner Mongolia increasing by 205,071 km^2, Qinghai increasing by 27,105 km^2, Gansu increasing by 31,238 km^2, and the northeast place nearly unchanged. There is a decrease in the southern planting area such as Anhui, Jiangsu, Guizhou, and Fujian. The general increasing trend is moving to inland and northern part. Under future climate, the total suitable planting area is 3,344,202 km^2, which has a little decrease when compared with the current suitable planting area of 3,583,131 km^2. But the most unsuitable planting area is 57,730 km^2, which has an obvious decrease when compared with the current suitable planting area of 522,459 km^2 [18]. With the global warming, the suitable planting area of spring soybean within China is becoming wider and wider.

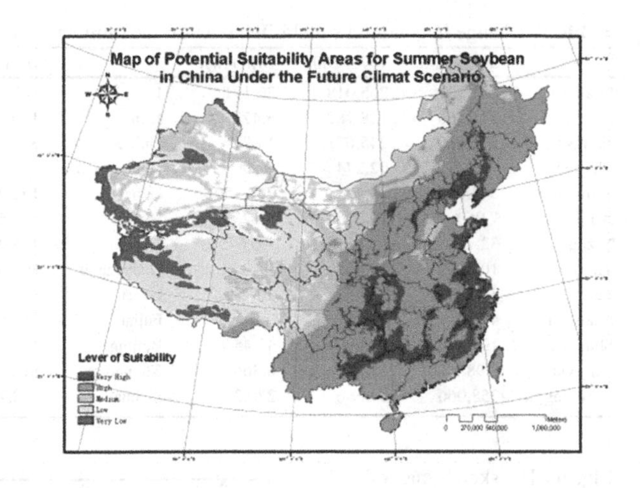

Fig. 5 Map of potential suitability areas for summer soybean in China under the future climate scenario

3.2 The Changes of Potential Suitable Planting Areas for Summer Soybean in China Under Future Climate Conditions

If very high, high, and medium suitable planting areas were chosen, the suitable planting area for summer soybean in China under future climate conditions is 6,118,013 km^2, which takes 81.47 % of the total area (Fig. 5).

From Fig. 5, the most suitable planting areas for summer soybean under future climate conditions focus on western Jilin, western Liaoning, parts of Hebei, Beijing, eastern Shandong, southern Jiangsu, Zhejiang, central Fujian, parts of Anhui, southern Hunan, northern Guangxi, southeast of Guizhou, Chongqing, Hubei, and parts of Shannxi.

We can learn from the comparison between Fig. 5 and GIS-based Evaluation of summer Soybean Growing Areas Suitability in China [18] that the change in most suitable planting area for summer soybean is relatively scattered without obvious regularity to the north, extend to Liaojing, Jilin and northern Hebei; to the east, extend to Shandong peninsula and to the south, extend to Yangtze River region. Most of the places around the country can grow summer soybean and the areas not suitable for summer soybean planting is decreasing.

Choose very high and high two levels as the more suitable planting area and extract the suitable planting area of each province with the national provincial boundary map and make comparison of the planting areas for summer soybeans of each province under the future climate conditions and current climate conditions [18] (Table 5, Fig. 6).

From Table 5, under future climate change, the potential suitable planting areas which have great increase include Inner Mongolia, Yunnan, Shanxi, Liaoning, Jilin, Heilongjiang, Guizhou, Gansu, Chongqing and Sichuan. The Inner Mongolia increased by 270,380 km^2, Shanxi by 47,454 km^2, and Heilongjiang by 72,987 km^2. Under the change in future climate, the planting areas of summer

Table 5 Change in potential suitability areas for the summer soybean in each province units: km^2

	2071–2100	1971–2000	Difference		2071–2100	1971–2000	Difference
Shandong	285,209	265,018	20,191	Liaoning	142,452	128,331	14,121
Anhui	134,260	128,782	5,478	Jilin	180,240	134,702	45,538
Jiangsu	247,500	245,671	1,829	Hubei	380,354	357,752	22,602
Zhejiang	96,873	92,274	4,599	Heilongjiang	362,846	289,859	72,987
Yunnan	344,033	82,474	261,559	Henan	152,033	163,191	−11,158
Xinjiang	7,060	19,706	−12,646	Hebei	163,652	144,904	18,748
Xizang	32,758	39,457	−6,699	Guizhou	175,988	135,713	40,275
Tianjing	10,869	10,868	1	Guangzhou	397,023	372,644	24,379
Taiwan	34,325	23,537	10,788	Gansu	95,973	51,830	44,143
Shanghai	4,910	4,910	0	Fujian	117,523	109,451	8,072
Shannxi	191,363	143,909	47,454	Beijing	16,124	15,933	191
Ningxia	7,988	1,622	6,366	Chongqing	81,724	70,433	11,291
Neimenggu	358,066	87,686	270,380	Sichuan	208,760	153,886	54,874

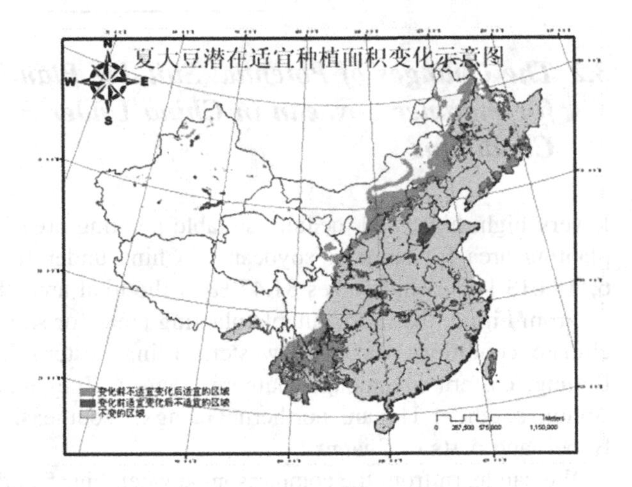

Fig. 6 The sketch map of change in potential suitability areas for the summer soybean

soybean in most of the provinces of China have increased. The areas of northeast and Inner Mongolia have increased in overall and only that of Xinjiang has decreased. The overall increasing trend is moving to inland and northern parts. The total planting area under future climate is 4,259,635 km^2 and has great increase when compared with the current suitable planting area of 3,306,593 km^2. The unsuitable planting area is 350,769 km^2, which has decreased a little when compared with the current suitable plating area of 695,312 km^2. With the global warming, the suitable planting area of summer soybean within China is becoming bigger and bigger.

4 Discussion

Compared with present, the overall most suitable planting areas for spring soybean under future climate conditions move to the northwest and the most unsuitable planting area for spring soybean are obviously decreased. The total suitable planting area under future climate is 3,344,202 km^2 and has deceased a little when compared with the current suitable planting area of 3,583,131 km^2 [18]. But the most unsuitable planting area is 57,730 km^2 and is greatly decreased when compared with the current suitable planting area of 522,459 km^2 [18].

Compared with present, the most suitable planting areas for summer soybean under future climate conditions are scattered without obvious regularity and extend to Liaoning, Jilin, and northern Hebei to the north and to Shandong peninsula to the east and to the Yangtze River region to the south. If the best and better planting areas were chosen, it can be seen that the suitable planting areas in inland and northern are increasing. The total suitable planting area in the future is 4,259,635 km^2, which is greatly increased when compared with the former 3,306,593 km^2. The unsuitable planting area is 350,769 km^2, which is decreased a little when compared with the current suitable planning area.

There are also some problems existed in this study. The data under future climate conditions, because of the limitations of the science itself of atmosphere and climate change as well as of computer technology, all the estimation about the future climate change trend has many kinds of uncertainty and it need further development of science and technology. Besides, this study makes the simulation by only presuming that the PRE and temperature change while other factors do not change. But in reality, there are many factors in the future climate will change because the interaction among various factors and each changing factor is not existed alone. For example, the global warming will make the evaporation of soil moisture increase and the organic matter and nitrogen lose, which will lead to further decrease in the PRE in arid regions, the increasing degree of soil erosion and the seriousness of salinity degree, so as to change the soil texture and pH, etc. Therefore, the conditions used in this study are relatively ideal conditions, so the result can only be used as a reference.

In general, global warming will cause the decrease in production potentiality of our major food crops and the increase in the instability. If the effective measures does not take actively against the climate change, the agricultural situation of China will be very challenging based on our existing production levels and security conditions. Therefore, it should take the climate change as the priority strategy to response to the climate change and take the promoting agricultural production and ensuring food safety as our main task to response to the climate change. This study is just from this aim and hoping the study results will have certain reference value for the macrodecision-making of the government.

Acknowledgments This work is financially supported by the Grants ISTCP 2010DFA92800 and 2012DFG31450.

References

1. Zhao, J., Yang, X., Liu, Z., Cheng, D., Wang, W., Cheng, F.: The possible effect of global climate changes on cropping systems boundary in China II. The characteristics of climatic variables and the possible effect on northern limits of cropping systems in south China. Sci. Agric. Sin. **43**(9), 1860–1867 (2010)

2. Rico, A.I.: Application of multicriteria evaluation to a spatial conflict in the urban fringe. ITC Msc thesis (2001)
3. Carver, S.J.: Integrating multi-criteria evaluation with geographical information systems. Int. J. Geogr. Inf. Syst. **5**, 321–339 (1991)
4. Wei, L.: Analysis of meteorological factors on soybean growth and yield. Heilongjiang Agric. Sci. **2**, 41–43 (2008)
5. Li, X.: The preliminary study of the relationship between climatic conditions and growing development of spring soybean. Crop Res. **4**, 15–17 (1998)
6. Yu, X., Guo, Y.: The effect of meteorological factors on the soybean output. Heilongjiang Meteorol. **2**, 3–4 (2002)
7. Zhou, J., Wang, C., Xu, W.: Comparison between transgenic soybean and non-transgenic soybean in resistance to stresses. J. Ecol. Rural Environ. **22**(2), 26–30 (2006)
8. Wang, Y., Feng, J., Yin, X., Xu, W.: Relationship of soil mechanical composition, pH value and main nutrients with soybean yield in planting areas of Jilin province. Jilin Agric. Univ. **26**(6), 653–658 (2004). Agric. Sci. Bull. **11**(4), 58–59 (2005)
9. Yan, P., Jiang, L., Wang, P., Nan, R., Wang, C., Wang, C.: Analysis of the illumination effect on the soy output from May to July. Heilongjiang Meteorol. **4**, 22–23 (2005)
10. Xiaoming, L.: Study the rate of soybean absorbing the phosphorus in different soil types. Soybean Sci. **2**, 75–77 (2002)
11. Han, X., Qiao, Y., Zhang, Q., Wang, S., Song, C.: Effects of various soil moisture on the yield of soybean. Soybean Sci. **11**, 269–272 (2003)
12. Qiu, J.: Affect of high temperature on soybean production. Soybean Bull. **1**, 12 (2003)
13. Fu, H., Liu, D., Guo, F., Wang, C., Zhao, L.: Effects of different soils on growth, development, yield and quality of small-seed soybean. Jilin Agric. Univ. **24**(5), 21–25, 38 (2002)
14. Chen, Y., Chen, R., Su, Y., Zhong, Z., Zhong, Q., Ji, Y.: Analysis of the meteorological factors affecting summer soybean growth in Huaibei area. J. Anhui Agric. Sci. **32**(4), 745–748 (2004)
15. Wan, F., Wang, Z., Li, F., Cao, H., Sun, G.: GIS-based crop support system for common oat and naked oat in China. Adv. Inf. Commun. Technol. **1**, 209–221 (2009)
16. Chen, Y., Wan, F., Zhou, W., Jia, C., He, W., Guo, R., Sun, G.: Crop mapping in China: the potential distribution of winter wheat. J. Lanzhou Univ. (Nat. Sci.) **45**(FO6), 12–18 (2009)
17. Li, Z., Zhu, Q.: Digital Elevation Model. Wuhan University Press, vol. 10 (2003)
18. He, W., Yang, S., Guo, R., Chen, Y., Zhou, W., Jia, C., Sun, G.: GIS-based evaluation of soybean growing areas suitability in China. Comput. Comput. Technol. Agric. IV–4th, 372–381 (2010)

Construction and Implementation of the Three-Dimensional Virtual Panoramic Roaming System of Hainan Ecotourism

Li-hua Wu, Jian-Ping Feng and Shu-qian He

Abstract The three-dimensional panoramic roaming system reproduces three-dimensional scene in panorama mode. The system can be the virtual scene roaming in the browser and has good interactivity and realism. In this chapter, the goal is to construct three-dimensional panoramic roaming system and proposes the generation process of virtual three-dimensional panoramic space based on panorama technology. Then, the chapter's depth study of the process of using a fisheye lens and digital camera to capture panoramic images, 360° panorama stitching algorithm, and using landscaping software Pano2VR production methods generates three-dimensional scene. Finally, Hainan Ecotourism Environment panorama design as an example, it introduces the whole process for achieving the three-dimensional panoramic virtual roaming system in details. Experiments show that the roaming system construction method is not only better visual effects and application valuably, at the same time, they will provide a technology reference for the three-dimensional panorama used in three-dimensional model, scene roaming, and interactive mode of Hainan ecotourism.

Keywords Three-dimensional panoramas · Image stitching · Panoramic roaming system · PanoV2R landscaping software

Panorama is to shoot 360° camera ring; one or more groups photos together into a panoramic image by computer technology to achieve real scene to restore full display mode and interactive viewing.

L. Wu · J.-P. Feng (✉) · S. He
School of Computer Science and Technology, Hainan Normal University,
HaiKou 571158, China
e-mail: fjp888@163.com

B.-Y. Cao et al. (eds.), *Ecosystem Assessment and Fuzzy Systems Management*,
Advances in Intelligent Systems and Computing 254, DOI: 10.1007/978-3-319-03449-2_31,

1 Introduction

The virtual panorama is also known as three-dimensional panoramic virtual reality; they are the real scenes of virtual reality technology based on panoramic images. Panorama is to shoot 360° camera ring; one or more groups photos together into a panoramic image by computer technology to achieve real scene to restore full display mode and interactive viewing. Supporting player's plug-in, we can look around with the mouse to control the direction of the viewer immersive feel. Also, in recent years, there have been many expressions based on three-dimensional panoramic images of real-case scenarios, such as Google Earth, Baidu Streetscape, Mengniu 360, panoramic passenger, moving scenery, and so on.

Three-dimensional panorama technology show in the tourist attractions mainly refers to the application on the Web site that provide tourist scenic spots and historical sites that show a three-dimensional panoramic effect. A high-resolution three-dimensional display scenic panorama with beautiful surroundings can give the tourists an immersive scene experience. While combining scenic tourist navigation map, the tourist easy click of a mouse can be achieved through a 360° viewing, as well as the entire area of virtual scene roaming. Currently, the three-dimensional panorama roaming system is a best practice of tourist attractions and tourism product innovation and promotion.

2 Overview of Three-Dimensional Panorama Technology

Three-dimensional panorama is also known as 360° panoramic views. It is a three-dimensional panoramic display technology, fist shooting look around to the existing multi-angle scene by a digital camera, after then to complete post-stitching images.

The main features of three-dimensional panorama technology are as follows:

1. The dimensional panorama image directly produced by shooting pictures, and the image quality is high, has a strong sense of reality.
2. The dimensional panorama roaming technology guide and interactive operability are very strong.
3. The amount of data is small, and it is suitable for network viewing.
4. The development cycle is short, and low production costs.

According panorama projection display of different ways, three-dimensional panorama technology implementations usually have three ways: cylindrical panorama, spherical 360° panorama, and cube panorama, as shown in Fig. 1.

These are the three methods that panorama have been projected onto cylinders inner surface, cubes inner surface, and the spherical inner surface. Cylindrical panorama is just along the horizontal direction of the scene around the shooting, and then split up panoramic, so it can only move horizontally browsing around.

Fig. 1 Cylindrical panoramic, cube panoramic and spherical panoramic

Table 1 Common authoring tools of panorama

Authoring tool	Function description
Panorama tools	Java-based development of a set of panoramic player software, support for roaming spherical panorama, cylindrical panorama roaming, and high dynamic range (HDR)
Ulead COOL 360	Provides advanced photo stitching, deformation, alignment, and blending tool, a series of photos into 360° panoramic painting or image, and through e-mail or Web page to send
Arcsoft panorama maker	High-quality professional puzzles program can quickly and easily be a series of overlapping photos together into a beautiful panoramic picture. Users can use this procedure to adjust the picture and edit the last generation, including modifying splicing point, splice location and the picture's brightness, contrast, and so on
Panorama studio	From a series of photos to create a seamless 360° wide-angle panorama. Provides automated stitching, image enhancement, and mixing functions. Can detect the correct focal length or lens
ADG panorama tools	Pictures can be created from a variety of network 360° panorama. Automatic mixing and correcting the color and brightness of the panorama
Pano2VR	A panorama scene converts to three-dimensional dynamic applications. This software can generate flat panorama, cylindrical panorama, spherical panorama, panoramic cube, and cross-shaped panorama
Object2VR	Software using the picture is a set of consecutive pictures; these pictures are of the object using a digital camera 360 for averaging the shooting angles, which need to shoot auxiliary turntable
QuickTime VR	Developed by the Apple's cross-platform multimedia packages, supports a variety of formats videos, pictures, streaming media, animation, sound, virtual reality, and interactive effects of virtual reality technology file
FSP viewer	Designed to show high-resolution full-screen panorama, it uses a new algorithm, the image is not only clear but also running smoothly
Pano2QTVR	A panoramic picture will be converted into QuickTime VR (QTVR) or Macromedia Flash (swf) format software

Spherical 360° panorama when shooting in both directions along the horizontal and vertical multi-angle shot looking around, stitching suture can be achieved through vertical and horizontal direction 360° wide-viewing angle display. After a three-dimensional panorama software to edit, add hot links that can be achieved by scenic hotspot, multi-virtual roaming between scenes, radar azimuth navigation, and other functions. The common authoring tools of panorama are also as shown in Table 1.

3 Generation Process of Virtual Three-Dimensional Panoramic Space

Three-dimensional panorama roaming refers to the panoramic space switch of the constructed panoramic image, its purposes is to navigate through different scenes. Three-dimensional panorama roaming system is needed to the appropriate hardware and software combination. First need the camera, fisheye lens, PTZ, tripods, and other hardware to capture fisheye photo, then use the panorama stitching software to put together photographs, released into a format that can play and browse, specifically generation process of virtual three-dimensional panoramic space, as shown in Fig. 2.

3.1 Rendering Method Based on Panorama

Currently, a variety of methods can be used to obtain a panoramic image. Shooting methods mainly include the following three ways:

The first, use the panoramic camera to directly capture a cylindrical panoramic image. Since this method requires the use of special camera equipment, it is expensive and complicated to operate, and therefore, the method is suitable for professional camera crews.

Second, use a lens with a larger horizon (fisheye lens) to shoot. In this way, there is a big captured image distortion. And generation of spherical panorama must be calibrated before and transformations.

Third, the use of ordinary cameras, camcorders, and other handheld devices for image sequence capture, or using the fisheye lens and single electric digital camera image acquisition method. Here, using a fisheye lens is an ultra wide-angle lens; the lens angle reaches or exceeds 180°. Because digital cameras use a single electronic panoramic shot of a 360° panorama head with a fisheye shot making, just need to take two that can easily synthesize an image to fight high-resolution 360° panorama. This is a quick and easy, yet as well as a high-cost solution.

Specific shooting skills is here, when a digital camera is with the focal length of 35 mm, a 360° panoramic images normally takes 7–9 photos, and when the focal length of 105 mm, when a 360° panoramic images generally requires 24 photos, the focal length is set at 35 mm. Fisheye lens choice, now widely used in a circular fisheye are drum fisheye and full-frame fisheye. Among the full-frame fisheye because the image to fill the screen, making the use of such fisheye panorama, the image quality can be high. In this study, panorama of the acquisition is a full-frame fisheye lens and a single electric digital camera method. Certainly, using full-frame fisheye lens when shooting also needs some skills, depending on the focal length of the resulting image will be shot entirely different.

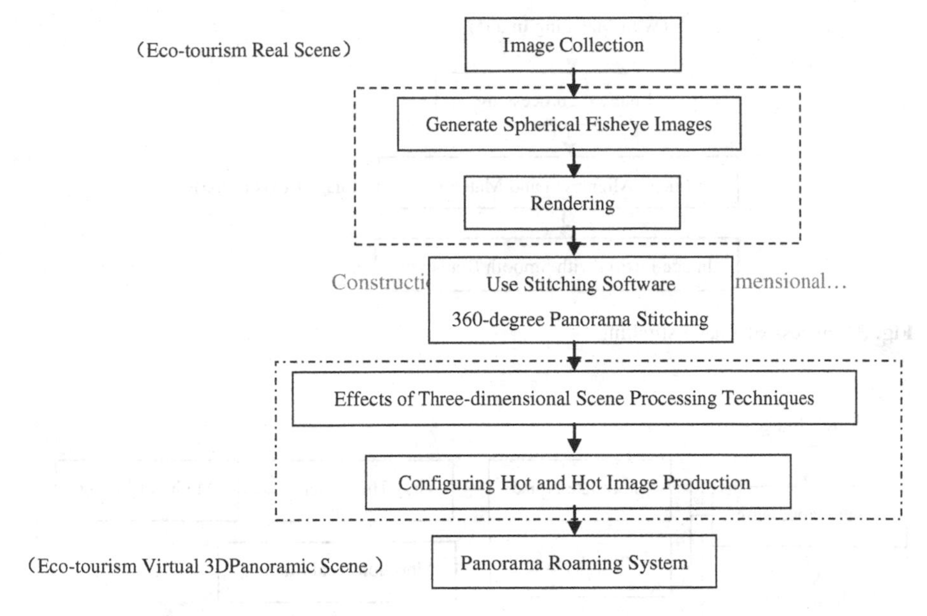

Fig. 2 Generation process of virtual three-dimensional panoramic space

3.2 360° Panorama Stitching

Panorama stitching is an important branch in the field of computer vision. It is a multiple related to the overlapping area and having a seamless image, to obtain a wide-viewing angle panoramic image technology. Currently, a great panorama stitching is theoretical significance and practical value of the research. It is to use ordinary handheld device to get the image, and then through the stitching software to correct the image distortion and to achieve automatic image stitching, to obtain high resolution and large field of view images.

1. **Image Stitching**

Panorama image mosaic generation technology is the most crucial step. Image stitching technology has three main steps: image preprocessing, image registration, image fusion with smooth boundary, as shown in Fig. 3.

(a) **Image Preprocessing**

Image preprocessing main methods are image smoothing, image enhancement, and image format conversion. Image enhancement refers to attenuate noise for improving the image sharpness. In this study, the two-dimensional image is pre-treated by image-smoothing techniques which eliminates the noise, gray enhancement, and binarization for improving the quality of two-dimensional image, and to obtain easy identification and processing images. The process of image preprocessing is shown in Fig. 4.

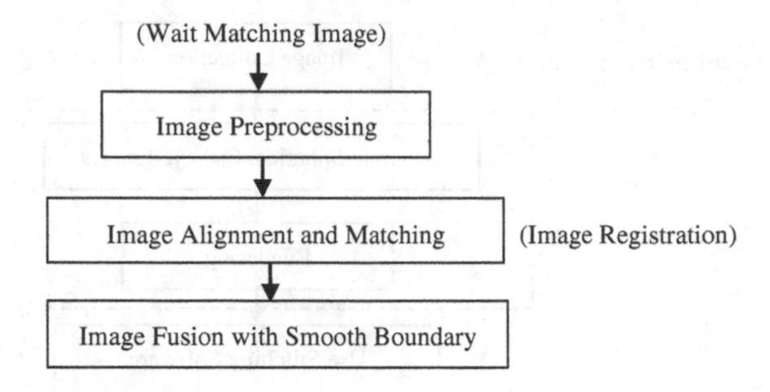

Fig. 3 Process of image stitching

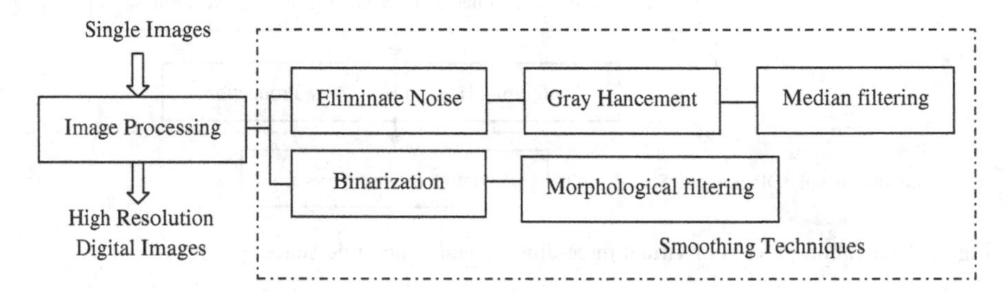

Fig. 4 The process of image preprocessing

(b) **Image Stitching and Fusion**

Image stitching core work is to study how to accurately obtain the two highest degrees of similarity in the image pixel coordinates. Existing panoramic image mosaic generation algorithm has mainly three types: region-based method, a method based on feature points, and phase-based approach. This study on the region-based and feature-based point stitching technology and splicing technology, in combination with the existing classic gray and column template matching algorithm based on the ratio of the algorithm, is given based on two-dimensional fractal image stitching algorithm, which efficiently finds image overlap and improves image stitching accuracy and computational speed, making the image to achieve a smooth seamless stitching.

2. **Stitching Software**

When using stitching software to create panoramas, generally, the process passes several processes: form import to simple editing, then automatic identification, then splicing process, then generated picture. Stitching a series of pictures of the software can be combined quickly, so that each image can be overlapped with each other up in place form a new image. Even in a fuzzy focal length, the camera does not need any data from the original pictures to calculate the focal length and the distortion parameters can be accurately measured to each pixel. In addition, each photo-stitching software can automatically adjust with the color edge, so as to provide a consistent image of the panorama tone; On the other hand, it can also be automatically adjusted to remove the adjacent picture deformation possibly and get high-quality and high accuracy results. Common stitching software is shown in Table 1.

3. **Panorama Stitching Methods**

In this study, a single power fisheye digital camera has acquired two panoramic photographs, which requires two photos stitched together to form 360° panorama. On cylindrical panorama, since the two images overlap region to be spliced larger, generally more feature-based matching method, such as feature-based curve matching, based on feature matching and feature-based contour matching, and so on. Direct method can also be used directly in the overlapping area of the image similarity to achieve this.

3.3 Construction of 360° Panorama Roaming System

After establishing a scene, the roaming project sets interactions and jumps between multiple scenes, and its contents include two parts of location and roaming. In cylindrical panorama after production, the resulting view has only a single viewpoint and cannot call for a real virtual reality tour; you must make a good panorama for a reasonable space to edit and organize. In this study, production software will guide people after some parameter settings that can be modified according to the actual needs of image quality, display size, and playback frame. For automatic rotation panorama, you can click the right turn autorotation feature, and then the general choice after loading begins to rotate. Finally, select the output, an swf file format, produced in 360° panoramic site; it will be embedded in the Web site panoramic display files inside a page, publish, and then, you can browse for the user.

4 Production of Panoramic Virtual Roaming System

Below, Hainan Ecotourism Environment panorama image design as an example is used to introduce the whole implementation process with three-dimensional panoramic virtual roaming system.

4.1 Tools and Software Selection

Shooting hardware devices: digital camera (wide-angle lens, fisheye lens), turntable, lighting lamps, and panoramic head.

Production software: Pano2VR and Object2VR are both three-dimensional panorama technology softwares; they are the embodiment of the three-dimensional static picture panorama technology. Among all softwares, Pano2VR software can generate flat panorama, cylindrical panorama, spherical panorama, panoramic cubes, etc; it is a 180 panorama of the scene into three-dimensional dynamic applications. In the main design of the skin on the scene editing and manipulation, for the use of custom skin edit mode, you can use different ways to produce different skin-editing effects, interactions, and presentation.

4.2 Panoramic Image Acquisition

While establishing a three-dimensional scene, first select an observation point, setting cameras, then certain angle per rotation, then an image exposure, and then stored into the computer. On the basis of image stitching, that is, about the same point in the object space, adjacent images are aligned with the corresponding pixel. On the implementation of cutting and splicing, a good image compression and storage is used to form panoramas. This study focuses mainly on cylindrical panoramas and cubic panorama.

4.3 Three-Dimensional Scenes by Pano2VR Software

In three-dimensional panoramic technically, Pano2VR mainly focuses on aspects of the object landscaping, this software uses two ways to generate panoramic scenes, one is through independent pictures and other is picture sequences, and export the swf or html generate a three-dimensional panoramic video files. And Photoshop designed and produced in this process is essential in picture processing and panorama stitching, as Fig. 5 shows the Boao Forum, Lingshan, Haikou West Coast, and Kangle Garden panorama stitching produce results.

Deal panorama mainly used for the production of landscaping software Pano2VR generates three-dimensional scene, in the text; the text information collection is completely unified text information on the selection and induction, so that each attraction details to be introduced will be collated with text messages for proper layout, through the skin of the production process in Fig. 5, into the panorama roaming system.

Figure 6 shows the production process:

First, in the custom skin (new skin) into the user-interface-related components. Second, svg format handled well, "arrow" through the "Add a picture" approach into the skin editor interface, and for each of the "arrow" double clicking to bring up the settings for action, purpose of doing so is to make use of by treatment interaction via mouse shortcuts, you can also choose to use the direction buttons to interact. Third, the "container" elements edit code "<b> $ ut </b>" which refers

Fig. 5 Effects of panorama production

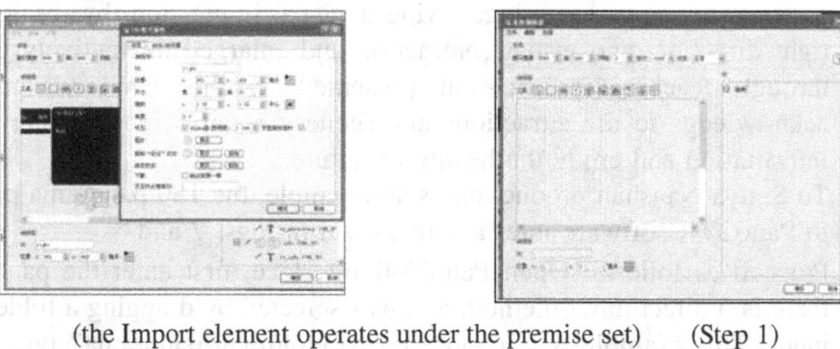

(the Import element operates under the premise set) (Step 1)

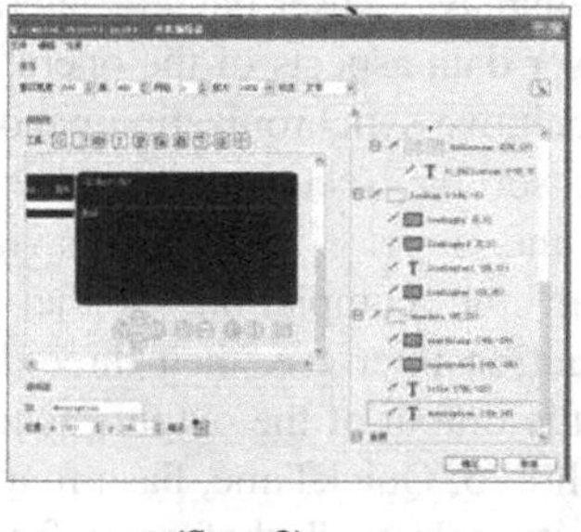

(Step 2) (Step 3)

Fig. 6 Production process of the skin

to the "Title" of the call, which will be called the operation of a user data interface Pano2VR inside information, which is text commentary information. "$ Ud" refers to the "user data" in the "Description" information for the call; the operation of these two codes is to make the production process become less cumbersome. These two codes are two relatively independent "container" to edit and adjust to the appropriate size of the container; the container can overlap between the skins of the element and can be edited through the container composition containing certain subsidiary relationship, so the "container" function is very powerful.

Step 1: Tourist Attraction Landscaping.

Summary: In this project producing a panoramic scene, tourist attractions in the landscaping areas mainly focused on two types of production, 360° and 360° cylindrical panorama panoramic ring. In this study, the production is single window size of 640×480. Ring cylindrical panoramic panorama is generated on the basis of the main difference in the angle of the settings that are different; cylindrical panorama is controlled at less than 360° in the case, and the resulting panorama is not fully engaged, while the ring is a panorama angle of 360° panoramic view of the front and rear fully engaged. In the three-dimensional panoramic visual display effect, ring-type three-dimensional panoramic effect will be more strengthened few. For plants constructed using landscape panorama panoramic scenes and scenes, the kind of scenario is based on a visual way to present surveying browsing, giving a visual immersion, by the left and right direction, and visual contraction and enlarged interactively browse through teaching information presented in this way; visitors can acknowledge to the attractions and scenery of an overall picture of the information and enjoy the beauty of nature.

To Sanya Nanshan production as an example, the 180 panorama pictures to Pano2VR software interface re shown in Figs. 7 and 8.

Proceed as follows: Open Pano2VR interface, first enter the panorama, there is a direct input method, and was selected by dragging a folder path input. This example is to choose the "cylindrical panorama" type. After importing panoramic pictures on the "display," "user data," "interactive hot spots" of the parameters set, the user data aspects of the operation to add the text message are generated to achieve synchronization panoramic view graphic effects. Then, you can set the focus to the interactive attractions panoramic panorama files and directory files via navigation links in interconnected way. Confirm the completion of the above parameters, return Pano2VR the production interface, then three-dimensional panorama is ready to output, in the output of the output process; to set the output format, it is divided into html5, QuickTime, flash four types of deformation, the project design unity selects flash format from the "Add" button to enter the output parameter and setting window is mainly produced by the skin hung to previous final confirmation generated

Fig. 7 Production interface of Pano2VR

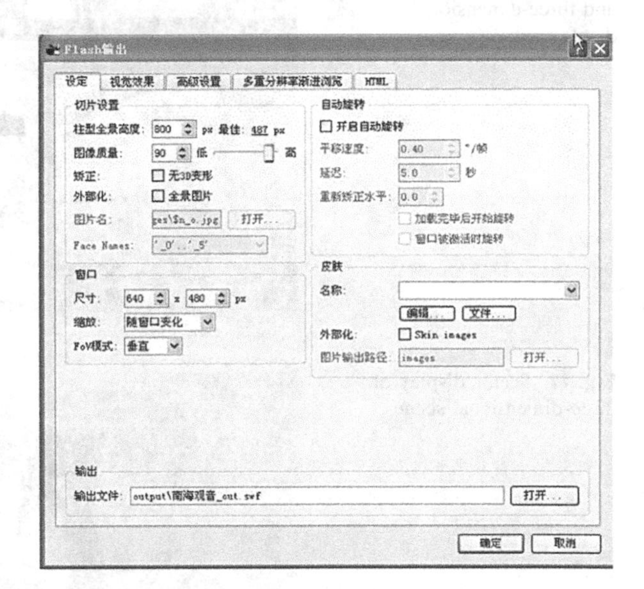

Fig. 8 Parameter settings

panoramic swf video files, html files, and Web project source files. It can be a simple choice of perspective and interactive demonstration to showcase their working behavior to detect the effect of the production work to be successful.

Step 2: Panorama sites

The shooting will be 100 separate three-dimensional panorama through dynamic Web site database, which is the study of "virtual panorama ecotourism site in Hainan," the formation of attractions panorama roaming through an interaction, making the site visitors a taste of the rich tourism resources in Hainan, advocacy Hainan to attract potential tourists to Hainan play. Technically, it was designed to add a richer perspective.

Fig. 9 Home page and of ecotourism virtual roaming system

Fig. 10 Scenic photographs and three-dimensional display entry

Fig. 11 Scenic display of three-dimensional scene

Diverse panoramic scenes can make presentation of Hainan's tourism resources; Diversification horizon of diverse subjects is shown in Figs. 9, 10 and 11 special Web site results.

Step 3: System Test and Evaluation

Hainan digital ecotourism three-dimensional virtual scene items tested in IE uses 360 secure browser, Baidu browser, Maxthon browser and makes use of tests that are using normal browsing.

5 Conclusion

Interactive 360° panoramic virtual roaming technology is a very energetic and has great potential for development of practical skills. In this chapter, three-dimensional panorama panoramic jigsaw roaming system and technology as the foundation, which describes the use of fisheye lens and a single electronic digital camera panorama stitching techniques for effectiveness, and panorama combined with roaming changes the original online monotonous, and poor interaction shows a two-dimensional plane, to achieve a three-dimensional real environment interaction with the viewer panorama roaming system. With Panoramic Virtual Tour technology, development and popularization by the means of achieving panorama roaming system will be more abundant, more and more sites use Virtual Tour technology development of Ecotourism sites.

Acknowledgments This study was supported by the Hainan Natural Science Foundation (No. 612121, 612126), supported by the key Science and Technology Projects of Haikou City in 2010 (No. 2012-017), and was supported by disciplines project of Hainan Normal University, expressed his thanks in this together.

References

1. Xu, H., Wang, P., Wang G.-S., et al.: Panoramic Jigsaw Implementation Techniques, **30**(4), 417–422 (2004)
2. Zhang, X., Li, L.: Panoramic puzzles technology research and application. Graphic Images **8** (2011)
3. Wu, H.-M.: Use of fish-eye panorama roaming technology to build a systematic approach to explore. Comput. Knowl. Technol. **6** (2009)
4. Lv, X.: Based on interest points matching panorama stitching algorithm. Comput. Program. Skills Maintenance **16** (2011)
5. Guo, C., Cao, F.: Dimensional panorama technology introduced in the area of application. Geospatial Inf **2** (2009)
6. Huang, J.: Dimensional digital technology in the museum website application. Digital Technol. Appl. **9** (2011)
7. Lvxiao, F.: Network dimensional interaction techniques (web3D). Technical Overview. Scientific information in Nov 2010 (2010)
8. Wu, F., Zhang, X.: Open Inventor single image-based three-dimensional reconstruction application. Shanghai Electric Inst. **12**(3), 216–219 (2009)

The Application of RS and GIS in Forest Park Planning: A Case Study of Diaoluoshan National Forest Park

Zhi Lin, Cong-ju Zhao and Peng-shan Li

Abstract The remote sensing images of Landsat TM and ETM+, June 2000 and 2010, were used to extract the land use information of Diaoluoshan National Forest Park and its surrounding areas by using ERDAS Imagine 9.0. Ten factors, elevation, slope, geological structure, land cover, settlement, the location of Diaoluoshan Nature Reserve, and so on, were chosen to evaluate ecological suitability. Applying the Delphi method to assess the grade and weight of the evaluation indicators, and then on the basis of the 1:50,000 digital topographic map to create digital raster graphics of different assessment factors using Arc GIS 9.1. The regional ecological suitability index (RESI) of each assessment unit was calculated by overlapping of each assessment grid. The RESI value of multifactor weighted superposition was 1.35–4.85 in the research areas, and the land was classified into 5 types, high suitability, suitability, basic suitability, unsuitability, and high unsuitability zone, by using the *K*-means clustering method. After that, generating spatial distribution map of regional suitability was based on the RESI reclassification, which provided a scientific method for the ecological conservation, resource exploitation, utilization, and management of Diaoluoshan National Forest Park.

Keywords RS · GIS · Ecological suitability assessment · Forest park planning · Diaoluoshan National Forest Park · Hainan

Z. Lin · C. Zhao (✉)
School of Geography and Tourism, Hainan Normal University, Haikou, 571158, People's Republic of China
e-mail: congjuzh@hainnu.edu.cn

P. Li
Department of Tourism, Hainan Technology and Business College, Haikou 570203, People's Republic of China

B.-Y. Cao et al. (eds.), *Ecosystem Assessment and Fuzzy Systems Management*, Advances in Intelligent Systems and Computing 254, DOI: 10.1007/978-3-319-03449-2_32,

1 Introduction

With the development in society and economy and improvement in people's living standard, tourism has become an important part of modern society lifestyle. As an emerging travel means, forest tour is getting increasingly important in the tourism industry; public needs for the functions and properties of forest parks are diversified. In order to meet the needs of tourists, as well as to protect and promote the local environment and ecology, scientific planning and design of forest parks makes an indispensable means to enhance the forest park quality. Most existing planning methods are based on subjective analysis [1–3] and visual judgment [4] rather than quantitative analysis; therefore, the scientific basis for such planning can be doubtful. RS and GIS feature fast, real-time spatial information acquisition and analysis capabilities, which make it very suitable for large area, complex terrain, high ecological sensitivity forest park planning and design. Currently, GIS, RS, and GPS technology is widely used in national forest parks [5], nature reserves [6], and other types of landscape planning and design and has achieved good social, economic, and ecological benefits [7].

This research is a case study of Diaoluo National Park, adopts the two-phase (years 2000 and 2010) Landsat TM/ETM+ remote sensing images from the region as data source, and provides classifying, mapping and evaluation for the Park based on RS and GIS technology to achieve objectives of reasonable planning and design, and scientific management to provide a scientific basis for the conservation, rational exploitation of resources and management of Diaoluoshan National Forest Park.

2 Methods

2.1 Survey of Research Area

Located in the southeastern part of Hainan Island, Diaoluoshan National Forest Park is the only national forest park on the eastern tourism industrial belt and an important tourist destination in the Greater Sanya Tourist Area in the Island. Construction for its core area was completed in 2008, and it was then listed as a national nature reserve.

Diaoluo National Forest Park is situated at 109°41′38″–110°4′46″E and 18°38′42″–18°50′22″N, featuring a medium mountain landform with an average elevation of 500 m above sea level and the highest peak of 1,499.2 m. The annual average temperature is 24.4 °C, while the mean temperature of coldest month is 15.4 °C. Annual rainfall is ample between 1,870 and 2,760 mm. Climate in the summer months (May–October) is typically humid and rainy, especially in high altitude areas where the rain season is lengthy and misty. Winter months (November–April) are characterized by cool and dry climate, with occasional and short cold fronts.

2.2 Research Methods

2.2.1 Data Source and Pretreatment

The basic data of the research area come from Landsat TM and ETM+ remote sensing images of July 2000 and July 2010. Images of the research area are cropped out, so this research does not include outlying regions. These remote sensing images are then pretreated by means of format conversion, band selection, geometric correction, georeferencing, image mosaic, cropping, image enhancement, and so on. ERDAS 9.0 is used for remote sensing image geometric precision.

2.2.2 Ecological Suitability Assessment Matrix

Based on a 1:50,000 digital topographic map, an assessment matrix is constructed to evaluate the grades and weights of 10 environment assessment indicators by Delphi technique, including geological safety, hydrological safety, ecological safety, human activities, and environmental management. Meanwhile, ArcGIS 9.1 is applied for the space overlay analysis after rasterizing the thematic maps of various assessment parameters, and then, regional ecological suitability index (RESI) in each assessment unit (grid) is calculated.

2.2.3 Ecological Effect Evaluation of Land Use Planning

Most scholars tend to evaluate the ecological effects of land use planning from the aspects of landscape structure and ecosystem service function value. With the adoption of patch-level indices such as number, area, area ratio, and density of patch- as well as landscape-level indices, including diversity, dominance, and evenness of landscape as analysis indicators, this research evaluates whether landscape changes caused by planning still harmonize with original landscape structures in the surrounding areas.

3 Results and Analysis

3.1 Ecological Suitability Assessment Indicator System

Grades and weights of each assessment indicator are evaluated using Delphi technique. Ecological zones such as nature reserves, lakes, and wetlands are absolutely unsuitable for construction land and are marked according to “zeroing principle” in the overlay process to ensure that the areas are not for construction use (Table 1).

Table 1 Ecological suitability assessment indicator system

Environmental factor	Weight	Factor	Sub-weight	Property	Score
Geological safety	0.35	Slope	0.15	>25°	1
				15–25°	2
				10°–15°	3
				5°–10°	4
				<5°	5
		Elevation	0.15	>1,000 m	1
				600–1,000 m	2
				300–600 m	3
				120–300 m	4
				<120 m	5
		Geological structure	0.05	Within 200 m on both sides	1
				200–800 m on both sides	3
				Above 800 m on both sides	5
Hydrological safety	0.15	Slope	0.10	<5°	1
				5°–15°	3
				>15°	5
		Distance	0.05	<20 m	1
				20–60 m	3
				>60 m	5
Ecological safety,	0.40	Buffer zone away from nature reserve	0.20	<100 m	1
				100–300 m	2
				300–700 m	3
				700–1,500 m	4
				>1,500 m	5
		LUCC	0.20	Mountain coppice	1
				Mountain rainforest	2
				Lowland rainforest	3
				Agricultural and forestry land	4
				Construction land and bare land	5
Human activities	0.10	Urban settlement	0.05	>1,000 m	1
				500–1,000 m	3
				<500 m	5
		Road traffic	0.05	>500 m	1
				200–500 m	3
				<200 m	5
Restriction factor	Nature reserve, core zone, buffer zone, and lake and wetland				0

3.2 Single-Factor Ecological Suitability Assessment

Five environmental factors are applied for assessing the ecological suitability in the research area, respectively, geological safety, hydrological safety, ecological safety, human activities, and environmental management.

3.2.1 Geological Safety

The research area is characterized by a medium mountain landform. Low mountains, waterfalls, and reservoirs distribute in the outskirts of the nature reserve. Landforms with elevation between 300 and 1,000 m above sea level occupy the largest area, accounting for 71.36 %. The research area features high terrain and steep slope, in which the area with slope >25° accounts for 41.93 % (Table 2).

Adverse geological structures in the research area consist of the Jianfeng-Diaoluo fault traversing the research area from west to east, the Wanning-Maling anticline located in the northeast, and the fault depressions lying in the southeast. According to "Earthquake intensity zoning map of China" promulgated by the China Earthquake Administration in 1990, most parts of the research area recorded a basic seismic intensity of 6° and a locally 7°. Therefore, the planning seismic design should go with the 7° (Fig. 1).

3.2.2 Meteorological and Hydrological Safety

Diaoluo Mountain sits along the typhoon path, and typhoon often brings flood disasters; moreover, its steep terrain leads to rapid flood drain, which is prone to landslide and mud-rock flow. According to the flood investigation in the past years, flood buffer zones can be divided into three-tier areas, respectively, within 50 m, between 50 and 100 m, and 100 m away. Areas further away are in generally free from flood threat (Fig. 2).

Table 2 Assessment results of some environmental factors

Elevation	Percentage	Topography	Percentage	Vegetation types	Percentage
<120	3.07	<5°	8.01	MBLT	2.20
120–300	14.92	5°–10°	5.02	MR	22.08
300–600	31.87	10°–15°	11.54	LR	66.47
600–1,000	39.49	15°–25°	33.49	AFL	5.79
>1,000	10.64	>25°	41.93	CBL	2.12
				Waters	1.35

MBLT Montane broad-leaved thicket
MR Mountain rainforest
LR Lowland rainforest
AFL Agricultural and forestry land
CBL Construction and bare land

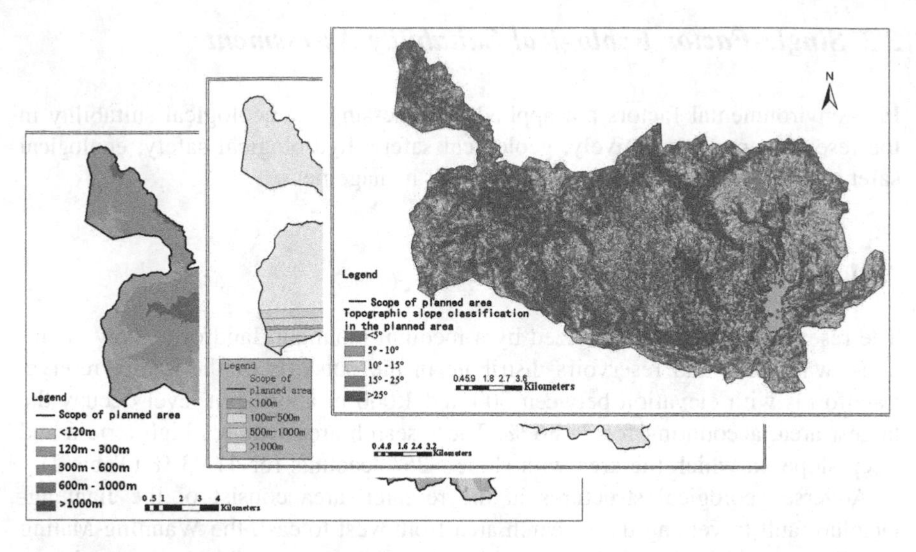

Fig. 1 Elevation and slope classification and fault buffer zones in the study area

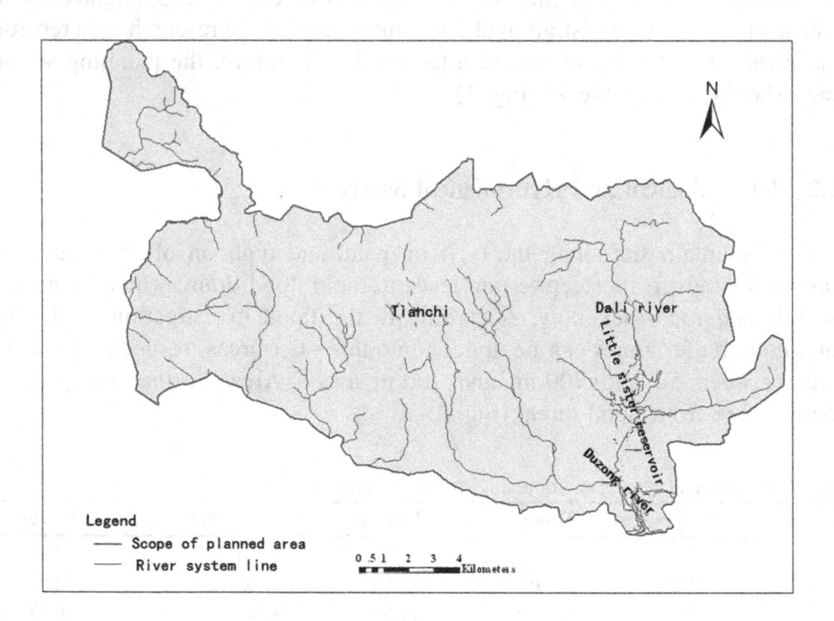

Fig. 2 Water system and its buffer zones

3.2.3 LUCC and Ecological Safety

Corresponding with the elevation changes, land cover in the research area varies from bottom to top, specifying tropical valley rainforest (altitude below 300 m), tropical lowland monsoon rainforest (300–700 m), montane rainforest (700–1,300 m), and

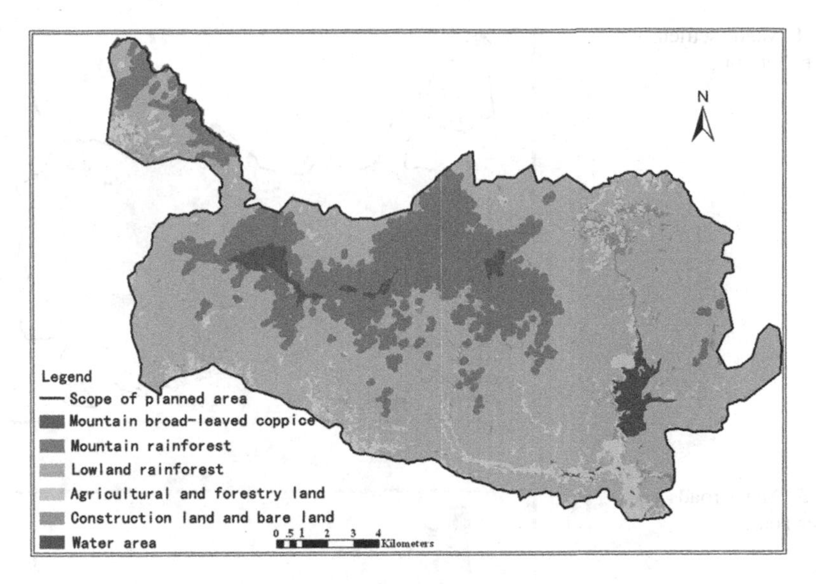

Fig. 3 Land cover and vegetation distribution in the research area

montane evergreen coppice on the 1,000–1,300 m isolated peak or the ridge zone adjacent to the peak. There are also tropical monsoon forest and tropical montane evergreen broad-leaved forest. LUCC type and area ratio and distribution are shown in Table 2 and Fig. 3.

3.2.4 Human Activities and Ecological Safety

Potential impact of urban settlement, road traffic, and major construction projects on the surrounding ecological environment is studied. Buffer zones of urban settlement and major construction projects can be divided into three-tier areas, respectively, within 500 m, between 500 and 1,000 m, and 1,000 m away; and buffer zones for main road traffic can be divided into three-tier areas, respectively, with 200 m, between 200 and 500 m, and 500 m away (Figs. 4, 5).

3.3 Comprehensive Ecological Suitability Assessment Scales

ArcGIS 9.1 is applied for the space overlay analysis after rasterizing the thematic maps of various assessment parameters, and the RESI in each assessment unit (grid) is calculated. The formula is RESI $= \sum S_i W_i$, with *S* being the property value of all the assessment indicators and *W* the weight value. A comprehensive ecological suitability assessment value *S* at 1.35–4.85 for the multifactor weighted superposition of the research area is hence obtained, which can be further

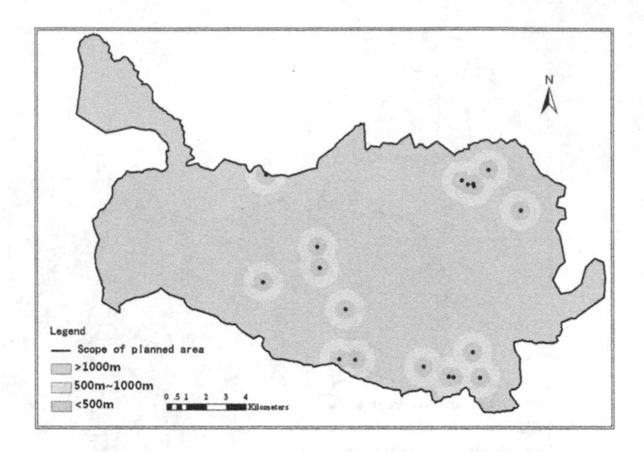

Fig. 4 Major settlements and buffer zones

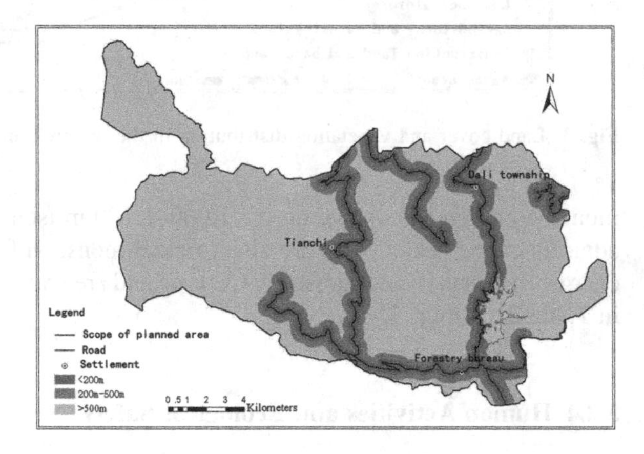

Fig. 5 Major roads and buffer zones

categorized into five types by the *K*-means clustering method: high suitability, suitability, basic suitability, unsuitability, and high unsuitability zones. Moreover, a regional suitability spatial distribution map is formulated using RESI reclassification (Table 3, Fig. 6).

As shown in Fig. 6, the classification results conform to the actual situation. Unusable land and land not suitable for construction are mainly located in the core areas and buffer zones; peripheral area to Little Sister Lake in the east and the high, undulating terrain area in the South are the ones not suitable for construction use; the pieces of land to the south of Dali Township and adjacent to the lake and the geological structure are not suitable for construction use either. The land suitable for construction use scatters.

In respect of the spatial distribution, most of the research area is natural forest land, while some secondary artificial forest and agricultural land sit in the south and east. Land suitable for construction use or those with higher scale are mainly distributed in the Forestry Bureau Service Area, north-central Dali Township, and eastern shore of Little Sister Lake. The terrain here is relatively low, far away from

Table 3 Assessment results for ecological suitability

Land use scale and score	Sensitivity	Land use orientation	Ecological characteristics	Area/ km^2	Percentage
Unusable (0)		Construction strictly prohibited	Core, water area, and wetland in the reserve	164.34	65.04
Highly unsuitable (1.35–2.39)	Extremely sensitive	Construction forbidden	High topography, wavy terrain, close to core zone	13.28	5.26
Unsuitable (2.40–2.85)	Highly sensitive	Not suitable for construction use	Comparatively high topography and wavy terrain, almost close to core zone and geological faults	24.54	9.71
Basically suitable (2.86–3.25)	Moderately sensitive	Limited construction use	Comparatively low topography, less wavy terrain, relatively far from core zone, and secondary artificial economic forest	25.01	9.90
Suitable (3.26–3.96)	Slightly sensitive	Suitable for construction use	Low and flat topography and terrain, mainly agricultural land, and good existing infrastructure	19.89	7.87
Highly suitable (3.97–4.85)	Insensitive	Very suitable for construction use	Low and flat topography, mainly construction land, convenient transportation and good infrastructure	5.62	2.22

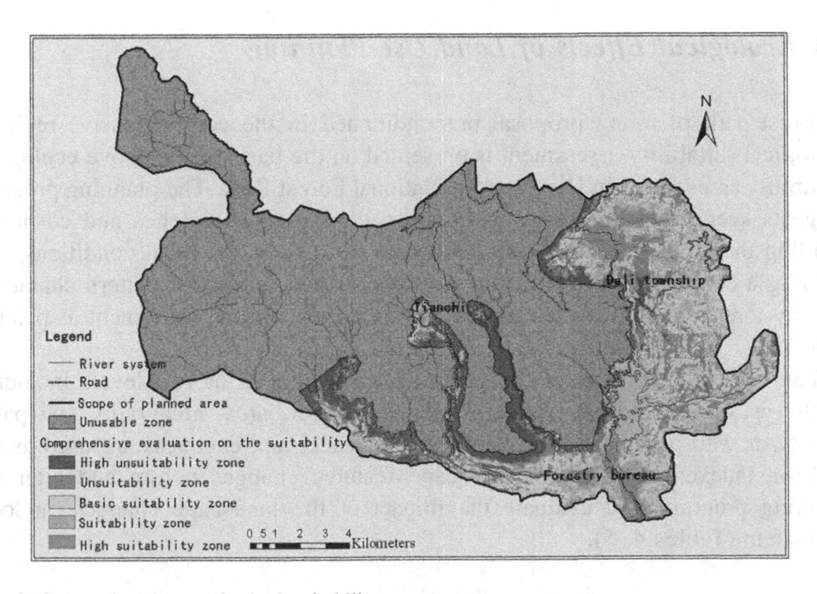

Fig. 6 Comprehensive ecological suitability assessment

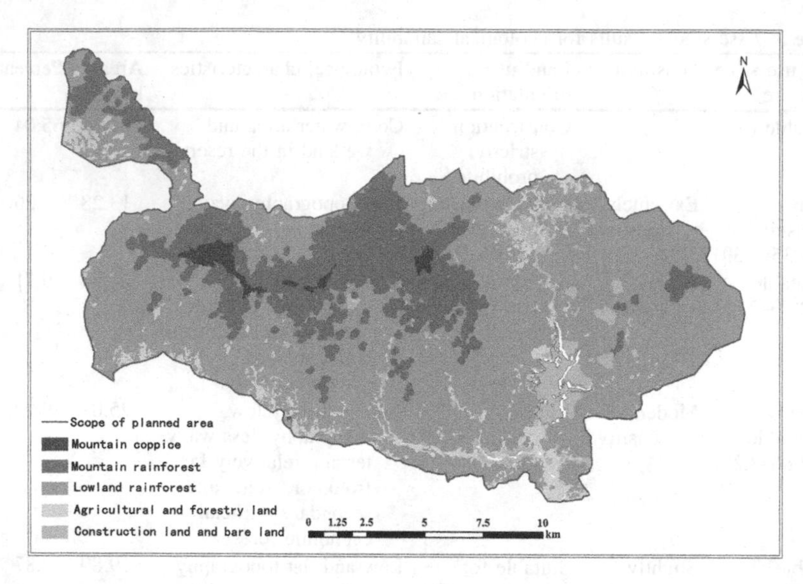

Fig. 7 Land use planning for Diaoluoshan National Forest Park

the core zone, and original human activities are frequent; transport infrastructure is good, the population density is low, township area is large, and building density is comparatively low. Hence, they can be used to build park themed for vacation, leisure, and sightseeing tourism.

3.4 Ecological Effects of Land Use Planning

A forest park planning proposal preconditioned in the comprehensive regional ecological suitability assessment is presented on the basis of the above ecological suitability assessment in Diaoluoshan National Forest Park. The planning program suggests keeping the existing landscape pattern such as patches and corridors, dividing the land plots in the research area in light of the local conditions, and forming a correlated yet comparatively independent landscape pattern characterized by forest as matrix, river as corridor and reservoir and settlement as patches (Fig. 7).

The ecological effects of the planning are evaluated at two levels including patch types and landscape patterns. Patch number, area, area ratio, and patch density as well as landscape diversity, dominance, and evenness are used as the analysis indexes to represent landscape features changes before and after the planning program and evaluate the impact of the landscape changes on local ecosystem (Tables 4, 5).

Table 4 Patch characteristics analysis

Patch type	Number		Area/km^2		Area percentage		Density/piece/km^2	
	CS	PS	CS	PS	CS	PS	CS	PS
MBLT	51	51	5.99	5.99	2.20	2.20	8.51	8.51
MR	754	775	60.23	59.96	22.08	21.98	12.52	12.93
LR	632	612	181.36	177.98	66.47	65.23	3.48	3.44
AFL	792	775	15.81	14.95	5.79	5.48	50.09	51.84
CBL	412	439	5.78	10.32	2.12	3.78	71.28	42.54
Waters	30	28	3.67	3.64	1.35	1.33	8.17	7.69

CS Current situation
PS Planned situation
MBLT Montane broad-leaved thicket
MR Mountain rainforest
LR Lowland rainforest
AFL Agricultural and forestry land
CBL Construction and bare land

Table 5 Landscape characteristics analysis

Index	Landscape diversity index	Landscape evenness index	Landscape dominance index
Status quo	1.03	0.57	0.76
Planning	1.07	0.60	0.72

It can be seen that there is a slight increase in the number of construction land and barren land patch from 412 to 439 after the planning; meanwhile, there is a significant drop in patch density from 71.28 pieces/km^2, dropped to 42.54 pieces/km^2. The patch density of other types go through little change before and after the planning, in which montane rainforest and lowland rainforest type patches account for about 88 % of the total area, in an absolutely dominant position. Scattered settlements are more densely concentrated after the planning, and the balanced expansion of such settlements reduces the division of natural type patch by human activities, as well as the impact on natural ecosystem.

An increase in landscape diversity index and landscape evenness index is also observed in the planning program, while the landscape dominance index is slightly reduced. Therefore, from the perspective of landscape ecology, the influence of planning on landscape structure and function is not obvious, yet the negative impact on the ecological environment is even less.

4 Conclusion and discussion

4.1 By applying the Delphi technique to screen the IF (Impact Factor) in forest park planning and land use, and assessing the grade and weight of the evaluation indicators, the single-factor ecological suitability assessment of the ten

IFs is then obtained. "Zeroing principle" is adopted to ensure that ecological zones such as nature reserves, lakes, and wetlands, which are absolutely unsuitable for construction, are not used for construction land. Assessment results conform to the actual situation, which indicates that the Delphi technique, or expert consultation method, plays an important role in screening IFs, grading and weighing evaluation factors in national forest park planning.

4.2 RS and GIS are applied for the overlay analysis of the thematic maps of various assessment factors. A comprehensive evaluation index of ecological suitability in the research area is hence obtained to be 1.35–4.85. The further use of the *K*-means clustering method helps categorize the research area into five types: high suitability, suitability, basic suitability, unsuitability, and high unsuitability zones, of which land of high suitability, suitability, and basic suitability account for 19.99 %, mainly distributed in the Forestry Bureau Service Area, north-central Dali Township, and eastern shore of Little Sister Lake, which are far away the core zone, and where transport infrastructure is good and the original human activities are relatively frequent.

4.3 A forest park planning preconditioned in the comprehensive regional ecological suitability assessment scarcely ever invade lowland rainforest and montane rainforest, which are important to the research area for their ecological conservation function, and has little negative impact on natural landscape. Besides exploitation and construction away from the areas evaluated in the comprehensive suitability assessment as land below basic suitability has even less negative impact on the ecological environment. Thus, on the one hand, the current ecological landscape and its function are well maintained, disturbance from human activities to natural landscape is lessened, and on the other hand environment quality of the scenic region is improved by integrating natural ecological forest land with residential and tourist land.

Applying RS, GIS technology, and RESI reclassification to produce Diaoluoshan National Forest Park planning on the premise of regional ecological suitability evaluation breaks through the traditional planning mode based on subjective perception and visual judgment; it scientifically constructs the suitability spatial distribution in the research area. Compared with the sole DEM generation using 3S technology [8] and park landscape investigation [9, 10], this research puts more emphasis on the comprehensive use of RS, GIS technology in ecological suitability assessment, landscape ecological analysis and landscape planning and designing, which embodies the idea of respecting natural ecology and existing with nature, and therefore is an important means to achieve the harmonious coexistence between man and nature.

Acknowledgments Financial support was received from the National Natural Science Foundation of China (No. 40961033 & 41361006) and International Science and Technology Cooperation Program of China (No. 2012DFA11270).

References

1. Yu, D.M., Li, X.L.: Planning and constructing Lingshan bay national forest park based on regional development. J. Qingdao Inst. Archit. Eng. **26**(3), 37–41 (2005)
2. Dan, X.Q.: Exploring scene planning and experiencing design—taking Leigong mountain forest park planning as an example. Cent. S. Forest Inventory Plann. **24**(3), 35–39 (2005)
3. Zhang, X.H., Gou, X.D., Wang, Y.: A preliminary study on the overall planning of the forest parks on the north slope of Qinling mountains within Shannxi province. J. Northwest For. Univ. **17**(1), 80–83 (2002)
4. Deng, Y.H.: Thoughts on planning and design of Wulingshan forest park of Chongqing city. For. Inventory Plann. **27**(2), 61–64 (2002)
5. Baskent, E.Z.: Controlling spatial structure of forested landscapes: a case study towards landscape management. Landscape Ecol. **14**(1), 83–97 (1999)
6. Zeng, H., Sui, D.Z., Wu, X.B.: Human disturbances on landscapes in protected areas: a case study of the Wolong nature reserve. Ecol. Res. **20**(4), 487–496 (2005)
7. Zhang, Y.C., Wang, Z.Q., Qiao, L.F.: Study on natural-ecological sensitiveness appraisal system of forest parks in mountain area. J. Anhui Agri. Sci. **33**(10), 1902–1903 (2005)
8. Shi, J.N.: DEM production and it's application to planning for forest park. Cent. S. For. Inventory Plann. **20**(1), 36–41 (2001)
9. Chen, H.L., Wang, Y.H., Ding, G.P.: Ecological sensitivity analysis on landscape ecological planning in tourist resorts—a case of Nanhuashan national forest park. Hunan Province **2**, 66–68 (2005)
10. Li, X.J., Meng, X.K., Zhang, F.Q.: Application of handheld GPS in planning of forest park. Hebei J. Forest. Orchard Res. **20**(1), 50–65 (2005)

Analysis of Spatial Distribution Pattern of Main Tourism Resources in Hainan Island Based on GIS

Jie-hua Song, Ping Wang and Cong-ju Zhao

Abstract This chapter takes Hainan main tourism resources as research object, GIS as platform, and the nearest neighbor index (R), geographic concentration index (G), Gini coefficient (Gini), Equilibrium ratio (ER) as calculation model to quantitatively analyze spatial distribution characteristics of main tourism resources in Hainan Island. The result shows that the nearest neighbor index $R = 0.728$, illustrating main tourism resources in Hainan Island are of clustered distribution; geographic concentration index $G = 27.322$, illustrating main tourism resources in Hainan Island are spatially concentrated; gene coefficient Gini = 0.945 m, illustrating main tourism resources are of unbalanced distribution between administrative regions. Economic equilibrium ratio coefficient ER = 16.13, area balance ratio coefficient ER = 13.041, suggesting that economy and area play little influence on the distribution of main tourism resources in Hainan Island. To better compare the differences between all the cities and counties, the equilibrium ratio coefficient is decomposed, the result shows that the maximum value of area balance ratio coefficient is in Wushizhang, the minimum value is in Haikou, the maximum value of economic equilibrium ratio is in Sanya, the minimum value is in Ledong.

Keywords GIS · Hainan Island · Main tourism resources · Spatial distribution pattern

J. Song · P. Wang (✉) · C. Zhao
College of Geography and Tourism, Hainan Normal University, Hainan 571158, China
e-mail: wangping2129@126.com

J. Song
e-mail: jiehuasong@yahoo.com.cn

B.-Y. Cao et al. (eds.), *Ecosystem Assessment and Fuzzy Systems Management*, Advances in Intelligent Systems and Computing 254, DOI: 10.1007/978-3-319-03449-2_33,

1 Introduction

The research on spatial distribution characteristics of tourism resources has important significance for exploration of spatial distribution law of tourism resources, rational use and development of tourism resources, scientific planning tourism layout, optimizing tourism resources allocation, and providing scientific advice and decision making for tourism development. Hainan Island, an international tourism island, has extremely rich and unique tropical tourism resources, which spread all over 1,528 km of the coast and can be developed into a world-class tourist resort. Currently, scholars mainly focus on qualitative analysis of classification, evaluation, development and protection, and development strategy of Hainan tourism resources [1–7], but play little attention to quantitative research of the spatial structure of tourism resources in Hainan. Therefore, this chapter proposes to quantitatively analyze spatial structure of tourism resources in Hainan using survey data of Hainan tourism resources as basic data, ARCGIS as analysis platform, to reveal the spatial distribution of main tourism resources in Hainan, as well as provide decision-making basis for the evaluation and development of them.

2 Research Scope and Data Sources

Hainan province includes Hainan Island and Sansha Islands (islands and sea area of Xisha, Zhongsha, and Nansha). The scope of this study is Hainan Island; considering tourists tend to choose higher-class tourism destinations and take little attention to lower-class tourism destinations in large-scale tourism space [8], this chapter mainly takes those influential tourism resources (i.e., comprehensive tourism resources of national A-level scenic spots, main natural tourism resources, main humanities tourism resources, and national key cultural relics protection units) as research object to analyze the spatial structure of tourism resources. The research data includes Hainan Island administrative zoning map, relevant data of Hainan Island tourism resources, and the statistical yearbook of Hainan Province in 2010.

3 Analysis of Spatial Distribution Pattern of Main Tourism Resources in Hainan Island

In order to quantitatively show the distribution differences and pattern change of main tourism resources between administrative areas of Hainan Island, the administrative boundaries of the cities and counties in Hainan Island and the distribution center points of main tourism resources are extracted from Hainan Island administrative zoning map and tourism resource distribution map using

ARCGIS10.0 software. On this basis, the calculation models—the nearest neighbor index, geographic concentration index, Gini coefficient, and equilibrium ratio index—are used to analyze the spatial distribution characteristics of tourism resources in Hainan Island.

3.1 Distribution Types

Spatial distribution of points is usually divided into three types: "uniform distribution," "random distribution," and "clustered distribution." At present, the nearest neighbor index, neighborhood average, Lorenz curve, statistical number of cells are main methods for measuring spatial distribution pattern of points [9–12]. The most commonly used method is the nearest neighbor index method, whose formula is as follows:

$$R = \frac{\overline{r_1}}{\overline{r_E}}$$

$$\overline{r_E} = \frac{1}{2\sqrt{\frac{N}{A}}}$$

$\overline{r_1}$ is the average distance of each point with its nearest neighbor point, $\overline{r_E}$ is the nearest neighbor distance in theory, N is the number of points, A is area. R indicates the distribution type of points. When $R = 1$, points are of "random distribution," when $R > 1$, points are of "uniform distribution," and when $R < 1$, points are of "clustered distribution." The calculations for main tourism resources in Hainan Island are that $\overline{r_1} = 4775.436\,\text{m}$, $\overline{r_E} = 6555.0467\,\text{m}$, $R = 0.728 < 1$, which indicate that main tourism resources in Hainan Island are of "clustered distribution" in space. See from Fig. 1, the distribution of main tourism resources in Hainan Island can be divided into two dusters: Haikou, Wenchang, Dingan, Qionghai, Wanning, Sanya, and Danzhou are clumped together into one duster, which are rich in main tourism resources; the other 11 cities and counties are clumped together into one duster, which are relatively insufficient in main tourism resources.

3.2 Spatial Distribution Balance

3.2.1 Concentration Degree

Geographic concentration index is an important index to measure the degree of concentration of geographic objects. The formula is as follows:

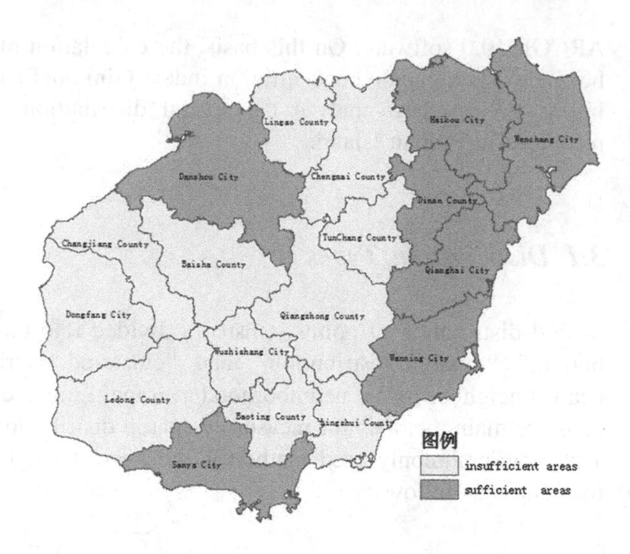

Fig. 1 Duster diagram of main tourism resources in Hainan Island

$$G = 100 \times \sqrt{\sum_{i=1}^{K} \left(\frac{N_i}{N}\right)^2}.$$

In the formula: G is the geographic concentration index, N_i is the quantity of main tourism resources in the region i, N is the total quantity of main tourism resources, and K is the number of regions. The value of G is more close to 100, main tourism resources' distribution is more concentrated; the value of G is more close to $G = 100 \times \sqrt{\frac{1}{K}}$, main tourism resources' distribution is more uniform. According to the above formula, concentration index of main tourism resources in Hainan Island is 27.322 (greater than 7.071, the value when main tourism resources in Hainan Island is uniform distribution), which indicates main tourism resources in Hainan Island are concentrated in spatial distribution.

3.2.2 Equilibrium Degree

The Gini coefficient is an important geography method to describe spatial distribution of discrete areas, and can be used to compare regional distribution differences between different subjects, so as to find the distribution regularity of the regions. In theory, the value of Gini coefficient is between 0 and 1, the higher the greater of the concentration degree. The calculation formula is as follows:

Pi is the proportion number of main tourism resources in *i* region accounted for total numbers of main tourism resources, *K* is the total number of regions, and *C* is the distribution uniform degree. According to the above formula, the Gini coefficient of main tourism resources in Hainan Island is 0.945, the value of *C* is 0.055, which show that main tourism resources in Hainan Island are of concentrated and spatially unbalanced distribution between administrative regions.

3.3 Spatial Distribution Difference

3.3.1 Analysis of Overall Difference

Because of the existence of some differences of administrative area, it is necessary to consider the differences in itself to better reflect the spatial distribution of tourism resources in Hainan Island. Based on this, the area equilibrium ratio index and economy equilibrium ratio index are used to calculate the equilibrium of spatial distribution, the formulae is as follows:

In the formula, ER is equilibrium ratio index; R_i is equilibrium ratio of main tourism resources in the i administrative area; X_i is the number of main tourism resources in i administrative area; X is the total number of main tourism resources; Ai is the area (or economic index) of the i administrative region; A is the total area (or total economic output) of the administrative regions; and K is the total number of administrative regions. If the distribution ratio of main tourism resources in the administrative areas is consistent with the area ratio (or economic ratio) of study areas, the equilibrium ratio is 0. Therefore, the closer to 0 the equilibrium ratio is, the more balanced the spatial distribution of main tourism resources is. According to the above formula, the economic equilibrium ratio coefficient for main tourism resources in Hainan Island is 16.13, area balance ratio coefficient is 13.041, indicating that the economy and area play small influence on the distribution of tourism resources in Hainan Island.

3.3.2 Analysis of Regional Differences

As it is difficult to find the internal differences between all cities and counties from the overall balance ratio, the equilibrium coefficient is broken down to get equilibrium ratios of all cities and counties (Figs. 2, 3), which is beneficial for analysis of their contribution degree to equilibrium and distribution coordination between tourism resources. As can be seen from Fig. 2, economic equilibrium coefficient of 13 cities and counties are greater than 0, indicating that this 13 cities and counties have the advantage in economy, where the maximum value is 2.765, in Wuzhishan City, followed by Baoting County, Ding'an County, Qiongzhong County, Baisha County, Wenchang City, Wanning City, Lingshui County, Danzhou City, Qionghai, Tunchang County, Sanya City, and Changjiang County, whose coefficients are 1.7585, 1.539, 1.451, 0.953, 0.932, 0.818, 0.775, 0.461, 0.351, 0.171, 0.0425, and 0.039; the remaining five cities' and counties' economic equilibrium ratio are less than 0, indicating that this five cities and counties are relatively lack advantage, of which the smallest value is −2.34, in Haikou. The area equilibrium

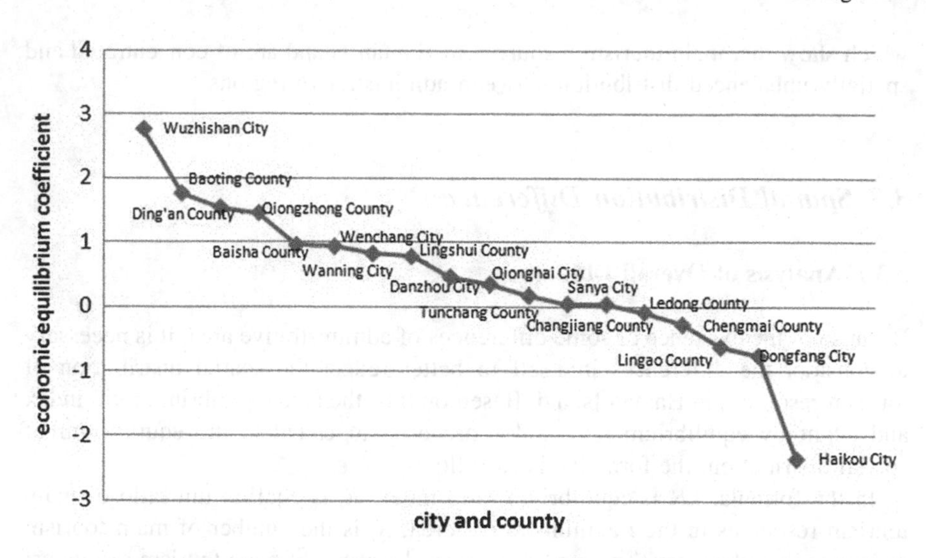

Fig. 2 Economic equilibrium coefficient of main tourism resources in each city and country

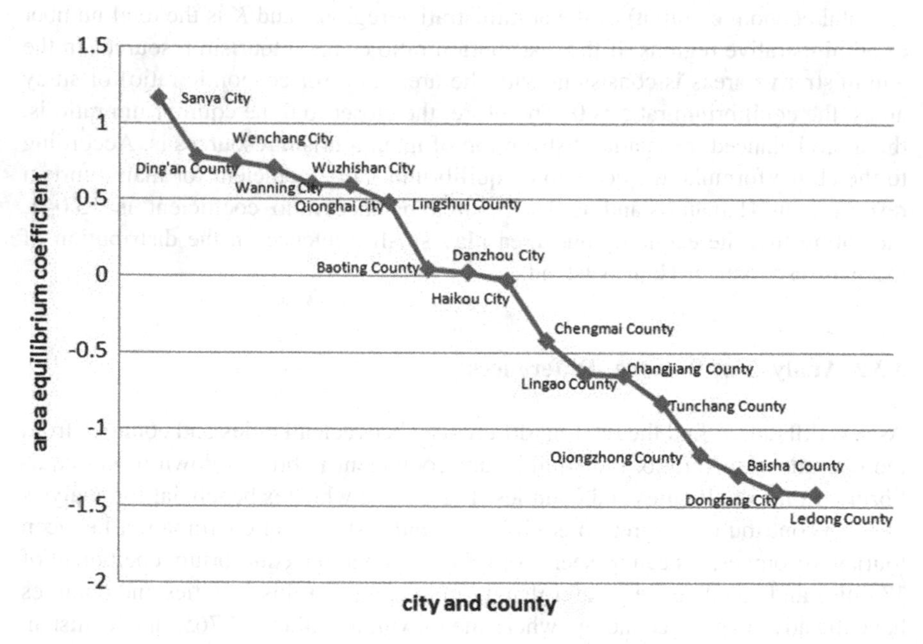

Fig. 3 Area equilibrium coefficient of main tourism resources in each city and country

coefficients of nine cities and counties are greater than 0, indicating that this nine cities and countries have the advantage in area, where the maximum value is 1.169, in Sanya, followed by Ding'an County, Wenchang City, Wanning City, Wuzhishan City, Qionghai City, Lingshui County, Baoting County, and Danzhou

City, whose coefficients are 0.7859, 0.748, 0.708, 0.6038, 0.594, 0.483, 0.0436, and 0.022; the other nine cities' and counties' area equilibrium ratio are less than 0, indicating that these nine cities and counties relatively do not have advantage, of which the smallest value is −1.425, in Ledong County.

4 Conclusion

The nearest neighbor index R of main tourism resources in Hainan Island is 0.728, explaining that main tourism resources in Hainan Island are of clustered distribution pattern in space; Sanya has a clear lead in the number and intensity of main tourism resources, which is closely related with Sanya's advantaged geographic condition, climate, regional culture, and others; the second city is Wenchang, one of Hainan tourism powerhouse, integrating eight tourism resources—sunshine, sea, beach, vegetation, air, island, flirtatious expressions, and countryside. So, it is rich in main tourism resources; the one also rich in main tourism resources is Danzhou City, which is a county-level city with the largest land area and the largest population of Hainan Province, and has the world of Lanyang hot springs, Shihua water tunnel, Guanyin cave, and so on.

From the perspective of spatial distribution balance of tourism resources in Hainan Island, the geographic concentration index $G = 27.322$, explaining main tourism resources in Hainan Island are concentrated in the spatial distribution; gene coefficient Gini = 0.945 m, explaining the distribution of main tourism resources in Hainan Island is concentrated and spatially unbalanced between administrative regions.

From the perspective of the balanced ratio of main tourism resources, economic equilibrium ratio coefficient ER = 16.13, area balance ratio coefficient ER = 13.041, suggesting that economy and area play small influence on the distribution of tourism resources in Hainan Island. The decomposition of equilibrium coefficient by cities and counties shows that the biggest area balance ratio coefficient is in Wuzhishan City, the smallest is in Haikou city, the biggest economic equilibrium ratio is in Sanya, and the smallest is in Ledong.

Acknowledgments This work is supported by International Science & Technology Cooperating Program of China (2012DFA11270), Hainan International Cooperation Key Project (GJXM201105), National Natural Science Foundation (61163042), and Young Teacher Research Fund of Hainan Normal University (QN1249)

References

1. Du, Y.: Research on strategy of tourism destination marketing of Hainan, pp. 1–86. Hainan University, Hainan (2011)
2. Yu, Z., Bi, H., Zhao, Z., et al.: A study on resources features and development of cultural tourism in Hainan Province. Res. Agric. Modernization **30**(05), 552–556 (2009)

3. Chen, X.: Investigation and analysis on the application of public relation in the promotion of Hainan tourism image. Humanit. Soc. Sci. J. Hainan Univ. J. Hannan Norm. Univ. (Nat. Sci.) **28**(3), 16–20 (2010)
4. Fu, G.: Investigation, classification and evaluation of natural tourism resources in Hainan. Nat. Sci. J. Hainan Univ. **28**(01), 52–58 (2010)
5. Jianping, Y., Du, N., Yu, T., et al. The tourist commodity market in Hainan: analysis & countermeasures for development. **18**(1), 84–88 (2005)
6. Liu, X., Tang, H.: A study on sustainable utilization of tourism resources in Hainan. Trop. Geogr. **22**(02), 152–156 (2002)
7. Xinjun, W.: Structural characteristics and development evaluation of tourism resources in Hainan Island. Trop. Geogr. **16**(02), 175–182 (1996)
8. Kai, Wang: Analysis of regional differences of tourism resources in China. Geogr. Land Res. **15**(3), 69–74 (1999)
9. Zhang, C., Yang, B.: Measurement Geography Foundation. Higher Education Press, Beijing (1984)
10. Lin, B.: Introduction of Measurement Geography. Higher Education Press, Beijing (1985)
11. Renzhong, G.: Spatial Analysis. Wuhan University of Geomatics Science and Technology Press, Wuhan (1997)
12. Xu, X., Zhou, Y., Ning, Y.-M.: City Geography. Higher Education Press, Beijing (2001)

Research of 3D Virtual Scene Generation and Visualization Based on Images

Jian-ping Feng, Li-hua Wu and Sheng-Quan Ma

Abstract Recently, virtual reality technology provides a new way for spatial information visualization; it has become a popular research topic in the field of computer graphics, computer visual and image processing, and other areas. This chapter first discusses the image-based modeling and rendering techniques (referred IBMR) and detailed analysis plenoptic functions, which describes technical theory basis of IBMR. And then in the current domestic, the three typical methods of 3D virtual scene generation based on panorama, concentric mosaics, and light field are studied. Finally, the chapter gives the principles of 3D virtual scene generation and processes based on panorama, those processes including projection transformation, image registration, image fusion, and inverse reprojection transformation. Because the above-studied technical methods can quickly and efficiently generate the panorama and can achieve a smooth transition between images coherent, they have been successfully applied to ecotourism site of Hainan virtual panoramic sights show. This method has better visual effects and application.

Keywords Plenoptic function · Panoramas · 3D virtual scene · Hainan ecological tourism

1 Introduction

Recently, along with the constant development of remote sensing technology, geographic information systems, virtual reality technology, and digital city promote the development of three-dimensional visualization technology. Specifically,

---

J. Feng (✉) · L. Wu · S.-Q. Ma
School of Computer Science and Technology, Hainan Normal University,
HaiKou 571158, China
e-mail: fjp888@163.com

B.-Y. Cao et al. (eds.), *Ecosystem Assessment and Fuzzy Systems Management*,
Advances in Intelligent Systems and Computing 254, DOI: 10.1007/978-3-319-03449-2_34,

virtual reality technology provides a new way for the visualization of spatial information; it has become a hot topic of the computer graphics, computer vision, and image processing research fields.

Three-dimensional modeling of complex models and display is the foundation of real-time virtual reality technology. Now, it can be divided into two categories. One is geometry-based rendering (referred as GBR); the other is image-based rendering (referred IBR).

Traditional geometry-based modeling method is to establish a complete geometric information and topology for the object and then generate three-dimensional images according the computer modeling results; the images are generally geometric model with polygons. Its main advantage is the observation point, and the direction of observation can change without restriction, to allow people to immerse simulation modeling environment, to meet 3I requirements of virtual reality technology, referred as immersion, interactive, and imagination. The most commonly used geometric modeling software are AutoCAD, 3dMax, Maya, and so on. But this modeling method is time consuming and laborious, and its modeling efficiency is not high. From the mid-1990s, IBR technology started to become a hot research in computer graphics, how to space any two-dimensional images taken in full three-dimensional display is really one of the latest technologies in today's world.

2 Modeling and Rendering Technology of Image-based

Modeling and rendering techniques of image-based can be defined as an image from a limited, known image from arbitrary viewpoint that can be seen on the new image. The newly created image is known as Get Image that is referred from the image to the image. The IBMR technique to eliminate the entire chain of geometric modeling part (the largest part of unnatural factors), combined with computer vision and computer graphics technology to achieve real images based on virtual modeling process. The image-based modeling and rendering techniques express explanation is as shown in Fig. 1.

Compared with the traditional GBR techniques, IBMR technology has modeled easy, fast rendering, realistic strong, and less computation advantages. In recent years, many scholars use IBR techniques to implement geometrical and optical model. And this is the modeling of virtual reality technology based on image.

This modeling method is based on the theory of plenoptic function; it is particularly used for the applications that use the method to very difficultly establish realistic geometry model of the natural environment, the environment and the need to reproduce the true original style applications. Now, they have begun to show its commercial value in the construction and the urban landscape, virtual tours, arts and culture, business and industrial technology, and other fields.

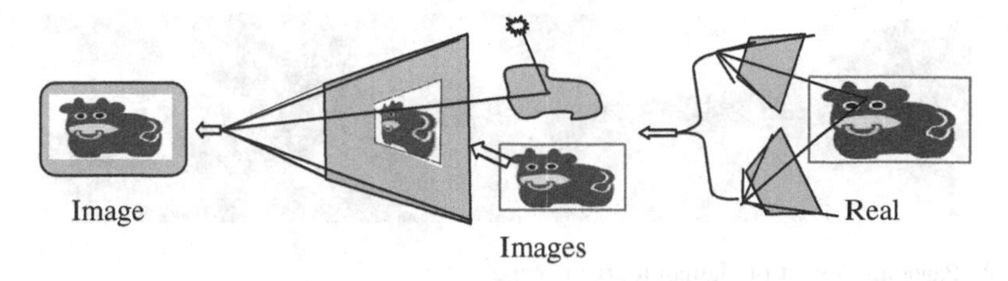

Fig. 1 The expression of interpretation of IBMR technology

3 Plenoptic Function

In the 1991, Professor Adelson from the American MIT Institute presented the plenoptic function concept and used it to describe the light collection can be seen from the space of a moment, from the viewpoint of any space, and any arbitrary azimuth and angle and wavelength range. General equation to describe the one parameter, the equation is as follows:

$$P_7 = P(V_X, V_Y, V_Z, \theta, \psi, \lambda, \tau) \tag{1}$$

where (V_X, V_Y, V_Z) means three-dimensional coordinate parameter viewpoint, spherical angle θ and ψ defined any one line of sight from the viewpoint, λ is the wavelength, and τ means for a moment. Plenoptic function as the scene image is a function, an image which is given by way of an accurate description of the scene. Plenoptic function depicts a given scene shines all possible environments.

The first application is the seven-dimensional plenoptic function $P_7 = (V_X, V_Y, V_Z, \theta, \psi, \lambda, \tau)$, to get the complete plenoptic function; all-seeing-dimensional modeling requires at least five-dimensional plenoptic function (light scenes can be assumed constant, nor change over time) $P_5 = (V_X, V_Y, V_Z, \theta, \psi)$; When the function simplifies to two parameters $P_2 = (\theta, \psi)$, it is the panorama; when the function simplifies to three parameters $P_3 = (\theta, r, h)$, it is concentric mosaics; when the function simplifies to four parameters $P_4 = (u, v, s, t)$, it is light field.

Plenoptic function presents an accurate description of problem for IBR technology; it became the theoretical basis of IBMR techniques. To this end, IBR problem can be described as follows: given a set discrete sampling set of plenoptic functions, IBR purpose is to build a continuous representation of plenoptic function. IBMR technology can be broken down into the following three processes: (1) sampling of plenoptic function; (2) reconstruction of plenoptic function; and (3) resampling of plenoptic function. This expression issues further study and provides many ways, for example, how to choose the best sampling point and how to reconstruct a continuous plenoptic function by the sampling points and so on.

Fig. 2 Panorama effect of Hainan tourism scene

4 Virtual Scene Generation Methods Based on Images

Currently, the typical image-based 3D virtual scene rendering methods are panorama, concentric mosaics, and light field. These methods can produce very realistic visuals, but you need to collect a large number of base images.

4.1 Rendering Method based on Panorama

In 1993, Professor EricChen first raised the method of panorama in the computer image processing. This method was later applied to Apple's Quick Time VR system software. Panorama is to shoot 360° camera ring in one or more group photos, and those photos are stitched into a panoramic image through computer technology to achieve interactive, all-round view of the real scene. Figure 2 shows the panorama effect of Hainan tourism scene. When Player plug-in support, use the mouse to control the direction of looking around, can be left to the right or can be near or far, viewers have the feeling of proximity to real scene. Details about this part of the description are as shown later.

4.2 Rendering Method based on Concentric Mosaics

He Liwei and Shen Xiangyang, president of Microsoft research Asia, proposed the rendering method based on concentric mosaics. It is different with the traditional panoramic view; concentric mosaics is combined into photographs by a camera at different viewpoints, and these viewpoints are continuously distributed. In fact, it is located on a concentric series of slim camera picture splicing.

The basic process of rendering method based on concentric mosaics: sampling, constructed concentric puzzle, and resampling.

1. **Sampling**: The sampling system of concentric mosaic consists of concentric number frame fixed cameras in a horizontal rotating rod, with each frame camera do along the continuous motion, and each camera is a circle, shown in Fig. 3.

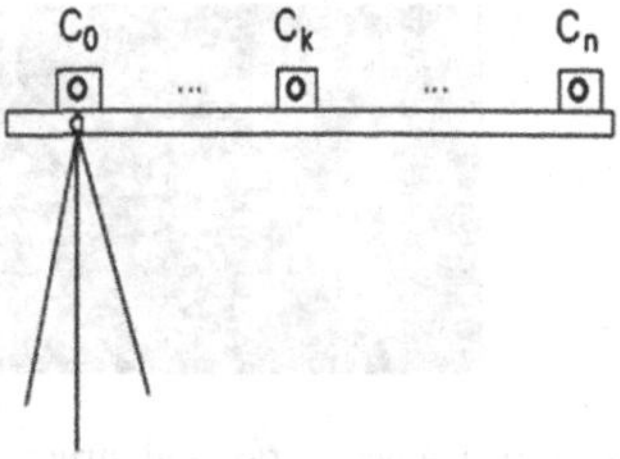

Fig. 3 Sampling by camera

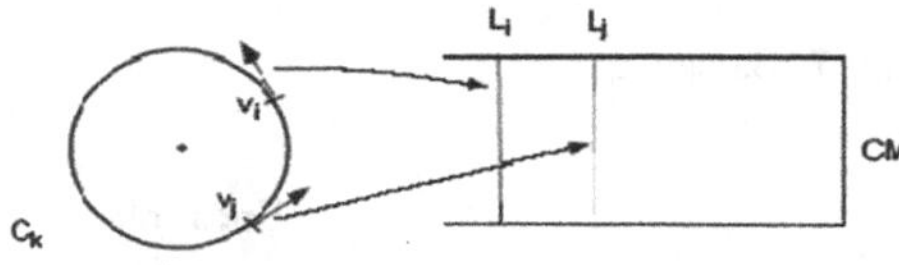

Fig. 4 Constructed concentric puzzle

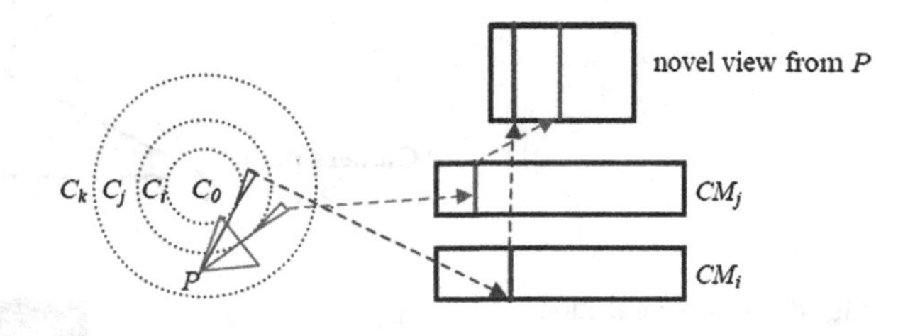

Fig. 5 Resampling

2. **Construct concentric puzzles**: Assuming that each frame cameras are slit camera, that at any point of only a vertical light into the camera, this light and round in the same plane, and at this point tangent to the circle. Different points on the circle line taken all together, you can get a concentric mosaic. Along each circle has two opposite tangential direction of light, the camera can be set back to back two samples. In this way, we were able to capture all the rays along the circumference, at any point in the region to establish optical function, shown in Fig. 4.
3. **Resampling**: When draw a new view, simply to determine the level of the viewpoint within the concentric circles of light belongs puzzle, or by interpolating adjacent concentric puzzle to get. For example, in a new light at the viewpoint P, it can be obtained at the point of V_j concentric puzzle CM_j, as shown in Fig. 5.

4.3 Rendering Method Based on Light Field

In 1996, experts Levoy and Hanrahan proposed rendering method based on light field. Optical flow is the apparent movement grayscale image, which is a two-dimensional vector field, which contains information that is the image point of the

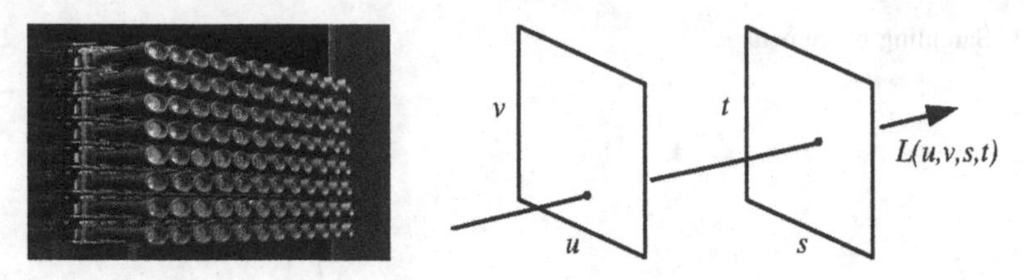

Fig. 6 Camera array and structured parameters of two planes

Fig. 7 Sampling

Fig. 8 Data organization

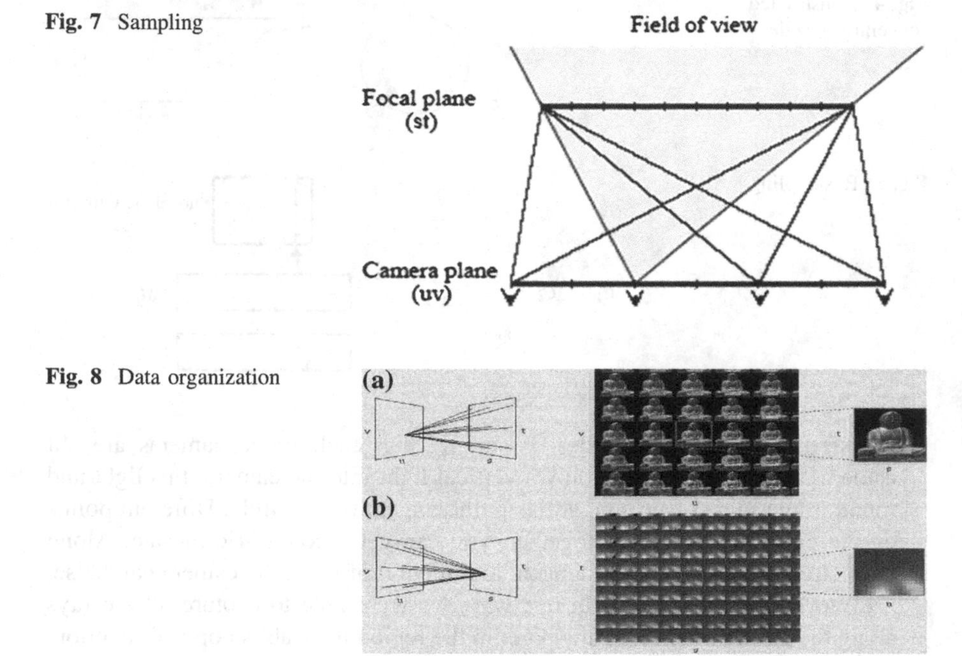

instantaneous velocity vector information. The research purpose of light field is to calculate approximate light flow fields that cannot be obtained directly from image sequences.

The basic process of rendering method based on light field: sampling, representation, and resampling.

1. **Sampling**: Sampling array used in the camera is shown in Fig. 6. The plenoptic function simplifies to 4-dimensional $P_4 = (u, v, s, t)$. They are structured parameters of two planes and sampling, as shown in Fig. 7.
2. **Representation**: Corresponding data organization is shown in Fig. 8.
3. **Resampling**: Resampling of generating a new viewpoint is a simple linear process, which is a small amount of computation. However, this technique relies on the high sampling frequency and requires a very large storage capacity to store a large number of sample images.

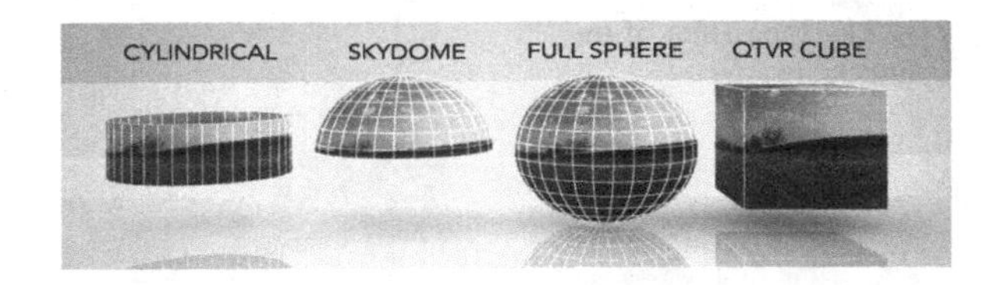

Fig. 9 Panorama body model

Fig. 10 Spherical panorama unfold effect

5 3D Virtual Scene Generation Method Based on Panorama

Panorama production process include the following: panorama model selection, image acquisition, image stitching, and image fusion. Panorama body model used can be divided into three patterns cube, cylinder, and sphere, as shown in Fig. 9. The three patterns were already 180 panorama of the projected cube, cylinder, and sphere's inner surface. Shown in Fig. 10 is a spherical pattern panorama unfold surface effect.

Panorama's formal description based on plenoptic function is the use of discrete images or continuous capture of a video image sequence as basic data, to form a panoramic image by processing and then put organized into multiple panoramic image space through a suitable space model. In this space, the user can move forward, backward, 360° circular looking up, looking down, close look, and distance operations such as virtual panoramic image design.

5.1 Panorama Shooting and Production Tools

Common hardware equipment of panorama shooting are as follows: digital camera (wide-angle lens, fish-eye lens), turntable, lighting lamps, and panoramic head. The basic production method is as follows: image acquisition with a camera in a fixed viewpoint according to a certain way (usually in accordance with uniform angular pivoting 360°), then image after collecting input computer, and image stitching, blending, and other processing to generate seamless panoramic image, and the finally reuse computer projector displayed and provide limited local roaming capabilities. Currently, the major application of the panoramic view is in

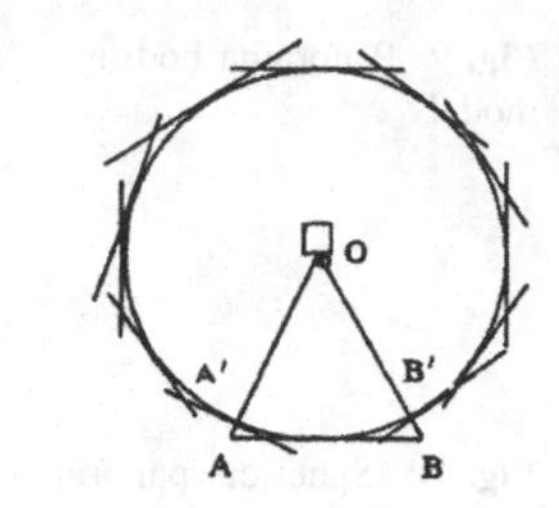

Fig. 11 The radius of the cylindrical surface

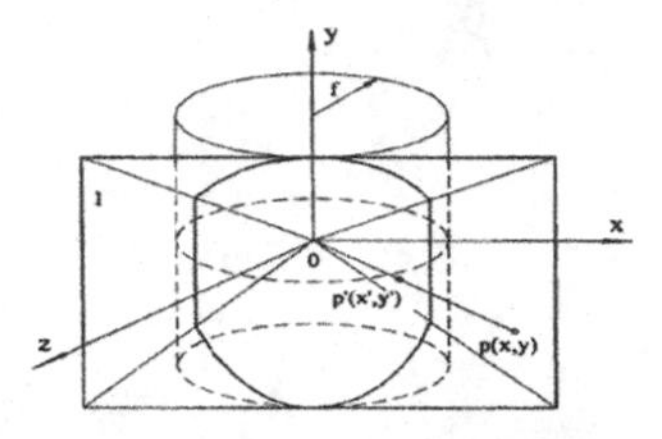

Fig. 12 Projection transformation

a virtual environment, game design, film special effects, and virtual museums. In the commercial areas, there are some more famous shooting and production tools, and they are Apple's QuickTime VR, IPIX Viewer, Live Picture, IBM's Hot Media, and other systems.

5.2 3D Virtual Scene Generation Process

Panorama-based virtual scene generation process includes the following: projection transformation, image registration, image fusion, and reprojection transformation.

1. **Projection Transformation**: The point in space of each camera transforms into an unified cylindrical surface coordinate. The radius of the cylindrical surface is camera's focal length (in pixels as the unit). Camera focal length can be estimated by the number of shots, as shown in Figures 11 and 12.
2. **Image Registration**: Existing generation algorithm of panoramic image mosaic can be divided into three categories: feature-based approach, flow-based method, and phase-based correlation method. After the image has been transformed, to find similar regions by overlapping the two images, looking for similar regions, the general algorithms are based on human visual characteristics defined as a pattern vector, respectively, for the two image similarity distance calculation, based on distance to find similar images' similar parts, as shown in Fig. 13.
3. **Image Blending**: After getting a good image stitching, you also need to overlap the image fusion process in order to achieve a seamless mosaic image.

Fig. 13 Found two photos overlap splicing

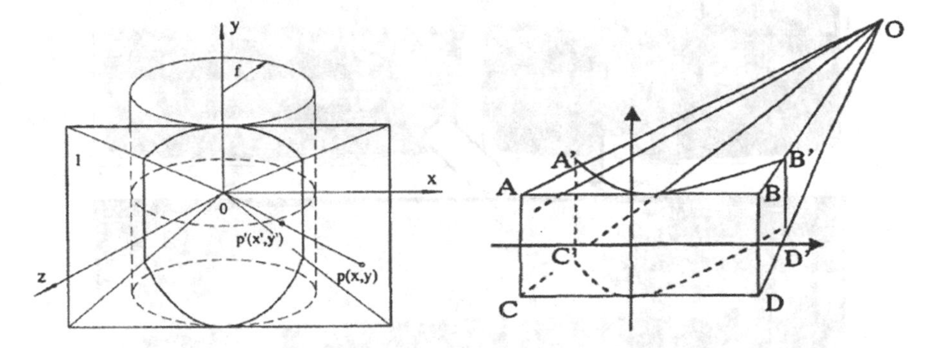

Fig. 14 Reprojection transformation

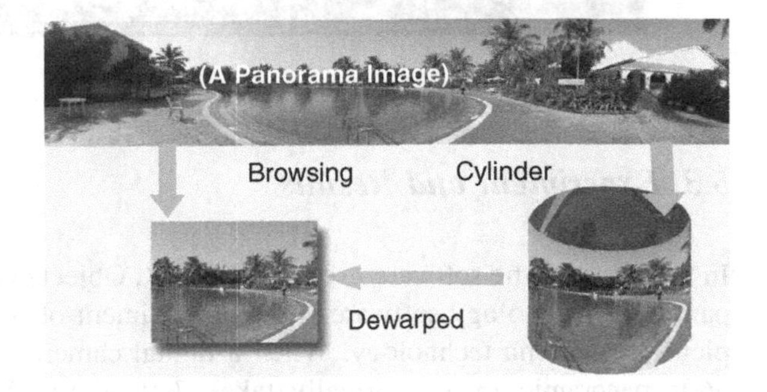

Fig. 15 Panoramic images of cylinder

Main fusion methods are fade fusion method, finding the minimum energy path method, and multi-resolution spline fusion method, as shown in Fig. 13.

4. **Reprojection Transformation**: According to the orientation of the viewpoint, calculated part of the image should be displayed, and in other beneficial OpenGL graphics library, this step can be done automatically, as shown in Fig. 14.

Fig. 16 Reprojection

Fig. 17 Fused panorama

5.3 Experiment and Results

In this chapter, the software used is Pano2VR, Object2VR, both three-dimensional panorama technology software is the embodiment of the three-dimensional static picture panorama technology. When a digital camera's focal length is 35 mm, a 360° panoramic image normally takes 7–9 photos. When the focal length is 105 mm, a 360° panoramic image generally requires 24 photos, the focal length set at 35 mm.

In this chapter, the three-dimensional virtual scene panorama generation method based on panoramic images for authentication is discussed. Experimental images using a digital camera ecotourism in Hainan as photos of the scene, the same layer adjacent to the angle of rotation between the two pictures 30° between the upper and lower image rotation angle is 60°. Figure 15 is a computer after shooting a cylindrical panoramic image stitching process.

The panorama projection transformation is shown in Fig. 16. Panorama fusion is shown in Fig. 17. The experimental results show that large numbers of images, this study used the virtual scene generation method can generate cylindrical panoramic image, such as spherical panoramas, photos, high precision registration method, fusion method can achieve better coherence smooth transition between images, with good visual effect.

6 Conclusion

Hainan has a unique natural environment, vast territory, and abundant resources. To effectively protect the ecological environment and the tropical island resources in Hainan ecotourism activities, the image-based 3D virtual scene can be reproduced in the field of ecotourism attractions roaming to achieve interactive tour feature. This method has been successfully applied to the ecotourism attractions of Hainan virtual panoramic display by a 360° panorama way to achieve ecotourism attractions of Hainan 3D virtual panorama roaming. Because it reproduces the real three-dimensional virtual scene, it has good travel Website interactivity and realism, thus greatly increasing the visibility and scenic spots in Hainan Ecological Website hits.

Acknowledgments This study was supported by the Hainan Natural Science Foundation (No: 612121, 612126), supported by the key Science and Technology Projects of Haikou City in 2010 (No: 2012-017) and was supported by disciplines project of Hainan Normal University expressed his thanks in this together.

References

1. Zhang, X., Li, L.: Panoramic puzzles technology research and application. Graph., Images **8**, 34–37 (2011)
2. Liu, D.: 3D virtual scene generation image and visualization. Syst. Simul. **17**(1), 73–75 (2005)
3. Were, Y.: Resonance VRML-based virtual scene roaming implementation. Comput. Eng. Design **29**(14), 3748–3751 (2008)
4. Liu, S., Gu, Y.: 3D build efficient virtual scene. Comput. Eng. Design **27**(2), 303–306 (2006)
5. Li, Z.: Domestic and foreign research status of virtual reality technology. Liaoning Tech. Univ. Nat. Sci. Ed. **23**(2), 238–240 (2004)
6. Gong, Y., Pu, X., Zhang, X.: Virtual scene roaming technology and system realization. Comput. Eng. Appl. **43**(15), 89–91 (2007)
7. Jun, H., Zhou, J.: 3D virtual scene dynamic interactive technology research. Comput. Eng. Sci. **29**(7), 55–57 (2007)
8. Xu, L., Jiang, Y.: Roaming the virtual scene modeling and implementation. Syst. Simul. **18**(1), 120–124 (2006)
9. Xu, H., Wang, P., Wang, G., et al.: Panoramic jigsaw. Implementation Techn. **30**(4), 417–422 (2004)

10. Wu, Q., Zhou, D., Caixuan, P.: Based on image sequences and roaming virtual scene reconstruction. Chin. J. Image Graph. **11**(1), 113–118 (2006)
11. Wu, F., Zhang, X.: OpenInventor single image-based three-dimensional reconstruction application. Shanghai Electr. Inst. **12**(3), 216–219 (2009)
12. Liu, G., Peng, Q., Bao, H.: Review and prospect of image-based modeling. J. Comput. Aided Design Comput. Graph. **17**(1), 18–27 (2005)
13. Levoy, M., Hanrahan, P.: Light field rendering. Proceedings of the ACM SIGGRAPH, pp. 31–42. ACM, New York (1996)
14. Lv, X.: Based on interest points matching panorama stitching algorithm. Comput. Program. Skills Maintenance (16) (2011)
15. Guo, C., Cao, F.: Dimensional panorama technology introduced in the area of application. Geospatial Inf. (2) (2009)
16. Huang, J.: Dimensional digital technology in the museum website application. Digital Technol. Appl. (9) (2011)

Part IV
Tourism Culture, Development and Planning

International Tourism: Geopolitical Dimensions of a Global Phenomenon

Olivier Dehoorne, Kevin Depault, Sheng-Quan Ma and Huhua Cao

Abstract Tourism is not apolitical. The development of this field has diverse, complex, and contradictory economic, political, social, cultural, and environmental impacts. The revenues derived from international tourism continue to break records: 1.035 billion people traveled for leisure in 2012. Thus, tourism has become a mass practice. The history of international tourism in the last 50 years can be divided into three stages: (1) the period of 1950–1980, which were characterized by a gradual democratization of tourism supported by sustained economic growth; (2) the 1990s, which were marked by an euphoric tourist market that pushed for increasing access across the globe for international tourism; and (3) the years following 2001, which are marked by maturity in the field of tourism. The ecstatic vision of a world without borders was met with harsh reality after the tragic events of September 11, 2001, leading to increased awareness of the complexities for tourism in a world fraught with conflict. Selective diffusion of global tourist flows is the current economic and political logic driving the tourism industry and translates into sophisticated business relations. The climates of insecurity that affect some fragile tourist destinations in the periphery are more sensitive to the expansions and contractions of international tourist markets.

Keywords International tourism · History of tourism · Center–periphery · Geopolitics · Security · Power

O. Dehoorne (✉) · H. Cao
Faculty of Law and Economics, CEREGMIA-Université des Antilles et de la Guyane, Guyane, France
e-mail: dehoorneo@gmail.com

H. Cao
e-mail: caohuhua@uottawa.ca

K. Depault · H. Cao
Department of Geography, University of Ottawa, Ottawa, Canada
e-mail: kdepault@gmail.com

S.-Q. Ma · H. Cao
School of Information and Technology, Hainan Normal University, Haikou, China
e-mail: mashengquan@163.com

B.-Y. Cao et al. (eds.), *Ecosystem Assessment and Fuzzy Systems Management*, Advances in Intelligent Systems and Computing 254, DOI: 10.1007/978-3-319-03449-2_35,

1 Introduction

Hazards such as natural disasters, military activity, recurring conflicts, and social unrest remind us that international travel is not without risk. The field of international tourism exists in a complex and potentially unstable environment, with dangers such as the explosion of a balloon flying over Luxor (19 dead, February 2013), the kidnaping of five tourists by an armed commando during the Holy Week in Acapulco (2013), the rape of a tourist in a minibus in Rio de Janeiro (April 2013) or that of a Swiss camper on the road to the Taj Mahal (March 2013). Gone are the days when the world was viewed as a vast playground open to tourists.

The flow of international tourism continues to break records: 1.035 billion people traveled for leisure in 2012, representing a growth rate of 4 % [World Tourism Organization (WTO)]. In 2000, 687 million people traveled, up from 278 million only 20 years earlier. Thus, tourism has become a mass practice [1] in the space of half a century: The WTO recorded only 25 million international tourists in 1950. Selective diffusion of global tourist flows corresponds to an economic logic and political tradition in sophisticated reports and evolutionary forces.

Tourism studies take on a new dimension in addressing power structures, often overlooked, and historical forces essential to understanding the internal and external sources of tourism development.

2 Tourism is Not Apolitical

Tourism is understood as a development choice with complex and contradictory political, social, cultural, environmental, and economic impacts [2]. Tourism can also be understood as an "engine of growth, prosperity and well-being" (WTO). It generates trade across the world and its export earnings (including tourist transportation) exceeded $1,200 billion USD in 2011, a rate of $3.4 billion per day (WTO).

Having acknowledged its overall economic impact, the international tourism industry is not without some fundamental problems. First, revenues are disproportionately captured by more developed countries, which act as both the primary transmitting and receiving locations for tourists.[1] Furthermore, these countries possess the bulk of international logistics infrastructure (tour operators, air transport companies, etc.). This concentration, within a context of increasing

[1] Receipts from international tourism amounted to $1,050 billion USD in 2011. A total of 64.6 % of expenditures are made in countries with advanced economies (classification based on the International Monetary Fund), divided between the European Union (36.6 % of the total market), North America (14.1 %), and North-East Asia (13.9 %). In terms of emissions of international tourists, the positions are identical: Europe emits more than 50 % of international tourists, followed by Asian economies (China, now the world leader with 83 million in 2012, followed by Japan and South Korea) and North America. Over 75 % of international tourists choose destinations within their region of origin.

international competition, limits the transmission of tourism knowledge and skills to economies of the south. Developing countries are forced to incur significant costs in order to reap the benefits of tourism through importing essential products and services demanded by international tourists, resulting in the famous "leakage" [3] that limits the economic benefits of tourism on economies in the global south. In addition, there is also the issue of diffusion and distribution of income, and as a result the extent of the benefits that reach the poor.

These dilemmas regarding tourism revenues fuel accusations from local populations that the government is prioritizing the accommodation of foreigners at the expense of the needs of the country. The percolation of tourism revenues across different social strata is slow to respond to side effects of the tourism industry, such as the rising cost of living, inflation and dollarization of the local economy, and the privatization of space.

It is in this context that we must analyze the evolution of tourist flows and assess economic and political issues regarding tourism in their international, national, and local dimensions, according to the rivalries of power [4]. And in the margins and peripheries of the world—in developing, emerging, and least developed countries—the tourist is at the heart of political strategies: For some, it is to show their success through the window of the modernity embodied in this place, in contrast, other common opponents, Democrats or extremist ahead will focus their destabilization of that power (the construction of discourse of ordinary hatred of foreigners to possible action terrorist in its international dimension).

3 The Advent of Mass Tourism: A Half-Century History of Tourism

The short history of tourism can be divided into four periods: (1) before 1950; (2) 1950–1990; (3) 1990–2000; and (4) since 2001. The concept of tourism can be traced to the beginnings of industrialization, when the romanticization of nature among the wealthy, coupled with a concern for health and well-being, triggered a self-induced shift among communities away from industrial centers. A rediscovery of shorelines [5] and artificial climates led to a search for exoticism toward the Middle East, Asia, and the Far East. These travelers were decidedly aristocrats and economic elites (bankers, industrialists, etc.) [6]. Over time the famous process of Boyer's "runoff along the social pyramid" (1999) occurred, and demand for leisure increased to wider social strata. Gradually, the concept of mandatory, paid holidays became democratized in industrialized countries [7], although the World Wars temporarily interrupted this phenomenon.

From 1950, robust economic growth favored changing patterns of consumption in society (such as the need for recognition and self-esteem, see Maslow's pyramid) leading to the advent of the "leisure society" [8] and the institutionalization of free time. The introduction of transport, both individual (cars were used in 41 %

of all international tourist trips in 2011) and collective (most notably in the form of air travel) accompanied the spread of global tourism. The development of large aircraft (Boeing 747, 1969) and charter flights and the deregulation of air space made international travel more affordable for the masses (51 % of trips in 2011 were taken by air travel). Distant destinations thus became more accessible: Marrakech or Iceland time a weekend, Bali or the Maldives for a week-long holiday.

All these elements contributed to the advent of mass tourism. This is reflected by the massive concentration of visitors at a few cramped reception areas. The development of cruise tourism is particularly significant in this regard: The latest boats put into service have a capacity of more than 6,000 passengers, and it is not unusual to see 12,000 people landed simultaneously at small Caribbean destinations. Democratization and the mass spread of tourism facilitated the development of new venues, and a progressive structuring flows, where offer conditions increasingly demand, like the concept of holiday villages (such as Club Med).

Brief but euphoric, the last decade of the twentieth century was characterized by an extraordinary expansion of the tourist sector. The fall of the Berlin Wall and the disintegration of the Soviet Union in 1989 signaled the end of a bipolar world and triggered curiosity among Western youth to discover Eastern Europe, an area formerly entrenched behind the Iron Curtain. The unification of the world under a capitalist system created a promising geopolitical context for the tourism industry. This period also saw the opening of China (17,877 foreign visitors recorded in 1965, compared with 31.2 million international tourists in 2000 and 57.6 million in 2011), the end of apartheid in South Africa (the second most popular African tourist destination behind Morocco, with 8.4 million in 2011), Vietnam (1.4 million in 2000 and 6,000,000 in 2011), not to mention the positioning of new Caribbean destinations (such as Cuba and the Dominican Republic) as seaside destinations for mass tourism.

The early twenty-first century, and the events of September 11, 2001, began the age of maturity for international tourism. The 2002 Sari Club bombing in Kuta (Bali), in which nearly 200 people were killed on the "blessed Island of the Gods," highlighted the vulnerability of tourist places and their relation to geopolitical issues.

These tragic events ended the euphoric approach to tourism and remind us that tourism is not apolitical: It cannot escape the social, economic, and political context of international destinations. There have been significant social events that have influenced the trajectory of tourist mobility and demand. These include attacks in Egypt (Cairo, 1996, Luxor, 1997), the Middle East and the Mediterranean (Djerba, Casablanca, Amman, the Turkish resorts of Marmaris and Antalya), on the Sinai coast (Taba, 2004 Sharm el-Sheikh, 2005; Dahab, 2006), and to the Indonesian and Philippine shores, through to piracy near the Horn of Africa. A citizen might imagine tourists to be neutral as a result of the temporary nature of their stay in a location, but tourists are particularly vulnerable targets. Specific policies are implemented to ensure the security of tourists, such as the deployment of the army in popular tourist destinations (i.e., the pyramids along the Nile), with substantial punishments for potential attackers.

The actual volatility of tourists is not increased by "channel surfer" behavior, but is rather the result of a maturity that develops according to the travel experiences of tourists and the representation of their desired world [9]. The tourist industry emerged in a secure and peaceful world and has developed in a context of economic uncertainty (the 2008 financial crisis), political crisis (the state of emergency declared in the USA on September 14, 2001 is still in force), and environmental hazards (especially in the wake of the Fukushima disaster on March 11, 2011).

4 Security is Key

Tourism in more peripheral locations remains fragile: The film Bronzed (1978), shot in Club Med Assine, Côte d'Ivoire, is a case in point. Once a promising destination, Côte d'Ivoire and the city Abidjan have since slipped in popularity both for international tourism and for weekend visits.[2] The same fate has befallen the Club Med in Haiti, where the dictatorship of the Duvalier family[3] guarantees economic stability and generous investment, particularly from the United States. The palaces and luxury hotels had globally renown until the spread of AIDS in the 1980s in Haiti.[4]

In the search for a strictly controlled otherness, removed from the stresses of everyday life and global geopolitics, tourists and investors alike favor confined places devoted exclusively to tourism. Investments are concentrated in areas conducive to isolation and tourism specialization such as islands, coastal archipelagos, peninsulas, and barrier beaches around which clubs and tourist complexes are grouped, economies of scale that also facilitate safety. An example of this is the recent tourism boom on Honduras' Archipelago Bay Islands, situated in the coastal towns of Tela and Puerto Cortes away from gangs, drug trafficking and other illegal activity. Another is the Venezuelan archipelago of Los Roques, which remains spared from the petty crime, social unrest, and political instability that affects the rest of the country. Social and political disruption, however, weaken marginal tourist destinations. A three-month long state of emergency was declared in the capital of Trinidad and Tobago in 2011 in response to gang activity in the neighborhoods of Laventille, Morvant, and Beetham. In Jamaica, it is advisable for

[2] The latest figures provided by the WTO for Côte d'Ivoire report 301,000 international tourists for 1998.

[3] In a referendum held on June 14, 1964, 99 % of voters approved a constitutional amendment that gave the presidency to François Duvalier for life (because according to the constitution, the president of the republic could not be re-elected).

[4] Several medical studies at the time believed that the virus originated in Africa was spread in the United States via Haiti.

tourists to avoid public transportation and public beaches, as well as main neighborhoods in the cities of Kingston, Montego Bay, and Spanish town (Ministry of Foreign Affairs). Additional examples could be provided to understand the logic of tourist resorts, which were imposed in emerging tourist destinations in the late 1990s. Installed at coastal locations, such "transplants" hardly accepted, these resorts concentrated on tourist consumption (and logically income) in peace, with the presence of local populations limited to authorized workers only. Such resorts exist along the northern coast of Jamaica, the Dominican Republic, and the Cuban archipelagos of Cayo Coco and Cayo Largo.

The current development of cruise tourism operates on the same logic: The ship is not just a simple means of transport from one stop to another; it becomes the principal destination for cruise passengers.[5] The cruise ship is a floating resort, a secure environment of festivity passing by scenic landscapes [10].

5 When the Tourist Becomes a Potential Target

Four levels of insecurity can be defined: (1) tourists as victims of petty crime; (2) tourists as bargaining chips; (3) tourists as symbolic targets; and (4) environmental hazards of inadequate accommodation.

First, the tourist can be exposed to an immediate predation without any significant premeditation. Due to economic discrepancies between tourists and local populations and the financial liquidity that tourists possess, they are easy targets for petty crime. Tourists can be grabbed during popular rallies, when they are not paying attention or at nightlife locales specializing in extortion and prostitution (such as nightclubs in Mauritius). Petty crime occurs in strategic locations such as between downtown hotels and the waterfront (as in Cape Town, South Africa), or the interior, with attacks by bandits (as in Burkina Faso). For all these actions which are petty crime, the tourist is only the most lucrative victim.

Then, there are premeditated acts that target tourists for their economic value and/or for political reasons. They can fall victim to kidnaping for ransom by organized crime (such as in Mexico and Haiti). Tourists also have political value as victims in guerrilla zones or in areas experiencing communal tension (such as in the southern Philippines or in the regions of Mandalay and Pegu in Burma). Such politically motivated actions may attract the attention of the media and subsequently reach international outlets.

The tourist, by nationality, is a representative of their home country, a nation that may engage in war. Thus, tourists can be the victims of improvised explosive

[5] The average annual growth of the cruise industry is 7.5 % over the period 1980–2011. The number of cruise passengers increased from 3.7 million in 1990 to 7.2 million in 2000 to 14.8 million in 2010 and 16.65 million in 2011, with a total of 105 million overnight passengers. Growth prospects for the industry are favorable, with 25 additional vessels expected between 2013 and 2015, with a total of 360,000 beds; a $10 billion investment over three years.

devices (IEDs), suicide attacks, or terrorist infiltrators operating in international networks. Areas of instability in the Middle East, North Africa, and the Sahel-Saharan region of Africa are particularly common locations for this security risk.

Finally, tourists' security may be compromised if the level of development and services are not consistent with the requirements for international tourism. Less developed countries face infrastructure limitations, such as at the archipelago of Cape Verde, which lacks a helicopter to assist in the ascent of the volcano Pico Grande, rescue boats capable of intervening in a stormy sea, and a decompression chamber in the event of diving accidents. Other limits may include a lack of hygiene, deficient medical supervision or prolonged exposure to health risks.

6 Conclusion

Despite the economic instability and current global crises, international tourism continues to consolidate its growth. While the purchasing power of Western middle classes (which supply the bulk of mass tourism) [11] increases, emerging powers such as China are supporting market growth.

Global tourism is dominated by major poles that structure their regional space. On its margins, emerging tourist destinations fit together, vulnerable, more or less stabilized, then forgotten and dead-angles margins. In these uncertain times, the tourism industry is aware of its own limits [12]. The daily deterioration of fragile zones, lack of hygiene, food hazards, urban unrest, police or paramilitary brutality, and state weakness are all factors that need to be considered within the confines of the global tourism industry. Tourism movements occur in a secure environment that shifts and transforms the borders of tourism in favor of destinations that guarantee security. A few emerging destinations appear capable of meeting the current safety requirements of the global market (such as Cuba, Vietnam, and Cambodia), while many unstable areas (like the majority of African states) have only marginal potential for tourism.

Acknowledgments The authors would like to thank the National Natural Science Foundation of China (No. 40961033) and International Science & Technology Cooperation Program of China (No. 2012DFA11270) for the support given to the study.

References

1. Boyer, M.: Histoire du tourisme de masse, coll. PUF, Paris (1999)
2. Cazes, G.: Tourisme et tiers-monde, un bilan controversé. Les nouvelles colonies de vacances. Edition Harmattan, Paris (1992)
3. Logossah, K.: L'industrie de la croisière dans la Caraïbe: facteur de développement ou pâle reflet de la globalisation? Téoros **26**(1), 25–33 (2007)
4. Giblin, B.: Le tourisme: un théâtre géopolitique ? Hérodote **127**, 3–14 (2007)

5. Corbin, A.: Champs. Le territoire du vide. L'Occident et le désir du rivage, 1740–184. Flammarion. Coll. 218 (1990)
6. Urbain, J.-D.: Sur la plage, mœurs et coutumes balnéaires au XIXe et XXe siècles. Edition Payot, Paris (2002)
7. Lanquar, R.: Sociologie du tourisme et des voyages. Presses universitaires de France, Paris (1985)
8. Dumazedier, J.: Vers une civilisation du loisir? Seuil, Paris (1962)
9. Michel, F.: Désirs d'Ailleurs. Essai d'anthropologie des voyages. Éditions Histoire & Anthropologie, Strasbourg (2002)
10. Dehoorne, O., Petit-Charles, N.: Tourisme de croisière et industrie de croisière. Etudes Caribéennes, 18 (2011)
11. Hoerner, J.-M.: Géopolitique du tourisme. Armand Colin (2008)
12. Dehoorne, O., Saffache, P., Tatar, C.: Le tourisme international dans le monde : logiques des flux et confins de la touristicité. Etudes Caribéennes, 9–10 (2008)

Data Inputs and Research Outputs: Social Network Analysis as a Tool in Sampling and Dissemination Strategies

Matthew Gaudreau

Abstract This paper examines the social aspects related to data collection and findings dissemination in participatory modeling and research. Two issues are examined, namely: (1) when collecting volunteered data from human sources in a specific community, the potential for sample bias is strong, and (2) the uptake of research results by policymakers and stakeholders is not guaranteed. The author explores the use of social network analysis and existing theory in social science research to improve volunteered data collection methods and uptake of research findings.

Keywords Data · Volunteered information · Research dissemination · Social science methodology · Social network analysis · Science-policy interface · Power relations

1 Introduction

Some of the preceding articles have focused on a variety of different research projects using complex models such as ecosystem assessment and management (EAM), geographic information systems (GIS), and environmental warning systems [1]. Each of these platforms relies on various sources of data, including census information from various levels of government, environmental indicators, and agricultural output, among others. The novelty of many of these systems is in the use of unique data inputs, received from human participants in the systems of study. For example, using crop yield data received from local farmers, and using local knowledge in Hainan's ecological warning system.

M. Gaudreau (✉)
School of International Development and Global Studies, University of Ottawa, Ottawa, Canada
e-mail: mjrgaudreau@gmail.com

B.-Y. Cao et al. (eds.), *Ecosystem Assessment and Fuzzy Systems Management*, Advances in Intelligent Systems and Computing 254, DOI: 10.1007/978-3-319-03449-2_36,

The inclusion of such volunteered information and local knowledge is an important factor in grounding these models, and their results, in the communities they serve. It is also necessary, as the micro-level data required of the models would be largely unattainable without the assistance of local populations. On the other hand, as management-based models, the output of related research has the intention of informing stakeholders and policymakers in order to better manage ecological and agricultural systems.

These two points—collection of volunteered data from local populations and using the model outputs to inform policy—are issues that are commonly dealt with in social science research [2, 3]. They are issues that exist at the science-policy interface and are subject to the power relationships that characterize the world of social relationships [4, 5]. Conceptualizing the context of participatory data collection and the provision of information to inform policy as social challenges will assist the efficacy of these modeling exercises.

The science-policy interface is a "social processes which encompass relations between scientists and other actors in the policy process, and which allow for exchanges, co-evolution, and joint construction of knowledge with the aim of enriching decision-making" [3]. This conception of scientist-stakeholder/policymaker relationships accurately describes the idealized processes that are undertaken in the participatory modeling activities typical of ecosystem assessment and management (EAM) and ecological/agricultural warning systems. It also allows for us to understand the social nature of these processes, which requires careful consideration of the how scientists and stakeholders go about collecting and exchanging information.

There are in fact many science-policy interfaces, and social processes, that take place in any decision-making scenario [6]. This article will proceed with an examination of issues that arise at two instances of the science-policy interface. These two instances are at the stages of information inputs for scientific modeling, and research output to affect policy. This will be followed by a discussion of measures that can be taken to improve research at each of these instances of the science-policy interface.

2 Data Inputs and Research Outcomes

One of the specific advantages of EAM models is that they are adaptable to specific contexts. This is both in terms of the issue that is being evaluated (e.g., agricultural production, ecological degradation) and the setting in which it is taking place (e.g., rural Gansu or Shaanxi province, or urban coastal cities such Liaoyang or Sanya). The appealing aspect for implicated communities is its use of local knowledge inputs related to issue at hand, and its practical orientation designed to process and provide information as a direct input into policy design.

On the input side, EAM systems use a complex mix of data with varying sources and levels of completeness. Geographic information systems (GIS) use relatively reliable sources of information such as local and national census data. This information is also, by nature, complete information, and therefore meets the standards of random sampling when a sample is drawn from a dataset. However, the use of volunteered information in EAM research, while a distinct advantage in involving local population and gaining access to data, is also the source of a potential pitfall.

On the research output side, EAM systems have goal of providing informational indicators to be used in management or policy decisions. The data inputs and models, tailored to context-specific cases, are specifically designed for these practical uses. However, EAM systems are subject to the common frustration faced by many scientific endeavors: achieving the application of indicators from models to inform on-the-ground management practices and policy design [7, 8].

These two related aspects of management-based modeling with volunteered information will be investigated individually to understand the social processes behind data gathering and research output.

2.1 Outsourcing Data Collection

There are multiple approaches to community participation in the social sciences, often involving the direct engagement of populations implicated in the subject of research. Typically, these approaches have involved varying degrees of including local populations or stakeholders in an iterative process of research design, data collection, and results consultation [9, 10]. However, within these approaches, the level of involvement of local populations or stakeholders can vary from the full inclusion that is characteristic of Participatory Action Research to simplify the provision of data points for use in modeling.

Most pertinently, recent research related to environmental governance and natural resource management has sought to include local knowledge as an informational input into devising effective management strategies [11–13]. This has been done in various ways, from collaborative modeling in which communities or stakeholders have input into the way predictive models are constructed [7, 14], to using local knowledge as a singular input into a model predetermined by researchers [15, 16]. With the advent of digital technologies increasing the potential connectivity between researchers and stakeholders, participation has been sought through accessing "crowd sourced" and "volunteered information" [16–18].

In the context of EAM, participation most closely resembles sourcing volunteered information, often containing local knowledge of environmental or agricultural conditions. The use of these methods to improve the accuracy and applicability of model outcomes makes this area of research susceptible to the social processes that make up the science-policy interface. This includes the power relationships that are embedded in social processes through uneven resources and uneven abilities to influence policy change [19].

In particular, sample selection bias may be an issue when different members of communities hold different social positions in the affected communities. When working in communities, this issue can be particularly problematic given the methods that may be needed to enter and work in communities for the purpose of data collection. This is a commonly cited issue in development studies [20]. These local social processes can prevent a representative sample of a particular information point from being gathered. Understanding the social relationships and power structures that may exist in stakeholder communities can give researchers the information necessary to understand the limitations of their samples or strategies for optimal access full stakeholder communities.

2.2 Research Results and Affecting Policy

Similar to understanding social relationships at the level of those volunteering information, it is also important that those undertaking EAM projects understand the power relationships that exist in relation to management and policy design. This is one of the issues that is intended to be addressed through collaborative modeling and has been a longstanding issue dealt with in public administration research. This ability for scientific research such as EAM to effect policy outcomes is precisely the focus of recent concepts such as "evidence-based policy," "evidence-informed policy," and "the science-policy interface" [3, 4, 21].

In particular, there is a noted issue in the relationship between the information that is generated by scientists, and the effect that this information has on policy output [21]. As many in the field of political studies have noted, it is not at all certain that findings in scientific study will translate directly into policy [4]. In fact, there are many other factors that go into policy decision-making, of which the findings from a given scientific study are only one [21, 22].

In the context of an early warning system, the evidence produced by a scientific study integrating multiple variables in a predictive model of ecological degradation may not necessarily lead to policy change [22, 23]. To put it simply: If an early warning system begins flashing a warning, who will listen? Similarly, if an agricultural warning system indicates management change is necessary, who will uptake new management practices?

Examples of climate change studies are pertinent to this question, serving as an example of the complexities involved in translating scientific research into public policy. Pielke [21] points to the US case of opposition to climate change policy and the agreement between two US politicians. One believes in the science behind anthropogenic climate change while the other does not; however, they both agree on not implementing taxation-based policy for fear of hurting economic growth [22].

The reality of the policy-making process is that scientific research is only one element that is incorporated into decision-making [4, 21]. As such, in the same way that understanding the social and political dynamics of communities that volunteer information, it is also important to understand the social and political dynamics related to any policy-output targets. That is, in taking on a management or applied orientation, the research results have the necessary goal of being used to inform policy. As policy making at the science-policy interface is a social process, having a clear picture of these social processes becomes pertinent.

Thus, understanding the social relationships between various policy networks and interest groups operating around the subject of research can give a more complete understanding of how to disseminate research results. From both the input and the output sides of EAM-related research, the social relationships in each instance of the science-policy interface become methodologically important and pertinent to research uptake.

3 Potential Solutions to Social Aspects of Research

Knowing how those volunteering information are (dis)connected, and understanding the relationships that exist between policymakers, are of primary importance. These factors help to understand the power dynamics that may or may not exist in relation to a given EAM research project. Given the issues that EAM systems face surrounding science-policy interfaces, we can now begin to assess ways forward. In particular, the use of social network analysis to uncover the social dynamics underlying input and output sources will be examined.

EAM research is a long-term process, often taking years to design. Because it is a long-term process, this gives a unique opportunity to use pre-study techniques to better collect (and collect better) data from communities and to have plans for dissemination of research outputs. As research at the interface between science and policy has shown, participation with stakeholders from the very beginning of work can enhance both community and policy-maker acceptance [7, 14]. Here, it is argued that using social network tools in this context can also give researchers an opportunity to improve their own work and dissemination of results.

Social network analysis is a structural approach to research that puts relationships between social actors at the center of its concern [24]. It has been described as both a theory and a methodology and has been applied in multiple disciplinary contexts [25, 26]. However, as a tool, it can also be applied for the purpose of understanding the layout of social relationships in a community with which research is being undertaken, or in deciding which organizations/people may be the most effective for getting EAM research heard.

By mapping social relationships, we can discover how people or organizations are (or are not) related to one another. This can shed light on the structure of relationships that were not previously known to the researcher (or even to the community), thereby increasing representation in data collection and/or the ability effect policy change. Below, each will be discussed with respect to the utility and potential applications of social network analysis to improve EAM research.

3.1 Volunteered Data: How is Your Sample Connected?

As discussed in Sect. 2.1, the methods involved in collecting volunteered data involve power relationships that are not always considered by those who usually rely on objective data.

Figure 1a gives a hypothetical example of how a “random sample” gained through volunteered information can actually appear in terms of social relationships. In this hypothetical example, a village of 20 people is being engaged to collect data about crop yields and fertilization methods. The data is collected through a partnership with the village leader, who helps to provide access to other villagers in order to collect data. A sample of 13 villagers (65 % response rate) will be used as data for input into an EAM model.

Under a random sample model, a 65 % response rate may seem highly desirable and a representative sample. However, given that the data sample is collected by researchers through partnership from the village leader, it is important to take a second look at how this data maps onto the social landscape of the village. Figure 1b demonstrates how a “random sample” might be skewed by the “gatekeeper” approach to collecting village-level data.

In this example, we can see that the village has a networked cluster represented by 15 villagers who all share friendship connections. This group can further be divided into three groups, the five on the top, five in the center, and four to the bottom. Further, there are five isolated village members who share connections with each other, but not to the other villagers. If we imagine that the village leader is represented by the node labeled “1,” we can see how the snowball sampling made its way through the friendship network. The result is that villagers 17–20 are not sampled as they do not share friendships with main group related to the village leader.

This kind of sample bias is one that is commonly faced when engaging human subjects; however, it may not be immediately dealt with by researchers who commonly use full census data. There are many hypothetical explanations for why villagers 16–20 are not connected to the rest of the villagers through friendship (or other relationship qualifiers). The point is that this difference (and underrepresentation in volunteered data) may skew the results of the model that depends on local information.

Figure 1 goes further than pointing out a previously understood bias in sampling. It also points to the usefulness in understanding the social networks that exist among partner communities. Of course, in practice, the geographic area in which EAM research is being conducted does not have a population of 20—but perhaps 2,000 or 200,000. This can make social network analysis a large undertaking. Despite the logistical difficulty, the data collected can lead to ensuring a much more representative sample of the desired population—which is necessary for meeting the assumptions of most models.

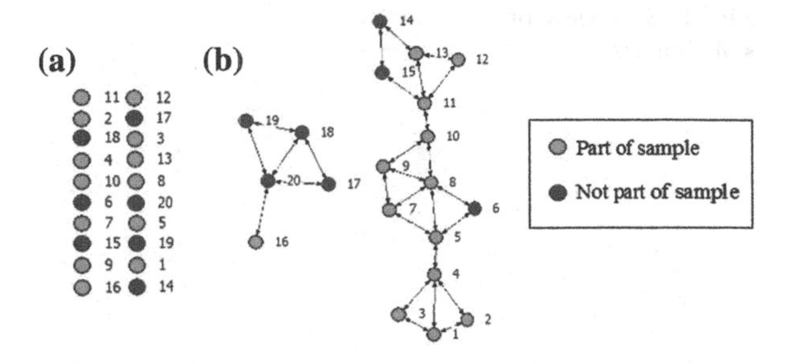

Fig. 1 Comparing assisted sample and SNA viewpoints

3.2 Collaboration at the Science-Policy Interface: Are You Connecting with the Right People and Organizations?

In terms of research output, as discussed in Sect. 2.2, it is not guaranteed that the results of EAM research will be transmitted for use in better management or policy. Having an understanding very early on in research about who the stakeholders are and who the existing players in decision making are, can help to determine the best way to approach research partnerships and prepare for research diffusion.

Similar to understanding social relationships at the level of informational inputs, an SNA view of the important research output networks is also useful. Figure 2 represents a hypothetical network of stakeholders (or alternatively, a network of policymakers). This is the same graph as displayed in Fig. 1b, but adjusted to represent certain nodes (which represent people) with numbers for ease of reference.

This graph helps to illustrate the type of information that a preliminary network analysis can provide. Where node (1) is the leader of a village or of a policy department, the other nodes represent various other actors involved in a hypothetical environment. A hierarchical logic may dictate that the most important person to contact would be the person in the highest position in an organization; however, as many network studies have shown, this is not always the case [27]. There are many alternative choice strategies within or between organizations that may lead to better informational uptake.

In Fig. 2, we can see that node (8) is a highly connected individual with five immediate relationships. In addition, he or she needs very few steps to reach most other actors in the network. Conversely, the person represented by node (14) has only two immediate connections and takes many steps to reach most other actors in the network. We can also see that, as before, nodes 16–20 do not share ties with the rest of the network.

This type of information can give an extremely valuable perspective on how best to disseminate research for the purpose of policy uptake (depending on the policy/ management goals of the researchers). For example, if the goal is to encourage villager uptake of an agricultural production practice, then the hypothetical network

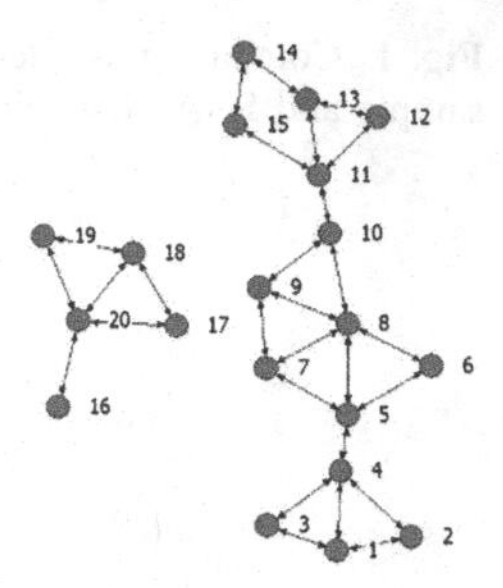

Fig. 2 SNA view of stakeholders

study has shown us that the isolated network (16–20) may not be in the loop when it comes to being informed by the main network. Thus, in order to encourage the use of EAM results among villagers, the network analysis will allow us to understand the best way to reach less networked groups.

Likewise, among the main network of villagers, it may be most beneficial to convince the person represented by node (8) to uptake a new production practice. If this person agrees, the network graph indicates that this person will be better placed to diffuse the new techniques to others. Of course, there may also be value in having the village leader agree as well, however what network analysis shows us is that the person in the "leader" position according to hierarchy may not be the one who has the most influence.

Having a perspective of the way that networks between people are structured can add the benefit of being able to target the most appropriate partners—whether at the level of the village, of a government department, or between organizations.

4 Conclusion

As can be seen in both the instances of collecting volunteered data and disseminating research on affecting policy/management change, each is a social process at the science-policy interface. Understanding these science-policy interfaces as social processes can help researchers to approach data collection and research dissemination in a strategic way. It cannot be assumed that community partners will be able to reach a representative sample for volunteered information, nor can it be assumed that practically oriented research will simply be taken up by managers.

The hypothetical network examples have shown the potential for social network analysis as a tool to map and assess social relationships. Its use facilitates the understanding of social structures that exist in partner communities or stakeholder networks. Having this knowledge at an early stage in EAM research can help to ensure appropriate representation among partner populations and effective targeting of individuals or organizations for the purpose of research dissemination.

Please to see other papers in this volume for examples of volunteered data as an input in modeling.

Acknowledgments Thanks to Professor Shengquan Ma and Professor Huhua Cao for providing a stimulating and interdisciplinary venue for discussing these important issues and for providing funding support for this research. Thanks also go to Professor Mark Saner of the Institute for Science, Society and Policy who offered guidance on issues of science-policy.

References

1. Bodin, Ö., Crona, B.I.: The role of social networks in natural resource governance: What relational patterns make a difference? Global Environ. Change **19**(3), 366–374 (2009)
2. Van den Hove, S.: A rationale for science–interfaces. Futures **39**(7), 807–826 (2007)
3. Watson, R.T.: Turning science into policy: challenges and experiences from the science–policy interface. Philos. Trans. Royal Soc. B: Biol. Sci. **360**(1454), 471–477 (2005)
4. Elwood, S.: Critical issues in participatory GIS: deconstructions, reconstructions, and new research directions. Trans. GIS **10**(5), 693–708 (2006)
5. Saner, M.: A Map of the Interface between Science and Policy. Council of Canadian Academies, Staff Papers (2007)
6. Cockerill, K., Daniel, L., Malczynski, L., Tidwell, V.: A fresh look at a policy sciences methodology: collaborative modeling for more effective policy. Policy Sci. **42**(3), 211–225 (2009)
7. Schneider, S.H.: Integrated assessment modeling of global climate change: transparent rational tool for policy making or opaque screen hiding value-laden assumptions? Environ. Model. Assess. **2**(4), 229–249 (1997)
8. Whyte, W.F.E.: Participatory Action Research. Sage Publications, Chicago (1991)
9. Jakeman, A.J., Letcher, R.A., Norton, J.P.: Ten iterative steps in development and evaluation of environmental models. Environ. Model. Softw. **21**(5), 602–614 (2006)
10. Blaikie, P., Brown, K., Stocking, M., Tang, L., Dixon, P., Sillitoe, P.: Knowledge in action: local knowledge as a development resource and barriers to its incorporation in natural resource research and development. Agric. Syst. **55**(2), 217–237 (1997)
11. Berkes, F., Colding, J., Folke, C.: Navigating Social-Ecological Systems: Building Resilience for Complexity and Change. Cambridge University Press, Cambridge (2003)
12. Raymond, C.M., Fazey, I., Reed, M.S., Stringer, L.C., Robinson, G.M., Evely, A.C.: Integrating local and scientific knowledge for environmental management. J. Environ. Manage. **91**(8), 1766–1777 (2010)
13. Hare, M., Letcher, R.A., Jakeman, A.J.: Participatory modelling in natural resource management: a comparison of four case studies. Integr. Assess. **4**(2), 62–72 (2003)
14. Tripathi, N., Bhattarya, S.: Integrating indigenous knowledge and GIS for participatory natural resource management: state of the practice. Electron. J. Inf. Syst. Dev. Countries **17** (2004)
15. Lee, T., Quinn, M.S., Duke, D.: Citizen, science, highways, and wildlife: using a web-based GIS to engage citizens in collecting wildlife information. Ecol. Soc. **11**(1), 11 (2006)
16. Flanagin, A.J., Metzger, M.J.: The credibility of volunteered geographic information. Geo J. **72**(3–4), 137–148 (2008)
17. Goodchild, M.F., Glennon, J.A.: Crowdsourcing geographic information for disaster response: a research frontier. Int. J. Digit. Earth **3**(3), 231–241 (2010)
18. Ernstson, H., Sörlin, S., Elmqvist, T.: Social movements and ecosystem services—The role of social network structure in protecting and managing urban green areas in Stockholm. Ecol. Soc. **13**(2), 39 (2008)
19. Leach, M., Mearns, R., Scoones, I.: Environmental entitlements: dynamics and institutions in community-based natural resource management. World Dev. **27**(2), 225–247 (1999)
20. Keller, A.C.: Science in Environmental Policy: The Politics of Objective Advice. The MIT Press (2009)

21. Pielke, R.A.: The Honest Broker. Making Sense of Science in Policy and Politics. Cambridge University Press (2007)
22. Koetz, T., Farrell, K.N., Bridgewater, P.: Building better science-policy interfaces for international environmental governance: assessing potential within the Intergovernmental Platform for Biodiversity and Ecosystem Services. Int. Environ. Agreements: Polit. Law Econ. **12**(1), 1–21 (2012)
23. Hanneman, R.A., Riddle, M.: Introduction to Social Network Methods. University of California, Riverside Press (2005)
24. Borgatti, S.P., Halgin, D.S.: On network theory. Organ. Sci. **22**(5), 1168–1181 (2011)
25. Marin, A., Wellman, B.: Social network analysis: an introduction. In: Scott, J., Carrington, P.J., (eds.). The SAGE Handbook of Social Network Analysis, pp. 11–25. Sage Publications (2011)
26. Krackhardt, D., Hanson, J.R.: Informal networks: the company behind the charts. Harvard Bus. Rev. **71**(4), 104–111 (1993)

Economic Valuation of Terrestrial Ecosystem Services: Present and Future Planning Perspectives for Ledong County, Hainan Province, China

Li-qin Zhang, Hu-hua Cao, Jiang-hong Zhu and Li-ping Qu

Abstract Economic valuation of land ecosystem services is a quantitative method for analyzing the ecological impact of land use that provides scientific conclusions for land-use decisions. This paper conducts a land ecosystem service value (ESV) assessment based on the present land-use situation during 1996 and 2008 and land-use planning between 2008 and 2020 in Ledong Li Autonomy County, Hainan province, China. The study concentrates on the reclassification of sub-ecosystems according to land-use types and the revision of ESV coefficients for different sub-ecosystems based on previous results obtained by Costanza et al. (Nature 387: 253–260, 1997) and Xie et al. (J. Nat. Res. 23(5):911–919, 2008). Implications of ESV development trends on land-use optimization are analyzed.

Keywords Ecosystem service value · Land use · Land-use planning · Value coefficient · Ledong

1 Introduction

Ecosystems produce different uses for human beings, directly or indirectly, from ecosystem functions [1]. How to quantitatively measure the economic value of ecosystem services is a key scientific question for decision-makers based on the balance between ecosystem services and economic benefits. Since the well-known article of economic valuation of global ecosystem services by Costanza et al. [1], the number of publications studying ecosystem service value (ESV) has grown exponentially, with almost 80 % of works on the subject being published after

L. Zhang (✉) · H. Cao
Department of Geography, University of Ottawa, Ottawa, ON K1N6N5, Canada
e-mail: lzhan143@uottawa.ca

L. Zhang · J. Zhu · L. Qu
Faculty of Public Administration, China University of GeoSciences, Wuhan 430074, China

B.-Y. Cao et al. (eds.), *Ecosystem Assessment and Fuzzy Systems Management*, Advances in Intelligent Systems and Computing 254, DOI: 10.1007/978-3-319-03449-2_37,

2007 (statistical data from the DirectScience database from a search for "ecological service" in "abstracts, key words, and article titles").

In the initial stages, there was much debate on the meaning of valuation of ecosystem services. Costanza et al. (1998) clarified the meaning for decision making [2]—"we (humans—both as a society and as individuals) are forced to make choices and trade-offs about ecosystems every day. These imply valuation. To say that we should not do valuation of ecosystems is to simply deny the reality that we already do, always have and cannot avoid doing so in the future". To evaluate the economic value of ecosystem, services are the essential basis of land and other natural resource plan [3, 4].

Research on this subject expanded from a global [5, 6] to regional and community scales [7]. Studies of land use and ecosystem service concentrate on land use/cover change and its impacts [8–9], the impact of both land-use change and socioeconomic development on ESV [10, 11], spatial expansion or landscape homogenization and ESV impacts [12–13], and so on. Relationships between ecosystem services, economy, and social sciences are specially published in the journal of Ecological Complexity 2010, for the basic identification which is the prerequisite for further research. Valuation model progress and its applications are examined [6, 11, 14]. A special issue in 2010 Ecological Economics (2013) has been published for the corresponding research in China.

Costanza has classified terrestrial ecosystem into nine subtypes: Forest, grasslands, wetlands, lakes/rivers, desert, tundra, ice/rock, cropland, and urban. Subsequent literature has adopted the nine sub-ecosystem classification, while more local research provides more detailed classification. In 2008, Biological Conservation issued a call for multiple classification systems for ecosystem services research that focuses on further division of sub-ecosystems.

The valuation model is another field researchers are interested in. The research of ESV by Costanza et al. [1] has concluded the values globally, which significantly influenced subsequent research. Xie et al. [15] studied the value coefficient for China, using the Delphi method with surveying of 200 professionals, based on the coefficient by Costanza [15]. Xie reclassified the ecosystem service categories into nine types according to the 17 types of R. Costanza (Table 1), making the ecosystem services much easier to be identified and calculated.

R. Costanza's coefficients have relatively abundant information for the ESV sub-ecosystems of forest, wetlands, and lakes/rivers, but there is lack of available information for urban, desert, tundra, and ice/rock, as well as for regulation and supporting services of cropland, lakes/rivers, and wetlands. G. Xie has revised and expanded information for unused land ecosystem, as well as for other ESVs, but information for urban ecosystems is still unavailable. In this paper, the ESV coefficients of paddy land, horticultural land, and construction land including urban are studied, which is complements former research.

The following questions are posed in this paper: (1) Did the last decade's land-use impact its ESV positively or negatively (1996–2008) in Ledong Li Autonomous County, Haina? How much has changed? (2) How will the land-use planning impact future ESVs during 2006–2020 and what are the implications for land-use plans?

Table 1 Classification of ecosystem services

Service type	9 types by G. Xie	17 types by R. Costanza
Provision service	Food production	Food production
	Raw materials	Raw materials
Regulation service	Gas regulation	Gas regulation
	Climate regulation	Climate regulation, disturbance regulation
	Hydro regulation	Water regulation, water supply
	Waste regulation	Waste regulation
Support service	Soil conservation	Erosion control, Soil formation, nutrient cycling
	Biodiversity conservation	Pollination, biological control, habitat/refuge, Genetic resources
Cultural service	Aesthetic landscape	Recreation, cultural

2 Method

2.1 Study Area

Ledong Li Autonomous County, with a jurisdiction of 2,763.22 km^2 and population of 458,876, 38.61 % of which are minorities (2010), is located in the southwest of Hainan island (18°24′–18°58′N and 108°39′–109°24′E) (Fig. 1). It is a very typical tropical region with dry and humid seasons. Ledong is an agricultural county with a relatively low economic growth rate (2.81 % per year between 2000 and 2005). Ledong has 84.3 km of coastline and 1,389 km^2 of sea in the south. It is located west of the famous "3S (Sun, Sand, and Sea)" tourism city of Sanya. Tropical rain forests are in the northern area of Ledong. The biodiversity in Ledong is very abundant, with diversified topographies such as beaches, coastal plains, alluvial plains, and hilly and mountainous areas. The land-use types in Ledong have very typical tropical characteristics. The elevation of Ledong is higher in north and lower in south, with the Limu mountain Ridge in the north and Wuzhi mountain Ridge in the east. In the middle area is a basin, and in the southwest has a coastal plain. The mountainous area occupies 40 % of the whole region. The precipitation in Ledong is uneven in different seasons with a greater frequency of droughts in spring. The spatial differentiation of soils in Ledong is obvious with eight types and 14 subtypes, for example, salinization swamp soil in coastal areas, paddy soil in basin areas, and yellow soil in mountainous areas above 800 m of elevation. In Ledong, Yinggehai has the famous and one of the biggest salt fields in southern China. Also, oil gas fields in Yinggehai area are now explored in the south sea region.

2.2 Data Collection and Preparation

In this research, data for present land use in 1996 and 2005 are collected from the Ledong land-use survey database and data for land-use planning between 2006 and 2020 are collected from the land-use planning database. Socioeconomic information

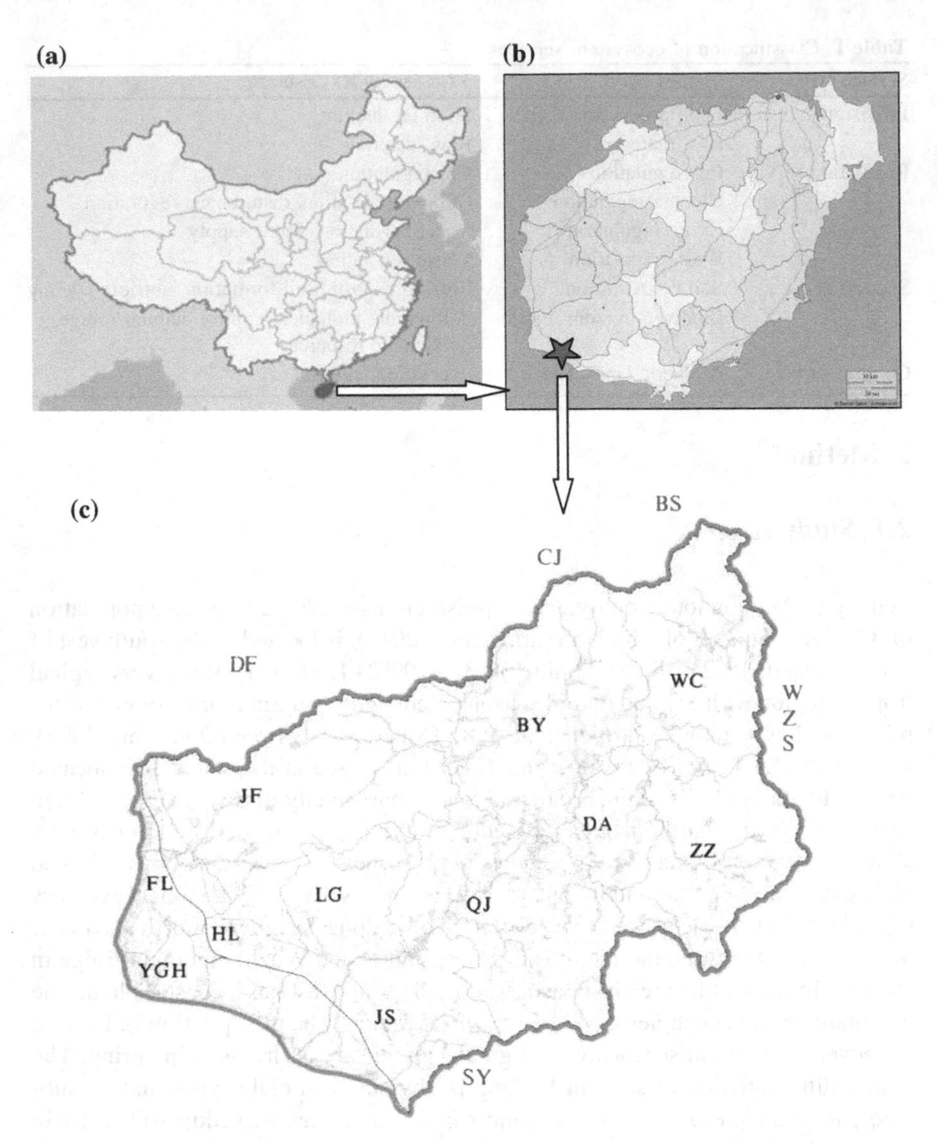

Fig. 1 Ledong location map in China and Hainan (the base maps of **a** and **b** are from http://d-maps.com/). **a** Hainan in China. **b** Ledong in Hainan province. **c** Ledong jurisdiction map

is collected from the annual statistical database of Ledong County. Since land-use classification is different from what the ESV research required, data conversion was conducted. Seven types of terrestrial sub-ecosystems are studied in Ledong County, with the conversion of Ledong land-use types into the sub-ecosystem types defined by R. Costanza for service value computation, in which tundra and ice/rock are eliminated, paddy land has been separated from wetland or cropland, and urban has been expanded into construction land (Table 2).

Table 2 Land ecosystem types in Ledong

Serial No.	Ecosystem	Land-use types
1	Dry cropland	Irrigable cropland, dry cropland, facility agricultural land
2	Paddy land	Paddy field
3	Horticultural land	Orchard land, tea plantation land, mulberry field
4	Forest land	Forest land, open forest land, shrub land
5	Grassland	Natural grassland, human-made grassland, other grassland
6	Wetland	Pond, inland beach, coastal beach
7	River and lake	Reservoir, river, lake
8	Construction land	Urban land/town, rural residential land/village, transportation land, mining land, scenic site, special industrial land, ditch, water conservation land
9	Unused land	Sand land, bare land

2.3 Assignment of Ecosystem Service Value Coefficient

The ESV coefficients for different sub-ecosystems are assigned according to the result studied by Xie et al. [15], with additional revisions. The revision is computed for the additional sub-ecosystems, including horticultural land, paddy land, and construction land. For other sub-ecosystems, the coefficients by G. Xie are adopted directly (Table 3).

2.3.1 General Function for Total ESV

The formula of ESV coefficients for sub-ecosystem i is:

$$\mathrm{ESV}_i = \sum A_i * \mathrm{ESV}_{ij} \tag{1}$$

In formula (1), ESV_i indicates the service value of sub-ecosystem i; A_i is corresponding land-use area of ecosystem i; and ESV_{ij} indicates the value of ecosystem service j of the sub-ecosystem i.

The total ESV is computed by:

$$\mathrm{ESV}_t = \sum \mathrm{ESV}_i \tag{2}$$

In formula (2), ESV_t indicates the total ESV of the study area; and ESV_i indicates the ESV of sub-ecosystem i.

2.3.2 Horticultural Land ESV Coefficient

Compared with forest land, the production or direct service value of horticultural land is higher, while the values of regulating, supporting, and recreation service value are lower.

Table 3 X-coefficient of land ESV in Ledong

Ecosystem service		Forest	Horticultural land*	Grassland	Cropland	Paddy land*	Wetland	River and lake	Construction land*	Unused land
Provision service	Food production	0.33	0.33	0.43	1	1.50	0.36	0.53	0.017	0.02
	Raw materials	2.97	1.485	0.36	0.39	0.39	0.24	0.35	0.012	0.04
Regulating service	Gas regulation	4.32	2.16	1.5	0.72	1.21	2.41	0.51	0.026	0.06
	Climate regulation	4.07	2.035	1.56	0.97	6.78	13.55	2.06	0.049	0.13
	Hydro regulation	4.09	2.045	1.52	0.77	6.72	13.44	18.77	0.039	0.07
	Waste regulation	1.72	0.86	1.32	1.39	7.20	14.4	14.85	0.066	0.26
Supporting service	Soil conservation	4.02	2.01	2.24	1.47	1.00	1.99	0.41	0.021	0.17
	Biodiversity	4.51	2.255	1.87	1.02	1.85	3.69	3.43	0.051	0.4
Cultural service	Cultural landscape	2.08	1.04	0.87	0.17	2.35	4.69	4.44	0.170	0.24
Total		28.12	14.22	11.67	7.9	28.98	54.77	45.35	0.45	1.39

*Coefficients are computed based on the coefficient by Xie et al. [15]
Other coefficients are adopted directly from the coefficient by Xie et al. [15]
1 Unit = US $54

$$ESV_{hef} = 50\% * ESV_{fef} \tag{3}$$

In formula (3), ESV_{hef} is the ESV of horticultural land except food production service value and ESV_{fef} is the corresponding ESV of forest land.

$$ESV_{fh} = ESV_{ff} \tag{4}$$

In formula (4), ESV_{fh} is food production value of horticultural land and ESV_{ff} is the food production value of forest land.

2.3.3 Paddy Land ESV Coefficient

Paddy field is a type of agricultural wetland, which has higher ESV than dry cropland but lower than natural wetland. Most rice is produced in paddy fields in Ledong. Paddy fields occupy 30,301 ha (2008) in Ledong, which is about 5.6 times greater than the area of wetland (4,595 ha in 2008). To identify and differentiate the ESV of paddy fields and those that dry lands and wetlands offer are essential in the study area. Ecosystem services provided by paddy fields include groundwater recharge, climate change mitigation, water purification, culture and landscape, and support of ecosystems and biodiversity [16].

$$ESV_{pefr} = 50\% * ESV_{wefr} \tag{5}$$

In formula (5), ESV_{pefr} means ESV of Paddy land except food production and raw material service value and ESV_{wefr} means corresponding ESV of wetland.

$$ESV_{fp} = 150\% * ESV_{fc} \tag{6}$$

In formula (6), ESV_{fp} is food production service value of paddy land and ESV_{fc} is the food production service value of cropland.

$$ESV_{mp} = ESV_{mc} \tag{7}$$

In formula (7), ESV_{mp} means raw material provision service value of paddy land and ESV_{mc} refers to the raw material provision service value of cropland.

2.3.4 Construction Land ESV Coefficient

In order to compute the ESV of urban areas, Bolund divided the urban ecosystem into six subsystems, including street tree lawns/parks, urban forest, cultivated land, wetland, stream, and lakes/sea [17]. We compute the ESV of construction land through an estimation of the proportion of its green space and water areas. According to town and village plan, construction land in Ledong is composed of at least 5 % greenland and water areas, so the ESVs of construction land (except cultural service value) equal to 5 % of the minimum value of those of the other sub-ecosystems, with the exception of unused land. Since construction land, especial urban land has a high recreation and cultural service function, especially in some indigenous local communities, the cultural service value of construction land is equal to the minimum (hereby equal to the cultural service value of cropland), which might be lower estimated.

$$ESV_{cec} = 5\% * ESV_{comin} \quad (8)$$

In formula (8), ESV_{cec} means the ESV of construction land except the cultural service value; ESV_{comin} refers to the minimum value of other types of land except unused land.

$$ESV_{cc} = ESV_{cmin} \quad (9)$$

In formula (9), ESV_{cc} means the cultural service value of construction land and ESV_{cmin} means the minimum cultural service value of other types of land.

3 Result

3.1 Land-Use Changes

During 1996 and 2008, only constriction land including urban land and mining, scenic site and special industrial lands have increased in area, while the areas of other types of lands have decreased. Between 2008 and 2020, this tendency will change according to land-use plans, while dry cropland, especially that of facility agricultural land will increase, as will those of forest land, reservoir (which is included in the type of river and lake), rural residential land, mining, scenic site, and special industrial land (Table 4).

Between 2008 and 2020, the changing scope is much larger than that during 1996 and 2008. The area of dry cropland will increase 1,169 ha, while the area of facility agricultural land is planned to increase 1,380 ha and that of the other dry cropland will decrease. The area of paddy land, horticultural land, and grassland will decrease. The area of forest land is planned to increase 859 ha, while in contrast, the area of wetland will decrease 1,601 ha during the planning period, which is markedly different from its nearly stable status during the past 12 years. The area of urban and transportation lands is planned to decrease, but within it the area of rural residential land, scenic sites and special industrial land will increase.

The planned changing of land-use structure is the basis of the changes of sub-ecosystem types. Wet land has a relatively high ESV than other types of land, and so the total ESV might decrease while wetland converts into other types of uses.

3.2 Land Ecosystem Service Value Changes

The total terrestrial ESV in Ledong is around $340 million USD. Among the total value, the direct ecological service (provision service) occupies only 10.37 % of the total, including food production service (2.18 %) and raw materials supply

Table 4 Ledong land-use changes between 1996–2008 and 2008–2020 ha

	1996	2008	2020	1996–2008	2008–2020
Dry cropland	18,558	18,544	19,713	−15	1,169
#Facility agricultural land	37	37	1,417	0	1,380
Paddy land	30,325	30,301	30,182	−24	−118
Horticultural land	68,986	68,932	68,480	−54	−452
Forest land	131,079	130,977	131,835	−103	859
Grassland	3,188	3,186	2,925	−2	−260
Wetland	4,598	4,595	2,993	−4	−1,601
#Pond	3,121	3,118	1,488	−2	−1,630
#Inland beach	275	275	385	0	110
#Coastal beach	1,202	1,201	1,121	−1	−81
River and lake, reservoir	6,664	6,659	6,732	−5	73
Construction land	12,905	13,111	13,442	206	330
#Urban	1,829	2,018	1,833	189	−185
#Rural residential land	6,034	6,030	6,124	−5	94
#Transportation land	645	645	541	−1	−103
#Ditch and Water conservation	109	109	109	0	0
#Mining, Scenic site, and special industrial land	4,288	4,310	4,835	23	524
Unused land	18	18	18	0	1
Sum	276,322	276,322	276,322	0	0
Total	276,322	276,322	276,322	0	0

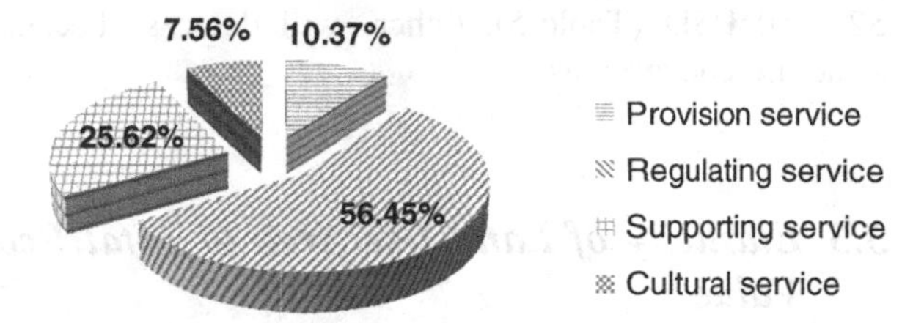

Fig. 2 Structure of ecosystem service in Ledong, 2008

(8.20 %). The regulating service, including gas, climate, and hydro and waste regulations has the highest proportion of the total with 56.45 %, while the supporting service composes 25.61 % of the total, in which soil conservation service is 11.80 % and biodiversity conservation service occupies 13.80 %. Cultural service has the proportion of 7.55 %, which lies not only in nature, but also in human environment such as urban and rural residential areas, tourism sites, transportation land, etc.(Fig. 2).

From 1996 to 2008, the value of all kinds of ecological services has remained nearly stable, ranging from $339,621 USD to $339,361 USD, and within it all the different types of service values have decrease slightly, from −$5,589 USD for the food production service value to −$45,503 USD for the hydro regulation service value. But between 2008 and 2020, three types of ESVs will increase: The food production service value will increase by $26,021 USD, while the raw material

Table 5 Land ESV in Ledong

Ecosystem service		Total ESV (10^3\$USD)			Changes of ESV (\$USD)		(%)
		1996	2008	2020	1996–2008	2008–2020	2008
Provision service	Food production	7,389	7,384	7,410	−5,589	26,021	2.18
	Raw materials	27,840	27,818	27,918	−21,642	99,381	8.20
Regulating service	Gas regulation	42,377	42,345	42,303	−32,859	−41,642	12.48
	Climate regulation	52,864	52,823	51,796	−40,796	−1,027,513	15.57
	Hydro regulation	58,725	58,679	57,716	−45,503	−963,254	17.29
	Waste regulation	37,755	37,726	36,622	−28,774	−1,103,304	11.12
Supporting service	Soil conservation	40,087	40,056	40,078	−31,124	22,250	11.8
	Biodiversity	46,875	46,839	46,715	−36,081	−124,178	13.8
Cultural service	Cultural landscape	25,638	25,620	25,290	−18,076	−330,373	7.55
Total		339,621	339,361	335,918	−260,500	−3,442,149	100

Table 6 Elasticity of land-use type to ESV

Coefficient type	Forest	Horticultural land	Grassland	Cropland	Paddy land
X-coefficient	5.84	1.54	0.04	0.21	1.38
Coefficient type	Wet land	River and lake	Construction land	Unused land	
X-coefficient	0.38	0.46	−0.01	−0.02	

provision service value will increase by \$99,381 USD and soil conservation by \$22,250 USD (Table 5). Others will decrease because of the related land-use structure conversions.

3.3 Elasticity of Land-Use Area to Total Ecosystem Service Value

The ESV resilience of land use is expressed with the changing rate of ESV to the changing rate of land-use area.

$$E_i = (1 - E_{i0}/E_{i1})/(1 - A_{i0}/A_{it}) \qquad (10)$$

In formula (10), E_i is the ESV elasticity of land type i; E_{i0} is the ESV of land type i at time t_0; E_{i1} means the ESV of land type i at time t_1; A_{i0} means area of land type i at time t_0; and A_{i1} means area of land type at time t_1.

Table 6 shows the ESV elasticity of forest is the biggest (5.84) and that of horticultural land and paddy land rank the secondary and tertiary (1.54 and 1.38, respectively). Construction land has the lowest ESV elasticity (−0.01).

4 Conclusion and Discussion

An ecosystem provides human beings with direct provisions such as food and materials, but these services might not fully correspond to socioeconomic development, i.e., during the agricultural era, the direct provision service should be the most important and biggest part; but with the development of agro-technology, the direct provision service value decreased while regulating and supporting function increased. The structure of land ecosystem service in Ledong in 2008 illustrates that the regulating service occupies the highest proportion with 56.45 % of the total in Ledong in 2008, and the supporting service ranks secondary with 25.62 %, while the provision service is 10.37 %, and the cultural service occupies the least with 7.56 %.

In the results by Costanza et al. [1], some ESVs are unavailable because of the lack of available information, for example, recreation and cultural services of urban land, regulating and supporting services of cropland, lakes/rivers, and wetland, ecosystem services of desert and other unused land, and so on. Those unavailable coefficients require further attention from researchers in the future. Also, there are some more details for the differentiation between different landscapes (land-use types) that require further study, especially in local areas. In this study, sub-ecosystems are reclassified according to Ledong land-use characteristics, and the service value coefficient is revised according to ecosystem functions from different sub-ecosystems, especially for the additional three systems including horticultural land, paddy land, and construction land, according to the results studied by Xie et al. [15]. Cropland and wetland have been divided into dry cropland, paddy field, and wetland (excluding paddy field). Unused land is considered for the ecosystem service. We also differentiated the coefficient of ecosystem service of forest and horticultural land, as well as the creation and cultural service from urban/construction land based on the average green area (including green space, surface water area, etc.) in construction areas. All these revisions are essential to the whole ecosystem service itself, even though the study is still in the probing stages.

The ESV elasticity hints at a new method for land-use structure optimization, i.e., from the objective of maximization of the terrestrial ESV, forest land is the first to be developed, followed by horticultural land and paddy land, area shrinking of those types of land should be very carefully and restrainedly. On another hand, since the lack of available literatures and information, valuation of ecosystem services of construction land and unused land is still not very clear, thus may weaken the significance of their contributions to the whole ecosystem service. To avoid the possibility of a decrease in the terrestrial ESV, construction land should be expanded very carefully and restrainedly. And the same attitude should be paid to unused land changes.

Since different methods generate different coefficients, how to make the research conclusion more persuasive with comparable parameters is actually the scientific question for ESV. This research is trying to use comparable parameters based on previous results, with some updates and revision, in order to make a conclusion that is not only scientific, but also practical. Furthermore, mapping of the ESV [3, 7] related to land-use changes in Ledong should be pursued for land-use planning or related resource planning decisions.

Further questions for local ESV research include multi-classification according to local land-use types and patterns; models for marketing and nonmarketing ESV, respectively; and marketing value based on comparable benchmarks.

Acknowledgments Thanks to the support by the Fundamental Research Funds for the Central Universities, China University of Geosciences (Wuhan), and the project of Construction of an Early Warning Information Platform/Module based on Ecotourism Assessment and Management (EAM)—Its Application in the Building of Hainan International Tourism Island of China (2012DFA11270).

References

1. Costanza, R., d'Arge, R., Groot, R., et al.: The value of the world's ecosystem services and natural capital. Nature **387**, 253–260 (1997)
2. Costanza, R., d'Arge, R., Groot, R., et al.: The value of ecosystem services: putting the issues in perspective. Ecol. Econ. **25**, 67–72 (1998). (After years' research and discussion)
3. Egoh, B., Reyers, B., Rouget, M.: Mappig ecosystem services for planning and management. Agric. Ecosyst. Environ. **127**, 135–140 (2008)
4. Raymond, C.M., Bryan, B.A., Macdonal, D.H.: Mapping community values for natural capital and ecosystem services. Ecol. Econ. **68**, 1301–1315 (2009)
5. Goot, R., Brander, L., Ploeg, S., Costanza, R., et al.: Global estimates of the value of ecosystems and their services in monetary units. Ecol. Serv. **1**, 50–61 (2012)
6. Sutton, P.C., Costanza, R.: Global estimates of market and non-market values derived from nighttime satellite imagery, land cover, and ecosystem service valuation. Ecol. Econ. **41**, 509–527 (2002)
7. Plieninger, T., Dijks, S., Oteros-Rozas, E., et al.: Assessing, mapping, and quantifying cultural ecosystem services at community level. Land Use Policy **33**, 118–129 (2013)
8. Zhao, B., Kreuter, U., Li, B., et al.: An ecosystem service value assessment of land-use change on Chongming Island China. Land Use Policy **21**, 139–148 (2004)
9. Wu, K., Ye, X., Qi, Z., et al.: Impact of land use/land cover change and socioeconomic development on regional ecosystem services: the case of fast-growing Hangzhou metropolitan area. China. Cities **31**, 276–284 (2013)
10. Dong, X., Yang, W., Ulgiati, S.: The impact of human activities on natural capital and ecosystem services of natural pastures in North Xinjiang. China. Ecol. Model. **225**, 28–39 (2012)
11. Ma, S., Swinton, S.M.: Valuation of ecosystem services from rural landscapes using agricultural land prices. Ecol. Econ. **70**, 1649–1659 (2011)
12. Estoque, R.C., Murayama, Y.: Examining the potential impact of land use/cover changes on the ecosystem services of Baguio city, the Philippines: scenario-based analysis. Appl. Geogr. **35**, 316–326 (2012)
13. Laterra, P., Orue, M.E., Booman, G.C.: Spatial complexity and ecosystem services in rural landscapes. Agric. Ecosyst. Environ. **154**, 56–67 (2012)
14. Costanza, R., Fisher, B., Mulder, K., et al.: Biodiversity and ecosystem services: a multi-scale empirical study of the relationship between species richness and net primary production. Ecol. Econ. **61**, 478–491 (2007)

15. Xie, G., Zhen, L., Lu, C., et al.: Expert knowledge based valuation method of ecosystem services in China. J. Nat. Resour. **23**(5), 911–919 (2008). (in Chinese)
16. Anan, M., Yuge, K., Nakano, Y., et al.: Quantification of the effect of rice paddy area changes on recharging groundwater. Paddy Water Environ. **2**, 41–47 (2007)
17. Bolund, P.: Ecosystem services in urban areas. Ecol. Econ. **29**, 293–301 (1999)

Hainan Resources and Characteristics of Cultural Tourism

Jing Ma, Sheng-Quan Ma and Olivier Dehoorne

Abstract Hainan Island, which is located in the southern part of China, is influx of immigrants and is multicultural meeting point, which is rich in cultural tourism resources and local cultural characteristics. Meanwhile, Hainan has a unique history of the development and unparalleled geographical location, and the development of Tuehai Path and Eastern channel high-speed railway, the establishment of SkyCity and Sansha City, the convocation of Boao Forum for Asia and the establishment of International Tourism Island, all gave the Hainan tourist industry a unique opportunity for development. In another words, Hainan cultural tourism has great potential for exploratory. This chapter will present the current status of cultural tourism resources in Hainan, detailed classify Hainan cultural tourism resources, analyse the cultural characteristics of Hainan tourism and provide a reference basis for the ecological tourism development of Hainan International Tourism Island.

Keywords Cultural resources · Tourism culture · Tourism development · Hainan

J. Ma (✉)
School of Tourism, Bournemouth University, Bournemouth BH12 5BB, UK
e-mail: majing0802@outlook.com

S.-Q. Ma
School of Information and Technology, Hainan Normal University, Haikou 571158, China
e-mail: mashengquan@163.com

O. Dehoorne
CEREGMIA-Université des Antilles et de la Guyane, Guyane, France
e-mail: dehoorneo@gmail.com

B.-Y. Cao et al. (eds.), *Ecosystem Assessment and Fuzzy Systems Management*, Advances in Intelligent Systems and Computing 254, DOI: 10.1007/978-3-319-03449-2_38,

1 Introduction

Hainan Province, which is located in the southernmost of China, is only tropical island province in China. Administrative area of Hainan includes Hainan Island, Sansha Islands (Xisha Island, Zhongsha Island and Nansha Island), reefs and its sea area. The land area of Hainan is about 35,400 km^2, the sea area of Hainan is about 2 million square kilometres; and thus, Hainan has the smallest land area and the largest ocean area among all Chinese Province.

The highest peak Wuzhi Mt located in the centre of Hainan Island, the elevation of the land reduces gradually from central mountain area down to cost line. All the rivers originate from central part, so that the river system looks like a radiation. On this sunny island, you can see coconut trees everywhere. The coastline of the island is 1,528 km long, which contains beautiful white and pristine sandy beach with clear blue water and high-end resort hotels and communities. Hainan is blessed with broad variety of tourism resources, endowed with wondrous island physiognomy, pleasant tropical island climate, beautiful ecological environment, charming beach, unique ethnic customs, rich island cuisine and various specialty and other tourism resources, it is a vibrant destination and wonder land of many tourists.

Hainan Island, since the establishment in 110 BC, experienced the history of 2,000 years, has formed the urban system structure which includes three district cities, six county-level cities and 10 counties. The total population reached 9,078,200 in 2011; the majority of the population belongs to the Han ethnic group, followed by Li, Miao, Zhuang, Hui and other ethnic minorities. Ethnic minorities in Hainan have a long history, profound culture and verity folk-cultural forms. Rich cultural tourism resources balanced stunning natural environment and created a unique charm of Hainan tourism culture. The capability of a university decides its competitive advantage and management performance in the essence [3]. Therefore, it does make sense for universities to accumulate, develop, evaluate and utilize their capabilities.

2 The Current Status and Classification of Hainan Cultural

Hainan cultural tourism resources refer to all kind of material or moral wealth created by human [1], from the motivation which could stimulate tourists to travel to Hainan, and be used by tourism to producing social, economic and environmental values [4]. This chapter, based on differences of landscape forms, divides Hainan cultural tourism resources into natural tourist resources and humane tourist resources:

2.1 Landscape Cultural Tourism Resources in Hainan

2.1.1 Cultural Tourism Resources of Physiography Sights

Mountains: Topographically, Hainan Island features hills, mountains and alluvial plains. Therefore, the height of the mountain in Hainan would not be high, the altitude is mostly around 800 m or less and only few peaks are over 1,500 m (the highest peak of Wuzhi Mountain is 1,867 m above the sea level). The hilly area of Hainan Island, generally called 'Wuzhi hilly area', divided into two parts—Wuzhi Mountain (southwest) and Limu Ridge (northeast), by Wanquan River and Changhua River fault valley. Wuzhi Mountain, located southeast of the island, named because it shaped like fingers, is the highest peak on the island. Limu Ridge is located in central part of island and contains clouded canyons and rolling hills which include Yingge Ridge (1,812 m), Limu Ridge (1,412 m), and other peaks. It also include Nanshan, Dongshan Ridge, Bawang Ridge, Jianfengling, Diaoluo Mountain, Tongguling, Houmi Ridge, Baishiling, Hainan Hill, Nanfeng Mountain, Gaoshan Ridge, Qixianling, Ji Gong Shan, Shengui Mountain, Exian Mountain and Baohu Ridge.

Caverns: The famous caverns of Hainan include Sanya Dongtian, Leiqiong Global geopark, Pen Dropping Cave, stone flower tunnel, Maogan Qianlong Cave, Houmi Cave, Emperor Cave. Among them, Sanya Dongtian is the Taoist cultural attraction which located in southernmost of China, reputed as the 'first Scenic Spots of Qiongya over 800 years' from ancient times due to its beautiful view of sea, mountains, rocks and caves, and has become first national 5A-class tourist attractions.

Fossils: Hainan unique fossils include Hainan Ebony, crab fossils and Yellow Agalmatolites. Ebony, known as 'Oriental tree' and 'plant mummy', is the trees buried in mud after earthquakes, floods, landslides and other natural phenomena, under hypoxia, high pressure and affection by microorganisms conditions, were carbonized after 1,000 years. Crab fossils, belonging to the class Cephalopoda, were buried underground billion years ago due to environment changings or crustal movements; only few prehistoric crabs, which have not decayed, could form fossils under favourable conditions. Yellow Agalmatolites are distributed mainly over Wuzhi Mt and peripheral mountains, mostly in Wuzhi Mt and Qiongzhong, Lingshui, Changjiang, Dongfang, Ledong have a small amount of distribution.

2.1.2 Water Landscape and Cultural Tourism Resources

Rivers: Rivers of Hainan Island rises in the central southern mountainous and hilly areas, Nandu River flows northward, Changhua River flows westward, Wanquan River flows east, Lingshui River and Yacheng River flows south. Nandu River, length of 314 km, is the longest river of island and rises in south-west part of island—Jakarta Grand Ridge, flows north-east through Baisha, Danxian, Chengmai, Qiongshan, etc., and runs into the sea in Haikou City. Changhua River is the second longest river on the island, length of 224 km, rises in Wuzhi Mt, flows towards south-west to Ledong then turn to north-west, and runs into the sea in western part of Changhua. Wanquan River, the famous river, is only 100 km long, has two sources: from east part of Wuzhi Mt and south part of Limu Ridge, mingles at Qionghai, and runs into the sea in Boao.

Lakes: Hainan Island is high in the middle and low on all sides, which means it is difficult to form a relatively large natural lake under geographical conditions. The largest lake is an artificial lake—reservoir which formed by intercepting in the middle of the river. There are two major reservoirs, Songtao Reservoir and Guangdaba Reservoir. In addition, the famous lakes also include Nanli Lake, Yunyue Lake, Muse Lake, Wanquan Lake, Dongshan Lake, Haikou People's Park East and West two lakes, and Yu Longquan.

Beach: Hainan Island has more than 1,500-km-long coastline, beaches takes about 50–60 % of all; most of the bays are calm with clear water and white and soft sand. The famous Gulfs include Yalong Bay, the East Sea, Perfume Bay, Shimei Bay, Yunlong Bay, Spring Garden Bay, Boao Harbour, Mulan Bay, Hougang harbour, and Yingge Sea.

Springs: Hainan has varied springs—34 springs have been proven. Spring tour is a speciality of Hainan; Wanning Xinglong Spring, Danzhou Blue Ocean Springs, Qionghai Guantang Spring, Baoting Qixianling spring, and Sanya Nantian Hot spring are the six famous springs among them.

Falls: Hainan Island has charming waterfalls, most of them hidden in the dense primeval forest, for example, Taiping Mt Fall, Baihua Ridge Waterfall, Whitestone Creek Fall, Fengguo Mt Waterfall, Hongkan Waterfall, Yajia Fall, Wanquan River Fall, Lumu Waterfall, Longfen Fall, Dakangling Fall, and Houmiling Waterfalls.

2.2 Hainan Cultural Tourism Resources

2.2.1 Historical and Cultural Tourism Resources

Hainan cultural tourism resources are relatively abundant and unique. Hainan cultural tourism resources include as following aspects:

Cultural Sites: Hainan has rich historical culture and many cultural sites, which include Dongfang Fulongyuan Neolithic shell sites, Xinlongcun Neolithic sites, Shuinan Village of Yazhou Ancient City, old Changhua Town, Zhonghe Ancient City, old Yazhou Town, Ming Dynasty Shuihui Residence Site, Xiuying Fort, meteorite crater and undersea villages.

Museums: The more famous museums in Hainan are Hainan Ethnic Museum and Lingshui County Museum. Hainan Ethnic Museum, located in the north of Yaxu Ridge of Wuzhi Mt, is diversified provincial museum which displayed historical relic of Hainan history and culture of Li, Miao, Hui ethnic groups and studied in Hainan ethnic group's culture. Hainan Lingshui County Museum located in the 'Qiongshan Hall' mainly exhibited the early historical revolutionary remains of Lingshui County.

Cliffside: Inscriptions: A place of deep cultural significance, Hainan boasts many Cliffside inscriptions by historical figures, for example, Ultima Thule Cliffside Inscriptions, Bijia Hill Cliffside Inscriptions, Nanshan Cliffside Inscriptions and Wuzhi Mt Cliffside Inscriptions. Among them, Bijia Hill Cliffside Inscriptions, by combining with the Tang, Song, Yuan, Ming and Qing Dynasty calligraphy and poetry, are the largest Inscriptions in Hainan Province and have more than 100 inscriptions.

Historical figures: Throughout history, there have been a number of historical figures come to Hainan, which include Hairui, Wang Zuo, Zhang Yuesong, Qiu Jun, Su Dongpo, Ma Yuan, Zhao Ding, Soong Ching Ling, Feng Baiju, Yang Shanji, Zhang Yunyi and Bai Mashan. They have been invaluable in the social development and progress of civilization of Hainan.

Colleges: Ancient colleges, undoubtedly, are cultural tourist resort of Hainan. The Dongpo Academy (national), Qiongtai College and Sibei College are the typical representatives of among Hainan Colleges.

2.2.2 Tourism Resources of Architectural Culture

Temples and pagodas are as follows: Chengmai Golden Temple, Boao Temple, Huafen Temple, Lingao Temple, Wenchang Temple, Xitian Temple, Yachengkong Temple, Koushui Temple, Meilang Pagodas, Jianlong Pagodas, Jieyuan Fang, Kuixing Pagodas, Qingyun Pagodas, Lingao Pagodas and Doubin Pagodas.

Cemetery and tombs are as follows: Xisha cemetery of naval battle martyrs, Ancient Islamite Graves, Xing Yu tomb, Former Residence of Hairui, Hairui Tomb, Qiujun Tomb, Zhang Yuesong Tomb and Zhaoding cenotaph, Five Saints Temple.

2.2.3 Artistic Culture Tourism Resources

Hainan Li culture has a history of 3,000 years. The legends, folk music, brocade, dance and sculpture of Li, Miao and other ethnic groups are valuable for development of cultural tourism in Hainan. Ship house, salt producing techniques, Qiong Opera, folk music of Li, and other 21 local folk culture are listed in national nonmaterial cultural heritage lists.

Artistic handicrafts: Hainan local culture is represented by Li culture, which is rich in cultural relics, such as traditional textile, dyeing and embroidery skills, Changjiang Li group ceramics skills with circled clay bar, Li style of fabrics making with tree bark. Nine out of the following handicrafts, Li style's brocade, paper cutting, silver ornaments, straw, bamboo and rattan plaiting, carved bone hairpin, horn carving, coconut carving, shell carving, shell painting, pearl jewellery, Miao style batik, butterfly painting, Lingao puppets, etc., have been included in national nonmaterial cultural heritage.

Drama and Opera: Hainan has a unique Qiong Opera. Lingao puppet drama and other arts forms, including 'The Story of Huaiyin', 'Pipa', 'The Yang Family', 'Apotheosis of Heroes', 'The Hollow', 'Gelsemium Elegans', 'Autumn poem', 'Zhang Wenxiu', 'Searching College', 'Dog Street Concubines' and other classic operas, are invaluable.

Fiction: Local characteristics works are 'Red Detachment of Women', 'Dada Se', 'Luhuitou', 'Huangzaopo' and other influential literary works and comedy.

TV shows Famous TV shows, which are based on local life or historical events, include 'The Liberation of Hainan Island', 'End of the World with Weaver', 'Island Without Winter', 'Su Dongpo in Hainan', 'The Sunset Glow on the Sea' and 'South Island'.

2.2.4 Food Culture Tourism Resources

Hainan food culture has a long and unique history.

Typical dishes: Due to its coastal geography, Hainan has abundant aquatic products. With these aquatic resources as ingredients, numerous appetizing dishes can be easily cooked. Among them, pepper mantis shrimp and grilled oysters are classic specialties in Hainan.

Traditional Cuisines: traditional dishes and special snacks in Hainan, including Wenchang chicken, Jiaji duck, grilled Dongshan goats, Hele crab, glutinous rice cake with coconut shreds, Hainan chicken rice, Hainan fish balls, Hainan four treasures claws, rock grilling lambs, Qiong Shan tofu, Jinhua seafood rolls, green lobster balls, Hainan rice noodle, Hainan pancakes, Li style bamboo rice and Hainan coconut buns.

2.2.5 Folk Culture Tourism Resources

Hainan folk custom, as a special cultural resources, is showing its unique charm in the modern tourism industry.

Traditional Festivals: Hainan has a lot of festival with ethnologic features, such as the Flower Exchanging Festival, Military Slope Day, Dragon-Head-Raising Festival, Li Miao March Day, Hainan International Coconut Festival, Danzhou Folk Song Festival, Hainan Joy Festival, Baoting Water Playing Festival and Li Building Fire Festival. Li Miao March Day is definitely one among them, for pursuing love, celebrating harvest and remembrance of one's departed ancestors.

Temple fairs: By the folk culture, temple fairs gradually become a new model of New Year travel in Hainan. The most unique temple fairs in Hainan include Yuchan Palace New Year fair, Haikou Red Lake Park temple fair, Dingan temple fair, Nanshan temple fair. 'With culture paving the way for tourism activities to promoting traditional culture and succeeding folk art' is the essence of temple fairs culture.

Folk entertainment: Li people's character is unrestrained; they are good at singing and dancing. Li's dance has its own unique charm, for example, Li group's 'bamboo dance' has become the most distinctive dance performances in Hainan. Moreover, the Li group's firewood dance has also become one of the most popular dances. Furthermore, Danzhou ballad, Yazhou Octave instrumental folk songs, Hainan gongs and drums, literature lanterns, etc. have also become the typical representatives of Hainan folk culture.

2.2.6 Religious Culture Tourism Resources

Religion and tourism culture are inseparable. The major religions in Hainan include Buddhism and Taoism.

Buddhism: The famous temples of Hainan include Nanshan Buddhist Cultural Park, Boao Temple and Huafeng Temple. Nanshan Temple is located in Nanshan Cultural District of 'Buddhism Cultural Park', which is 40 km west of Sanya. It is an antique-Tang-style temple in the mountain, facing the sea, known as the 'Buddhist Kingdom on the Sea'. It is the largest newly built temple for the past 50 years and the largest temple in Southland. Boao Temple is a mark of continuation and development of the Buddhism history of Hainan. In AD 748, the monk Tang Jianzhen went to Japan and met the wind drift to Hainan, lived for a year and a half, built a Buddhist temple on the island, and began to spread Buddhist culture. Boao Ancient Town, which is located on the east coast of Hainan Island, left a lot of historical sites and Guanyin Buddhist legends. Huafeng Temple, which is located in Wanning, is an ancient cave based on a large lying concave stone as the wall, giant stone as the top. According to historical records, the Tang Dynasty monk Lei Zhenhai repaired this temple, renamed as the Huafeng Temple.

Taoism: Taoism is combined with Li group's primary believes, become a mainstream religion in Li area, and still affected the Li folk customs and social life. The most famous Yuchan Palace, which is located on piedmont of Wenbi Ridge of Dingan, was the retiring place of Bai Yuchan who was revered as the Fifth Founder of the Golden Pubic region sect of the Southern Taoism branch.

2.2.7 Trade Culture Tourism Resources: Business District 'Arcade'

Arcade derived from the need for the local climate and the commercial form, its existence is related to the surrounding business environment and humane environment. Arcade is not only an architectural culture, but also has cultural and historical significance and commercial economic value. Haikou Arcade is centred at Boai Road, Zhongshan Road, Xinhua Road, Deshengsha Road, Jiefang Road and other five streets, it covers about 2 km^2, and there are more than 600 arcade buildings, big and small. Nowadays, the unique Southeast Asia-style carvings on the arcade are still legible, one shop followed another, there are still have wood floor for walking, with a block of sheltered sun roof, strung together as a long hallway. Haikou arcade is a reflection of early ‘sales in front of workshop’ business model, that contains specific climatic conditions, cultural identity and economic factors.

2.2.8 Academic Culture Tourism Resources

Patriotic education bases: Red tourism has become another trend of the tourism development, Comparing to other tourism forms, it has more significance for patriotic education. Thus it has become patriotism education base. Hainan patriotic education bases including Xisha cemetery of naval battle martyrs, Baima Mountain Memorial Park, Zhang Yunyi General Memorial Hall, the Memorial Park of the Red Detachment of Women, Mrs Xian Memorial Hall, Baisha Uprising Monument, Qiongya Public School Monument, former site of Hainan Qiongya Column headquarter, Murui Mountain Revolutionary Bases and Pavilion, Yang Shanju Former residence, Feng Baiju former residence.

Regional characteristics: The unique geographical location gave Hainan Island the unique charm of the regional characteristics of cultural tourism resources. Such as marine culture, island ecology culture, rainforests, volcanoes culture, ethnic customs and culture, longevity culture, etc. These cultural resources with rich local characteristics and strong vitality, are the fountains of cultural resources for future tourism.

Technology Parks: In recent years it formed a number of preferential policies in Hainan to strengthen the construction of science and technology parks, which mainly include Danzhou National Agricultural Technology Park, Hainan Rare Plant Science and Technology Park, Hainan National 5A-class Eco-technology Park.

College town: Hainan College town is located in Guilinyang, including Qiongtai Teachers College, Hainan College of Economics and Business, Hainan Normal University and Haikou College of Economics.

3 Hainan Tourism Cultural Characteristics

3.1 Natural Eco-friendly Culture, to Enhance the Taste of Hainan Tourism Resources

Since the establishment of ecological province, Hainan has achieved remarkable results in ecological and environmental protection, industrial development, living environment development and ecological culture cultivation. With the rapid economic and social development of Hainan, ecological environment quality still remains the leading level. Hainan provincial government focused on the protection of original ecological situation of Hainan Island, and took a flexible approach in light of local conditions in the process of developing cultural tourism, and actively protected the existing topography and vegetation, built harmony characteristics with the surroundings, created a coordinated islands ecotourism atmosphere, enhanced the taste of Hainan tourism resources, giving passengers comfortable and relaxed enjoyment [2].

3.2 Unique Red Culture, to Enhance the Spirit of Cultural Connotation of Hainan Tourism

The revolutionary history of persisting 'red flag does not fall in 23 years', by Qiongya Special of Chinese Communist Party, the revolutionary stories and songs in film and ballet 'The Red Detachment of Women', makes many people inviting to Hainan Red Tourism. Hainan took advantage of this selling point, with the Red resources of Wuzhi Mt and Murui Mt, established the Memorial Park of the Red Detachment of Women, Yunlong Former Residence, Tankou Anti-Japanese First Shot Blocking Action Sites, Feng Baiju Former Residence, the Major Battlefield of the Inundation Battle sites and other revolutionary tradition education bases, and created a group of red brand to increase unique cultural elements of Hainan tourism.

3.3 Tropical Island Marine Culture, to Characterize Hainan Tourism Products

Hainan Island adjoins the South China Sea, belongs to tropical maritime climate. The sunny climate with atmospheric circulation is ideal for biological growth. Therefore, island is rich for biological and marine resources and marine culture. Shells, pears, coconuts, coffee, tea and costume become unique cultural elements. Hainan took there as a selling point, only coconuts products has coconut powder, coconut candy, coconut carving, ornaments and several kinds of tourism products. These tourism products with characteristics of Hainan, presenting visitors pleasure, and resulting in a strong desire to buy, almost every visitor came to the island cannot be refused to pay for it.

3.4 Modern Fashion Culture, to Colour Hainan Tourism Activities

With the tourism development, Hainan emphasis on realizing the full potential of natural resources and achieving optimal distribution of resources, while injecting in film and television projects, sports, fitness, performing arts and entertainment, and forming a unique "Hainan charm".

3.5 Ethnic Minorities Culture, to Present Tourists with a High Aesthetic Experience

Ethnic Culture in Hainan has done a lot of discovering and serving, such as build international entertainment and world-class performing arts programmes through the introduction, organizing the primitive singing and dancing show which reflected the Li and Miao history and culture, create a number of famous tourist brands, to bring tourists a high aesthetic experience.

3.6 Religious Culture Tourism in Hainan, to Inject New Vitality

During the development of cultural tourism, Hainan clearly recognized that religion culture is distinctive, attractive tourism resources with great tourism value. In recent years, Hainan religious and cultural resources development and utilization has opened up a new tourism market, attracted a large number of tourists. Development of Nanshan Buddhist Culture is a typical example, it not only could compare with the Great Wall, Terracotta Warriors and other scenic spots among the first national 5A-class tourist attractions, but also become the benchmark and symbols in Hainan. Nanshan scenery spot also plans to establish Buddhism institution, Buddhist research institute, Buddhist culture exhibition, to become Buddhist cultural centre with a unique style at home and abroad.

4 Conclusion

Culture is the soul of tourism; the charming environment and unique cultural resources of Hainan Island are substance of building international tourism island. Hainan tourism resources exhibit a specific structural feature—the tropical island scenery and land resource as the based level, beaches, hot springs, mountain and forest as the core level, Li Miao customs, cultural historical sites, other social and cultural tourism resources and other natural resources as the auxiliary level. Rich cultural tourism resources, strong local culture, unique history of the development and location, development of international tourism island have created historical opportunities for Hainan tourism development. The main advantage of eco-tourism resources of Hainan Island is embodied in good ecological value, high quality of the environment, higher scientific, leisure and healthcare value, as well as higher Beach ornamental value, which makes the development of cultural tourism in Hainan has great potential to explore.

Acknowledgments This work is supported by International Science & Technology Cooperation Program of China (2012DFA11270) and Hainan International Cooperation Key Project (GJXM201105).

References

1. Fu, G.J.: Investigation, classification and evaluation of natural tourism resources in Hainan. J. Hainan Univ. (Nat. Sci.) **28**(1), 52–58 (2010)
2. Liao, L.N., Zou, G.Y.: Essential Synthesis of Tourism and Culture. Guangxi Economics, pp. 55–57 (2012)
3. Wang, P., Song, J.H.: Hainan Tourism Geography. Hainan Publishing House, Haikou (2013)
4. Zuo, X.: Assessment and development researches on eco-tourism resources of Hainan island. Chin. Agric. Sci. Bull. **20**(1), 217–223 (2004)

Sustainable Tourism and Ecotourism: Experiences in the Lesser Antilles

Sopheap Theng, Xiao Qiong, Dorina Ilies and Olivier Dehoorne

Abstract This study examines new tourism strategies in small island destinations in the Caribbean. Heavily dependent on a tourism sector that favours seaside activities and faced with socio-economic limits and outdated environmental models, these islands seek to develop policies which will maximize economic activity in a context of sustainable tourism. Beyond the scope of the Convention on Sustainable Tourism Zone of the Caribbean (the STZC Convention), which aims to establish the first sustainable tourism zone in the world, this article analyzes the development of ecotourism on the islands of Dominica (an emerging destination) and St. Lucia (which is in a consolidation phase).

Keywords Ecotourism · Sustainable tourism · Mass tourism · Island · Lesser antilles · Island of Dominica · St. Lucia

S. Theng (✉) · O. Dehoorne
Faculty of Law and Economics, CEREGMIA—Université des Antilles et de la Guyane, Schoelcher Cedex, France
e-mail: sopheaptheng@gmail.com

O. Dehoorne
e-mail: dehoorneo@gmail.com

X. Qiong
Southwest Universtiy for Nationalities of China, Chengdu, China
e-mail: xiaoqiong71@hotmail.com

D. Ilies
University of Oradea, Oradea, Romania
e-mail: iliesdorina@yahoo.com

B.-Y. Cao et al. (eds.), *Ecosystem Assessment and Fuzzy Systems Management*, Advances in Intelligent Systems and Computing 254, DOI: 10.1007/978-3-319-03449-2_39,

1 Introduction

In the tropical islands and coastal areas, the limits of mass tourism are evident. International tourism stimulates economic activity in less developed regions. But these mass flows of tourism and consumption patterns tend to disregard sustainable development in these small countries [1, 2]. The impact on physical and human environments, including rapid urbanization and rising inflation, need to be taken into account in tourism development models. This model of tourism development tends to permit the exit of revenue outside the Caribbean, leaving too little income locally [3]. Through questions on ecotourism and models of alternative tourism more generally, the aim is to rethink tourism development in a sustainable perspective [4, 5]. Tourism experiences in the Caribbean are particularly interesting because there are many cases in which mass tourism in this region has led to economic and environmental failure. More recently, governments and local communities are mobilizing to promote forms of tourism best suited to their territories. The Caribbean has 20 million tourist visits and more than 20 million overnight stays in cruise tourism annually. The Caribbean has seen its revenues multiply by 5 in the last 30 years. These flows are relatively modest in the context of global tourism, but are significant in light of the characteristics of these island territories. The Caribbean islands are tiny (only 235,000 km^2) and support significant permanent populations (nearly 40 million people). Their production systems are much diversified and their natural resources are particularly vulnerable [1, 6, 7]. At the sub-regional level, the highly selective tourist flows primarily favour international locations that can offer the most famous tourist facilities and which meet the essential standards for safety and security [8].

In these island destinations, tourism development has substantial economic, cultural, environmental and social impacts. The contributions of tourism are significant for supplying employment, stimulating investment, improving services and infrastructure, and improving the quality of life in destinations. But the introduction of these new forms of consumption negatively impacts the environment (with an excessive density of buildings in sensitive areas, significant increases in waste, etc.) and host societies (inflation, lack of access to resources, social conflicts) [9, 10].

It is in this context that we must reflect on the state and potential of sustainable tourism in general and ecotourism in particular. First, Caribbean destinations are faced with the obsolescence of their product: Too standardized and targeting customers from traditional source markets, the industry needs to innovate to better meet the demands of environmentally-concerned citizens. Those countries that operate in a highly competitive regional context need to develop an area of cooperation and lay the foundation for a common framework for discussion on sustainable tourism development, facilitating exchange of expertise in the management of tourist places. Finally, ecotourism projects, undertaken as part of a global reflection on sustainable development, are being added to the different tourist strategies of Caribbean destinations.

2 Cruise Ship as Destination

2.1 Ecotourism to Promote Alternative Approaches

The appeal of tropical islands were initially constructed on the basis of the sun, sand and sea, with the stereotypical landscape of sandy beaches flanked by palm trees leading to a turquoise lagoon, in a tropical atmosphere with long sunsets lulled by the trade winds. This perpetual stereotype reflects the entire standardized logic of mass tourism and tourism consumption that has been imposed in the Caribbean. Today, the need to transform the tourism product in response to changing customer expectations in general and new client niches in particular. The search for alternative tourism policy also reflects a desire to redevelop these areas, which have gradually lost their allure due to high concentrations of tourist consumption and subsequent pollution. Redeployment flow is another approach that favours landlocked inland territories, which authorizes the use of tourism as a planning tool—as was done in some European countries [11]. This approach allows the integration of new resources by reconsidering the economic framework to improve benefits for the local population.

It is in this context that initiatives, more or less innovative, are being deployed in the interiors of some Caribbean islands. The inclusion of land resources facilitates the design of new tourist products, often categorized as ecotourism. Various theories propose forests as places both feared and desired (the "Green Hell"), through a variety of uses: for leisure (like raids and other courses in the trees), for research and education (identification of species) or ethno-tourism (meetings with traditional/indigenous groups that live in the forest); during day trips (as part of a beach-focused holiday) or for specific periods of time [12, 13].

More than ever, given the current limitations of mass tourism, island destinations are aware of the need to engage in alternative, more responsible forms of development. These insights explain the enthusiasm of ecotourism stakeholders, with the vague focus that accompanies the definition [14]. The most accepted definition in academia is probably that of Mexican environmentalist. Ceballos-Lascurain (1987), who defines ecotourism as "a form of tourism that involves visiting relatively intact and undisturbed natural areas with the specific aim of studying and admiring the scenery, the wild plants and animals it houses, as well as any cultural event (past or present) observed in these areas" [14].

Interest in approaches and ecotourism experiences is uneven, the theoretical constructs must adapt to the economic and social realities of the territories of experimentation. As highlighted by Ziffer [15] and Couture [16], this form of tourism "draws primarily on the natural history of a region, including its indigenous cultures, which also requires active management by the country or host region, which undertakes to establish and maintain sites with residents, ensure regulatory and affect the company's revenue to finance land management and community development will be part of the character of the host territory". Hence, as pointed out by Dehoorne and Transler [5], the concept of "encounter" is important. The success of ecotourism, which is based on the empowerment of tourists and host societies, lies in a form of sustainability that can only be realized through encounters with the companies and individuals operating in these environments. These meetings contribute to the establishment of more equitable and inclusive relationships.

Enhancement of natural and cultural heritage, sustainable production systems, improving the living conditions of host communities, integrating decision-making processes, increasing visitor accountability and strengthening relationships between tourists and local companies are all concerns that motivate ecotourism. This development model allows to extricate local development processes from the income generated by tourism consumption more thoughtful and organized.

And more broadly, as noted Lequin [17], ecotourism is a way to travel that "represents a new wave of thinking about development and the tourist experience that is based on the principles of sustainable tourism i.e., a form of eco-tourism, in the broad sense, meaning both resource protection, respect for cultural identity and empowerment of stakeholders". The challenge for ecotourism is to balance economic development and sustainability of ecosystems, taking into account the needs of host communities and building the conditions of a meeting between hosts and guests [5, 16, 18].

2.2 A New Perspective of Tourism Resources

Through the perspective of ecotourism, a new assessment of resources extends the tropical atmosphere to the sugar canes, bananas and volcanic landforms of interior island landscapes. This focus on island interiors affords new perspectives on developing land resources in the volcanic islands such as Guadeloupe (the area of Basse-Terre, the Soufrière Hills volcano, rises to 1,467 m above sea level) Dominica (1,445 m from Morne Diablotins), Martinique (1,397 m at the summit of Mount Pelée) and in the southern Caribbean Arc, St. Lucia, St. Vincent and Grenada. Some volcanoes remain active, permitting fumaroles and sulphur springs (like those on St. Lucia) to escape. The island of Dominica is advantageously positioned compared to the Lesser Antilles with eight volcanoes, including one of which houses the second-largest boiling lake in the world.

These small territories offer unique landscapes that evolve with the altitude, changes in precipitation amounts and exposure and profile slopes. The staging is asymmetrical between the windward and leeward coasts. On the same island, formations ranging successively xerophytic thorny scrub, dry forest and the rainforest hygrophile and sometimes even nebelwald can be witnessed [1]. Endemism is particularly developed and the islands demonstrate a great wealth of biodiversity. For example, more than 1,700 species of flowers can be found on the island of Martinique, including bougainvillea, hibiscus, Gabon tulips, and wild *anthuriums* and *heliconias*. These environments are also conducive to the birds as on the island of Dominica, which has about 170 species. This includes permanent species, the most notable of which are the hummingbird, imperial parrot or 'Sisserou' (Amazona imperialis), the red neck 'Jaco' parrot (Amazona arausiaca) and the "malfini" eagle or mountain plover. They are also large frigates (with glowing male crop) as well as various herons and pelicans. The rich fauna also includes lizards, iguanas and reptiles with the boa constrictor of Dominica (which can be between 4 and 5 m long) or the trigonocephalus (or "iron spear" triangular head) of Martinique and St. Lucia.

This biodiversity allows for innovation in tourism and recreation planning in general, although development thus far has been slow. However, given a new profile of tourists that are more curious, more athletic and better equipped, the realization of adequate facilities is essential to both to make these inland features accessible and to protect them from deterioration (and the loss of economic value that would result). The Caribbean states in the framework of regional cooperation intend to equip themselves with common tools for tourism management, both for formerly frequented areas and new sites of exploration, as part of a sustainable management perspective.

3 A General Framework: The Sustainable Tourism Zone of the Caribbean

Reflection on the sustainability of tourism among the Association of Caribbean States (ACS) led to the signing of the Convention on Sustainable Tourism in 2001 at the Second Summit of Heads of State and Government on the island of Margarita (Venezuela). This agreement lays the foundation for the creation of the Sustainable Tourism Zone of the Caribbean (STZC), which is the first area of sustainable tourism in the world. The aim of the 26 member states and associated signatories is to provide a legal framework for the promotion of sustainable tourism in the Greater Caribbean, focusing specifically on the promotion of heritage and ecotourism [19].

3.1 Political Positioning

Born out of the concept of sustainable development, sustainable tourism is understood as covering all levels of politics and tourism practices at global and local levels. The definition used in the Declaration of the STZC specifies that "sustainable tourism must respond adequately to the challenges of rising unemployment and currency flows, the protection and preservation of the environment and natural resources, heritage conservation and the preservation of cultural values". Moreover, sustainable tourism should support "community participation and involvement of local interest groups in the process of tourism development, such as policy formulation, planning, management, ownership and distribution of profits from this activity."

This convention recognizes the scope of the tourism sector as a considerable source of employment (particularly on islands facing social conflict), foreign direct investment and foreign exchange earnings. Accordingly, under the STZC, it is not a matter of limiting the flow of visitors, but rather to circumscribe tourist programs with the aim of sustainability. The convention also aims to address increasing levels of pollution caused by uncontrolled tourism development that is causing the degradation of heritage resources and loss of wealth and, in the medium term, will destroy the heritage of Caribbean countries [18].

This approach to tourism development in the perspective of sustainability should define adequate policies to protect natural areas and local cultures and additionally, the contributions of tourism to the preservation of the environment and societal development. In this sense the concept of sustainability is understood as encompassing the protection of biodiversity as well as of culture and the environment. Within the context of tourism development, sustainability identifies the human being as the centre of focus in order to facilitate a more equitable distribution benefits.

"Sustainability" is thus approached from several different angles. While supporting the development of the tourism market, the objectives are to integrate environmental protection and Caribbean cultural identity into the industry, stimulate the participation of local communities and promote technological solutions to waste management, including alternative energy production. Lastly, sustainability aims to develop integrate these approaches in a legal and regulatory framework in order to adopt appropriate policy instruments.

3.2 From Theory to Practice: Defining Indicators

The methods for implementing sustainable tourism policy requires the definition of common indicators and tools to identify trends in the context of territorial diagnosis and support local processes, adapted to the realities and objectives of the territories (see Table 1). Normative indicators are common to all parties and comply with the Convention establishing the STZC, while local codes are defined by the policies of tourist destinations.

The indicators are based on three dimensions: Heritage (for all aspects of the environment, culture, way of life), economic (the management of financial and material resources), and social (human aspects related to tourism). These indicators include one or more sensor elements, based on qualitative and quantitative criteria, which are likely to evolve over time as a function of tourism in the concerned area. An ethical approach is introduced in work undertaken at the local level and particularly in the social dimension where the emphasis is on community involvement, transparency in decision-making and access to information, and

Table 1 The dimensions of sustainability in tourism

Heritage dimension	Economic dimension	Social dimension
Criteria		
• Renewal of resources	• Profitability, territorial integration of tourism	• Community participation (decision making, planning and distribution of tourism benefits)
• Environmental preservation and management	• Community interest	• Well-being, dignity and respect
• Valuation of assets (natural, historical, artistic and local/cultural)	• Flexibility, adaptability and innovation	• Education, awareness and training
• The quality of the living environment	• Control of tourism development	• Aid initiatives
		• Accessibility and inter-regional transport systems
Normative indicators (common to all destinations)		
• Energy consumption and management	• Integration of the local economy	• Employment origins in the tourism sector
• Water consumption	• Tourist satisfaction	• Child prostitution
• Water quality		• Tourist security
• Access to safe drinking water		
• Environmental preservation and management		
• Control of environmental impacts		
• Effectiveness of solid waste treatment systems		
• Effective wastewater management and treatment		
• Valuation of identity and culture		
Local indicators (determined by destination)		
• Inventories of flora and fauna	• Tourist traffic/frequency	• Labour laws in business
• Vulnerable areas	• Length of the season	• Information transfer
• Valuation of traditional architecture	• Sustainable Tourism Network–Research	• Corporate communication

(continued)

Table 1 (continued)

Heritage dimension	Economic dimension	Social dimension
• Landscape valuation	• Cooperation–Coordination	• Level of training
• Scents and sounds	• Observational apparatus	• Training system
• Traffic	• Estimating device	• Network and sales distribution
• Agricultural activities	• Marketing strategy	• Support for business creation
	• Pluriactivity	• Air and sea links
	• Planning	• Support for the promotion of sustainable development in enterprises
	• Public authority	• Support for the promotion of traditional knowledge
	• Economic power	• Access to people in need
	• Taxation	• Diversity of supply
		• Commercial activities and service

(*source* CEA)

solidarity and cooperation among all stakeholders, for example in the definition of local objectives.

The drivers selected for the implementation of tourism sustainability indicators under the STZC can be grouped into three categories:

1. Significant levels of tourism development. The concentration of tourist activity in small areas occasionally impacts the quality of resources and thus tourist revenue. These destinations must diversify their tourism product beyond the beach and open new areas for tourism consumption as part of a policy of controlled development oriented toward sustainability. The most significant examples are the beach resorts in northern Dominican Republic, Cancun and the Riviera Maya, and the Jamaican coast of Ocho Rios as well as smaller islands like St. Lucia.
2. Destinations which have not yet attained a position on the international market. They intend to use this opportunity to develop the tourism sector using the concept of sustainability and ecotourism, in the manner of Costa Rica in the late 1970s. These countries are in a prototype phase, as the tourist islands of Dominica, the Guatemalan department of Izabal, and Trinidad and Tobago (through the development of the national parks of Chaguaramas and Matura).
3. Declining destinations which are no longer able to compete against the larger mass tourism destinations of the moment. This category mainly consists of the French West Indies, where one of the outcomes of the current crisis could lead to conversion to ecotourism and green tourism more generally.

4 Ecotourism at the Heart of a New Strategy

4.1 Ecotourism as a Basis for Development in Dominica

The island of Dominica, located between the French territories of Guadeloupe and Martinique, is an emerging tourist destination (see Fig. 1). The late entry of this small state, independent since 1978, on the tourist scene is partly explained by the absence of the traditional resources enjoyed by other tropical islands that have had success with tourist resorts [20, 21]. Dominica, a relatively isolated country, is sparsely populated (less than 80,000 people in an area of 750 km^2 in 2005) and is among the poorest in the region. This island, which has experienced very little anthropization, is the most mountainous of the Windward Islands and possesses only a few black sand beaches. Since the 1990s, the government of Dominica has developed a tourism strategy based on the uniqueness and quality of its natural and human resources [22]. Dominica will never be "the island of 365 beaches "(as advertised by the island of Antigua), but instead it strives to become an "island of 365 rivers". The aim is to mobilize all local resources under a matrix of tourism development with an emphasis on ecotourism.

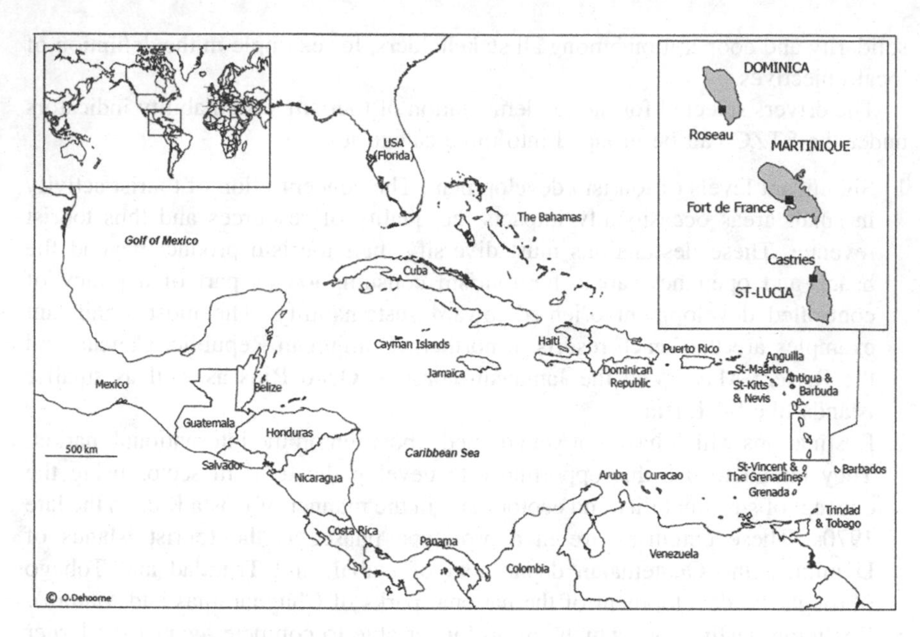

Fig. 1 The Caribbean and the Lesser Antilles

Official communications stress the "virginity" of the island, with its lush forests (which dominate 62 % of the island), the fumaroles and hot spring of its volcanoes, and its many rivers and waterfalls. Promotions for the island focus on some major sites such as the Boiling Lake, Emerald Pool, Trafalgar Falls and the Soufrière Sulphur Springs, and planning is under way to form a network of protected areas and natural parks which will extend over one fifth of the country. Dominica is therefore the epitome of a sustainable development model in action, relying on intelligent resource management. Ecotourism supports the creation of micro-enterprises based on family and community, supporting specific jobs (guide, crafts) within local communities and improving habitats through small accommodation units with an appreciation of the role of women.

Tourist flows in Dominica remain modest (less than 85,000 tourists in 2006, CTO, 2007), but growth has been steady despite the complexity of air paths and the high cost to reach the destination without charter flights. Dominica has developed a reputation in the field of sustainable tourism, as evidenced by the Green Globe 21 certification obtained in 1997.

Under these conditions, ecotourism is a first step to initiate economic development of the country. Tourist consumption has a significant impact on training, other sectors of production (such as small-scale farming) and allowing revenue flows to remain among the local population. Sustainable tourism projects contribute to awareness of environmental and cultural challenges and consolidate territorial cohesion [23, 24]. And beyond the tourism sector alone, it is part a strategy for sustainable development of the whole island.

4.2 *From Mass Tourism to Ecotourism*

Caribbean tourist destinations, including those with classic seaside-based sectors, are diversifying their services through ecotourism. Some destinations do this by customizing the tourism product or by emphasizing territorial realities, such as forest landscapes and volcanic terrain. These destinations must continue to strengthen their position and undertake strategies to increase revenue and improve the distribution of that revenue in their areas by opening new areas of tourism activity. The environmental commitment can, in such circumstances, prove to be superficial.

The problem of the island of St. Lucia is similar to those of other destinations in the Greater Antilles. This small country (165,000 people in an area of 616 km^2) experiences consistent tourist flows (302,000 tourist stayovers and 360,000 cruise passengers in 2006) that are confined to the north-western coast (Rodney Bay and vicinity). The tourism sector, which developed after the island achieved independence in 1979, is experiencing a decline in agricultural exports (the agricultural sector makes up 15 % of GDP and employs more than 10,000 people). However, economic, social and environmental limits necessitate a reconsideration of the island's approach to tourism development: Areas of tourist concentration belong to multinational groups with objectives that differ from those of the government. Local job creation in tourist accommodation is essential, but the economic contribution, given the unstable nature of tourist accommodation and consumption, is far from satisfactory [25]. The concentration of tourists on beach resorts is not conducive to the diffusion of tourism revenue throughout the whole territory. The government has begun planning the St. Lucia Heritage Tourism Programme in order to change the tourism sector so that it corresponds with the economic and cultural needs of Saint Lucians. The program focuses on the spatial distribution of tourism practices through ecotourism initiatives, for example along the Caribbean coast (as in the Soufrière Marine Reserve) and the volcanic mountains at the center of the island. It is an approach that is meant to complement the mass tourism practiced in Rodney Bay. This model of sustainable tourism reflects a compromise between, economic imperatives and environmental and cultural preservation.

5 Conclusion

Tourism is essential for economies in the Caribbean, a region which faces fierce competition in the sector, characterized by a plethora of identical products with sales pitches based primarily on the cost of benefits. Faced with the current limitations of mass tourism, Caribbean states are developing a conceptual framework to support a sustainable tourism policy through the promotion of alternative practices often categorized as ecotourism.

Development projects and resources are defined in terms of economic imperatives in these small independent states, which face significant rates of underemployment and debt. Sustainable development initiatives attempt to reconcile tourist practices with development and conservation but are impeded by financial limitations, especially in small independent islands such as Dominica and St. Lucia. The question of income for sustainable tourism development is essential.

Ecotourism can offer a unique tourist product that is synonymous with environmental quality and authenticity of place. Emerging destinations are marketing themselves as "green islands" to assert their originality in the international tourism market. As such, ecotourism is a potential first step in the development of tourism in these countries. The circumstances are completely different for areas characterized by mass tourism which measures certain limits of their approach to development. Whether or not significant tourism flows are necessary to be able to generate significant employment and the revenues in order for further development to be realized on these island territories, the flows must nevertheless be part of a rational development project to be able to generate real spillover effects.

References

1. Dehoorne, O.: Le tourisme dans la Caraïbe. De La Nécessité de Sortir du Tourisme de Masse. Téoros **26**(1), 3–5 (2007)
2. Hiernaux, D.: Le tourisme de masse au Mexique : un virage? Téoros **26**(1), 15–20 (2007)
3. Hillali, M.: Le tourisme international vu du sud:essai sur la problématique du tourisme dans les pays en développement. Presse de l'Université du Québec, Québec (2003)
4. Breton, J.-M. (ed.): Ecotourisme et développement durable (2001)
5. Dehoorne, O., Transler, A.-L.: Autour du paradigme de l'écotourisme. Etudes Caribéennes **6**, 13–26 (2007)
6. Weaver, D.: Ecotourism in the Caribbean basin. In: Cater, E., Lowman, G. (eds.) Ecotourism: A Sustainable Option, pp. 159–176. Wiley, New York (1994)
7. Island Resource Foundation (IRF): Tourism and Costal Resources Degradation in the Wider Caribbean. Island Resources Foundation, St. Thomas (IVA) (1996)
8. Hall, C.-M., Timothy, D.-J., Duval, D.-T.: Safety and security in tourism: relationship, management and marketing. Co-published J. Travel Tourism Mark. **15**(2/3/4) (2003) (The Haworth Hospitality Press, New York)
9. Stonich, S.: The Other Side of Paradise: Tourism, Conservation and Development in the Bay Islands. Cognizant Communication Corporation (1999)
10. Dehoorne, O.: Le tourisme dans l'espace caribéen: logiques des flux et enjeux de développement. Téoros **26**(1), 6–14 (2007)
11. Béteille, R.: Le tourisme vert. PUF, Paris (1996)
12. Augier, D.: L'écotourisme à la Martinique, Mém. Master économie. Université des Antilles et de la Guyane, p. 108 (2005)
13. Murat, C.: L'écotourisme comme alternative touristique. Eléments de réflexions à partir de l'exemple dominicais, Mém. Master Economie, Université des Antilles et de la Guyane, p. 127 (2007)
14. Lequin, M.: Ecotourisme et gouvernance participative. Presses de l'Université du Québec, Montreal (2001)

15. Ziffer, K.: Ecotourism: The Uneasy Alliance. Conservation International and Ernest & Young, Washington, DC (1989)
16. Couture, M.: L'écotourisme, un concept en constante évolution. Téoros **21**(3), 5–13 (2002)
17. Lequin, M.: L'écotourisme. Expérience d'une interaction nature-culture. Téoros **21**(3), 38–42 (2002)
18. Smith, V.L.: Hosts and Guests: The Anthropology of Tourism. University of Pennsylvania Press, Pennsylvania (1989)
19. Girvan, N.: Rencontre à Margarita. URL http://www.acs-aec.org/PressCenter/GreaterCaribbean/French/index13.htm, consulté le 27 décembre 2009 (2001)
20. Wilkinson, P.F.: Tourism Policy and Planning: The Case Studies from the Commonwealth Caribbean. Cognizant Communication Corporation (1997)
21. Weaver, D.B.: Managing ecotourism in the Island microstate: the case of Dominica. In: Diamantis, D. (ed.) Ecotourism: Management and Assessment, pp. 151–163. International Thomson Business Press, London (2004)
22. Patterson, T., Rodriguez, L.: The political ecology of tourism in the commonwealth of Dominica. In: Gössling, S. (ed.) Tourism and Development in Tropical Islands. Political ecology Perspectives, pp. 60–87. Edward Elgar Publishing, Northampton (2003)
23. Wiley, J.: Dominica's economic diversification: Microstates in a Neoliberal Era. In: Klak, T. (ed.) Globalization and Neoliberatism: The Caribbean Context, pp. 155–178. Rowman & Littlefield, Lanham (1998)
24. Patterson, T., Cousins, K., Kraev, E., Gulden, T.: Integrating environmental, social and economic systems: a dynamic model of tourism in Dominica. Ecol. Model. **175**, 121–136 (2004)
25. Wilkinson, P.-F.: Tourism policy and planning in St. Lucia. In: Gössling, S. (ed.) Tourism and Development in Tropical Island, pp. 88–120. Edward Elgar, Cheltenham (2004)

Cruise Tourism: Global Logic and Asian Perspectives

Olivier Dehoorne, Corina Tatar and Sopheap Theng

Abstract Cruise tourism is an industry with significant growth (17.6 million passengers in 2011). This economic sector, which is currently dominated by North American clients who prefer destinations in the Caribbean Sea, is diversifying throughout the world. The cruise industry is strengthening its position in the Mediterranean and Baltic seas and is increasing deployment in East Asia from the shores of the South China Sea to the major rivers. Between continuity and reconstruction, the cruise industry faces several challenges such as the limits of gigantism (more than 6,000 passengers), its relationship with the host territories (ports of call) and the renewal of product "cruise".

Keywords Cruise · Cruise industry · Product · International tourism · Mass tourism · Fordism · Gigantism · World

1 Introduction

Each year, the cruise ship industry's share of international tourism increases: More than 17.6 million international tourists chose cruise ships in 2013 (CLIA—Cruise Line International Association). It has had an annual growth rate of 7.5 % since 1980. This reflects substantial growth since the early 1970s, when there were just

O. Dehoorne (✉) · S. Theng
Faculty of Law and Economics, CEREGMIA—Université des Antilles et de la Guyane, France, France
e-mail: dehoorneo@gmail.com

S. Theng
e-mail: sopheaptheng@gmail.com

C. Tatar
University of Oradea, Oradea, Romania
e-mail: corina_criste_78@yahoo.com

B.-Y. Cao et al. (eds.), *Ecosystem Assessment and Fuzzy Systems Management*, Advances in Intelligent Systems and Computing 254, DOI: 10.1007/978-3-319-03449-2_40,

over 100,000 cruise ship clients. Demand is very strong in this sector, where a consolidation process undertaken in the early 2000s has led to the establishment of an oligopolistic market. Favourable growth prospects drive investment, which are polarized between three actors that dominate the market: Carnival Corporation, Royal Caribbean Cruises and Star Cruises Group [1].

If the first cruises began on transatlantic routes between Europe and the Americas in the late nineteenth century, the definition of the cruise ship product is accurate in the 1960s. The first official cruise ship was the Normandy (1935), which connected New York to the European continent in four days and three hours. After the Second World War, the number of maritime transatlantic passengers declined as air transport became more common. The number of cruise passengers fell from one million in 1956 to 132,000 in 1973. In the late 1960s, the cruise industry began to develop as a specific tourism product: A ship dedicated exclusively to transport tourists, with more room for immigrants on board and its own terms of use and specific time frames. With specific geographic scopes and circuits, the liner was designed as the first manifestation of the cruise ship [2–4].

The nature of cruise tourism as we know it today is relatively recent. Like all tourism practices, it gradually democratized from an activity reserved for the wealthy elite to a form of mass tourism [4–6].

This study proposes to measure the growth of cruise tourism and analyse the logic of the cruise industry and its deployment worldwide.

2 Cruise Ship as Destination

The cruise ship is not merely a means of transport connecting stops, more or less arranged for the occasion, which motivate the movement of the visitor. The ship is part of the pleasure and is even central to the tourist experience. This fundamental distinction determines the originality of the cruise ship: The means of transport is the main place of residence from which tourists might depart for day or half-day trips. Cruise ships are tourist destinations in themselves, which overshadow the host territories (ports of call) which they replace [3, 7].

The cruise ship is a floating resort with all the services and amenities required to meet client needs [3, 4, 8]. The cruise ship is a mobile resort [2, 6]. As a festive and secure place from which to view exotic landscapes, cruise ships have become the destination. Tourists do not travel on any ship; they board the "Boreal" (Compagnie du Ponant), Oasis of the Seas, Allure of the Seas Royal Caribbean International (RCI), etc. Marketing campaigns focus on the ship, its amenities and its multiple services [3, 4, 6] which make the ports of call more attractive.

A product of Fordism, the advent of mass cruise tourism developed in accelerated stages based on classic distinction and imitation [9] applied to tourism practice [5]. The growth of cruise tourism is based on a remarkable economic strategy that depends on the "desire of somewhere else" [10–12], more or less distant, comfortable and secure, approaching otherness without being overly exposed [17] in a festive environment.

With its standardization of consumption and product, the cruise industry is one of mass tourism. The sector's activity is controlled by two major groups: RCI, founded in 1968, and carnival cruise line (CCL), founded four years later. More recently, in 1994, the Walt Disney Company launched its own line called Disney Cruise.

3 A Booming Market

The primary organization developing statistical data to monitor developments in the cruise industry is the CLIA (founded in 1975,[1] CLIA monitors 26 cruise companies). The cruise industry appears to be booming, but attention should be paid to discrepancies in statistics between the numbers recorded by different cruise lines under CLIA and the total number of cruise passengers reported by each host country. In the latter case, cruise tourists are actually day-trippers (they stay in the host country for less than 24 h). The total number of cruise tourists reported by receiving countries (ports of call) should not be confused with the number of tourists reported by the CLIA, which corresponds to the actual number of cruise tourists.

Cruise lines reported a total of 350,606 beds in 2013. While their collective accommodation capacity represents only 6 % of tourist accommodations available in hotels around the world (WTO—World Tourism Organization), the occupancy rate for cruise accommodations is approximately 103 %. The occupancy rate is calculated on the basis of two passengers per cabin, but many of these cabins are occupied by families with three or four people, contributing to an occupancy rate which exceeds 100 %.

In this context, accommodation capacities in the cruise industry capacity have increased dramatically. In the early 1980s, the supply of cruise accommodations was fewer than 50,000 beds. This number has increased from 100,000 beds in 1992 to 150,000 beds in 1999. Accommodations have continued to increase from more than 200,000 beds sold in 2003, to 250,000 in 2006 and over 300,000 in 2010 (see Fig. 1).

From 2000 to 2013, 167 new cruise ships have been added to the industry. In 2012 alone, 14 ships, with a collective capacity of approximately 16,700 beds, were added.

Naturally, this demand has led to fierce competition among cruise companies and increased capacity in each vessel: Cruise companies began putting into service 220,000 tonne ships, with 6,600 beds in each [8, 10]. Reception capacity has increased from 2,500 passengers (as on the Grand Princess) to 3,360 passengers

[1] Founded in 1975, CLIA is a non-profit organization which represents the interests of 26 member companies. It participates in the regulatory processes and policies of the cruise industry and communicates across a network of roughly 14,000 travel agencies (about 14,000). The two main sources for this study are CLIA and the World Tourism Organization (WTO, 2003).

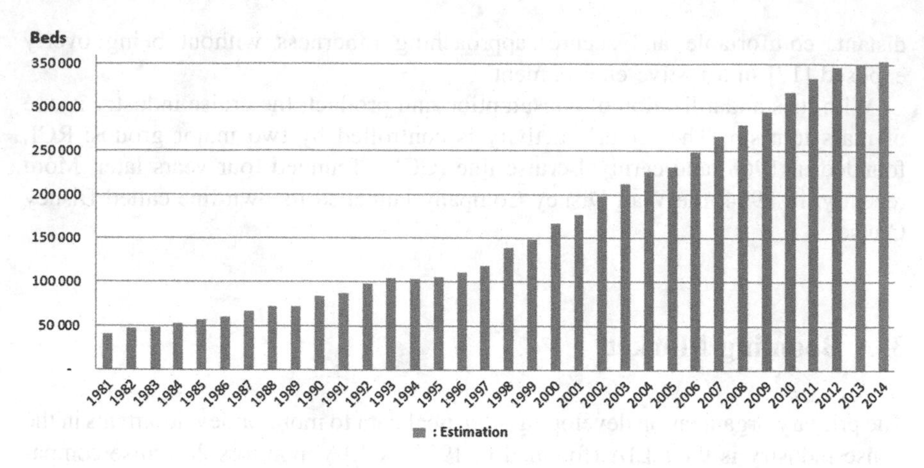

Fig. 1 Evolution of the cruise industry's accommodation capacity (1981–2014)

(Carnival Destiny) to stabilize around 6,000 beds (Oasis of the Seas and Allure of the Seas) [4].

These gigantic ships (with a 6,000 bed capacity) have allowed a decline of the traditional cruise product in favour of new customer segments. The economies of scale realized on such vessels have democratized the cruise industry and contributed to the timely development of mass cruise tourism: 10,000–12,000 people disembark on the same day on the same time slots at a stop on small Caribbean islands. This massive phenomenon, focused on short passages of time, necessitate a reconsideration of the construction of ports of call and the types of consumption that logically evolve from the uniformity and standardization that characterizes the cruise industry.

Between 1980 and 2012 cumulatively, the cruise industry carried 225 million tourists (including 188 million North Americans). As Fig. 2 demonstrates, the number of cruise passengers has exploded since 1990. During the 1990s, the total number of tourists onboard grew annually by 80 % to reach 7.214 million people in 2000. Cruise tourism growth was 106 % during the following decade, with an annual output slightly less than 15 million in 2010.

This sector has not been affected by the economic recession. On the contrary, crises and political tensions seem conducive to cruise tourism as it provides a secure and safe environment. The international tension and events following the terrorist attacks of September 11, 2001, which negatively affected many tourist destinations, actually strengthened the cruise industry, which experienced a 19.5 % growth in the number of passengers during the period of 2000–2002.

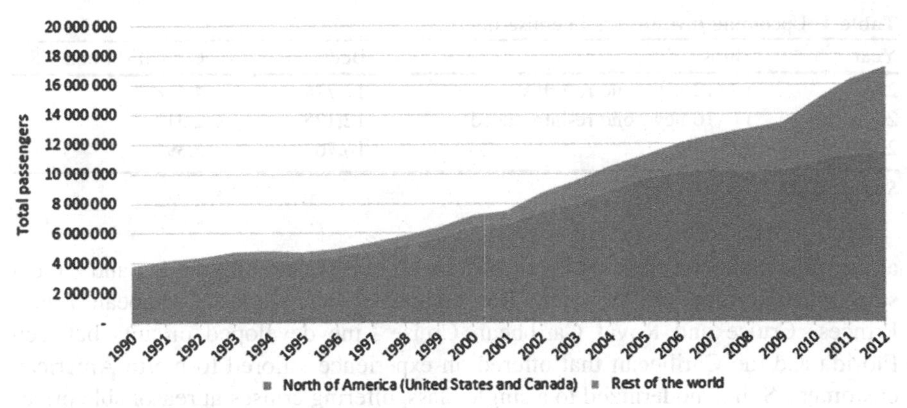

Fig. 2 Trajectory of cruise tourism (1990–2012)

4 Cruising Routes Globally

Cruise destinations remain concentrated in the Caribbean and Mediterranean basins, while American clientele continue to dominate demand on global cruise tourism [4, 8, 11]. However, the potential for emerging client markets are considerable, as demonstrated by spatial reconfigurations currently taking place that serve to renew and diversify the cruise industry.

4.1 The Inescapable Weight of North America

North American customers (the USA and Canada), who represented 67 % of cruise clientele in 2013, are fundamental for the cruise industry. In a context of global growth in cruise passengers, their relative share is shrinking from 91 % in 2000 (7,214,000 individuals) to 83.9 % in 2006 (12,010,000 individuals) and 72.8 % in 2010 (14.82 million individuals). The weight of North American clients remains significant but is decreasing due to the expansion of the cruise industry in Europe and the South China Sea.

The domination of the North American market can be explained by several factors. The first cruises, which were organized in the northern Caribbean (Bahamas, Puerto Rico, Cuba), were driven by clientele in New York. This demand for cruise ships doubled during Prohibition (the Volstead Act, 1919–1933), as cruise ships allowed for the legal consumption of alcohol in international waters. At the same time, development of the first regional circuits also began in the Baltic and Mediterranean seas.

The development of mass cruise tourism in the North American market emerged in the 1960s, with destinations concentrated in the Caribbean Sea. It was

Table 1 Upcoming investments in cruise industry

Year	Ships	Beds	Capital (billion US$)
2012	13 (12 new, one refurbished)	17,774	4,418
2013	11 (10 new, one re-introduced)	12,125	2,317
2014	13	16,702	3,395

Source CLIA

at this time that entrepreneurs decided to invest in this fledgling market and build a substantial commercial product. The companies Norwegian Caribbean Lines, Princess Cruise and Royal Caribbean Cruise Line developed circuits between Florida and the Caribbean that offered an experience tailored to North American customers. Ships modernized to a single class, offering cruises at reasonable prices in a relaxed environment that was in stark contrast to the more conservative atmosphere of pre-war cruises. Robust economic growth and increased demand for leisure among the retiring baby boomer generation situated cruise ships at the centre of the North American tourist market. In North America, the facilities required for cruise ships are accessible, including airline service between ports of departure and clientele cities and an impressive hotel infrastructure to group tourists on the eve of their departures. It is also easy for North American clients to purchase a cruise for a weekend. The cruise industry is able to offer North American clients the Caribbean sun in the middle of the northern winter.

The perception of cruises as stable and accessible products is facilitated by the media. One example of this is the TV series "The Love Boat" in 1970, which depicted life on a cruise ship and popularized cruise practices. Now, 6.4 % of US nationals who go on vacation choose cruises.

The total number of cruise passengers originating in other parts of the world, from Europe to Asia, has not exceeded 40 million people in the last 22 years. But recent progress in European and East Asian markets has contributed to the global diffusion of the cruise industry. Global growth in the cruise sector (10.4 %) is more sustained than in the North American market alone (4.5 % see Table 1).

In 2011, 6.2 million Europeans were cruise tourists, accounting for 30 % of all cruise passengers globally (an increase from 23 % in 2006). European cruise ports took on in 5.6 million passengers (not exclusively Europeans). For its part, the booming Asian market has a total of 1.7 million cruise passengers on board or 10.4 % of global activity and is projected to reach 7 million cruise passengers by 2020 (Asia Cruise Association) (Table 2).

Overall, the expansion of the cruise industry on various national markets highlight its growth prospects in North America (where the rate of market penetration in the United States and Canada are, respectively, 3.26 and 2.01 %), Europe (with a rate of 1.49 % in Germany) and the rest of the world.

Table 2 Penetration of the cruise industry by country (2011)

World rank	Country	Population (millions)	Cruise revenue (millions)	Penetration rate (%) (cruise revenue/ population)	Penetration rank
1	USA	312.10	10.448	3.348	1
2	UK	63.07	1.700	2.695	3
3	Germany	81.80	1.388	1.697	6
4	Italy	60.80	0.923	1.518	8
5	Canada	35.00	0.763	2.180	4
6	Spain	46.20	0.703	1.522	6
7	Australia	22.62	0.623	2.754	2
8	France	65.40	0.441	0.674	11
9	Scandinavia	25.50	0.306	1.200	10
10	Benelux	28.00	0.159	0.568	12
11	Switzerland	7.90	0.121	1.532	7
12	Austria	8.40	0.104	1.238	9
Total	Global total		19.400		

Source CLIA

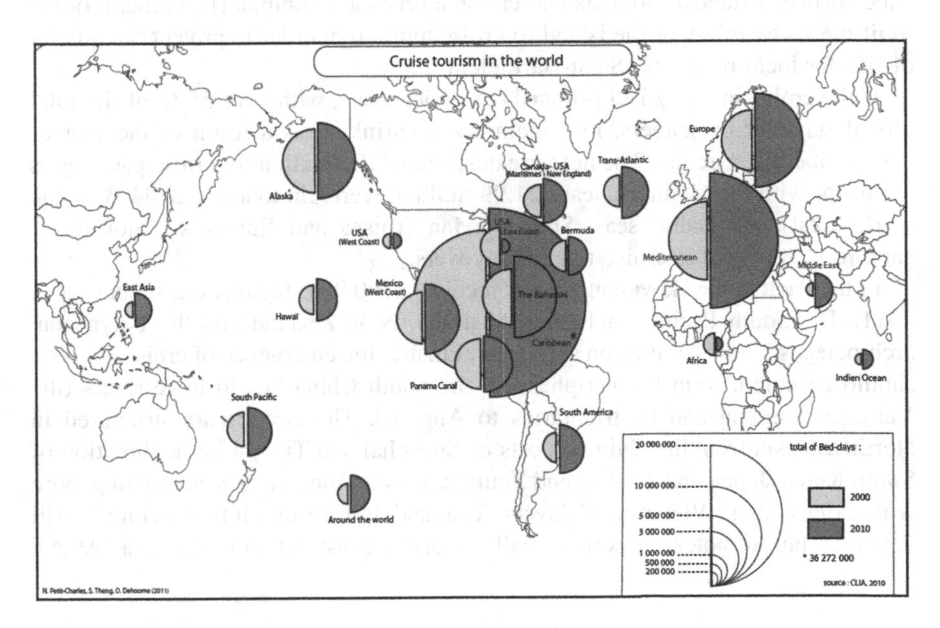

Fig. 3 Cruise tourism globally (2010)

4.2 Expansion of the Cruise Industry: Since its Historic Basins to the Confines of the Tourist World

In 2011, cruises brought a total of 107.5 million overnight tourists to locations across the globe. The preferred stayover destination remains the Caribbean, with

36.2 million tourists in 2011 (33.7 % of the total). With the addition of the Bahamas (6.5 million), the Caribbean area dominates almost 40 % of cruise tourist overnight activity (see Fig. 3). This strengthened global position over the past decade is logically explained by close proximity to the North American centre. The regional activity of the cruise continues through Mesoamerica (2.5 % of overnight stays focused on pan-American packages) to the Pacific coast of Mexico (3.27 % of nights) and back up towards Alaska (6.18 % of nights).

In the Caribbean area, cruise tourism is gradually becoming as important as stayover tourism, as noted by Logossah [3], in "13 des 34 pays de la Caraïbe, le flux de croisière excède celui de tourisme de séjour en 2004". Several destinations, such as the Mexican island of Cozumel, St. Maarten and Bahamas, specialize in cruise tourism. In practice, this specialization is incompatible with tourism stayovers because of occasional impacts caused by the influx of cruise passengers on specific popular places. The overconcentration of visitors generates nuisances for the local population. Host jurisdictions must make strategic choices [12]: (1) exclusively benefit from cruise tourism (Cayman Islands and St. Martin), e.g. implementing service taxes; (2) appropriately sharing space so that interferences between stayover tourists and cruise tourists are minimal (Dominica); or (3) limit the accessibility of the island to cruise tourism in order to protect the quality of life for local residents (Saint-Barthélemy).

If the cruise industry is a primarily American one, with over 55 % of the total global stayover passengers, its hegemony is shrinking as a result of the globalization of cruise circuits. The next greatest spatial destination for cruise passengers is Europe with the Mediterranean (21.99 million overnight tourists, 20.44 % of the world total) and Baltic seas. Scandinavian cruises and Europe's major rivers capture 8 % of global cruise tourist stayovers.

In the Pacific, the Hawaiian Islands receive 2,190,000 stayover cruise tourists in 2011. The South Pacific, including Australia, New Zealand and the Polynesian archipelagos, totals 2.9 million stayovers. Finally, the emergence of cruises in Asia should be noted, from the periphery of the South China Sea to large rivers (the Yangtze, Mekong and its tributaries to Angkor). The circuits are structured in Northeast Asia from the Chinese ports of Shanghai and Tianjin in the direction of South Korea, Japan and Taiwan and Southeast Asia, from Hong Kong to Singapore with extensions to Vietnam, Malaysia, Thailand and Burma. Infrastructure is still deficient, but 80 potential ports of call are under construction in the area (ACA).

4.3 Diversification of Cruise Ship Destinations

The spatial diffusion of the cruise industry is taking place alongside the emergence of new luxury cruise lines targeting wealthier customers. Thus, the company Paul Gauguin Cruises deploys low-capacity ships (332 passengers with 217 crew members) in the South Pacific between Tahiti and French Polynesia. The company

Hurtigruten, with 120 years of experience operating in the Baltic, is expanding its cruise lines from the North Sea to Greenland and Antarctica.

River cruises are the fastest growing segment of the industry, with an annual growth rate of over 10 % in the last five years. The development of river cruises has been dominated by a few large owners as Avalon Waterway (with 45 routes in the Danube Mekong through the Nile), the Uniworld Boutique River Cruises (more than 40 destinations in Russia and Southeast Asia) and the American Cruise Line (mainly on the Mississippi).

Finally, the opening of the Shanghai cruise line in 2006 and the investment of the company Costa in Chinese markets resulted in 300,000 passengers boarding for cruises in the South China Sea in 2010. The Chinese tourism market has considerable potential, having received a total of 57 million international tourists in 2010.

5 Challenges for the Cruise Ship Industry

The cruise industry has considerable potential, especially in terms of access to new markets that favour selective deployment in stable ports of call around the world. However, some limitations should be highlighted, such as the capacity to manage gigantic cruise lines, relations with host countries and prospects for innovation. The current system of cruise tourism is very profitable, but is ultimately too standardized.

5.1 Diversification of Cruise Ship Destinations

Economies of scale and mass production explain the emergence of gigantism in cruise lines, with mega-ships capable of carrying 5,400–6,000 passengers, onto which must be added crew members who, depending on the company and vessel capacity, range from 1,000 to 2,100. Larger vessels, such as the Oasis of the Seas and Allure of the Seas, have up to 8,400 people on board.

This gigantism is not without hassle for the ports of call, which are forced to implement construction of impressive harbours and dredging sites for the exclusive use of cruise ships. These liners are also of concern for social and environmental reasons. The enormity of the ship necessitates the rigorous organization (during excursions or meal services) which contradicts ideas of consumer freedom of choice, which is often dictated by identified groups (e.g. on the basis of nationality) that must follow the program as designed.

In addition to these modest stresses, there are the inevitable stressors which occur among groups of people in confined spaces, such as alcoholism, sexual assault and theft. These high densities of individuals with purchasing power also manifest itself in tension during stopovers, with local residents viewing tourists as easy prey for theft, leading to increases in security measures and occasionally a change in stopover destinations from maritime to mainland locations [17].

High densities of individuals moving between air-conditioning and swimming pools in the open air, in a large vessel containing water stocks that are subject to changes in temperature, represents a ripe environment for health complications. The confined spaces of cruise ships at sea render disinfection impossible and serve as a favourable environment for sea viruses such as Legionella [18]. Sick passengers may be quarantined, but remain contagious, and outbreaks of gastroenteritis are fairly common [19,20].

Gigantism also heavily contributes to environmental degradation [21, 22]. A ship of 3,000 passengers generates 800,000 L of waste water (from toilets), four million litres of grey water (showers, sinks, laundry), 90,000 L of bilge water (mixture of fluids, hydrocarbons, lubricants, degreasers, etc.), 40 tons of non-hazardous solid waste (plastic, glass, food waste) and 1–2 tons of hazardous solid waste (see site association Robin Hood[2]). Non-submersible waste can be incinerated onboard. These environmental challenges are not irresolvable, and their resolution within a "green growth of the cruise industry" is possible, but there are direct costs that must be shouldered by cruise companies. The integration of this parameter that can impact on product costs of mass cruise tourism as conceived in the gigantism as currently practiced.

5.2 Relations Between Cruise Lines and Ports of Call[3]

In 1972, the CCL launched the concept of "fun ship", advancing the idea that the ship is the central product of the cruise tourism. The steamer "Destination" is a floating resort, with all the desired amenities that could possibly be imagined (pools, spas, beauty salons, theatres, night clubs, casinos, shopping malls, jogging trails, mini-golf), as well as a feeling of security. Weaver [23] analyzes the "Disneylandization" of the cruise with its normalization of a staged, festive atmosphere. We must reconsider the attraction ashore in ports of call, in comparison with the services and amenities available on cruise ships. Interest stops become quite secondary, while time passed aboard ship increases, resulting in a reduction in money spent at ports of call, while consumption on board is increasing and diversifying. At the end of the cruise, the average amount of money spent on drinks on the cruise ship per passenger is at least 30 % of the price of the entry ticket that must be paid on leaving the ship [4, 7].

The current logic of the cruise industry, which places the ship at the centre of the product, allows for the maximization of profits on the ships at the expense of

[2] http://www.robindesbois.org/.

[3] Between the tax per passenger, wharfage laws and possible lease of reserved space, on the basis of an average 200 boats arriving annually with 3,500 passengers on each.

land destinations (ports of call) [24–26]. These financial realities lead the authorities of the host country to reconsider their position in the cruise industry. The debate focuses on the extent of specific improvements to be made in light of economic realities.

Certainly, financial projections by the association of cruise companies are encouraging: An American cruise tourist spends an average of $1,770 USD for a week's vacation (CLIA). During the 2008–2009 season, the cruise industry generated $2.2 billion USD in the Caribbean, directly supporting 56,000 jobs with total sales of $720 million USD (CLIA). In fact, spending levels are highly variable, and the most popular purchases are luxury and duty-free products. If the expenditure per passenger reaches $193,22 USD per passenger during a stopover in the Virgin Islands over the course of several stops, the average expenditure is hardly the price of a souvenir T-shirt.

Next, the argument that the locations at which a cruise ship disembarks is a potential tourist destination for holiday reads as rather tourist propaganda. We must distinguish situations. Tourists who are accustomed to standardized mass consumption and divide their time between leisure malls and major sporting events choose destinations that offer substantial tourist amenities (such as Cancun, Montego Bay, Punta Cana). Frequent stops at isolated, poorly urbanized areas are too unfamiliar and thus are inadequate for this tourist demographic [4, 7]. They prefer to experience local cultures for short periods of time from the balcony of a ship or dock space, between exoticism and voyeurism: Near enough to photograph poverty without having to experience it too intimately or for too long. For many destinations, cruise tourism is not synonymous with traditional tourism stays: the consumption patterns, temporary relationships and satisfaction criteria are not compatible.

Therein lies the dilemma for the home territories: they must invest tens of millions of dollars into an industry for which they can expect between $750,000 and one million dollars USD in annual revenue. To accommodate cruise ships and thus maintain financial benefits for their entire economy, public authorities must invest in ports of call at a loss. It is in this context that negotiation occurs between the cruise lines, which are concerned with maximizing their profits with clientele that is practically captives and host territories, which are in competition with each other and are faced with economic imperatives and exclusive investment opportunities. It is for these reasons that ports of call become, in the eyes of the cruise industry and cruise passengers, only postcard landscapes, more or less exotic and interchangeable [4, 7].

5.3 Towards a Renewal of the Cruise Industry

The cruise industry has experienced mass production in the Caribbean, driven by North American consumers [27–29]. The "Caribbean cruise" product embodies the golden age of Fordism applied to the cruise industry. The product is standardized, and economies of scale are achieved through the development of mega-liners that increase productivity. The Caribbean cruise tourism market is stabilizing, and the growth prospects of penetration of the North American market are interesting. The product is quite attractive to inexperienced clients, but it shows signs of obsolescence among more sophisticated clientele. The cruise industry, with these practices and mass consumption, has expanded towards Europe and East Asia.

Companies diversify their products by targeting different customer segments, such as Princess Cruises, which has six vessels that provide 115 different routes through a network of 350 destinations in the world. Its cruises last between 7 and 107 days, up to around the world with the crossing of the oceans, and visit five continents (the Caribbean in Senegal, through the Panama Canal, Patagonia and Hawaii).

Besides mass tourism, the cruise industry appears to be shifting away from gigantism, focusing instead on specific customer segments such as Disney Cruise Line's Dream (2,500 passengers, commissioned in 2011). Ship owners target more upscale clientele through renewed offers, upscale luxury, in line with ships like Oceania Cruises' Marina (1,250 passengers, commissioned in 2011), the Uniworld Boutique River Cruises 'Victoria River and Avalon Waterways' Panorama (commissioned in 2011, with a reception capacity of no more than 200 beds).

The cruise industry is globalizing from consumption based on mass production and is committed to a gradual product differentiation that includes offering superior quality. Diversification of the supply, the redefinition of spatial circuits of cruise lines coupled with increasing specialization is not necessarily innovation. It is rather rearranging and creating new versions of the same cruise ship product. The range displayed is more the result of a supply strategy that aims to win market share in a competitive environment. As such, the foundations of the cruise industry remain attached to a mass production base.

6 Conclusion

Far from exclusively offering gigantic cruise lines, the cruise industry has shifted, selectively diversifying the targeted customer demographics and destination circuits. This relatively young industry remains a booming sector that helps to define the new contours of global tourism in a context of political and economic instability. Safety, fun and controlled otherness remain the hallmarks of success for global cruise tourism.

The ship has become a "floating village resort" that travels in the middle of a postcard landscape, where the stops are decidedly secondary to the ship itself. Repeated experiences in the Caribbean highlight the economic and business logic of the cruise industry beyond the home territories [30–32], which are unable to negotiate the economic conditions needed for their development.

The cruise companies construct and direct the cruise product. They gauge potential cruise developments and the maximization of profits against desirable qualities of home territories. In accordance with economic growth in home territories and their potential for tourism development, the cruise industry continues to extend from its historical base in North America and Europe to Asian markets, mainly Chinese.

References

1. Dickinson, B., Vladimir, A.: Selling the Sea: An Inside Look at the Cruise Industry. Wiley, New Jersey (2008)
2. Wood, R.: Cruise Ships: Deterritorialized destinations. In: Lumdon, L., Page, S.J. (eds.) Tourism and Transport: Issues and Agenda for the New Millenium, pp. 133–145. Elsevier, Amsterdam (2004)
3. Logossah, K.: L'industrie de la croisière dans la Caraïbe : facteur de développement ou pâle reflet de la globalisation? Teoros **26**(1), 25–33 (2007)
4. Dehoorne, O., Murat, C., Petit-Charles, N.: Le tourisme de croisière dans l'espace caribéen : évolutions récentes et enjeux de développement. Etudes caribéennes. (2009). URL, http://etudescaribeennes.revues.org/3843
5. Boyer, M.: L'invention du Tourisme. Gallimard-Découvertes, Paris (1996)
6. Dowling, R.K.: Cruise ship tourism. Cabi international, Australis (2006)
7. Dehoorne, O., Petit-Charles, N., Theng, S.: La croisière dans le monde permanences et recompositions. Etudes caribéennes (2011). URL, http://etudescaribeennes.revues.org/5629
8. Grenier, A.: Le tourisme de croisière Téoros **27**(2), 36–48 (2008)
9. Bourdieu, P.: La distinction: critique sociale du jugement. Éd. de Minuit, Paris (1979)
10. Saïd, E.: L'Orientalisme. L'Orient créé par l'Occident, Le Seuil (1980)
11. Michel, F.: Désirs d'Ailleurs. Armand Colin, Paris (2000)
12. Amirou, R.: Imaginaire du Tourisme Culturel. Éditions. PUF, France (2000)
13. Augé, M.: Non-Lieux. Introduction à une Anthropologie de la Surmodernité Le Seuil, Paris (1992)
14. Charlier, J.: Les nouvelles frontières géographiques et techniques du marché mondial des croisières. FIG (2009). URL, http://archives-fig-st-die.cndp.fr/actes/actes_2009/charlier/article.html
15. Fournier, C.: Le tourisme de croisières en Méditerranée. Géoconfluences (2011). URL, http://geoconfluences.ens-lyon.fr/doc/typespace/tourisme/TourScient6.htm
16. Bresson, G., Logossah, K., Pirotte, A.: L'industrie de croisière aux Caraïbes tend-elle à évincer l'industrie du tourisme de séjour? Document de travail. ERMES, University Paris 2. (2007). URL, http://ermes.u-paris2.fr/doctrav/0707.pdf
17. Jaakson, R.: Beyond the tourist bubble? cruiseship passengers in port. Ann. of Tourism Res. **31**(1), 44–60 (2004)
18. Klein, R.A.: The cruise sector and its environmental impact. In: Schott, C. (ed.) Tourism and the Implications of Climate Change: Issues and Actions, pp. 113–130. Emerald, Bingley (2010)
19. Miller, J., Tam, T., Maloney, S., Fukuda, K., Cox, N., Hockin, J., et al.: Cruise ships: high-risk passengers and the global spread of new influenza viruses. Clin. Infect Dis. **31**, 433–438 (2000)
20. Cramer, E., Gu, D., Durbin, R.: Diarrheal disease on cruise ships, 1990–2000-the impact of environmental health programs. Am. J. Prev. Med. **24**(3), 227–233 (2003)
21. Butt, N.: The impact of cruise ship generated waste on home ports and ports of call: a study of southhampton. Mar. Policies **31**, 591–598 (2007)
22. Jones, R.: Chemical contamination of a coral reef by the grounding of a cruise ship in bermuda. Mar. Pollut. Bull. **54**, 905–911 (2007)

23. Weaver, A.: The Disneysation of Cruise Travel. In: Dowling, R.K. (ed.) Cruise Ship Tourism, pp. 389–396. CABI, Wallingford (2006)
24. Wilkinson, P.-F.: Caribbeancruisetourism : Delusion ? Illusion ? Tourism Geographies **1**(3), 261–282 (1999)
25. Weaver, A.: Spaces of containment and revenue capture: super sized cruise ships as a mobile tourism enclave. Tourism geographies **7**(2), 165–184 (2005)
26. Chin, C.B.N.: Cruising in the global economy. Profits, Pleasure and Work at Sea. Ashgate, Washingtong (2008)
27. Hall, J., Braithwaite, R.: Carribean cruise tourism-a business of transnational partnership. Tourism Manage. **11**(4), 339–347 (1990)
28. Wood, R.: Carribean cruise tourism—globalization at sea. Ann. Tourism Res. **27**(2), 345–370 (2000)
29. Weaver, A.: The McDonaldization thesis and cruise tourism. Ann. Tourism Res. **32**(2), 346–366 (2005)
30. Klein, R.A.: Troubled seas: social activism and the cruise industry. In: Dowling, R.K. (ed.) Cruise Ship Tourism: Issues Impacts Cases, pp. 377–388. CABI Publishing, Oxfordshire (2006)
31. Petit-Charles, N., Marques, B.: Répartition du tourisme de croisière dans la Caraïbe : quels déterminants ?. CEREGMIA, Document de travail No. 2010-09 (2010). URL, http://www2.univ-ag.fr/RePEc/DT/DT2010-09_Petit-Charles_Marques.pdf
32. Petit-Charles, N., Marques, B.: Determining factors for the distribution of cruise tourism across the Caribbean. Tourism Econ. **18**(5), 1051–1067 (2012)

Ecotourism as an Alternative to 'Sun, Sand, and Sea' Tourism Development in the Caribbean: A Comparison of Martinique and Dominica

Leah Weiler and Olivier Dehoorne

Abstract The Caribbean islands face ever-increasing challenges to their tourism industries, particularly the decline in tourist demand for the traditional 'sun, sand, and sea' tourism offered in this region. Given most Caribbean islands' overwhelming dependence on tourism for economic growth, this decline is alarming. This article compares tourism development on the islands of Martinique and the Commonwealth of Dominica to further explore reasons for the decline of tourism and to examine the potential success of ecotourism to revive flagging Caribbean tourism industries.

Keywords Tourism · Caribbean · Ecotourism · Sustainable development · Martinique · Dominica

1 Introduction

Tourism development is an increasingly popular option for developing countries that wish to lessen their economic dependence on primary agricultural products. This is particularly the case for Caribbean islands, whose isolated location, limited cropland, and prevalence of natural disasters render them particularly vulnerable. Tourism is also an industry which, given the wide range of stakeholders involved, involves a large degree of conflicts of interest and related problems. The traditional form tourism development has taken in the Caribbean has been 'sun, sand, and sea'

L. Weiler (✉)
School of International Development and Global Studies,
University of Ottawa, Ottawa, Canada
e-mail: lweil083@uottawa.ca

O. Dehoorne
CEREGMIA—Université des Antilles et de la Guyane, Guyane, France
e-mail: dehoorneo@gmail.com

B.-Y. Cao et al. (eds.), *Ecosystem Assessment and Fuzzy Systems Management*,
Advances in Intelligent Systems and Computing 254, DOI: 10.1007/978-3-319-03449-2_41,

mass tourism, consisting of large-scale, foreign-owned facilities, environmental degradation, repatriation of profits outside of the host location, and loss of cultural identity [1–3]. Given this reality, it is clear that a more sustainable alternative to mass tourism is required for the Caribbean that promotes economic development while taking into account local environments and cultures.

Ecotourism is an alternative to mass tourism that allows for tourism development and economic growth while protecting local environments and cultures. The relationship between tourism, socioeconomic development, and the environment, however, complicates sustainable tourism development. For some, ecotourism is associated with antidevelopment, and often in the early stages of tourism development, 'the social costs and economic risks are pooh-poohed as the whines of no-growth naysayers' [4]. Others suggest that, if not managed appropriately, ecotourism may develop into mass tourism in response to increased client demands [5]. It is an approach that, while addressing many concerns of mass tourism, is not without problems.

This chapter will compare tourism development in the Commonwealth of Dominica and Martinique, a French overseas department. Both are Caribbean islands with a wealth of natural features that attract tourists. However, they each possess very different histories, cultures, and tourism industries. This chapter will provide brief overviews of the general state of the tourism industries in Dominica and Martinique, respectively, before examining the literature on ecotourism in the Caribbean as an alternative to mass tourism. The trajectory of tourism development in Dominica provides an encouraging example of successful ecotourism that may be beneficial to other Caribbean islands such as Martinique that are facing various problems with the tourism sector.

2 Tourism Development in Dominica

The Commonwealth of Dominica is an island nation and a former British colony in the Lesser Antilles region of the Caribbean Sea. It is nicknamed the 'Nature Isle of the Caribbean' for its wealth of natural features, including mountainous rainforests, hot springs, and ongoing geothermal activity. Dominica is home to numerous rare bird and plant species that are protected by an extensive natural park system. Dominica contains Morne Trois Pitons National Park, a tropical forest with scenic volcanic features that attained UNEP World Heritage Status in 1997. Within it is Boiling Lake, the second largest, thermally active lake in the world. The island also contains Marine Reserves at the submerged volcano at Prince Edward's Bay and the submerged volcano at Soufrière/Scotts Head (Fig. 1). Tourism benefits Dominica economically by shifting the island away from overdependence on agricultural exports, especially bananas, the principal crop [6].

Tourism has developed more slowly on this island as compared to the rest of the Caribbean, with prospects for large-scale tourism hindered by mountainous physical geography and vulnerability to hurricane activity [7–9]. The Kastarlak

Fig. 1 Soufrière/Scotts head marine reserve (photo taken by author 2013)

Report [10] recommended a tourism policy based on environmental assets, and so, Dominica developed nature-based tourism as a development strategy [11]. France and Wheeller [12] described Dominica as a place visited by small numbers of adventurous travellers, as opposed to the client behaviour typified by mass tourism and resort destinations.

Issues have arisen with the degradation of Dominica's natural resources in the past. One tourist fell to her death during the hike to Boiling Lake as a result of the erosion of the path from frequent tropical downpours [13]. Such degradation of tourist places makes them less attractive to clients and can lead to a decline in tourist arrivals [14]. Additionally, Dominica captures less than 1 % of tourists to the Caribbean [15]. In more recent years, however, Dominica's ecotourism industry, characterized by small-scale, locally owned facilities, has seen increasing growth as an alternative to mass tourism [16]. Since 1986, stay-over tourist arrivals to the island have grown annually by almost 7 % [17]. This island nation is able to distinguish itself from other Caribbean destinations by offering an alternative tourism experience to the international market.

Dominica is the only Eastern Caribbean island that is with a community of pre-Columbian native Caribs (or Kalinago), who were largely exterminated or driven from neighbouring islands. Approximately 3,000 Caribs remain, most of whom live in eight villages on the east coast of Dominica in the Carib Territory on the northeast coast of Dominica. The Waitukubuli Trail is named after the Carib peoples' name for Dominica, which means 'tall is her body' [18]. Boxhill and Severin [19] found that the social impacts of ecotourism on the Caribs are largely positive. Thus, it appears as though tourism development on Dominica has included active participation by the indigenous Carib people in decisions regarding their territory.

A 1993 Caribbean Tourism Organization survey indicated that most tourists visiting Dominica do so to enjoy its natural beauty [20]. Additionally, Slinger [21] suggests that ecotourism in Dominica has contributed to a renewed interest in Caribbean culture. There is evidence indicating that Dominicans benefit substantially from the tourism industry: 96 % of employees in the island's tourism labour force were Dominican [22]. Local communities' views are incorporated into decision-making processes regarding ecotourism on the island. One example of this is the Waitukubuli Trail, which extends throughout all of Dominica and has a system comanagement of resources that allow for the participation of local communities [23].

In 2007, hotels accounted for only 29 % of travel accommodations available on Dominica [24], reflecting a preference for smaller-scale accommodations such as guesthouses. The predominance of local tourism businesses suggests that repatriation of profits outside of Dominica is limited [25]. Ultimately, it appears that the small-scale nature of ecotourism in Dominica has benefited the island nation by increasing linkages between tourism and other sectors, such as the economy, environment, and culture.

This largely positive assessment should be tempered with a caveat, however. It appears as though more recent policy decisions by the Dominican government could jeopardize the ecotourism industry on the island. These actions include a lack of impact assessment reports at new tourist sites [26]; hydroelectric development at Trafalgar Falls and other sites [27]; and the construction of a model Carib village, which was rejected by the Carib people due to a lack of consultation with local people [28]. While some past actions by the Dominican government are concerning in which they may indicate a shift towards mass tourism, policy choices by the island's authorities respecting tourism have largely remained faithful to the concepts of environmental protection and local participation.

3 Tourism Development in Martinique

The island of Martinique, or Madinina (the 'island of flowers'), is a French overseas department that is also located in the Lesser Antilles region southeast of Dominica. As a part of France, Martinique is also part of the European Union and thus has an economic advantage over its neighbours in the form of financial assistance from the French metropole [29]. The island is located along a fault line between the North American and Caribbean Plates and thus possesses several centres of volcanic activity. Ash from the active volcano Montagne Pelée has resulted in black sand beaches along the northern coasts, in contrast to the white sands of les Salines in the south. The Natural Regional Park of Martinique (le Parc naturel régional de la Martinique—PNRM) consists of several areas, including the rain forests around Mont Pelée, the Caravelle peninsula, and the area around les Salines and le Diamant [30]. Tropical conditions have contributed to lush vegetation on the island, including rain and mangrove forests, as well as a wealth of beaches.

Martinique's economy is heavily dependent on both agricultural exports and tourism. Many goods must be imported, contributing to a trade deficit that requires large transfers of aid from France. Martinique's status as a department of France has resulted in a much higher standard of living than on most other Caribbean islands. Tourism revenue in Martinique is concentrated in the south, where there is a wealth of beaches, and is mostly derived from traditional mass tourism in the Caribbean [31]. As of 2011, tourism accounted for about 9 % of Martinique's GDP [32].

Steps have also been taken to improve environmental regulation. For example, the conservation of mangrove forests and similarly fragile areas has been increased [33]. As of December 2011, Martinique had 1 national park, 2 regional parks,

2 nature reserves, and 19 protected areas [34]. However, the Martiniquian population does not appear to care very much about environmental issues [35]. This is partially evident in the decline in the quality of bathing water, with a 63.8 % seawater quality and a reported quality of 33.3 % for freshwater [36]. It would appear as though a certain amount of indifference among the population is to blame for thwarted efforts at environmental protection. This conclusion should be taken with a grain of salt, however, as there is currently a public campaign by the General Council in Martinique to reduce the disposal of waste in the water and on the ground [37].

In contrast to Dominica, tourism is declining in Martinique. In 2008, the number of registered tourists visiting the island was 40 % less than it had been a decade previously [38]. Martinique's local population associates the tourism sector with servitude in the old plantation economy and so regards the industry with a certain amount of distaste [39]. There are several other factors which explain the decline of tourism in Martinique, including socioeconomic tensions, competition from neighbouring islands, and an overdependence on French clients [40, 41]. Dehoorne and Augier [42] point out that neglect of tourist sites results in deterioration that further reinforces a decline in tourist visitation. There are multiple factors contributing to the decline of tourism in Martinique, including environmental degradation. Future efforts towards tourist development in Martinique must take into account environmental protection in order to ensure the successful growth of the industry.

One proposed solution to the flagging tourism industry is increased development of ecotourism in Martinique [43, 44]. Such developments could spread the distribution of tourist attractions more evenly across the island, relieving strains on coastal areas [45]. It is one method in which the mutually reinforcing aims of environmental protection and sustainable economic development can be realized within the tourism sector.

4 Ecotourism as Authenticity: The Sustainable Way Forward in Tourism Development

Islands face unique challenges within the tourism industry, including geographic isolation, susceptibility to natural disasters, and vulnerable ecosystems [46]. Nonetheless, their unique location and identity allows for differentiation from other tourist destinations. This is particularly the case for islands in the Caribbean, whose reliance on tourism as a driver of economic growth and development is fairly substantial. In 2012, almost 25 million tourists visited the Caribbean, a 5.4 % increase from 2011 [47]. However, the extent to which local economies and environments benefit from mass tourism is questionable, as the dominance of foreign tourism companies renders populations vulnerable to the whims of international markets [48]. Traditional tourism as practiced in the Caribbean has led to unsustainable exploitation of sand, limestone, and other materials due to limited supply of building materials [49].

Carlet [50] argues that the host population, tourism clients, and tourism organizations have a mutual interest in ensuring sustainable tourism development. However, the idea of a 'win–win' scenario in which environmental protection occurs alongside income growth in the tourism industry is more of an ideal than a realistic expectation [51]. As such, compromise is required between economic growth, environmental protection, and local cultures.

Ecotourism is an emerging concept that is used in multiple ways, contributing to a vague definition. It includes sustainable approaches to development, protection of natural areas, inclusion of local populations, and poverty reduction [52]. It is an approach to tourism that has a lower impact on the physical and cultural environment. For islands in the Caribbean, ecotourism allows for differentiation and economic growth at the same time as retaining local control and protecting the natural environment.

Ecotourism has numerous potential benefits for Caribbean islands seeking to diversify their economies while simultaneously protecting the natural environment and local cultures. It is different from mass tourism in which, rather than relying on a narrow range of client countries, it aims to diversify the market base as much as possible [53]. Moreover, because ecotourism is not dependent on summer weather, it is less vulnerable to the 'deluge–drought' cycle from summer to winter common to many Caribbean tourist destinations [54]. Ecotourism provides an alternative to mass tourism that allows for sustainability both in environmental regulation and economic growth in a manner that benefits local communities.

5 Conclusion

Tourism development presents substantial opportunities for Caribbean islands, whose geographic isolation, primary commodity dependence, and environmental vulnerability leave them susceptible to economic instability. However, the trajectory of traditional sun, sand, and sea tourism in the Caribbean has resulted in environmental degradation and of expatriation of profits outside the region, reinforcing the very conditions that made tourism development crucial to these islands in the first place. Evidence from the tourism industry in Dominica suggests that ecotourism can provide substantial benefits to Caribbean islands in the form of regional development and economic growth while encouraging the protection of both the environment and local/indigenous cultures. However, diligence must be exercised in ensuring the active inclusion of local populations and in limiting as much as possible the negative impacts of tourist activities on the environment. To this end, attention must be paid towards overutilization of natural resources and the maximum tourist capacity of small islands.

References

1. Blommestein, E.: Sustainable tourism in the Caribbean: an Enigma? In: Griffith, Mand., Persuad, B. (eds.) Economic Policy and the Environment: The Caribbean Experience, pp. 191–220. University of the West Indies, Mona, Jamaica (1995)
2. Brohman, J.: New directions in tourism for third world development. Ann. Tourism Res. **23**(1), 48–70 (1996)
3. Klak, T., Flynn, R.: Sustainable development and ecotourism: general principles and Eastern Caribbean case study. In: Jackiewicz, E., Bosco, F.J. (eds.) Placing Latin America, pp. 115–136. Rowman and Littlefield, Lanham, MA (2008)
4. Richter, L.K.: The politics of tourism in Asia, p. 21. University of Hawaii Press (1989)
5. Ioannides, D.: Planning for international tourism in less developed countries: toward sustainability? J. Plan. Lit. **9**(3), 249 (1995)
6. Bonnerjea, L., Weir, A.: Commonwealth of Dominica Poverty Assessment: Report Prepared for the Government of the Commonwealth of Dominica. British Development Division in the Caribbean, Roseau, Dominica, Bridgetown, Barbados
7. Fermor, P.L.: The Traveller's Tree: A Journey Through the Caribbean Islands. Harper and Row, New York (1950)
8. Ward, F.: Golden Islands of the Caribbean. Crown, New York (1972)
9. Blume, Helmut: The Caribbean Islands. Longman, London (1974)
10. Kastarlak, B.: Tourism and its Development in the Third World, p. 25. Routledge, London (1975)
11. Christian, C.S., Lacher, T.E., Hammit, W.E., Potts, T.D.: Visitation patterns and perceptions of national park users—case study of Dominica, West Indies. Carib. Stud. **37**(2), 83–103 (2009)
12. France, L., Brian, W.: In: Barker, D., McGregor D.F.M. (eds.) Environment and Development in the Caribbean: Geographical Perspectives, pp. 59–69. University of the West Indies, Jamaica (1995)
13. Cater, E.: Environmental contradictions in sustainable tourism. Geogr. J. **161**(1), 25 (1995)
14. Butler, R.: The concept of a tourist area cycle of evolution: implications for management of resources. Can. Geogr. **24**(1), 5–12 (1980)
15. World Trade Organization: Regions: the Americas. UNWTO World Tourism Barometer **6**(1). 21–25. www.unwto.org/facts/eng/pdf/barometer/UNWTO_Barom08_1_en.pdf
16. Patterson, T., Rodriguez, L.: In: Gossling, S. (ed.) Tourism and Development in Tropical Islands, Political Ecology Perspectives, pp. 60–87. Edward Elgar Press (2003)
17. Slinger-Friedman, V.: Ecotourism in Dominica: studying the potential for economic development, environmental protection and cultural conservation. Island Stud. J. **4**(1), 6 (2009)
18. Dominica Hotel and Tourism Association: Welcome to Dominica. Destination Dominica 2000, pp. 18–22 (2000)
19. Boxhill, I., Severin, F.: An exploratory study of tourism development and its impact on the Caribs of Dominica. Int. J. Hospitality Tourism Adm. **5**(1), 1–27 (2004)
20. Slinger-Friedman, V.: Ecotourism in Dominica: studying the potential for economic development, environmental protection and cultural conservation. Island Stud. J. **4**(1), 10 (2009)
21. Slinger, V.: Ecotourism in the last indigenous Caribbean community. Ann. Tourism Res. **27**(2), 520–523 (2000)
22. Slinger-Friedman, V.: Ecotourism in Dominica: studying the potential for economic development, environmental protection and cultural conservation. Island Stud. J. **4**(1), 10 (2009)
23. Sarrasin, B., Tardif, J.: Écotourisme et resources naturelles à la Dominique: la congestion comme pratique novatrice. Téoros **1**, 38 (2012)

24. Euromonitor International: Travel and Tourism in Dominica. www.euromonitor.com/Travel_And_Tourism_in_Dominica (2007)
25. Slinger-Friedman, V.: Ecotourism in Dominica: studying the potential for economic development, environmental protection and cultural conservation. Island Stud. J. **4**(1), 11–12 (2009)
26. Watty, W.R.F.: Feature Address W.R. Franklin Watty, Chairman of Diaspora Affairs, Dominica Academy of Arts and Sciences at the Annual Awards Gala Dominica Hotels and Tourism Association. http://da-academy.org/Watty_Address_DHTA.pdf (2008)
27. Evans, P.G.H., Williams, D.: Development and Management of Nature Sites—Integrating Conservation with Ecotourism in Dominica. Project Report No. 4. Brussels, European Community Project No. B7-5040-24. Ecosystems Ltd (1997)
28. Boxhill, I., Severin, F.: An exploratory study of tourism development and its impact on the Caribs of Dominica. Int. J. Hospitality Tourism Adm. **5**(1), 23–24 (2004)
29. Logossah, K., Céliméne, Fred: Évaluer l'impact économique du tourisme: Un exemple de modélisation macro-sectorielle de l'économie martiniquaise. Téoros **26**(1), 53–62 (2007)
30. Parc naturel regional de la Martinique (PNRM): Les missions de PNRM. PNRM. http://pnr-martinique.com/les-missions-du-pnrm/ (visited 5 Aug 2013) (2013)
31. Sheller, M.: Consuming the Caribbean: From Arawaks to Zombies. Routledge, London and New York (2003)
32. Euromonitor International: Travel and Tourism in Martinique. http://www.euromonitor.com/travel-and-tourism-in-martinique/report (2012)
33. Plantin, C.: L'évolution géo-environnementale de la Martinique entre les années 1956 et 2006. Les Cahiers d'Outre-Mer, pp. 551–564 (2011)
34. INSEE: Espaces naturels faisant l'objet d'une protection réglementaire au décembre 2011. INSEE Martinique. http://www.insee.fr/fr/themes/tableau.asp?reg_id=23&ref_id=tertc01301 (2011)
35. Martouzet, D.: Fort-de-France: Ville Fragile?. Anthropos, Paris (2001)
36. INSEE: Qualité des eaux de baignade en 2011. INSEE Martinique.http://www.insee.fr/fr/themes/tableau.asp?reg_id=23&ref_id=tertc01304 (2011)
37. Daniel, P.: Gestion des dchets: bataille contre les prospectus. DOMactu. http://www.domactu.com/actualite/138616583087874/martinique-gestion-des-dechets-bataille-contre-les-prospectus/ (7 Aug 2013)
38. Dehoorne, O., Furt, J.M., Tafani, C.: L'écotourisme, un 'modèle' de tourisme alternative pour les territories insulaires touristiques français? Discussion à partir d'expériences coisées Corse-Martinique. Études caribéennes 19: Par. 5 (2011)
39. Dehoorne, O., Furt, J.M., Tafani, C.: L'écotourisme, un 'modèle' de tourisme alternative pour les territories insulaires touristiques français? Discussion à partir d'expériences coisées Corse-Martinique. Études caribéennes 19: Par. 7 (2011)
40. Augier, D.: L'écotourisme forestier: pour unrapprochment entre tourisme et environnement à la Martinique. Études caribéennes: Par. 18 (2007)
41. Desse, M.: Guadeloupe, Martinique, LKP, crise de 2009, crise économique, déclin économique: de crises en crises: la Guadeloupe et la Martinique. Études Caribéennes 17 (2010)
42. Dehoorne, O., Augier, D.: Towards a new tourism policy in the French West Indies: the end of mass tourism resorts and a new policy for sustainable tourism and ecotourism. Études caribéennes 19: Par. 15 (2011)
43. Augier, D.: L'écotourisme forestier: pour unrapprochment entre tourisme et environnement à la Martinique. Études caribéennes: 1–8 (2007)
44. Dehoorne, O., Augier, D.: Towards a new tourism policy in the French West Indies: the end of mass tourism resorts and a new policy for sustainable tourism and ecotourism. Études caribéennes **19**, 1–16 (2011)
45. Augier, D.: L'écotourisme forestier: pour unrapprochment entre tourisme et environnement à la Martinique. Études caribéennes: Par. 22 (2007)

46. Vellas, F: Les spécificités de l'insularité et le développement touristique. Le tourisme et les îles, pp. 19–32. L'Harmattan, Paris (1997)
47. Nicholson-Doty, H.B.: State of the Industry Report. Caribbean Tourism Organization. St. Michael, Barbados. http://www.onecaribbean.org/content/files/StateofIndustryFeb2013.pdf. 1–2 (14 Feb 2013)
48. Britton, R.A.: International tourism and indigenous development objectives: a study with special reference to the West Indies. Ph.D. Thesis, University of Minnesota, Minneapolis (1978)
49. Jackson, I.: Carrying capacity for tourism in small tropical Caribbean Islands. UNEP Ind. Environ. **9**(1), 7–10 (1986)
50. Cater, E.: Environmental contradictions in sustainable tourism. Geogr. J. **161**(1), 21–28 (1995)
51. Cater, E.: Environmental contradictions in sustainable tourism. Geogr. J. **161**(1), 22 (1995)
52. World Tourism Organization: Sustainable Development of Ecotourism: A Compilation of Good Practices in SMEs. WTO, Madrid (2003)
53. Weaver, D.B.: Alternative to mass tourism in Dominica. Ann. Tourism Res. **18**(3), 424 (1991)
54. Weaver, D.B.: Alternative to mass tourism in Dominica. Ann. Tourism Res. **18**(3), 426 (1991)

On the Archiving Conservation of Hainan Tourism Culture

Jin-xia Zheng

Abstract The development of Hainan tourism industry needs to exert the unique charm of Hainan tourism culture; therefore, the exploration, inheritance, and protection of Hainan tourism culture become extremely important. This article draws forth archiving conservation of Hainan tourism culture by analyzing comprehensively the current situation of Hainan tourism culture; it also explains further the connotation of archiving conservation of Hainan tourism culture. In the end, it puts forward constructional methods of living environment of archiving conservation of Hainan tourism culture and implementation plan of archiving protection of Hainan tourism culture based on principle of archives protection.

Keywords Archiving conservation · Hainan · Tourism culture

1 Introduction

Hainan Island has attracted plenty of tourists coming for sightseeing since the beginning of the construction of Hainan International Tourism Island in 2009, the tourism industry of Hainan therefore has been growing as well. In 2012, Hainan received 33.2037 million trips from home and abroad, total revenue of tourism reached RMB37.9 billion, which is 17 % increase in that of previous year, the international tourism policy effect really benefits Hainan [1].

However, in some experts' opinion, the economic growth of Hainan relies on deeply on real estate industry and the development of Hainan international tourism

J. Zheng (✉)
School of Information and Technology, Hainan Normal University,
Haikou 571158, China
e-mail: zhengyinzhi323@163.com

B.-Y. Cao et al. (eds.), *Ecosystem Assessment and Fuzzy Systems Management*,
Advances in Intelligent Systems and Computing 254, DOI: 10.1007/978-3-319-03449-2_42,

island is becoming a feast of real estate industry. While the resource of the real estate is limited, the sustainable development of Hainan tourism industry cannot rely on real estate only. *Several opinions of the state council on boosting Hainan International Island Construction* makes a point of making Hainan an open, green, civilized, and harmonious island with beautiful ecological environment, unique cultural glamour and civilized, peaceful, and harmonious society. In this case, injecting Hainan local culture into tourism industry with the help of the natural beautiful ecological environment to exert Hainan unique tourism cultural glamour is the only way of development.

In the recent years, the conservation and heritage of Hainan tourism culture has become the common topic of government and residents that some local cultures are endangered and disappearing has aroused concerns on the conservation of Hainan tourism culture from all society. There is no doubt on the importance of Hainan tourism culture, however, how to protect Hainan tourism culture, in what way to dig and protect Hainan tourism culture, and how to fully, comprehensively, factually inherit it and with deepened meaning has become pending problem.

2 On the Archiving Conservation of Hainan Tourism Culture

The tourism culture of Hainan originates from Hainan culture and is an indispensable part of Hainan local culture. The tourism culture of Hainan consists of ecological tourism culture, tropical custom tourism culture, folk custom tourism culture, overseas Chinese culture, religion culture, and humanistic tourism culture forming from Hainan historical culture, etc. The combination of Hainan unique natural ecological resource and special ethnic groups, history, and humanity forms peculiar tourism culture [2].

For the past few years, China and Hainan government has increased the exploitation on Hainan local culture which greatly enriched the connotation of Hainan tourism culture and lifted the quality of Hainan tourism culture; however, there are still some questions as follows.

2.1 Inadequate Exploitation and Protection

Hainan government has done quite a lot of work regarding digging and spreading Hainan tourism culture in recent years and has repaired and protected many cultural sites like the renovation of famous 'Qiongtai Ancient Academy' and 'Dongpo Ancient Academy' which shows the salvage of government to 'academic culture'. However in general, it is still not enough in regard to protection and exploitation on existing culture, for instance, the declining and broken ancient streets that have arcade buildings, there are not many ancient streets left, most of them are turned to lifeless modern arcades which is really sad loss of 'arcade culture'; the protection and exploitation of 'arcade culture' are far from enough.

2.2 *Improper Construction Way of Hainan Tourism Culture*

At present, the excavation, protection, and exploitation of Hainan tourism culture are managed by many divisions, there involved many government sectors run by themselves. This multiple management results in severe waste and overlapping of resources. In the meantime, some local governments value application while despise protection and management. For example, the protection and management of some tourism cultures are dismissed or excessively used and destructively exploited, even use them as cash cows distorting folk art under the name of inheritance and innovation after they are applied for non-material cultural heritage, resulting in distorting or soulless of non-material in heritage, which fully illustrate the decentralized protection way has become hindrance that gets in the way of healthy development of Hainan tourism culture, we must find a way that emphasizes keeping the authenticity of the precious tourism culture also more suits sustainable development.

2.3 *Insufficient Awareness of Protection and Exploitation of Local Culture*

The public awareness of the government in charge on the excavation, protection, and exploitation is not enough, and many residents lack adequate knowledge of the importance and urgency of the protection of Hainan local culture, resulting in some cultures that need oral teaching and rote memory are disappearing and destroying and abandoning rare objects and materials, abusing and overexploiting cultural artifacts are happening time to time.

3 The Connotation of Archiving Conservation of Hainan Tourism Culture

It happens time to time that Hainan tourism culture disappears and dies due to its authenticity being lack of reliable guarantee and the carrier being unfixed, which demands us to find a way to protect Hainan tourism culture that suits Hainan tourism culture and be able to escort the protection of Hainan tourism culture. Archive has specific material carrier and is a solidified information; to archive Hainan tourism culture is a good plan to protect Hainan tourism culture [3].

3.1 The Archiving of Hainan Tourism Culture

The '-ing' as in 'archiving' means translation into certain quality or change to some form when used after nouns or adjectives. Archiving is to change some things to archives completely and process them according to requirements of archival management [3].

The archiving of Hainan tourism culture is to change Hainan tourism culture into archives and display them in the form of archives and manage them by the way of archival management.

3.2 The Connotation of Archiving Conservation of Hainan Tourism Culture

Conservation is the connection point between archival work and Hainan tourism culture. Archiving is the important means of Hainan tourism culture carrier which facilitates the protection and inheritage of Hainan tourism culture. Archiving is a process; archiving conservation of Hainan tourism culture is a systematic approach of digging and protecting Hainan tourism culture deeply. It translates Hainan tourism culture into archives, fits it into archival management, and protects it by utilizing systematized, normalized archival management methods, deepening, extending, and inheriting its connotation.

The tourism culture itself has the characteristics of archives, the archiving conservation of Hainan tourism culture stresses more on digging the cultural connotation of tourism culture itself, uplifting its cultural taste and making it the culture peculiar to Hainan that Hainan people are proud of, and inheriting it positively and extensively. In the meantime, archives itself have rich cultural connotation. The explicit knowledge and implicit knowledge contained in the politics, economy, and culture carried in archives are the inexhaustible source for tourism culture connotation.

4 Conclusion

The archiving conservation of Hainan tourism culture not only solidifies the dynamic tourism culture to certain carriers to form the tourism culture information resource through photographing, recording, videotaping etc., recording methods, but more emphasizes on filing Hainan tourism culture and bringing it into the range of archival management which is more than copying rigidly archival management methods for Hainan tourism culture protection, specific proposals as follows:

4.1 Living Environment Construction of the Archiving Conservation of Hainan Tourism Culture

The living environment construction of Hainan tourism culture mainly refers to legal basis, management mechanism, talent team, and social significance, etc. Hence, to do well the archiving protection of Hainan tourism culture, one must follow the following:

1. Change previous multiple management of culture protection to diversified management mechanism that features government being the leading guide, archival sector being the construction main body, other cultural, tourists, etc. sectors being the assistant bodies. Government grants archival sector the right of being the main body within its power, making the best use of the archival management system, advanced archival protection technology, good archival professional and technical personnel of the archival sector, and serving them as strong backup of Hainan tourism culture archiving conservation. Cultural and tourism sectors coordinate positively with archival sector, providing materials and utilizing its advanced means to protect and improve the living status and environment of tourism culture resource.
2. Set up sound, complete, and impeccable laws and regulations, build up favorable legal environment of Hainan tourism cultural archiving protection. The previous laws on cultural protection mostly laid stress on the rights that cultural and tourism sectors enjoy, while most laws that involve archival sector granted it only intervention role instead of dominant role, which cannot fully exert the advantages the archival sector has on cultural protection, esp., on intangible culture protection. So there should be considered strengthening the rights of archival sector, making archival sector the protective main body and having it justifiably join the protection of Hainan tourism culture.
3. Consolidate protection awareness, link application, protection, and publicity. Mind determines action, only that the general awareness of whole society is improved, can better protection on cultural resources be realized. So to develop Hainan local economy, both government and society need to strengthen protection awareness of tourism culture, all administration areas need to link application, protection, and publicity when actively applying for all kinds of cultural heritages, regardless of the application results, there should have subsequent protection and inherit on their tourism cultural resources, meanwhile, do good publicity on Hainan tourism culture, make whole society understand it and realize the urgency of cultural protection, making Hainan tourism culture protection deeply rooted in people's minds.
4. Found outstanding team. The team of Hainan tourism culture archiving protection should include archival research scholars, cultural research scholars, non-material culture inheritors, archive workers, culture workers. only through strengthening the training, cooperation of these professionals, Hainan tourism culture archiving protection can be better done.

4.2 The Implementation of Hainan Tourism Culture Archiving Protection

The work link of Hainan tourism culture archiving protection is to be managed according to the process of archival management.

1. The Collecting, Filing Work of Hainan Tourism Culture. First to do a general survey, investigating, registering, collecting, and recording the current situation of all kinds of Hainan tourism cultures and then organize relative experts and scholars to sort them out, identifying and confirming them based on the research and investigation, categorizing them according to different types and then place them on file. In addition, ceaselessly replenish and complete the files according to the trends of the culture in the future. After filing Hainan tourism culture, there should establish digitizing database of Hainan tourism culture and realize computer retrieval and remote retrieval.

2. The Custody Work of Hainan Tourism Culture Archives. Protection of Hainan tourism culture archives varies with the carriers of the archives. For tangible archives, not only should the entity and related materials of the archives be protected by using various technologies and methods, but digitalize them through multimedia technology forming multimedia archival materials. For example, display dynamic scenes of important scenic spots by using GIS system. As for intangible cultural archives, we should inherit and protect them by using all kinds of carriers. For example, for some rare or endangering intangible culture, except completing the collecting, sorting and custody work of the data, there is more to do on the training of the heritors and to file one by one afterward. The archivists not only are required having rich folk culture accomplishment or related training, but also be trained on archival management [4].
3. Utilization of Hainan Tourism Culture Archives. Hainan tourism culture archives are an important inheriting medium for the inheritance of the culture. Even for the disappeared culture, we can know it, learn it, and publicize it through the recording, videotaping, and multimedia materials. For the public, multimedia is an important platform for them to understand Hainan tourism culture. In this case, unfold rich, profound historical culture of Hainan through Hainan tourism culture.

References

1. Sheng-Zheng: The Twenty-first Century Economic Report, Hainan Reported, 20 July 2013. http://www.21cbh.com/2013/7-20/zONjUxXzcyNzIzOQ.html
2. http://bbs.szhome.com/commentdetail.aspx?id=130639402&projectid=130010. 1 May 2012

3. Lei-Chu: A Research on the Archival Method of Intangible Cultural Heritage Protection, China Masters' Theses Full-text Database
4. Discussion On How To Protect Intangible Heritages Through Building Archives Meijuan-Qin Archives Management, May 2007

Research on Classification and Evaluation of Tropical Sports Tourism Resources in China

Taking Hainan for Example

Tian Jianqiang, Xia Minhui and Song Jingmin

Abstract This chapter conducted classification and evaluation on tropical sports tourism resources in Hainan by applying document literature method, questionnaire survey method, field investigation method, expert interview method, and mathematical statistical method, according to the national standard of the People's Republic of China (GB/T 18972-2003) *Investigation and Evaluation of Tourism Resources*, *Tourism,* and *Tourism Geography* theories. Tropical sports tourist resources in Hainan are divided into 2 main categories, 7 sub-categories, 67 basic types, and 513 resource monomers. The author established evaluation index system of sports tourism resources, weight and scoring method, and evaluation ranking. Qualitative evaluation on tropical sports tourist resources in Hainan is as below: rich in resources, high aggregation index, complete types, and unique functions. This article set up the index system and weight and scoring evaluation grade for quantitative evaluation of sports tourism resources. By using expert scoring and calculating, the author conducted quantitative evaluation respectively on the participation type and ornamental type of sports tourism resources; the result is of practical and reasonable significance and has instructive significance on development of tropical sports tourist resources in Hainan.

Keywords Sports tourism resources · Classification and evaluation · Hainan

---

T. Jianqiang (✉) · X. Minhui
School of PE of Hainan Normal University, Haikou, China
e-mail: tiantian1978@yeah.net

S. Jingmin
Department of PE of Hainan University, Haikou, China

B.-Y. Cao et al. (eds.), *Ecosystem Assessment and Fuzzy Systems Management*, Advances in Intelligent Systems and Computing 254, DOI: 10.1007/978-3-319-03449-2_43,

1 Introduction

As a tropical region branch of sports tourism industry, tropical sports tourism is highly favored by people group from developed countries as well as those from the subtropical zone, temperate zone, and cold temperate zone. Tropical sports tourism resources serve as the basis for the development of tropical sports tourism. Hainan is the second largest island and also the only tropical island of China, covering 34,000 km^2 of land area. At the same time, Hainan is a big coastal province, covering the marine area of about 2,000,000 km^2, accounting for 2/3 of the total marine area of China. Hainan holds China's tropical location advantages with its unique tropical climate, tropical oceanic islands, tropical geomorphology, and ethnic folk customs. Meanwhile, its excellent ecological environment, the world's third largest tropical rainforest and the national only Li nationality amorous feelings have derived and created many sports tourism resources.

2 Research Object and Methods

2.1 Research Object

Tropical sports tourist resources in Hainan.

2.2 Research Methods

2.2.1 Literature Consultation Method

Access to the international social science database IBSCO HOST database, Canada SILVER, Sports Tourism International Council Website, search the theme words "sport tourism resources" by GOOGLE, and 47 related papers were searched finally.

In China National Knowledge Infrastructure (CNKI) Database System, the General Administration of Sport of China Information (SPORTS) database, type the key words: "development mode of tourism, sports tourism, Hainan tourism" to search; we can search a total of 244 related papers that were published during 1994–2013. We consulted and searched about 18 PhD dissertations relating to tourism development and Hainan tourism. In addition, we consulted with more than ten books about the study, including *Tourism economics*, *Introduction to Tourism Resources,* and other related theories.

2.2.2 Expert Interview Method

The author interviewed about 20 experts and scholars from General Administration of Sport of China, Scenery Tourism Department of Tongji University, Tourism Administration of Hainan Province, Department of Culture and Sports of Hainan Province, Higher Education Institutions and listened to the proposals on the establishment of index system, questionnaire survey, and research achievement and discussed with the experts.

2.3 Questionnaire Investigation and Statistical Analysis

2.3.1 Questionnaire Design

Questionnaire compilation: process flow diagram is as follows:

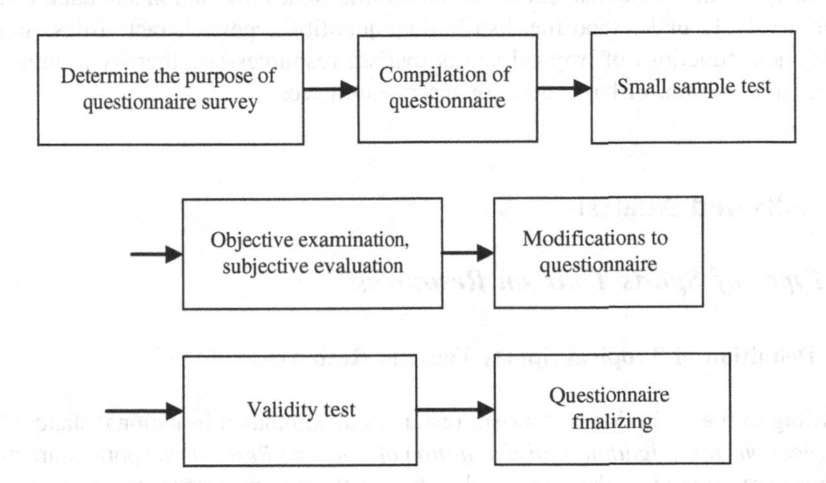

Questionnaire test: Experts shall examine content validity and structural validity of the questionnaire (Tables 1, 2).

2.3.2 Questionnaire Analysis

This thesis applies international general statistical software statistical product and service solutions (SPSS) to conduct statistical description of questionnaire and employs the combined method of quantitative analysis and qualitative analysis to evaluate the tropical sports tourist resources in Hainan.

Table 1 Expert's evaluation on content validity of the questionnaire

Contents	Rational	Basically rational	Irrational
The number of experts	11	5	0.0

Table 2 Expert's evaluation on structural validity of the questionnaire

Contents	Rational	Basically rational	Irrational
The number of experts	12	4	0.0

2.4 Field Investigation Method

Through the field investigation on 18 cities and counties in Hainan, the author personally experiences many sports tourism projects. Through the systematic investigation and study on the characteristics of tropical sports tourism resources in Hainan, this thesis extensively collected information, information and data, and comprehensively understood the distribution, quantity, types, characteristics, and development functions of tropical sports tourism resources and thereby accumulated a large amount of basic data for the research work.

3 Results and Analysis

3.1 Types of Sports Tourism Resources

3.1.1 Definition of Tropical Sports Tourism Resources

According to the definition of tourism resources as stipulated in national standard *Classification, Investigation, and Evaluation of Tourism Resources*, sports tourism resources are defined as the sum of all kinds of things and factors in nature and human society that can make people feel attractive to sports tourism and can be developed and utilized by sports tourism industry and can produce economic benefit, social benefit, and environmental benefit. Tropical sports tourism resources refer to the sum of all kinds of things and factors in the nature and human society of tropical region that can make people feel attractive to sports tourism and can be developed and utilized by sports tourism industry and can produce health benefit, economic benefit, and social benefit.

3.1.2 Classification of Tropical Sports Tourist Resources in Hainan

Main categories: According to the participation degree of sports tourism resources in meeting sports tourism demand, tropical sports tourism resources can be divided into two main categories: natural category and humane category, which can specifically be divided into participating sports tourism resources and ornamental-type sports tourism resources.

Sub-categories: According to the main categories of tourism resources as specified in *Classification, Investigation, and Evaluation of Tourism Resources*, the tropical sports tourism resources in Hainan can be divided into seven subcategories.

Basic types: According to the formation principle of sports tourism resources, tropical sports tourism resources in Hainan can be classified into 67 basic types, including 513 tropical sports tourism resource monomers comprising of rich physical factors (Tables 3, 4).

3.2 Evaluation of Sports Tourism Resources

3.2.1 Establishment of Sports Tourism Index System

Structure of the evaluation system of sports tourism resources is as follows:

Sports tourism activities	Popularity of sports tourism resources
	Appropriate travel periods or the service scope of
	The scale of sports contests and cycle of national sports activity
Element value of sports tourism resources	Use value, or recreation value or ornamental value of sports tourism resources
	Sports tourism resources contain sports culture, sports science, or sports aesthetic value
	Whether sports tourism resources are associated with rare and strange geological landscapes
	Whether sports tourism resource is reasonable
	Value orientation of sports physical exercises
	Whether sports tourism resources are subject to interference and destruction of natural and artificial power, and whether it is well preserved
Value-added	Environmental protection and environmental safety of sports tourism resources
	Safety of sports tourism facilities

Table 3 Classification sheet of tropical sports tourism resources in Hainan

Category	Segmentation of category
Main categories (2)	Natural category and humanistic category
Sub-category (7)	Geographical landscape type, hydrological landscape type, ecotype, folk custom type, festival celebration type, events type, and stadium-base type
Basic types (67)	Coastal type, island type, mountain type, canyon type, waterfall type, strange stone and caves type, geological landscape type, rivers type, lakes type, hot-spring type, tropical rainforest type, animal protection area type, nature protection area type, folk custom type of humanistic category, festival type of humanistic category, event type of humanistic category, golf type, training base type, sports center type, stadium and swimming pool type, etc.
Resources monomers (513)	Independent-type tropical sports tourism resources monomers (407) and collective-type tropical sports tourism resources monomers (106)

3.2.2 Determination of the Index Weight System

The author conducted investigation on 12 people including experts from tourism and sports science and sports tourism developers on the index system constructed in the study through Delphi method and then worked out the average and got the scale value for comparison between various factors and endowed 100 points to the general evaluation system; finally, the weights of sports tourism resources evaluation index system could be obtained as follows (Table 5).

3.2.3 Scoring Standard of Evaluation on Sports Tourism Resources

It is required to give score on the basis of sports tourism resources common-factors comprehensive evaluation system, resource factor value and resources influence have the score range 100 points in total; value-added: "environmental protection and safety" has positive and negative points, in the range −23 points ∼5 points. Each evaluation factor is divided into four grades, and its factor scores shall be correspondingly divided into 4 grades.

3.2.4 Scoring and Grading

1. Scoring: According to the evaluation on the sports tourism resources monomer, we can get the score given by the comprehensive evaluation on the monomer sports tourism resources common-factors. On the basis of the total score given by the evaluation on sports tourism resources monomer, the total score is divided into five grades from high to low: fifth-grade sports tourism resources have the score range >90; fourth-grade sports tourism resources have the score range >75–89; third-grade sports tourism resources have the score range

Table 4 Classification system of tropical sports tourism resources in Hainan

Main category	Sub-category	Basic types	Representative resources
Participating sports tourism resources	Geographical landscape type	Mountain, river valley, caves, beach, reef, and, etc.	Five Finger Mountain, Diaoluo Mountain, Jianfeng Ridge, Bawang Ridge, Limu Mountain, Baihua Ridge, Qixian Ridge. Wangxia Valley, Qiceng valley, Bronze Drum Ridge at Qizi Bay
	Hydrological landscape type	Waterfall, artificial lakes, rivers, hot springs, coastal lagoon harbor, tropical submarine landscape, etc.	Yalong Bay, Dadonghai Sea, sea scarp, Shimei Bay, Riyue Bay, Chunguo Bay, Tufu Bay, Fengjia Bay, Gaolong Bay, Coconut Bay, Wanquan River, Changhua River, Nanli Lake, Yunyue Lake, Songtao Reservoir, Daguang dam reservoir, Wanquan lake, dam
	Ecological sports tourism resources	Tropical primitive forests, valley rainforest, marine plants, marine aquaculture animals	Jianfeng Ridge, Diaoluo Mountain, Five Fingers Mountain, National Forest Park Five Fingers Valley, Jianfeng Ridge Canyon, Bawang Ridge Canyon, Changhua River Canyon, Wanquan River Canyon
	Sports tourism facilities	Stadium, sports club, gym, etc.	Haikou City Stadium, Gangfei table tennis club, Wenhua fitness center
	Folk custom and humanistic activities	Ethnic folk customs, folk sports tourism, folk festivals of ethnic minority	Bull racing, cross-bow shooting, bamboo pole dance, wrestling, martial arts, climbing vines and picking flowers, Yajia (tortoise-like backward tug-of-war), carrying-pole dance, spinning top
Ornamental-type sports tourism resources	Athletic sports events sports culture communication	The Olympic games, world championships, world cup, international competitions with great influence, sports architecture, sculpture, painting, photographing	Round-the-island bicycle race, round-the-island sailing race, mission hills golf race, rally racing, Nanshan Taijiquan health congress, winter training base, international women's tennis

>60–74; second-grade sports tourism resources have the score range >45–59; first-grade sports tourism resources have the score range >30–44; sports tourism resources under grade has the score range <29.
2. Rating: fifth-grade sports tourism resources are also known as "special grade sports tourism resources"; fifth-grade, fourth-grade, and third-grade sports tourism resources are collectively known as "excellent sports tourism resources"; second-grade and first-grade tourism resources are known as "ordinary sports tourism resources".

3.3 Evaluation on Tropical Sports Tourist Resources in Hainan

3.3.1 Qualitative Evaluation on Tropical Sports Tourism Resource System in Hainan

1. Tropical sports tourist resources in Hainan have comprehensive types and varieties, diversified sports tourism functions.
2. Tropical sports tourist resources in Hainan have marine characteristics and excellent quality.
3. Tropical sports tourism resources in Hainan have rational conditions for existence and ideal portfolio.
4. Tropical sports tourism industry in Hainan has lower degree of development and great potential in resources.

3.3.2 Quantitative Evaluation on Tropical Sports Tourist Resources in Hainan

According to the characteristics and development practice of tropical sports tourism in Hainan, evaluation on sports tourism resources according to the above research can be divided into participation type (including mountain type, coastal type, canyon type, canyon type, river valley type, river type, cliff type, hot-spring type) and ornamental type; according to the index weight system determined in this study, this thesis employs expert scoring approach to calculate and thereby obtain the comprehensive evaluation results of tropical sports tourist resources in Hainan (Table 6).

1. Evaluation on participating sports tourism resources (Table 7).
2. Evaluation on Ornamental-type Sports Tourism Resources (Table 8).

Table 5 Weight value of sports tourism resource index evaluation system

	Ornamental value	Esthetic value	Rare and exotic landscape	Scale abundance and probability	Interference and damage preservation	Popularity and influence	Appropriate travel period or the scope of use	Value-added weight
Total N	12	11	12	12	12	12	12	10
missing values	0	1	0	0	0	0	0	2
Average	24.25	23.27	14.00	10.75	4.67	16.25	7.08	−18.90
Median	25.00	22.00	14.00	10.50	5.00	12.00	5.00	−23.00

Table 6 Evaluation on participating sports tourism resources (I)

Coastal type	Total score	Grade	Rating	Mountain type	Total score	Grade	Rating
Yalong Bay	94	Fifth grade	Special grade	Jianfeng Ridge	90	Fifth grade	Special grade
Dadonghai	91	Fifth grade	Special grade	Qixian Ridge	90	Fifth grade	Special grade
West Island	92	Fifth grade	Special grade	Limu Ridge	77	Fourth grade	Excellent grade
End of the world	66	Third grade	Excellent grade	Yingge Ridge	90	Fifth grade	Special grade
Shimei Bay	84	Fourth grade	Excellent grade	Diaoluo Mountain	80	Fourth grade	Excellent grade
Perfume Bay	77	Fourth grade	Excellent grade	Macaque Ridge	76	Fourth grade	Excellent grade
Tufu Bay	76	Fourth grade	Excellent grade	Yajiada Ridge	70	Third grade	Excellent grade
Wuzhizhou Island	82	Fourth grade	Excellent grade	Bawang Ridge	90	Fifth grade	Excellent grade
Demarcation Islet	82	Fourth grade	Excellent grade				

Table 7 Evaluation on participating sports tourism resources (II)

Canyon, ravine type	Total score	Grade	Rating	Total score	Folk custom type	Total score	Grade	Rating
Five Fingers mountain canyon	95	Fifth grade	Special grade	95	Bull Racing	84	Fourth grade	Excellent grade
Wanquan river valley	91	Fifth grade	Special grade	91	Cross-bow shooting	90	Fifth grade	Special grade
Changhua river valley	86	Fourth grade	Excellent grade	86	Swinging	79	Fourth grade	Excellent grade
Diaoluo mountain valley	81		Excellent grade	81	Wrestling	71	Third grade	Excellent grade
Qicha river valley	62	Fourth grade	Excellent grade	62	Climbing vines and picking flowers	75	Fourth grade	Excellent grade
Limu vidge canyon	60	Third grade	Excellent grade	60	Bamboo pole dance	95	Fifth grade	Special grade
Jianfeng ridge canyon	76	Third grade	Excellent grade	76	Yajia (tortoise-like backward tug-of-war)	85	Fourth grade	Special grade
Bawang ridge	70	Fourth grade	Excellent grade	70	Carrying-pole dance	75	Fourth grade	Special grade
Yingge ridge	65	Third grade	Excellent grade	65				

Table 8 Evaluation on ornamental-type (sports events) sports tourism resources

Sports event	Recreation value	PE value	Exotic landscape	Scale abundance	Preservation degree	Popularity	Appropriate travel period	Environmental protection	Total score	Grade	Rating
Round-the-island bicycle race	24	21	13	10	5	11	5	5	94	Fifth grade	Special grade
Round-the island sailing race	21	18	10	9	5	9	5	6	83	Fourth grade	Excellent grade
Nanshan Taijiquan health congress	20	17	9	8	5	8	5	5	77	Fourth grade	Excellent grade
Iron man triathlon	19	19	8	7	5	7	5	5	75	Fourth grade	Excellent grade
Golf tournament	23	21	12	9	5	10	5	5	90	Fifth grade	Special grade
World Cup Surfing competition	22	19	11	8	4	9	5	6	84	Fourth grade	Excellent grade

4 Conclusions and Recommendations

4.1 Conclusions

1. Tropical sports tourism resources in Hainan can be divided into 2 main categories, 7 sub-categories, 67 basic types, and 513 resources monomers.
2. Qualitative evaluation on tropical sports tourist resources in Hainan are as follows: rich types and diversified functions; marine characteristics and excellent quality; better portfolio, lower development degree, and the resource potential is far from full play.
3. Quantitative evaluation on tropical sports tourist resources in Hainan.

This thesis established sports tourism resources index system by applying experts consulting method and further conducted categorized quantitative evaluation on tropical sports tourism resources in Hainan. After discussion and certification of experts, the evaluation results prove scientific and rational, and the conclusion has practical and reasonable significance, and it is of practical significance for the development of tropical sports tourism resources in Hainan.

4.2 Suggestions

1. Development of tropical sports tourist resources in Hainan shall be dominated by the protective development of ecological environment and reduce the damage on resources to a minimum degree. At the same time, we should advocate that sports tourists should protect environment as their sacred duty in their tourism activities.
2. Sports tourism resources evaluation index system established in the study does not only consider the characteristics of sports tourism resources, but also absorb the merits of other tourism resources evaluation index, but it still needs further research and improvement.
3. This study proves that this system is effective in the evaluation of sports tourism in Hainan, but whether it is also effective for sports tourism resources in other regions also needs to be further verified in practice.

Foundation Project The National Social Science Fund Project—Youth Project "Empirical Research on Cultivation of Tropical Off-season Sports Tourism Market in China" (13CTY015) stage outcome.

About the Author Tian Jianqiang (1978-), male, born in Xuzhou, Jiangsu Province, MA, mainly engaged in the study on sports, humanities and social science. Tel: 13278911438

References

1. Xia, M., Su, Y., etc.: Analysis of current development situation of sports tourism in Hainan and the study on special planning. Ind. Sci. Tribune (2) (2013)
2. Song, J.: Study on the development strategy of tropical sports tourism in Li Ethnic minority in Hainan Province. J. Hainan Univ. (Humanit. Soc. Sci. Ed.) (5) (2011)
3. Jia, J.: Study on present situation and sustainable development of sports tourism in Hainan Province. J. Hainan Univ. (Nat. Sci. Ed.) (3) (2012)
4. Zhang, J.: Development and Practice of Ecological Tourism, p. 6. China Tourism Press, Beijing (2001)
5. Li, X.: On sports tourism and fitness, p. 8. Beijing Sport University Press, Beijing (2003)
6. Bihu, W.: Principle of regional tourism planning, p. 5. China Tourism Press, Beijing (2001)
7. Wang, X.: On Hainan tourism development and investment trend. Doctoral dissertation of Beijing University (1995)
8. Wang, T.: On the characteristics and development principle of ethnic minority sports tourism. J. Xinjiang Univ. (5) (2000)
9. Ding, H.: Study on talents training model and curriculum system design of sports tourism in China. J. Tianjin Univ. Sport (1) (2002)
10. Zhenfang, H.: Study on Coastal Ecological Tourism Development Model. Doctoral Dissertation of Nanjing Normal University, June 2003
11. Chen, L.: Australia paying more attention to sports tourism. Foreign Sports Dyn. (27) (1998)
12. Feng, S., Han, L.: A Comparative Study of Chinese and Foreign Sports Tourism Market. Subject of General Administration of Sport of China, Project No. 170SS9827
13. Zhu, J.: Preliminary exploration on issues concerning development of sports tourism projects. Sports Sci. (5) (2000)
14. Wang, X.: On the development mode of tourism resources and tourism resources sustainable development concept in China. Geogr. Sci. (7) (2004)
15. Bao, J.: Study on Tourism Development—Principle, Method and Practice, p. 9. Science Press, Beijing (2003)
16. Caspersen, C.J., Powell, K.E., Christenson, G.M.: Physical Activity, Exercise and Physical Fitness: Definition and Distinctions for Health-related Research. Public Health Reports, March 1985

ASEB Analysis of Tropical Out-of-Season Sports Tourism Products in China

Taking Sanya for Example

Zhao Feng, Tian Jianqiang and Wang Xiaolin

Abstract Sanya, as tropical coastal tourism city in China, is well known both at home and abroad, and it boasts of rich and unique sports tourism factors. The thesis, by taking the sports tourism products in Sanya for example and from the perspective of tourists, employed grid analysis method of ASEB to conduct in-depth analysis of strength and weakness, opportunities, and threats of sports tourism products in tropical region in China and activities, environment, experience, and benefits of tourists in the tourism process, and on this basis, the author put forward the strategies of developing and promoting out-of-season sports tourism products in tropical region in China.

Keywords ASEB · Out-of-season · Sports tourism products · Sanya city

1 Introduction

Sports tourists have greatly varied interest in different sports tourism resources, this characteristic determines the seasonal change of sports tourism activities [1]. Out-of-season tourism is relative to the in-seasonal tourism, it is in the pursuit of the maximized difference between residential area and destination. From the point of view of economics, its essence is the maximization of the benefit; from a cultural point of view, it is in fact a cross-cultural behavior; from the psychological point of view, it is a breakthrough against people's fixed "off-season" thinking pattern and a change in people's thinking. Out-of-season tourism is wide but not universal

---

Z. Feng
Jilin Sports College, Changchun 13022 Jilin, China

T. Jianqiang (✉) · W. Xiaolin
School of PE, Hainan Normal University, Haikou 571158 Hainan, China
e-mail: tiantian1978@yeah.net

B.-Y. Cao et al. (eds.), *Ecosystem Assessment and Fuzzy Systems Management*, Advances in Intelligent Systems and Computing 254, DOI: 10.1007/978-3-319-03449-2_44,

because it has its own applicable scope [2]. Tropical regions in China cover about 72,000 km^2 in total. Hainan as the main tropical region in China accounts for about 50 % of the total tropical area in China [3]. Therefore, Hainan province has the advantage richly endowed by nature in development of out-of-season sports tourism products [4].

2 Summarization of ASEB Analysis of Sports Tourism Products in Sanya

Based on in-depth interviews, depth interview questionnaire and questionnaire, on the basis of the Sanya sports tourism experience information and interview information feedback questionnaire, through the use of statistical software for tourists different experience requirements and results of the collation and analysis of certain, in individual cells after induction summary in ASEB analysis in conclusion, the analysis of Sanya sports tourism product ASEB (Table 1).

3 Specific ASEB Analysis of Sports Tourism Products in Sanya

General conclusion sheet of ASEB analysis basically summarized strength, weakness, experience and earnings, and other internal and external manifestations in Sanya sports tourism industry. In order to better reflect the purpose of the study, this thesis, based on the analysis of questionnaires and interviews with tourists, carried out ASEB analysis of strength, weakness, opportunities, and threats of sports tourism products in Sanya from four aspects: activities, environment, experience, and benefits of tourists.

3.1 Strength

3.1.1 Strength of Activities

Rich tropical sports tourism resources in Sanya provide a favorable basis for developing diversified and unique sports tourism products. At present, Sanya has very rich sports tourism products, sports tourism products that have been developed can be divided into five categories (Table 2): sports events products, leisure sports tourism products, festival celebration tourism products, thrilling sports tourism products, and folk-custom sports tourism products. Sanya sports tourism products have multiple attributes, strong portfolio of products, higher function in experience and entertainment and thus are sufficient to attract tourists to revisit there.

Table 1 General conclusion sheet of ASEB analysis of sports tourism products in Sanya [5]

	Activities	Environment	Experience	Benefits
Strength	Sports tourism products have multiple attributes, strong combination performance; strong uniqueness (tropical climate); products have strong function in experience and entertainment	Dominated by seaside resort, with beautiful natural environment, simple local folk customs, higher satisfaction of tourists, longer coastline, unique culture in center city (Li and Miao ethnic minority culture and "end of the earth" culture)	Integrating sightseeing, culture, fitness, leisure, and vacation; products have strong participation ability, full of excitement	Increase the sensory experience of tourists of sports tourism projects, get mental and physical relaxation and as well as psychological satisfaction
Weakness	Tourism items are obviously restricted by season (the number of summer tourists falls sharply and changes occur to the market demand). Location deviates from Circum-Bohai-Sea metropolitan circle and the Yangtze river delta metropolitan circle	The development of some products will have a negative impact on the environment (such as diving, tropical rain forest adventure), most of the tourism projects have a certain risk	Part of the sports tourism products are greatly restricted by age, physical health, and other factors, and price of ticket and products in some part of scene areas are much higher than that in similar cities (such as motor boats, Sea Trek Parasailing, surfing), experiencers are prone to feel fatigue	Local residents have lower participation degree; tourism income is limited to specific groups or teams; due to residence time limit, part of the tourists have no chance to experience time-consuming sports tourism products (such as swimming, diving and surfing)

(continued)

Table 1 (continued)

	Activities	Environment	Experience	Benefits
Opportunities	Having attracted greater concern of the government, vigorously developing the yacht, sailing, windsurfing, and other water sports, paying equal attention to scenic area development and planning, optimized composition of product development and festival tourism	Thanks to strengthened efforts in regional environmental remediation, the environment has been greatly improved, local competent agencies have established distinctive tourism image, strengthened efforts in network information publicity, and thereby the market order has been obviously improved	Develop different product combination and preferential package, encourage visitors to participate in sports, meet the diversified experience needs of tourists	National economic income grows faster, the number of travelers sharply increases; it is necessary to take various measures to make the tourists get benefit therefrom, and thereby, further develop potential tourists
Threats	The eastern coastal cities (such as Dalian, Qingdao, Xiamen, Hongkong) also have the same sports tourism products; part of the sports tourism products is constrained by environmental capacity (such as diving, boat)	Similar coastal tourist cities have higher economic level and stronger competition ability; Sanya is still lack of relevant supporting tourism service facilities	Local residents have less understand and participation in some part of sports tourism products; sports tourism souvenirs are too simple and single; part of sports tourism products life cycle step into debility period	Some physically limited visitors cannot get psychological satisfaction; tourists' actual experience is different from their expectation

Table 2 List of sports tourism products that have been developed in Sanya

Types of sports tourism products	Sports tourism products
1 Sports events products	Round-the-island bicycle race, round-the-island sailing race, hot-air balloon race across the Qiongzhou Strait, golden coconut golf show, European women's occupation golf tournament, ITF international women's tennis tournament, Sanya open competition of the world beach volleyball tournament, China (Sanya) international tropical rainforest adventure race, national students orienteering, world surfing championship
2 Leisure sports tourism products	Diving, sailing, windsurfing, swimming, banana boat, sea bicycles, sea trek parasailing, flying fish, algin fishing boat, spa, sea fishing, sunbathing, fishing, cycling, motorboat
3 Festival celebration tourism products	Tianya international wedding festival, longevity culture festival, March 3rd day, Junpo festival, festival of Madam Xian, Hainan joy festival, dragon boat festival, coconut festival, double ninth festival
4 Thrilling sports tourism products	Island tourism, offshore powerboat, tropical rainforest adventure, sea trek parasailing, island survival training, tropical desert hiking adventure, climbing and underwater photography, golf tourism, forest tourism and mountain climbing, water drift and dive, underwater photographing, suspended flying, paragliding, cliff diving, cave exploration, canyon zip, cliff descending, descending above creek, car travel
5 Folk-custom sports tourism products	There are more than 300 folk-custom sports tourism projects in the whole country; Sanya has bamboo pole dance, Yajia (tortoise-like backward tug-of-war), cross-bow shooting, carrying-pole dance, hitting the coconut with stick, dragon boat racing, racing board shoes, etc.

3.1.2 Strength of Surroundings

Situated on the latitude very close to that of world famous tourist resorts such as Hawaii of the USA, Bali Island of Indonesia, Thailand's Phuket Island, etc., Sanya boasts of beautiful natural environment and has the rich characteristics of tropical coastal cities. The territory of Sanya covers the coastline as long as 209.1 km, and there are 19 large and small harbors. Sanya serves as the sports tourist destination dominated by tropical coastal beach and tropical rainforest, enjoying world first-class sea water quality, beach quality and atmospheric quality, domestic higher level in forest coverage rate, and city greening rate. Excellent ecological environment has laid the foundation for sustainable development of Sanya and also attracted a large number of tourists both at home and abroad to come here for tourism, residential healthcare and real estate investment. Sanya has become an "ideal second residence" of people. Sanya attracts tourists from all over the world with its unique Li and Miao ethnic minority culture and "End of the World" culture. The environment determines the tourists' satisfaction on the sports tourism products to a certain extent, and thus good natural environment, and simple, honest and unspoiled local residents in Sanya can directly improve the satisfaction of tourists.

3.1.3 Strength of Experience

Integrating sightseeing, fitness, and leisure and vacation resort, sports tourism products in Sanya are full of thrilling experience and strong participation and thereby have very great attraction to tourists. With beautiful sea and beach, Sanya has carried out a lot of experience sports tourism projects, such as thrilling diving, leisure sunbathing, competitive beach volleyball, folk-custom fishermen life experience, etc., different consumption types have been designed with different experience items.

3.1.4 Strength of Benefit

Currently, domestic sightseeing tourism occupies the dominant position of the tourism market, multiple-style development, and promotion of Sanya sports tourism products which create good conditions for visitors participating in sports tourism, increasing the tourist's sensory experience, relaxing visitors mentally and physically and thereby improving tourists' travel satisfaction.

3.2 Weakness

3.2.1 Weakness of Activities

From November to next March, in each year is the golden tourism period in Sanya. Compared with the peak tourist season, Sanya will step into the hot summer during May–August every year, which is the slack season of tourism in Sanya, facing the sharply declined number of tourists. Sports tourism products in Sanya are dominated by water sports and beach sports, which are not too vulnerable to the impact of season, but slack tourism season comes during May–August due to the travel restrictions for tourists.

3.2.2 Weakness of Surroundings

The development and utilization of sports tourism products that have higher dependence on the environment will have a negative impact on the environment, e.g., diving sports have unrecoverable destruction on coral reefs; some sports events such as: sea trek parasailing, diving, mountain climbing, and the like have a certain risk and higher requirements for sportswear prepared and even require full-time coach to provide guidance.

3.2.3 Weakness of Experience

Part of the sports tourism products are greatly restricted by age, physical health, and other factors, and price of ticket and products in some part of scene areas are much higher than that in similar cities (such as motor boats, Sea Trek Parasailing, surfing), experiencers are prone to feel fatigue.

3.2.4 Weakness of Benefit

Local residents have lower participation degree; tourism income is limited to specific groups or teams; due to residence time limit, part of the tourists have no chance to experience time-consuming sports tourism products, e.g., playing an 18-hole golf race takes 4 h in total, it probably takes 2–3 h for tourists experiencing diving from training to going underwater to experience.

3.3 Opportunities

3.3.1 Opportunity of Activities

Local government pays great attention to the development of sports tourism products; second-phase action plan for Sanya tourism internationalization points out that we should "speed up the process of planning and developing marine theme park dominated by water recreation projects, strengthen the project introduction efforts, launch in-depth development of diving, fishing, yacht, sailing board, sailing, motor boats, surfing, sea traction umbrella, sea Golf and other marine tourism projects, initially form the tourism product system themed by golf, spa health, sea fishing, luxury yacht, marine sports, and other tourism projects," vigorously develop yacht, sailing and sailing board, and other water sports, pay equal attention to the development and planning of scenic area, and optimized portfolio of product development and festival tourism.

3.3.2 Opportunity of Surroundings

At present, with the increasing impact of tourism on Sanya economy, the government noted the importance of environment for the tourism development and thus began to increase the efforts in regional environmental remediation. As a result, government has greatly improved local environment, established a distinctive tourism image, strengthened efforts in network information publicity, and the market order has been improved eventually to some extent. Sanya tourism boasts of higher visibility and reputation; environmental rectification has enhanced the influence of Sanya tourism and created a good social environment for the development of tourism industry in Sanya.

3.3.3 Opportunity of Experience

With the development of sports tourism and people's understanding of sports tourism and the pursuit of modern people on the health, leisure, characteristics, novelty, affordability, participative tourism projects with health value are more and more favored by tourists, Sanya has not only developed rich sports tourism projects, but also formulated different preferential package of product portfolio, encouraged tourists to participate in the sports events, and met varied experience requirements of tourists.

3.3.4 Opportunity of Benefit

Sanya local residents have insufficient understanding and participation in some part of sports tourism products, so we should seize this opportunity to create more and more sports tourism projects that are suitable for the participation of local people and can provide employment opportunities for them; sports tourism souvenirs that have been developed in Sanya are rather single; and part of sports tourism products developed in the early period has stepped into the decline stage in their life cycle as the aging of marketing strategy and facilities with the passage of time.

3.4 Threat

3.4.1 Threat of Activities

Sports tourism started late in Sanya, and it is still in the primary stage and lack in facilities, safety measures, staffing of professional talents, but some eastern coastal cities of our country (Dalian, Qingdao, Xiamen, Hongkong) also have the same sports tourism products, overseas cities on the same latitude have more mature development in sports tourism projects; some sports tourism products are prone to environmental capacity constraints due to the impact of water domain and other environmental conditions.

3.4.2 Threat of Surroundings

Compared with similar coastal tourist cities such as Dalian, Qingdao, Xiamen, etc., that have higher economic level and strong competitiveness, Sanya still lacks the related tourism service facilities. Earlier developed sports tourism projects excessively pursue the maximized transition of economic interests, resulting in the destruction of the environment, e.g., the coral can be rarely seen in submarine world because of contamination of seawater and excessive flow of people.

3.4.3 Threat of Experience

Safety guarantee measures are the premises of developing and establishing sports tourism projects; relevant laws and regulations serve as necessary conditions for smooth operation of sports tourism projects; and service industry system plays a restriction on the development of sports tourism. The development of tourism industry in Sanya has come to the front of the tertiary industry, but the stale industry system seriously restricts the internationalization progress of Sanya tourism.

3.4.4 Threat of Interests

The development of tourism industry has a great drive toward the economy, especially for Sanya that a city takes tourism industry as a pillar; the maximization of economic benefit is the essential investment choice for investors. Because now the city of Sanya stays in the great situation of building international tourism island, many preferential policies will promote the development of tourism industry. However, corresponding laws and regulations are still not perfect enough, many investors will exploit loopholes of laws to seek illegal benefit; standardized development of sports tourism industry will be severely restricted if things go on like this.

4 Strategies of Developing and Promoting Out-of-Season Sports Tourism Products in Sanya

4.1 Attaching Importance to Study on Product Attributes, Strengthen Analysis of Environmental Constraints

We should fully understand the attributes of sports tourism products, especially unique features, heterogeneous compatibility, and functional extension of sports tourism products and strengthen the publicization of unique characteristics of

Sanya sports tourism products. Based on the characteristics of Sanya sports tourism products, it is very necessary to strengthen the analysis of restriction factors of sports tourism environment. In short, analysis of the environmental constraints is the determination of tourism marketing subject on its scope of activities in a certain meaning. It is required to clearly mark the products categories that Sanya can provide, so as to create good conditions for developing sports tourism products.

4.2 Market Segmentation, Aiming at the Tourist Source, Focusing on the Target

We should provide data for feasibility study on the development of sports tourism products through market segmentation, target market selection, and study on tourist source market. By making full use of the Internet, exhibition and tourism marketing cooperation, and other promotion measures, we should convey and transfer promotion information of Sanya sports tourism products, create demand, and achieve "AIDA" target effects [6]. On the basis of scientific analysis of existing value and rationality and feasibility of sustainable development of sports tourism products in Sanya, we shall immediately launch a persistent series of subsequent sports tourism products and lock the target, and obtain the special benefits at long last.

4.3 Establish the Brand, Foster the Market

The brand of Sanya sports tourism should have distinct tropical characteristics, and be full of the color of this era, keep abreast with popular culture tone, play climate advantage, establish the brand advantage, i.e., I have what you lack, I have specialty in what you have, and I create uniqueness in your specialty.

Foundation Project The National Social Science Fund Project—Youth Project "Empirical Research on Cultivation of Tropical Out-of-season Sports Tourism Market in China" (13CTY015) stage outcome.

About the author Zhao Feng (1969–), male, born in Beijing, associate professor, is mainly engaged in study on sports, humanities and social science.

References

1. Yang, Q.: 20 years of research on sports tourism in China: Review and prospect. Chin. Sports Sci. Technol **47**(5), 90–97 (2011)
2. Li, J.: A few ideas on the construction of Hainan international tourism island. J. Hengshui Univ. **13**(1), 122–124 (2011)
3. Zhou, C., Miao, X., Dai, G.: Case Analysis of International Tourism Planning. Nankai University Press, Tianjin (2004)
4. Xia, M.: Study on the Development of Tropical Sports Tourism: Taking Hainan as an Example, p. 12. Beijing Sport University Press, Beijing (2012)
5. Wang, X.: On characteristic sports tourism of Hainan. J. Hainan Radio TV Univ. **4**(6), 78–80 (2011)
6. Zhang, R.: Development strategy of sports tourism products. Sports Sci. Technol. **23**(2), 11–13 (2002)

Author Biography

About the author Zhao Feng (1969–), male, born in Beijing, associate professor, is mainly engaged in study on sports, humanities and social science.

Study on Sino-Foreign Cooperation in Golf Education

Li Jia, Tian Jianqiang and Jiang Yanwen

Abstract Through analysis of Sino-foreign cooperation in running golf management specialty, combined with existing problems in golf education in our country, this thesis constructed a decision-making model suitable for Sino-foreign cooperative school-running in higher vocational colleges and provided reference proposals for promoting realistic, pervasive, operatable Sino-foreign cooperative school-running mechanism.

Keywords Golf · Sino-foreign cooperation

With the development of golf sports in China, the lack of high-quality management personnel become an important factor restricting healthy development of golf sports in our country. As an important field of training senior management personnel in golf industry—golf management education in higher education institute is still in its infancy in our country, the quality of talented people cannot meet the actual needs of golf industry, and its disciplinary affiliation, talent training mode and the curriculum are still in the exploratory stage [1]. Professional golf management specialty in foreign countries guides the new trend of world golf management and education with its high-quality talents training standards; its advanced talents training model and scientific curriculum settings have great significance for golf management education in China.

L. Jia · J. Yanwen
Sanya Aviation and Tourism College, Sanya, Hainan, China

T. Jianqiang (✉)
School of PE of Hainan Normal University, Haikou, China
e-mail: tiantian1978@yeah.net

B.-Y. Cao et al. (eds.), *Ecosystem Assessment and Fuzzy Systems Management*, Advances in Intelligent Systems and Computing 254, DOI: 10.1007/978-3-319-03449-2_45,

1 Current Status of Domestic Golf Education

Since reform and opening up, China's golf industry has developed rapidly. According to calculations of insiders, the total number of existing golf courses has exceeded 300 and rapidly increases with the annual rate of 20–30 %. According to the calculation conducted by concerning experts, a golf course under normal operation needs 350 professional operation and management personnel. According to the above data, the number of gold courses will attain about 1,000 in the next 5–6 years, requiring about 350,000 employees engaged in golf industry. However, less than 500 students receive golf major education in university currently in China, and Shenzhen University took the initiative to found Golf School in 1997. Subsequently, Jinan University, Beijing Forestry University, Beijing Agricultural University, Tongji University, and other more than ten universities opened golf-related specialties, even if all of the graduates from the school mentioned above enter the golf industry. However, a huge gap still exists in the demand of talents on golf sports.

Many colleges and universities have signed cooperation agreements with overseas golf educational institutions to directly aim at high-end management talents market. These universities have higher starting points, uneven quality of students, and more comprehensive curriculums established, but golf major education in such universities is constricted by higher cost, too much basic knowledge, and lack of practice. Golf management is a work requiring higher practical operation, but golf education in many schools is confined to golf course management, business operation, and other basic theoretical knowledge; talents trained in such mode are difficult to adapt to the demand on talents of golf industry [2]. In addition, golf talents training still stays in the college education level in many schools. Because college graduates have weak theoretical foundation, lower knowledge level in economics and management, which greatly restricts the development of their future occupation, cannot fully meet the practical requirements of golf course. Shortage of professional teachers is also an important factor severely restricting the development of golf education in China, and most schools in China have this defect. In addition, the lack of experience in running school, irrational teaching mode, education method still in the exploratory stage, and other shortcomings seriously restrict the development of golf education in China.

2 Analysis of Sino-Foreign Cooperative Mode of Golf Education

Scotland is recognized as the birthplace of golf sports; the United States is the world's largest golf market. In terms of both golf sports level and development level of golf-related industries, the USA and UK play pivotal roles in the development of golf course in the world and the two countries also rank top in the world in golf education. Golf sports achieved rapid development during the period

1960–1970s, which is also the stage in which the USA and UK started systematic training of golf talents; a large number of highly qualified professional golf management personnel provide an important talent reserve for healthy development of golf industry in their countries and the world [3]. Among domestic existing Sino-foreign cooperation projects in golf education, most are projects carried out in cooperation with national educational institutions in the USA and UK. Cooperation fields are focused on three aspects including golf course facilities management, golf course management, and golf sports skills.

2.1 International Certificate Cooperation in Golf Industry

International professional certificate cooperation in golf industry corresponding to golf academic education refers to the Sino-foreign cooperative project dominated by skills and certification training, mainly covering training items of internationally recognized and common professional skills, with the objectives of obtaining internationally recognized certificates as the goal. As academic education has more normative requirements on school-running conditions, curriculums, teaching contents and arrangements, and fresh source quality, etc., both cooperation partners are usually more cautious. Almost all Sino-foreign cooperative golf education in China is concentrated in the cooperation in golf professional skills certificate training. We usually adopt "3 +1" or "2 +1" model, that is, trainees can complete 2 or 3 years of study in our country and continue to study abroad, and students who have eligible credit can receive a corresponding certificate internationally recognized in golf industry. For example, School of Golf of Shenzhen University cooperates with Elmwood College Scotland in the mode that students can choose to study for some period in China and then go abroad to continue to further themselves; two institutions mutually recognize the credits obtained in the counterpart, and students who have eligible credit can eventually win a corresponding certificate internationally recognized in golf industry awarded by foreign institution. Many domestic institutions have extensive cooperation with such kind of foreign institutions, with the cooperation fields covering golf course turf maintenance, golf club management, golf professional sports skills, and other aspects, such as golf referee training project cooperated between China Golf Association (CGA) and R&A, and golf professional coaches training project cooperated between CGA and Australian Professional Golf Association PGA; these are all widely praised by insiders.

2.2 Dual-Diploma Education

Colleges and universities mostly adopt single-diploma education, and students can only get the diploma from one college in traditional single-diploma education mode. Dual-diploma education means that students can obtain two certificates awarded by tow institutions after completing their studies required, and this education mode has enormous appeal to students who will be able to get certificate awarded by overseas institutions without going out of the country [5]. Therefore, some domestic higher education institutions setting up golf professional education have taken the Sino-foreign cooperative mode by adhering to the educational philosophy of "certification education combined with academic education" and extensively implementing "professional quality training certificate + domestic and international professional certification + graduation certificate + project practical training certificate". We usually adopt "3 + 1" or "2 + 1" model, that is, trainees can complete two or three years of study in our country and continue to study abroad; those who have eligible credit can win bachelor's degree or master's degree awarded by abroad education institutions.

3 Teaching Quality System Monitoring

Sino-foreign cooperative golf education started too late, but has developed rapidly in recent years; it is a kind of interactive and high-level talent recruitment mode benefiting multiple parties including students, teachers, and cooperative education institutions and complying with specific conditions of our country. On the one hand, students can obtain overseas diploma with lower costs and lower starting point in English; on the other hand, our teachers can access the world's newest educational philosophy and have more opportunities for international exchange and thereby promote the transition in teaching ideas and teaching methods, escalate professional construction and optimize talents training so as to achieve true internationalization of school-running means. But we should also note that we should form complementary advantages to the maximum extent rather than total Westernization in terms of the introduction of teaching modes and curriculums by integrating with China's actual conditions, rationally arranging curriculums and scientifically distributing class hours. Therefore, we must establish a standard quality control system.

3.1 Emphasis on Teaching Quality Monitoring Process

It is a key link to implement teaching evaluation on foreign teachers for ensuring the quality of teaching. Teaching evaluation aims to promote the teaching management and ensure the quality of teaching. Language barrier makes it difficult for teaching work of foreign teachers; domestic universities have a common problem that teachers of specialized courses cannot express themselves fluently in English, but teachers undertaking supplementary work do not understand too much on the golf industry; such a contradiction caused a lot of obstruction for guaranteeing high-quality teaching [6]. In addition, because of differences in the education system, many Western countries do not require teachers to keep students' papers, assignments, papers, and other information. These factors led to that our teaching evaluation on foreign teachers can only stay in such levels as completing of teaching hours, classroom atmosphere, as well as the students' feedback. As the result, we are unable to give sufficient concern to their teaching ability, preparation conditions, lecturing contents, and other aspects.

1. Establishment of scientific teaching quality evaluation index: Domestic university can establish student-based teaching evaluation system, establish teaching inspection system, and implement initial evaluation at the beginning of semester, mid-term, and final evaluation and other conventional teaching inspection and inspection and evaluation on periodic teaching work of our partners.
2. Establishment of supervisory department, management personnel, expert, or peer class-auditing system to regularly attend course lecturing and give written feedback comments on teaching effect and quality.
3. Establishment of student feedback system: Through timely collection, compilation, and feedback of information about front-line classroom teaching from students, we can provide a reference for teaching management and quality monitoring and timely adjust foreign teachers who were found to have some problems in the teaching inspection process. Foreign partner shall send full-time supervisors to closely collaborate with Chinese side, and it is required to carry out at least twice quality inspection and monitoring per semester for the cooperative projects; any problem found shall be promptly settled by negotiations.

3.2 Implementation of Post-teaching Quality Control Measures

1. Consummating course examination (evaluation) system. Examination (evaluation) is one of the most effective means of evaluation on classroom "teaching" and "learning" effects. Sino-foreign cooperative education institutions shall establish a complete set of examination management system to implement uniform standards and strict management over the proposition, invigilation, marking, and other aspects.
2. Establishment of "performance analysis report" system. After the examination of end course, it is required to conduct a comprehensive analysis and summarization on the degree of difficulty of the examination papers or operation items as well as students' achievement.

4 Inspiration on Golf Education in China

4.1 Talents Training Model for Golf Management Specialty

Golf management has very strong specialty and needs to be provided with the education dominated by job skills, career development skills, and that belong to vocational education in terms of the nature of education. Emphasis on golf sports skills, events arrangements, and club management are very important aspects of golf education. Golf industry is a complex system based on golf culture, requiring that management personnel should have multidisciplinary knowledge background and comprehensive abilities. Some educational institutions simply focus on the "Order-style Education", in which vocational education becomes the pre-employment training center. These institutions may input still higher investments in hardware, and some even build their own golf course for students training and internships, but they only emphasize the employment rate, pushing the students into the employment circle without follow-up support and training; this mode is not conducive to establishing vocational education brand. These institutes should strengthen the establishment and teaching of such curriculums as management and economics, and even physical education courses to expand the students' knowledge background and enhance their career development ability.

4.2 Discipline Affiliation of Golf Management Specialty

Sino-foreign cooperative golf education mentioned above is not truly a professional discipline in strict meaning; it is actually just a professional direction under a discipline, and students after graduation can be awarded with appropriate bachelor's degree (including management science, horticulture, and physical education). Many domestic universities should take advantage of their academic strengths to cultivate and train golf talents with general golf professional capabilities and unique expertise. Golf industry is a complex industry group, which is in demand of a variety of compound elites characterized by "specialty and multi-skills" or "pluri-expertise and multi-capability". It really resembles emphasis on trifles but ignorance of essentials to ignore golf education specialty features in different colleges and universities, but entangle with the issue of discipline affiliation of golf management specialty.

4.3 Emphasizing Training Its Cultural Attainment

From a social and cultural point of view, golf is not only a lifestyle, but also an attitude to life. Golf is an outdoor sport with the perfect combination of man and nature, and it is one of the recreational sports closing to nature and relieving physical and mental fatigue, possessing profound traditional culture deposits in Western medieval and modern cultural elements. The USA and UK attach great importance to the training of golf cultural quality in terms of advanced management personnel cultivation, and such curriculums as golfer development plan, golfing career development, sports psychology, healthy living, and nutrition that are very strange in domestic universities appeared in gold education institutions in the USA and UK, so that students can first become the practitioner of golf cultural and golf culture communicator. It is a pity that domestic golf management neglects of the settings of such curriculums.

5 Scarcity of High-Quality Teaching Materials Professional Faculty Shortage in Golf Education Specialty

In terms of selecting teaching materials, we should focus on two aspects: on the one hand, play the function of original teaching materials in terms of training language ability, insist on the principle of imparting knowledge from the shallow to the deep, and proceeding in an orderly way; on the other hand, we should learn advanced concepts of original teaching materials to promote localization of English teaching materials. According to the requirements of cooperative education mode, foreign partner shall be responsible for providing a full set of professional teaching programs and materials. Due to the restrictions of fresh source level of students, teaching staff quality and teaching conditions, and other factors, it will be of great difficulty for us if we implement teaching practice completely in compliance with teaching program and teaching materials offered by overseas partners. Therefore, domestic universities should adopt the method of "grafting" original teaching materials, i.e., according to actual situations of domestic college and universities, we shall adopt original materials, teaching plans (lecturing plans) provided by foreign partners as the main teaching materials in specialized curriculums, and meanwhile supplement advanced research achievements in teaching materials of domestic-related specialties in actual teaching activities so as to complete the docking of Sino-Western cross-cultural communication and constantly improve the teaching resources [7].

Sino-foreign cooperative education teaching staff is constituted by foreign teachers and Chinese teachers, who should have the common features of higher level of expertise and international communication skills. Cooperation agreement of Sino-foreign education project shall clearly prescribe that expatriate professional teachers should be outstanding full-time teachers from headquarters of foreign cooperative institution so as to ensure the quality of teaching. Domestic cooperative institute shall choose young and competent teachers with solid professional knowledge, excellent English level, and holding medium–high technical titles to undertake the teaching task of specialized courses and try to implement bilingual teaching. Through attending the classroom teaching of relevant specialized courses, participating in relevant professional academic conferences and activities, professional curriculum development, and accepting multi-faceted supervision of foreign teacher advisory group and other forms to understand project details and transplantation model, and thereby fully experience the curriculum characteristics of foreign vocational education, teaching philosophy, teaching methods in order to improve the overall quality of professional teachers.

6 Conclusion

With the deepening of internationalization of our country, more and more domestic educational institutions will join the ranks of the Sino-foreign cooperative education. Through Sino-foreign cooperation, we can introduce overseas high-quality educational resources and advanced educational philosophy, school-running and management experience, stimulate domestic golf field professional talents education, improve the educational level and social reputation of domestic partner institutions, and meanwhile cultivate a large number of technical applied talents meeting the market demand and complying with international standards. We will, based on the available foundation, continue to explore new ideas in Sino-foreign cooperative education matching the actual conditions of our institute and thereby promote the cause of Sino-foreign cooperation in running schools to achieve rapid and healthy development.

Foundation Project The National Social Science Fund Project—Youth Project "Empirical Research on Cultivation of Tropical Out-of-season Sports Tourism Market in China" (13CTY015) stage outcome.

About the Author Li Jia (1976), male, born in Shanxi, master, lecturer, mainly engaged in research of golf club management and golf education management.

References

1. Fang, X., Chen, X.: Comparative study of undergraduate education of golf management in China. J. Phys. Educ. **13**(6), 71–74 (2006)
2. http://careernet.pgalinks.com/helpwanted/empcenter/pgaandyou/pro.cfm?ctc=1637EB/OL
3. http://careernet.pgalinks.com/helpwanted/empcenter/pgaandyou/pro.cfm?ctc=1641EB/OL
4. Qingminm, Z.: Internationalization: The only way for Chinese High-level Universities. J. High. Educ. Res. (11), 2001
5. Nie, G.: Inspiration of U.S. higher education accreditation standards on higher vocational education in China. J Guangdong Educ. Inst. **23**(3), 69–73 (2003)
6. Dongmin, Y.: Considerations on Sino-foreign cooperation in higher education. J Hunan Med Univ (Social Science Edition), **8**(1), 2006
7. Wu, K., Zhang, C.: Infrastructure construction and practices of golf higher education in China. Special Zone Economy. **2**, (2006)

Part V
Application of Operations Research and Fuzzy Systems

Ideas of Ancient Extensional Operational Decision-making and Its Expectation

Pei-Hua Wang and Bing-Yuan Cao

Abstract By analyzing an instance of Liu Yan's idea in the Tang Dynasty, and the wisdom solution to "no food" in heartland surrounding, the writers in the paper points out that Liu Yan's method of application, in essence, is an idea of decision-making through operations research in extension, and that the use of transformation can be summarized as a matter element one. Thus, the matter element transformation to problems existed even since the ancient. A serious look at problems like these, for the law of finding out "ideas," and coming up with "approaches," no doubt, becomes useful.

Keywords Extension · Operation research · The matter element transformation

1 Introduction

The methods to extensional decision-making through operations research have already existed since the ancient.

Liu Yan's method (purchasing grain) [1] means allowing the officials in the multi-grain and convenient-traffic states and counties to divide the local grain prices and purchasing quantity in decades into five grades before presenting to a shipping company. Whenever new grain price was determined, they began to

B.-Y. Cao (✉)
School of Mathematics and Information Sciences, Key Laboratory of Mathematics and Interdisciplinary Sciences of Guangdong Higher Educatio Institutes, Guangzhou University, Guangzhou 510006, People's Republic of China
e-mail: bycao@gzhu.edu.cn

P.-H. Wang
School of Mathematics and Information Sciences, Key Laboratory of Mathematics and Interdisciplinary Sciences of Guangdong Higher Educatio Institutes, Guangzhou Vocational College of Science and Technology, Guangzhou 510550, People's Republic of China
e-mail: phwang321@163.com

B.-Y. Cao et al. (eds.), *Ecosystem Assessment and Fuzzy Systems Management*, Advances in Intelligent Systems and Computing 254, DOI: 10.1007/978-3-319-03449-2_46,

purchase the grain instead of reporting their decision for agreement. If grain price was decided for the first class, the number of acquisitions was the fifth grade; prices for fifth class, then the acquisition was the first grade; prices for the second grade, then the acquisition of the fourth class; prices for the forth grade, then the acquisition of the second class. At the same time they should report local prices and purchase quantity immediately to the shipping company. Thus, the officials in places of cheap grain price would purchase food naturally according to the highest share volume in history, and other places would purchase according to the standard price so as to avoid blindness. The shipping company still put summary for calculation of the purchase food plans reported by states and counties. If its amount was too much, then they reduced the amount by the places in expensive price and from far distance; If the planned share amount was short, then they increased the amount of the place with low price and close distance.

Liu Yan (AD 715–780 years) in the Tang Dynasty has been recorded in this text, who was recommended for the "child prodigy" with great wisdom at his 8 years [2]. He uses his own ingenuity, clever ideas, for purchasing and transporting Jiangnan grain to solve the Central Plains in "no food" and made outstanding contributions to recover the declining economy because of "An Shi Rebellion" in Tang Dynasty. This is a model of ancient operational decision-making.

The chapter, based on historical records, interprets extension operational decision-making ideas in today. It provides not only a new method of quantitative management thinking to policy-makers for how to combine qualitative with quantitative for better management decision-making, but also a good inspiration for them.

2 Conflicting Analysis and Determination of Matter Element Main Extension Sets

It is easy to see that it is a crystallization of extension operational decision-making by carefully analyzing the methods used by Liu Yan. The basic idea is to apply a series of matter element transformation in Matter Element Analysis to resolve the incompatibility issue.

Following next, we put method of extension operational decision-making [3, 4] into combination of Liu Yan's (purchasing grain) methods, which are described below:

What is the extension and what is the matter element?

Definition 1 [5, 6] The description is called "extension" that we develop an incompatible and opposite problem, change incompatibility into compatibility from two aspects of qualitative and quantitative, and opposite into coexist; while making things, features and value about the features composed of three triples, is called the matter element, denoted by $R =$ (things, features, value) $= (P, C, Q)$.

Food transport is a very difficult problem in the Tang Dynasty. First, transport speed slowed; second is the high cost of transport miscellaneous; and third is amazement of losses and waste. In particular, after the "An Shi Rebellion" in Tang Dynasty, China's Central Plains were famine, people suffering from hungry. In order to resolve this crisis as soon as possible, it is necessary to break the shackles of various conventional procedures and to take strong measures in order to alleviate this conflict. Otherwise, the consequences would be disastrous. Thus, at that time, the two contradictory sides showed hungry and the court with focus of the conflict being a word "grain".

Determine n states and counties to be $A_1, A_2,\ldots, A_n$, the distance to be $S_1, S_2,\ldots, S_n$, constituting n sub-systems and the establishment of fifth-grade grain prices as P_1, P_2, P_3, P_4, P_5, whose purchases were C_1, C_2, C_3, C_4, C_5. Clearly, food in high or low prices, purchased quantity and transportation distance directly involves solutions to time, speed and economic cost in the crisis of "no food " in central plains. To say in now words " fast " refers to as fast as possible,"good" to excellent solutions, and "economical" to lowest costs for solution to these problems. This requires grasping the principal contradiction, and if we hold the main contradiction, all questions may be solved. The grasp of the principal contradiction can be mathematically considered to be the main matter element extension sets. Therefore, the problem here, based on two matter elements of food price and distance, can be determined as their primary extension set: $\tilde{H}_0', \tilde{H}_0''$.

3 Matter Element Transformation

Liu Yan ingenious solution to plains "no food" siege usually referred to people "trick," "ideas," "method," and "strategy." Why Liu Yan could think of these methods? Whether it can be feasible in law, and what method can be followed? If anything, it is the dream of mankind to use a computer to achieve these, which is also a major problems of human research in this century and the next century. In fact, Liu Yan's method contains the use of the following four kinds of matter element transformation:

Replacement Transformation:

Substituting a thing for another thing, a feature for another one, and quantity value for another one is called a replacement transformation.

Liu Yan first shipped x million tons of grain from areas with surplus grain to the Central Plains in order to suspend contradictions in its "no food". He applied this transformation in flexibility.

Transform in Addition or Substraction:

Determining the increase, reduction in the acquired material according to the needs of the target market reduction, is said to be a transformation in addition or subtraction.

Table 1 Purchase prices in food of A_i state

Purchases	Prices				
	P_1	P_2	P_3	P_4	P_5
Grade					
1		C_4			
2	C_5				
3				C_2	
4			C_3		
5					C_1

In his view of the whole situation, Liu Yan considered price and shipping, respectively, to be two factors for determination of the increase or reduction in the region's food purchases in order to reach the purposes of multi-purchasing with low cost and shipping.

Component Transformation:

Dividing a substance product into different parts, or products through a variety of substances forms of a new substance combination transformation is called a component transformation.

Liu Yan successfully used the decomposition component transformation. By economic statistical data, he divided the number of food prices and purchasing quantity within decades from states and counties into five grades, and by a quantitative relationship between the two, summed up the trend of inversely proportional to the share price and volume, and set a reasonable purchasing rule.

Scaling Transformation:

This namely expands or contracts transformation on the matter element.

Liu Yan's purchasing food was made in Table 1.

From the statistics Table 1, we know that there exists a nonlinear relationship between price and share volume (see the real curve of the next Fig. 1).

By scaling transformation, the nonlinear relationship is changed into supply of linear relationship (see the virtual straight line of Fig. 1), which greatly simplifies the operation.

4 Conducting Compatibility Test

Let

$$\Delta C_i = \frac{M - M^*}{P_i}, \quad (\text{at } M > M^*),$$

or

$$\Delta C_i' = \frac{M - M^*}{F_i}, \quad (\text{at } M > M^*),$$

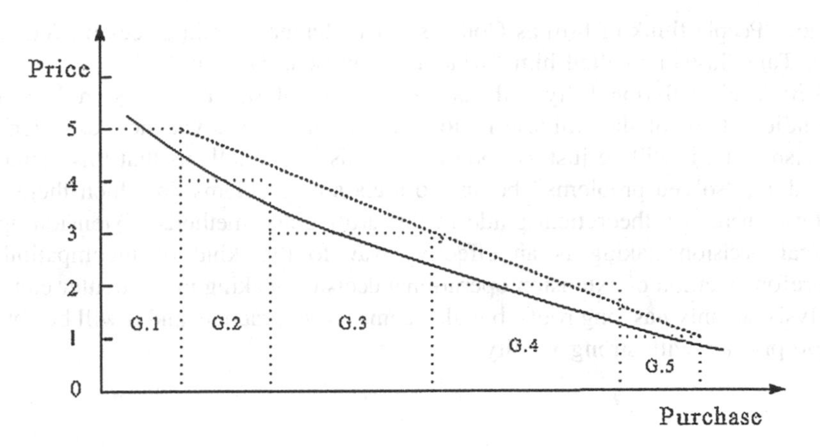

Fig. 1 The relationship between price and share volume

where M means the total amount of funds; M^* is a buying food fee reported; P_i represents grain prices by states and counties; C_i shows the amount of the purchase in low-price states and counties; F_i is transformation cost from their acquisition points; ΔC_i and $\Delta C'_i$ denote a cost increment.

Judgment:

The order quantity should increase if $\Delta CC_i > 0$; it should reduce or even not buy any grain if $\Delta C_i < 0$. When $\Delta C_i > 0$, according to the ΔC_i's size, we determine amount of the purchase of food.

Through the compatibility test, the transformation rationality and feasibility of the measures is also confirmed.

Liu Yan did use the above extension operational decision-making method, a rapid, economical, and complete solution to the Central Plains "no food" siege, admired by generations of people.

In fact, according to the "MengXi stature" records [7], Liu Yan did take the following three measures:

1. Actively rectify grain transportation;
2. The rational organization of transport, to take "local conditions and staging" approach, the expenses costs accounted for the summary accounts by steps;
3. Improvement of transport packaging, changed bulk of bagged, which greatly reduced its loss.

As a result of these measures, he not only accelerated the speed of transportation, but also greatly reduced transportation costs. For example, in the past, it took 9 months to transport grain from Yangzhou to Changan, the loss along the way reached to 20 %. Liu Yan spent only 40 days after improvement with no loss of semi-bucket; only costing 700 Wen fees (ancient money units) to transport per stone; “People think of him as God,” so that Changan’ grain prices moved steadily, Tang Su-tsung called him “Xiaohe” in present dynasty.

Liu Yan’s solution fully reflects the wisdom of our ancestors, reflects our splendid culture of the Chinese nation. If you simulate down their experiences, flourish them, it will be just as good thing. This article tells us that this “smart,” “wisdom, “solved problems” belong to a class of problems, in which there is a pattern; there is a theoretical guide in this artifice and methods. Extension operational decision-making is an effective way to this kind of incompatibility. Therefore, method of extension operational decision-making in the matter element analysis not only has long roots, but also comes from practice, and it will be able to guide practice with strong vitality.

5 Expectation and Example

With the rapid development of science and technology, research difficulty will continue to appear, research areas will continue to expand, and multidisciplinary cross-research is imperative. The extension sets research is combined with rough sets, and fuzzy sets will be likely to impetus artificial intelligence, information and control, economic management, and other fields as well. On December 6–7, 2005, China’s famous Xiangshan Science Conference was held in Beijing, where in No. 271 Colloquium, “extenics scientific significance and future development” appeared as the theme for its two days of discussion, whereby the extension upgraded to a new countertop. In China, a wide range of applications has appeared in the fields like the matter element and extension sets in the extension control, extension information theory, extension systems, extension programming, extension decision, extension logic, extension engineering and generalized force, matter element neural network, extension forecasting and evaluation, etc. [3, 8, 9].

Example (Transform bridge problem in Huanggang of Shenzhen in China) Suppose quaternary form to be 3-ary one $R = (P,C, Q)$ because the universe is the same, then

$R_1 =$ (Chinese inland, traffic rules, driving at right side) $= (P_1,C, Q_1)$,
$R_2 =$ (Hongkong, traffic rules, driving at left side) $= (P_2,C, Q_2)$,
$r =$ (come-and-go auto of Chinese inland and Hongkong, runtime, 24 h/day),
then this is a conflicting problem in R_1 and R_2 under the condition r: $(R_1 \odot R_2) \uparrow (r)$.
Let

$z_1 = (L_1,C,$ unilateralism moving from inland to Hongkong)
$z_2 = (L_2,C,$ unilateralism moving from Hongkong to inland),
and take $L = L_1 \oplus L_2$. Then, L is a link part with N_1 and N_2 where L_1 is a bridge orbit connecting right half highway in inland with the left half one in Hongkong

and L_2 is a_bridge orbit connecting right half highway in Hongkong with the left one in inland, $\oplus$ denotes integrable. Again, we take rough transform:

$$T_1R_1 = R_1' = (N_1 \oplus L_1, C, f_1(\omega)),\ T_2R_2 = R_2' = (N_2 \oplus L_2, C, f_2(\omega)),$$

where

$$f_1(\omega) = \begin{cases} \text{moving right}, & \omega = \text{inland}, \\ \text{unilateral moving to Hongkong}, & \omega = \text{orbit up in a bridge}, \end{cases}$$

$$f_2(\omega) = \begin{cases} \text{moving left}, & \omega = \text{Hongkong}, \\ \text{unilateral moving to inland}, & \omega = \text{orbit below in a bridge}, \end{cases}$$

such that $(R' \odot 1R'_2) \downarrow (r)$, that is, R_1 and R_2 coexist under the condition r, where $L = L_1 \oplus L_2$ is a connecting transition thing of R_1 and R_2. That is each one goes his or her own way, each gets what he or her needs, and it is compatible.

6 Conclusions

We may be similar to the introduction of fuzzy sets and fuzzy convex function [10], the introduction of extension convex sets and extension convex functions [11, 12], on the basis of most proposed extension optimization problem. So first in engineering and investment decisions to achieve [3, 7], we should research the success method of Liu Yan by using the theory of extension and conversion of the bridge thinking (i.e., the use of "side of the road, and perform" thinking), as soon as compiled computer program, computer advice, and strive to explore the "same idea," "anyway" the law. As soon as the design and manufacture of robots simulate the human brain thinking, today's vision is a reality of tomorrow, in order to benefit mankind.

References

1. Shen, G.: Liu Yan buy up grain surgery (method). Meng Xi Bi Tan. http://read.dayoo.com/book/content-10824-636986.htm/
2. Yu Y.H.: Datang Finance Minister: Liu Yan. Chinese Press. http://www.hongxiu.com/x/16350/
3. Cai, W., Yang, C.Y., Lin, W.C.: Method of Extension Engineering. Science Press in China, Beijing (2000)
4. Guéret, C., Prins, C., Sevaux, M.: Operations Research Case: Modeling, Solving. Dash Optimization Ltd (2002)
5. Cai, W.: The extension set and incompatible problem. J. Sci. Explor. **1**, 83–97 (1983)
6. Cai, W.: Extension theory and its application. Chin. Sci. Bull. **44**(7), 673–682 (1999)
7. Cao, B.Y.: Matter element analysis method and its application in the investment decision-making. Selected Papers of the Second Symposium on Matter Element Analysis of China's, Guangzhou (1986)

8. Cao, B.Y.: Rough posynomial geometric programming. Fuzzy Inf. Eng. **1**(1), 37–57 (2009)
9. Cao, B.Y.: The more-for-less paradox in fuzzy posynomial geometric programming. Inf. Sci. **221**, 81–92 (2012)
10. Cao, B.Y.: Interval and fuzzy (value) convex function with convex functional. J. Guangzhou Univ. **8**(4), 1–5 (2008). (Natural Science Edition)
11. Cao, B.Y.: Extension convex set. Acta Sci. Naturalium Univ. Norm Hunanenist **13**(1), 18–24 (1990)
12. Cao B.Y.: Further study of extension-convex-set-separate theorem. J. Changsha Normal Univ. Water Resour. Electr. Power **9**(3), 229–234 (1994). (Natural Science)

Fuzzy Comprehensive Benefit Evaluation of Different Land Management Mode: Semiarid Case Study in Fuxin North China

Ji-hui Yang, Xiao-yan Han and Jie Lv

Abstract Based on different land management mode, the comprehensive benefit of peanut planting is researched in Liaoning Fuxin area. Firstly, the independent variable of affecting comprehensive benefit is expressed by interval numbers. Secondly, the evaluation index of comprehensive benefit is given under different land management mode by the operation of interval number. Finally, the rank results of comprehensive benefit to peanut planting are obtained under different land management mode.

Keywords Comprehensive benefit · Land management mode · Interval numbers · Ranking

1 Introduction

Fuxin city is located at northwestern Liaoning. It borders Shenyang capital city of Liaoning on the east Chaoyang city on the west Inner Mongolia Autonomous Region on the north and Jinzhou and Yinkou port on the south. It is less than 500 km to Dalian. It is 220 km from east to west, north to south 120 km and total area of 10,355 km^2. It governs two counties of Fuxin Mongolia Autonomous County and Zhangwu County and five districts, and it is semiarid climate.

J. Yang · X. Han · J. Lv (✉)
College of Economics and Management, Shenyang Agriculture University, Shenyang ZIP 110866, People's Republic of China
e-mail: jieluesy@163.com

J. Yang
e-mail: yangjihui@163.com

X. Han
e-mail: xiaoyan1274@yahoo.com.cn

B.-Y. Cao et al. (eds.), *Ecosystem Assessment and Fuzzy Systems Management*, Advances in Intelligent Systems and Computing 254, DOI: 10.1007/978-3-319-03449-2_47,

The average frost-free period is 154 days, the date of the first frost in September 28 after the last frost date in April 28 after. The annual average temperature is between the 7.1–7.6 °C [1]. Its total population of 1.93 million people, 65 towns, 835 villages, 5,056 village groups, 290,000 households, 1.05 million agricultural population, and agricultural labor force 0.48 million people. Fuxin city has 52 million acres arable land, 36 acres per capita arable land. Peanut sown area reached 18.7 million acres in 2012; the peanut sown area makes up around 36 % of total arable land. The total output of peanut reached 0.6 million tons, which ranks first in Liaoning Province [2].

In December 2012, the research group of agricultural economics from College of Economics and Management, Shenyang Agriculture University, investigated peanut production and farmers' income issues in several towns of Fuxin Mongolia Autonomous County, including Fuxin town and Laohetu town.

The research objects include peanut-planting personnel under different land management mode. For example, cooperative, mutual-aid teams, peanut-planting large family, and the individual farmers.

However, the accurate data are difficult to obtain from the respondents on the basis of practical investigation. There is little doubt that the case is extremely complex. Most of the respondents cannot provide accurate data, perhaps some are not willing to provide accurate data, others often cannot provide accurate data due to lack of knowledge. In such cases, classics statistics, evaluation, and decision-making method almost had no effect. We must seek a new solution to these problems.

In a seminal paper written in 1965, Zadeh [3] described the properties of fuzzy sets. Recently, fuzzy set theory has been frequently applied to the field of agricultural economics; the comprehensive benefit evaluation and decision-making problem are becoming increasingly important in the fuzzy environment [4, 5]. One of the better summaries of why fuzzy theory is useful in decision making and is given by Yager in [6]: "It must be kept in mind that the use of a fuzzy set does not eliminate the subjective or fuzzy nature of the concepts with which we are dealing…, but it does give us a handle for dealing with subjective concepts in a rational way, in a similar manner as the method used by the Bayesian Decision Makers enables them to handle subjective probabilities and utilities."

In this research, a lot of information is presented in the form of interval numbers. In recent years, the research of interval numbers is relatively active, the literature is numerous [7–9]. However, the application of interval numbers' theory to the area of agricultural economy does not see more, this chapter is just a try.

This chapter will be organized as follows. Section 2 provides an overview of interval number and its operation. We introduce relative dominance degree of interval number in Sect. 3. Besides, ranking index and data preprocessing of fuzzy decision making will be described in Sects. 3 and 4, respectively. Section 5 gives the main conclusions and the research limitation.

2 Interval Number and Its Operation

Definition 1 [10] We call the real number close interval $a = [\underline{a}, \overline{a}] = \{x | \underline{a} \le x \le \overline{a}. \ \underline{a}, \ \overline{a} \in R\}$ a closed interval number, where $\underline{a}$ is called left endpoints, $\overline{a}$ is called right endpoints, R denote a real number set. While the degenerated close interval $[a, a]$ is seen as a real number a itself, $a = 0$ is especially example.

Similarly, we have the definition of other class interval number, for example, the open interval number $(\underline{a}, \overline{a})$, the semi-open, and closed interval number $(\underline{a}, \overline{a}]$, etc.

Definition 2 [11] Let $a = [\underline{a}, \overline{a}]$, $b = [\underline{b}, \overline{b}]$ is two closed interval numbers (Other types of interval numbers have similar definitions), then add, subtract, scalar multiplication, multiplication, and division operations of interval numbers can be defined as follows:

$$a + b = [\underline{a} + \underline{b}, \ \overline{a} + \overline{b}], \tag{1}$$

$$a - b = [\underline{a} - \overline{b}, \ \overline{a} - \underline{b}], \tag{2}$$

$$ka = \begin{cases} [k\underline{a}, \ k\overline{a}], & k > 0, \\ [k\overline{a}, \ k\underline{a}], & k \le 0, \end{cases} \tag{3}$$

$$a * b = [\underline{c}, \ \overline{c}]. \tag{4}$$

Here, $\underline{c} = \min\{\underline{a} \cdot \underline{b}, \ \underline{a} \cdot \overline{b}, \ \overline{a} \cdot \underline{b}, \ \overline{a} \cdot \overline{b}\}$, $\overline{c} = \max\{\underline{a} \cdot \underline{b}, \ \underline{a} \cdot \overline{b}, \ \overline{a} \cdot \underline{b}, \ \overline{a} \cdot \overline{b}\}$,

$$\frac{a}{b} = [\underline{c}, \ \overline{c}], \tag{5}$$

where $\underline{c} = \min\left\{\frac{\underline{a}}{\underline{b}}, \frac{\underline{a}}{\overline{b}}, \frac{\overline{a}}{\underline{b}}, \frac{\overline{a}}{\overline{b}}\right\}$, $\overline{c} = \max\left\{\frac{\underline{a}}{\underline{b}}, \frac{\underline{a}}{\overline{b}}, \frac{\overline{a}}{\underline{b}}, \frac{\overline{a}}{\overline{b}}\right\}$.

3 The Ranking Based on Relative Dominance Degree of Interval Number

The problem of fuzzy comprehensive benefit evaluation is generally related to the fuzzy decision making, this has more or less to do with the analytic hierarchy [12, 13]. If the studied problem has involved interval number, then the ranking of interval numbers is inevitable; the ranking problem of interval numbers has numerous literatures [14, 15], this article does not get into details.

Definition 3 [16] Let $a = [\underline{a}, \ \overline{a}]$, $b = [\underline{b}, \ \overline{b}]$ is arbitrary two interval numbers (including the degradation to the real number situation), $\theta = \frac{\overline{a} - \underline{b}}{\overline{a} - \underline{a} + \overline{b} - \underline{b}}$,

$$P(a > b) = \begin{cases} 1 - \frac{1}{2\rho^{\theta - \frac{1}{2}}}, & \theta \geq \frac{1}{2} \\ \frac{1}{2\rho^{\frac{1}{2} - \theta}}, & \theta < \frac{1}{2} \end{cases} \quad (6)$$

is called relative dominance degree of interval numbers a and b, where ρ is a parameter greater than 1.

Definition 4 Let $a_1 = [\underline{a_1}, \overline{a_1}]$, $a_2 = [\underline{a_2}, \overline{a_2}], \ldots, a_n = [\underline{a_n}, \overline{a_n}]$ is arbitrary n interval numbers (including the degradation to the real number situation), if $a_{ij} = P(a_i > a_j)$, then matrix $A = (a_{ij})_{n\times n}$ is called relative dominance matrix of interval numbers $a_1, a_2, \ldots, a_n$.

Definition 5 [16] Let $A = (a_{ij})_{n\times n}$ is relative dominance matrix of interval numbers $a_1, a_2, \ldots, a_n$ and $c_i = \sum_{j=1}^{n} a_{ij}$. If

$$c_{ij} = \frac{c_i - c_j}{2(n-1)} + 0.5, \quad i, j = 1, 2, \ldots, n, \quad (7)$$

then the matrix $C = (c_{ij})_{n\times n}$ is called fuzzy consistent judgment matrix of interval numbers $a_1, a_2, \ldots, a_n$.

Definition 6 [17] Let $A = (a_{ij})_{n\times n}$ is relative dominance matrix of interval numbers $a_1, a_2, \ldots, a_n$, $C = (c_{ij})_{n\times n}$ is fuzzy consistent judgment matrix of interval numbers $a_1, a_2, \ldots, a_n$,

$$w_i = \frac{1}{n} - \frac{1}{2\alpha} + \frac{1}{n\alpha}\sum_{j=1}^{n} c_{ij} \quad (8)$$

is called ranking index of interval numbers $a_1, a_2, \ldots, a_n$, where α is parameter and $\alpha \geq \frac{n-1}{2}$.

4 The Data Preprocessing of Fuzzy Comprehensive Benefit Evaluation

The data are imprecisely reported in Fuxin, it comes from imprecisely sources. Such data rarely are described by precisely defined mathematical models. In particular, the variable is assigned to the interval numbers, all kinds of classical statistics and comprehensive evaluation methods are difficult to be used. Therefore, there is a need to use more flexible methodology that allows the usage of imprecise description of data and analyzed quantities. Fuzzy statistics seems to be an appropriate tool to cope with such problems [17].

The research data were obtained through questionnaire to cooperative, mutual-aid teams, peanut-planting large family, and the individual farmers. The data were divided into three categories by economic, ecological, and social. The comprehensive benefit can be represented as

$$a_i = \lambda_1 b_i - \lambda_2 c_i + \lambda_3 d_i, \quad i = 1, 2, 3, 4, \tag{9}$$

where interval numbers a_1, a_2, a_3, a_4 are comprehensive benefit related to cooperative, mutual-aid teams, peanut-planting large family, the individual farmers, respectively. The interval numbers b_i, c_i, d_i are quantity related to economic, ecological, and social, respectively. The $\lambda_i (i = 1, 2, 3)$ is the weight of interval number b_i, c_i, d_i, respectively.

This chapter focuses on the application of the fuzzy ranking method, for the convenience of calculation, the weights $\lambda_1, \lambda_2, \lambda_3$ is taken as $\frac{1}{2}, \frac{1}{4}, \frac{1}{4}$, respectively. The value of parameter ρ is taken as e.

The research data from Fuxin are calculated by fuzzy statistical method [18], using operational formula (1–5) of interval number, the value of four dependent variable can be obtained as follows:

$$\begin{cases} \text{cooperative,} & a_1 = [7.58, 8.78], \\ \text{mutual-aid teams,} & a_2 = [7.12, 9.02], \\ \text{peanut planting large family,} & a_3 = [7.75, 8.13], \\ \text{the individual farmers,} & a_4 = [6.63, 7.82]. \end{cases}$$

Using formula (6), the above interval number is compared each other, and the following relative dominance matrix A of the above four interval numbers can be obtained:

$$A = \begin{bmatrix} 0.5000 & 0.5174 & 0.5705 & 0.6647 \\ 0.4826 & 0.5000 & 0.5277 & 0.6196 \\ 0.4295 & 0.4723 & 0.5000 & 0.6829 \\ 0.3353 & 0.3804 & 0.3171 & 0.5000 \end{bmatrix}.$$

The following fuzzy consistent judgment matrix C of the above four interval numbers can be obtained using formula (7):

$$C = \begin{bmatrix} 0.5000 & 0.5204 & 0.5280 & 0.6200 \\ 0.4796 & 0.5000 & 0.5075 & 0.5995 \\ 0.4270 & 0.4925 & 0.5000 & 0.5920 \\ 0.3800 & 0.4005 & 0.4080 & 0.5000 \end{bmatrix},$$

where $\alpha = 1.5$.

Using formula (8) to matrix C, the ranking index can be obtained as follows:

$$w_1 = 0.2781, \ w_2 = 0.2644, \ w_3 = 0.2594, \ w_4 = 0.1980.$$

The results of ranking are:

$$w_1 \succ w_2 \succ w_3 \succ w_4,$$

where the symbol "$\succ$" means "is preferred or superior to".

5 Conclusion

In this chapter, we have researched comprehensive benefit evaluation problem for peanut planting with incomplete information in different land management mode based on fuzzy statistical theory and fuzzy decision-making method [19]. The numerical results have shown that the best land management mode of comprehensive benefit is cooperative, the second is the mutual-aid group, the third is the peanut-planting large family, and the individual farmers have the worst comprehensive benefit. This result provides the theoretical basis for current land transfer and moderate scale management, that is to say, we must encourage and accelerate the work of land transfer and moderate scale management, which is beneficial for the improvement of the comprehensive benefit of land use. However, it is difficult to obtain the sufficient research data due to limited human and financial resources, and we cannot data mine more precise information based on these existing fuzzy data [20]. Determining the weight factor and parameter values is also not perfect, but it is a complex and important task in fuzzy decision making. The above shortcomings will be improved constantly in the further research.

Acknowledgments I am grateful to the experts of Fuzzy Information and Engineering Branch of ORSC and the referees for their valuable comments and support. This work is supported by the National Science and Technology Support Program Sub-topics (No. 2011BAD09B02-1) and the postdoctoral fund of Shenyang Agriculture University.

References

1. Fuxin Bureau of Statistics: Fuxin Statistical Yearbook 2013. China Statistics Press, Beijing (2013)
2. Liaoning Bureau of Statistics: Liaoning Statistical Yearbook 2012. China Statistics Press, Beijing (2012)
3. Zadeh, L.A.: Fuzzy Sets. Inf. Control **8**, 338–353 (1965)
4. Lai, Y.J., Hwang, C.L.: Fuzzy Multiple Objective Decision Making. Springer Press, New York, Berlin, Heidelberg (1996)
5. Mu, Y., Wang, W.H., Zhong, Q.: Fuzzy decision method in planting scheme for forms. Proceedings of International Conference on Agricultural Engineering, I89-92, Beijing, China (1999)
6. Yager, R.R.: Multiple objective decision-making using fuzzy sets. Intl. J. Man-Mach. Stud. **9**, 375–382 (1977)
7. Tong, S.: Interval number and fuzzy number linear programming. Fuzzy Sets Syst. **66**, 301–316 (1994)
8. Xu, Z.S., Da, Q.L.: Research on method for ranking interval numbers. Syst. Eng. **19**(6), 94–96 (1994)
9. Sengupta, A., Pal, T.K.: On comparing interval numbers. Eur. J. Oper. Res. **127**(1), 28–43 (2000)
10. Cao, B.Y.: Optimal Models and Methods with Fuzzy Quantities. Springer Press, Berlin, Heidelberg (2010)
11. Zimmermann, H.J.: Fuzzy Set Theory and Its Applications. Kluwer, Boston, MA (1985)

12. Wang, L., Xu, S.: Analytic Hierarchy Process Foreword. China Renmin University Press, Beijing (1990)
13. Zhang, J.J.: Fuzzy analytical hierarchy process. Fuzzy Syst. Math. **14**(2), 80–88 (2000)
14. Xu, Z.S., Da, Q.L.: Research on method for ranking interval numbers. Syst. Eng. **19**(6), 94–96 (2001)
15. Li, Z.L.: A ranking approach for interval numbers. Nat. Sci. J. Hainan Univ. **21**(1), 4–7 (2003)
16. Zhang, J.J.: Research on method for ranking interval numbers. Oper. Res. Manag. Sci. **12**(3), 18–22 (2003)
17. Lv, Y.J.: Weight calculation method of fuzzy analytical hierarchy process. Fuzzy Syst. Math. **16**(2), 79–85 (2002)
18. Wang, Z.Y., Wu, B.L.: Statistics of Fuzzy Data. Harbin Institute of Technology Press, Harbin (2008)
19. Xu, Z.: Uncertain Multiple Attribute Decision Making Methods and Its Application. Tsinghua University Press, Beijing (2004)
20. Klein, Y., Pery, R., Komem, J., Kandel, A.: Fuzzy data mining. Int. Ser. Intell. Technol. **15**, 131–152 (2000)